Computational Knowledge Discovery for Bioinformatics Research

Xiao-Li Li
*Institute for Infocomm Research, A*STAR, Singapore*

See-Kiong Ng
*Institute for Infocomm Research, A*STAR, Singapore*

Medical Information Science
REFERENCE

Managing Director: Lindsay Johnston
Senior Editorial Director: Heather A. Probst
Book Production Manager: Sean Woznicki
Development Manager: Joel Gamon
Acquisitions Editor: Erika Gallagher
Cover Design: Nick Newcomer

Published in the United States of America by
Medical Information Science Reference (an imprint of IGI Global)
701 E. Chocolate Avenue
Hershey PA 17033
Tel: 717-533-8845
Fax: 717-533-8661
E-mail: cust@igi-global.com
Web site: http://www.igi-global.com

Library of Congress Cataloging-in-Publication Data

Computational knowledge discovery for bioinformatics research / Xiao-Li Li and See-Kiong Ng, editors.
p. cm.
Summary: "This book discusses the most significant research and latest practices in computational knowledge discovery approaches to bioinformatics in a cross-disciplinary manner that is useful for researchers, practitioners, academicians, mathematicians, statisticians, and computer scientists involved in the many facets of bioinformatics"-- Provided by publisher.
Includes bibliographical references and index.
ISBN 978-1-4666-1785-8 (hardcover) -- ISBN 978-1-4666-1786-5 (ebook) -- ISBN 978-1-4666-1787-2 (print & perpetual access) 1. Bioinformatics. 2. Data mining. I. Li, Xiao-Li, 1969- II. Ng, See-Kiong.
QH324.27.C66 2012
572'.330285--dc23

2012002754

British Cataloguing in Publication Data
A Cataloguing in Publication record for this book is available from the British Library.

Table of Contents

Section 2
Gene Expression Data Analysis

Section 3
Protein Interaction Analysis

Section 4
Network Mining and Systems Biology

Detailed Table of Contents

Motif identification for DNA sequences has many important applications in biological studies, including diagnostic probe design, locating binding sites and regulatory signals, and potential drug target identification. There are two versions—the Single Group and Two Groups. Here, the occurrences of the motif in the given sequences have errors. Currently, most of existing programs can only handle the case of single group. However, most of the programs do not allow indels (insertions and deletions) in the occurrences of the motif. In this paper, the authors propose a randomized algorithm for the one group problem that can handle indels in the occurrences of the motif. Finally, an algorithm for the two groups' problem is given along with extensive simulations evaluating algorithms.

A probe mapping technique using a novel implementation of a persistent q-gram index was developed. It guarantees to find all matches that meet certain definitions. These include exact matching of the central 19 bases of 25 base probes, matching the central 19 bases with at most one or three mismatches and exact matching of any 16 bases. In comparison with BLAST and BLAT, the new methods were either significantly faster or identified matches missed by the heuristics. The 16 bp method was used to map the 342,410 perfect match probes from the Affymetrix GeneChip Rat Genome 230 2.0 Array to the genome. When compared with the mapping from Ensembl, the new mapping included over seven million novel matches, providing additional evidence for researchers wishing to further investigate the sources of signals measured in microarray experiments. The results demonstrate the practicality of the index, which could support other q-gram based algorithms.

A computational mutagenesis is detailed whereby each single residue substitution in a protein chain of primary sequence length N is represented as a sparse N-dimensional feature vector, whose M << N nonzero components locally quantify environmental perturbations occurring at the mutated position and its neighbors in the protein structure. The methodology makes use of both the Delaunay tessellation algorithm for representing protein structures, as well as a four-body, knowledge based, statistical contact potential. Feature vectors for each subset of mutants due to all possible residue substitutions at a particular position cohabit the same M-dimensional subspace, where the value of M and the identities of the M nonzero components are similarly position dependent. The approach is used to characterize a large experimental dataset of single residue substitutions in bacteriophage T4 lysozyme, each categorized as either unaffected or affected based on the measured level of mutant activity relative to that of the native protein. Performance of a single classifier trained with the collective set of mutants in N-space is compared to that of an ensemble of position-specific classifiers trained using disjoint mutant subsets residing in significantly smaller subspaces. Results suggest that significant improvements can be achieved through subspace modeling.

Annotating protein structures is an urgent task as increasing number of protein structures of unknown function is being solved. To achieve this goal, it is critical to establish computational methods for characterizing and classifying protein local structures. The authors analyzed the similarity of local surface patches from 609 representative proteins considering shape and the electrostatic potential, which are represented by the 3D Zernike descriptors. Classification of local patches is done with the emergent self-organizing map (ESOM). They mapped patches at ligand binding-sites to investigate how they distribute and cluster among the ESOM map. They obtained 30-50 clusters of local surfaces of different characteristics, which will be useful for annotating surface of proteins.

wavelet coefficients. Following thresholding, the coefficients are fed as an input vector into a two-dimensional Self-Organizing-Map clustering algorithm. Transformed data are then clustered by minimizing the Euclidean (L^2) distance between their corresponding fluctuation patterns. A multi-resolution analysis by Wave-SOM of expression data from the yeast *Saccharomyces cerevisiae,* exposed to oxidative stress and glucose-limited growth, identified 29 genes with correlated expression patterns that were mapped into 5 different nodes. The ordered clustering of yeast genes by Wave-SOM illustrates that the same set of genes (encoding ribosomal proteins) can be regulated by two different environmental stresses, oxidative stress and starvation. The algorithm provides heuristic information regarding the similarity of different genes. Using previously studied expression patterns of yeast cell-cycle and functional genes as test data sets, the authors' algorithm outperformed five other competing programs.

Regulatory network analysis and other bioinformatics tasks require the ability to induce and represent arbitrary boolean expressions from data sources. In this paper, the authors introduce a novel framework called BLOSOM for mining (frequent) boolean expressions over binary-valued datasets. Boolean expressions can be grouped into four categories: pure conjunctions, pure disjunctions, conjunction of disjunctions, and disjunction of conjunctions. The authors' main focus is on mining the simplest expressions (the *minimal generators*), but also to propose *closure* operators that yield *closed* (or unique maximal) boolean expressions. BLOSOM efficiently mines frequent boolean expressions by utilizing a number of methodical pruning techniques. Experiments showcase the behavior of BLOSOM for different input settings and parameter thresholds. Application studies on gene expression and gene regulation patterns showcase the effectiveness of this approach.

MicroRNAs (miRNAs) are short single-stranded RNA molecules with 21-22 nucleotides known to regulate post-transcriptional expression of protein-coding genes involved in most of the cellular processes. Prediction of miRNA targets is a challenging bioinformatics problem. AU-rich elements (AREs) are regulatory RNA motifs found in the 3' untranslated regions (UTRs) of mRNAs, and they play dominant roles in the regulated decay of short-lived human mRNAs via specific interactions with proteins. In this paper, the authors review several miRNA target prediction tools and data sources, as well as computational methods used for the prediction of AREs. The authors discuss the connection between miRNA and ARE-mediated post-transcriptional gene regulation. Finally, a data mining method for identifying the co-occurrences of miRNA target sites in ARE containing genes is presented.

Section 3
Protein Interaction Analysis

The genomes of many human pathogens have been sequenced but the protein-protein interactions across a pathogen and human are still poorly understood. The authors apply a simple homology-based method to predict protein-protein interactions between human host and two mycobacterial organisms viz., M.tuberculosis and M.leprae. They focused on secreted proteins of pathogens and cellular membrane proteins to restrict to uncovering biologically significant and feasible interactions. Predicted interactions include five mycobacterial proteins of yet unknown function, thus suggesting a role for these proteins in pathogenesis. The authors predict interaction partners for secreted mycobacterial antigens such as MPT70, serine proteases and other proteins interacting with human proteins, such as toll-like receptors, ras signalling proteins and immune maintenance proteins, that are implicated in pathogenesis. These results suggest that the list of predicted interactions is suitable for further analysis and forms a useful step in the understanding of pathogenesis of these mycobacterial organisms.

Inferring the relationship among proteins is a central issue of computational biology and a diversity of biological assays are utilized to predict the relationship. However, as experiments are usually expensive to perform, automatic data selection is employed to reduce the data collection cost. Although data useful for link prediction are different in each local sub-network, existing methods cannot select different data for different processes.

This paper presents a new algorithm for inferring biological networks from multiple types of assays. The proposed algorithm is based on transfer learning and can exploit local information effectively. Each assay is automatically weighted through learning and the weights can be adaptively different in each local part.

The authors' algorithm was favorably examined on two kinds of biological networks: a metabolic network and a protein interaction network. A statistical test confirmed that the weight that our algorithm assigned to each assay was meaningful.

The reconstruction of protein-protein interaction (PPI) networks from high-throughput experimental data is one of the most challenging problems in bioinformatics. These biological networks have specific topologies defined by the functional and evolutionary relationships between the proteins and the physical limitations imposed on proteins interacting in the three-dimensional space. In this paper, the authors propose a novel approach for the identification of potential protein-protein interactions based on the integration of known PPI network topology and transcriptomic data. The proposed method, Function Restricted Value Neighborhood (FRV-N), was used to reconstruct PPI networks using an experimental data set consisting of 170 yeast microarray profiles. The results of this analysis demonstrate that incorporating knowledge of interactome topology improves the ability of transcriptome analysis to reconstruct interaction networks with a high degree of biological relevance.

High-throughput techniques involve large-scale detection of protein-protein interactions. This interaction data set from the genome-scale perspective is structured into an interactome network. Since the interaction evidence represents functional linkage, various graph-theoretic computational approaches have been applied to the interactome networks for functional characterization. However, this data is generally unreliable, and the typical genome-wide interactome networks have a complex connectivity. In this paper, the authors explore systematic analysis of protein interactome networks, and propose a k-round signal flow simulation algorithm to measure interaction reliability from connection patterns of the interactome networks. This algorithm quantitatively characterizes functional links between proteins by simulating the propagation of information signals through complex connections. In this regard, the algorithm efficiently estimates the strength of alternative paths for each interaction. The authors also present an algorithm for mining the complex interactome network structure. The algorithm restructures the network by hierarchical ordering of nodes, and this structure re-formatting process reveals hub proteins in the interactome networks. This paper demonstrates that two rounds of simulation accurately scores interaction reliability in terms of ontological correlation and functional consistency. Finally, the authors validate that the selected structural hubs represent functional core proteins.

In this paper, the authors consider the problem of, given a metabolic network, a set of source compounds and a set of target compounds, finding a minimum size reaction cut, where a Boolean model is used as a model of metabolic networks. The problem has potential applications to measurement of structural robustness of metabolic networks and detection of drug targets. They develop an integer programming-based method for this optimization problem. In order to cope with cycles and reversible reactions, they further develop a novel integer programming (IP) formalization method using a feedback vertex set (FVS). When applied to an E. coli metabolic network consisting of Glycolysis/Glyconeogenesis, Citrate cycle and Pentose phosphate pathway obtained from KEGG database, the FVS-based method can find an optimal set of reactions to be inactivated much faster than a naive IP-based method and several times faster than a flux balance-based method. The authors also confirm that our proposed method works even for large networks and discuss the biological meaning of our results.

Searching for repeated features characterizing biological data is fundamental in computational biology. When biological networks are under analysis, the presence of repeated modules across the same network (or several distinct ones) is shown to be very relevant. Indeed, several studies prove that biological networks can be often understood in terms of coalitions of basic repeated building blocks, often referred to as network motifs.This work provides a review of the main techniques proposed for motif extraction from biological networks. In particular, main intrinsic difficulties related to the problem are pointed out, along with solutions proposed in the literature to overcome them.

Open challenges and directions for future research are finally discussed.

Protein classification can be performed by representing 3-D protein structures by graphs and then classifying the corresponding graphs. One effective way to classify such graphs is to use frequent subgraph patterns as features; however, the effectiveness of using subgraph patterns in graph classification is often hampered by the large search space of subgraph patterns. In this paper, the authors present two efficient discriminative subgraph mining algorithms: COM and GAIA. These algorithms directly search for discriminative subgraph patterns rather than frequent subgraph patterns which can be used to generate classification rules. Experimental results show that COM and GAIA can achieve high classification accuracy and runtime efficiency. Additionally, they find substructures that are very close to the proteins' actual active sites.

With rapid advances in genomics, phylogenetics has turned to phylogenomics due to the availability of large amounts of sequence and genome data. However, incongruence between species trees and gene

Preface

The completion of the sequencing of the human genome marked the beginning of the new era of the so-called *Fourth Paradigm* (Hey, Tansley, & Tolle, 2009) of data-intensive scientific discovery for biomedicine. In addition to the traditional study of living organisms in terms of their biological structure, function, growth, origin, evolution, distribution, and taxonomy (Avila, 1995), biologists have also significantly stepped up their efforts in uncovering the biological processes that underlie disease pathways in the clinical contexts to discover new drugs to benefit human beings. This has resulted in a flood of biological and clinical data ranging from genomic sequences, DNA microarrays, protein structures, and biomolecular interactions, to Gene Ontology, disease pathways, biomedical images, and electronic health records. We are in a scenario where our capability to generate biomedical data has greatly surpassed our ability to mine and analyze the data effectively. In this new era of data-intensive scientific discovery, the focus of biomedical research has shifted from data generation to knowledge discovery, or more specifically, ***computational*** *knowledge discovery*. The aim is no longer on enabling the biologists to generate more data rapidly, but on converting them into useful knowledge.

However, in order to translate the vast biomedical data into useful clinical applications, there are still fundamental data analysis difficulties that have to be overcome. Practical issues such as handling noisy and incomplete data (e.g. Protein-Protein Interaction data, or PPIs, having high false positive and false negative rates), processing compute-intensive tasks (e.g. large scale graph mining), and integrating various data sources (e.g. linking genomic and proteomic data with clinical databases) are new challenges faced by biologists in the data-intensive post-genome era.

The objective of this book is to disseminate the research results and best practice from cross-disciplinary researchers and practitioners working on computational knowledge discovery for bioinformatics. By doing so, we hope to bring better awareness of the interesting and challenging problems to inspire new computational solutions. The book is therefore aimed at three overlapping audiences: (1) researchers in the areas of bioinformatics, data mining, machine learning, and data structure; (2) practitioners in the industry in these areas; (3) instructors and postgraduate students in colleges and universities.

Upon reading this book, we hope that researchers in the life sciences will be convinced that *in silico* knowledge discovery tools can truly complement their work *in vitro* and *in vivo*. At the same time, we also hope that computer scientists, mathematicians, and statisticians will be attracted to the new data analysis challenges that are important for computational knowledge discovery in bioinformatics research, and join their life sciences colleagues in their impactful research to benefit humankind.

COMPUTATIONAL KNOWLEDGE DISCOVERY

Computational knowledge discovery is an increasingly important research area in the computer science domain that focuses on the methodologies and processes for automatically or semi-automatically extracting useful patterns or knowledge from large volumes of data. It is often also a cross-disciplinary practice, as the methodologies and processes are applied to discover knowledge in a discipline outside of the computer science domain that has become significantly data-rich—in our case, the life sciences. Its importance has clearly risen with the rapid growth of online information on Web as well as abundant databases used by industries (e.g. various biological or clinical databases), together with the increasingly urgent need to derive novel knowledge from the data generated.

Computational knowledge discovery is itself a challenging task, involving research from multiple computational disciplines such as machine learning, statistics, artificial intelligence, databases, pattern recognition, and data visualization. Generally, it consists of an iterative sequence of the following steps (Han & Kamber, 2005):

1. **data cleaning**: removal of noise and inconsistent data from the original data;
2. **data integration**: combining multiple data sources consistently;
3. **data selection**: identifying and retrieving only the data that are relevant to the analysis task from the database;
4. **data transformation**: transforming and consolidating data into a format that is appropriate for mining by performing summary or aggregation operations;
5. **data mining**: applying intelligent machine learning methods to extract data patterns—this is the key process in computational knowledge discovery;
6. **pattern evaluation**: identifying the truly interesting patterns that represent useful knowledge based on interestingness measures;
7. **knowledge presentation**: using intuitive visualization and effective knowledge representation techniques to present the discovered knowledge to the user.

Steps 1-4 are typically considered as *data pre-processing steps* for preparing the given data for the crucial data-mining task in Step 5 (Fayyad, Gregory & Padhraic, 1996). Steps 6 and 7 are often referred for *decision support* as they involve choosing interesting patterns/knowledge and presenting them to user in a user-friendly manner for users to do further studies as well as decision-making.

Let us look into more details of Step 5, which is the key process for computational knowledge discovery. Data mining typically involves one or more of the following tasks:

- **Association rule mining (Dependency modeling):** This task detects relationships between variables/features. One well-known example is *market basket analysis* in which patterns in supermarket customers' purchasing habits are extracted from their sales transactions ("market baskets") and translated for market campaigns. For example, the association rule {butter, bread}=>{milk} can be detected in most supermarket transaction data—it indicates that if a supermarket customer has bought butter and bread together, he or she is likely to buy milk as well. With such association rule mining, a supermarket can automatically detect which products are frequently bought together and use this information for marketing purposes, e.g., promotional pricing or product placements. In the biomedical domain, association rule analysis can be performed on gene expression values to mine for association rules where the antecedents could be features and their value ranges (cancer genes and corresponding gene expression values under different conditions) and the consequents the class labels (cancer or non cancer). The knowledge discovered can then be used for building a diagnostic system.

- **Cluster analysis or clustering**: This is the task of segmenting a set of objects into groups so that the objects in the same group are more similar to each other than to those in other groups. In machine learning, cluster analysis is considered a form of *unsupervised learning*, since there is no need for users of the clustering algorithms to provide training examples. One example use of cluster analysis in biomedical research is to cluster gene expression data into groups where genes in each group share similar gene expression profiles, to discover genes that belong to the same biological functions (Li, Tan & Ng, 2006).
- **Classification**: This is the task of assigning input data into one of a given number of categories. A classic example is spam filtering, in which the task is for an email program to classify a new email as a legitimate message or just spam. To build a classifier, a user must first collect a set of training examples that are labeled with the predefined known classes (for example, known spam messages and different types of legitimate email messages). A machine learning algorithm is then applied to the training data to build a classification model (classifier) that can be employed subsequently to assign the predefined classes to instances in a test set (for evaluation) or future instances (in practice) (Li, Liu, & Ng, 2007). An application of classification in biomedical research is to predict the biological functions of novel proteins or genes. Here, those proteins with known functions are used as training data to build a classification model for each biological function, which can be subsequently used to classify unknown proteins.
- **Regression analysis**: This task attempts to find a function, which models the data with the least error. It includes the techniques for modeling and analyzing several variables, where the focus is on the relationship between a dependent variable and one or more independent variables.
- **Anomaly detection (Outlier detection):** In this task, we attempt to detect data records, which do not conform to an expected or established normal behaviour. The mined results might be either interesting data records or error records, which require further investigation.

COMPUTATIONAL CHALLENGES

Let us now discuss in details the various data analysis challenges for computational knowledge discovery from biological data (Li & Ng, 2009).

Handling Noisy and Incomplete Data

Biologists have generated tremendous amounts of biological data with the newly available advanced high-throughput technologies, but many of the data sets are actually rather noisy and incomplete. For example, there is a high-level of noises in gene expression data detected by DNA microarrays and protein interactions detected by popular high-throughput assays such as yeast-two-hybrids. In addition, molecular structures (e.g. protein structures), molecular interactions (e.g. protein interactions), Gene Ontology (including the sub-ontologies, molecular function, biological process, and cellular component), and disease pathways are still largely incomplete due to relatively limited biological knowledge.

Let us take Protein-Protein Interactions (PPIs) as an example to illustrate that current high-throughput data sets can be very noisy (alarmingly high false positives) and incomplete (alarmingly high false negatives). Established high-throughput methods such as Yeast-Two-Hybrid (Y2H) (Fields & Song, 1989) and Tandem Affinity Purification-Mass Spectrometry (TAP-MS) (Puig, et al., 2001; Rigaut, et al., 1999) are popular methods that have enabled comprehensive detection of PPI data sets in various species. However, the quality of detected PPI data sets is far from satisfactory.

a. **Noisy data**: The experimental conditions in which the above detection methods are carried out may result in detecting interactions that do not occur under *in vivo* physiological conditions. In other words, the experimental data may not be very accurate and contain noisy false positive interaction data that do not occur in the cell.
b. **Incomplete data**: The high-throughput methods can also fail to detect various types of interactions. For example, researchers have reported loss of weak transient interactions, loss of post-translational modification, and bias against soluble or membrane proteins (Lalonde, et al., 2008; Tarassov, et al., 2008). These result in false negatives in the PPI data leading to incomplete coverage of the interactomes.

How serious is the data quality situation for protein interaction detection—specifically, what are the false positive and false negative data rates in current PPI data sets? Many researchers have attempted to answer this question (Bader & Hogue, 2002; Bader, Chaudhuri, Rothberg, & Chant, 2004; Gentleman & Huber, 2007; Hart, Ramani, & Marcotte, 2006; Von Mering, et al., 2002) by comparing the overlaps between the collected protein interactions from different large-scale biological experiments, or checking the consistency of the function or location information between the interaction partners, to estimate the accuracy and coverage of the PPI data. The results of all these investigations have consistently shown that the quality of the protein interaction data is indeed very problematic, with accuracy ranging from 10% to 50% (Gentleman & Huber, 2007; Sprinzak, Sattath, & Margalit, 2003; Von Mering, et al., 2002) and coverage lower than 50%, even for the most studied and curated interactome of *Saccharomyces cerevisiae* (yeast) (Hart, et al., 2006). One can only expect the accuracy and coverage of the PPIs for the other less-studied or more complex species to be even lower.

Can we employ computational knowledge discovery methodologies to help dealing with the experimental limitations of noisy and incomplete protein interaction data? To address the issue of noisy data or false positive interactions, we can computationally validate the experimental protein interactions using various robust reliability measures (Chen, Hsu, Lee, & Ng, 2006). To address the issue of incomplete interactome coverage or false negative interactions, we can computationally predict novel protein interactions to supplement the experimental data set of detected PPIs. One way to do this is to represent the known protein interactions (positive training examples) and the non-protein interactions (negative training examples) using various informative biological features/properties, such as protein domains, protein sequences (e.g., amino acids composition, etc.), protein functions, biological processes, cellular locations, structural information, and topological features extracted from the PPI network. Then, a supervised machine-learning model (such as support vector machines, Bayesian classifier, k-NN classifier, or decision tree) can be used to classify unknown protein pairs into interacting or non-interacting. Another approach is to develop network-cleansing techniques such as finding missing links and/or removing unreliable links, which can also be used to improve the quality of the entire protein interaction networks (Li & Ng, 2009).

A number of chapters in this book address the noisy and incomplete data issues in protein-protein interaction detection (Section 3). To address the incomplete data issue, Chapter 10, "Prediction of Protein-Protein Interactions between Human Host and Two Mycobacterial Organisms," predicts host-pathogen interactions to help biologists better understand disease processes. Chapter 12, "Incorporating Network Topology Improves Prediction of Protein Interaction Networks from Transcriptomic," introduces a novel approach Function Restricted Value Neighborhood (FRV-N) for the identification of potential protein-protein interactions based on the integration of known PPI network topology and transcriptomic data.

To address the noisy data issue, Chapter 11, "A Transfer Learning Approach and Selective Integration of Multiple Types of Assays for Biological Network Inference," proposes a transfer learning algorithm to integrate noisy interaction data to infer more reliable networks. Chapter 13, "Mining Protein Interactome Networks to Measure Interaction Reliability and Select Hub Proteins," describes a k-round signal flow simulation algorithm to measure interaction reliability from connection patterns of the interactome networks.

Processing Computer-Intensive Network Mining Tasks

In order to uncover the biological processes of how life functions in living organisms, it is important to investigate the biological interactions at the network level. Large-scale biological networks include gene regulatory networks, metabolic networks, protein-protein interaction networks, protein domain networks, as well as DNA-protein interaction networks. Given that proteins are the workhorses of biological processes and they function only through interactions, the most intensively studied networks are naturally the PPI networks.

Computationally, a PPI network is typically modelled as an undirected graph (sometimes with weighted links) where the nodes represent unique proteins and the links denote interactions between two proteins. Such a network is a very large graph with thousands of vertices and tens of thousands of edges even for a simple model organism such as yeast. One can only imagine the insurmountable complexity of the PPI networks for the more complicated species such as the human being.

Graph theory is an important tool to facilitate efficient analyses of the large-scale interaction networks (Barabasi & Oltvai, 2004). Some important graph mining techniques have been proposed and applied in discovering useful biological knowledge from PPI networks such as interaction motifs, network motifs (Alon, 2007), lethal proteins (Jeong, Mason, Barabsi, & Oltvai, 2001), protein complexes/functional modules (Li, Wu, Kwoh, & Ng, 2010; Li, Tan, Foo, & Ng, 2005; Xie, Kwoh, Li, & Wu, 2011), and disease genes (Yang, Li, Wu, & Kwoh, 2011). We can also investigate the evolution of various protein interaction networks by computing the alignment of the PPI networks. The computational challenge for such a task is clearly humongous, as we will need to handle multiple massive networks. Once the networks are aligned, we can transfer knowledge (e.g. protein interactions, functions, etc.) from a PPI network for well-studied species to other less well-studied species.

Clearly, mining the large protein interaction networks is a compute-intensive task. Techniques for reducing the search space and time complexity are therefore highly important, and we can obtain better graph mining results by exploiting biological knowledge coupled with the development of novel efficient graph mining techniques.

The chapters in Section 4 cover various topics in biological network mining. Chapter 15, "New Trends in Graph Mining: Structural and Node-Colored Network Motifs," provides a technical overview of effective computational methods for extracting motifs in three types of biological networks: gene regulatory networks, protein-protein interaction networks, and metabolic networks.

Chapter 14, "Finding Minimum Reaction Cuts of Metabolic Networks under a Boolean Model Using Integer Programming and Feedback Vertex Sets," introduces an integer programming-based method to find an optimal set of reactions to be inactivated in a large biological network, such as the *E. coli* metabolic network. The proposed method can find an optimal set of reactions to be inactivated much faster than a naive IP-based method and several times faster than a flux balance-based method.

Chapter 16, "Discriminative Subgraph Mining for Protein Classification," studies the graph classification problem for protein classification. Unlike existing methods, which use frequent subgraphs as features (which results in a large search space), this chapter presents two efficient discriminative subgraph mining algorithms: COM and GAIA, which directly search for discriminative subgraph patterns instead. Experimental results showed that COM and GAIA can achieve high classification accuracy and runtime efficiency.

Integrating Various Biological Evidences for System-Level Understanding

As mentioned, the recent advent in high-throughput experimental technologies has resulted in an increasing number of large datasets available at various biological levels from the genome to the proteome and metabolome. Many of the datasets are deposited in centralized databases that are publicly accessible by all the researchers. At the same time, there are also increasing efforts by biologists to provide annotations of biological knowledge, with projects such as the Gene Ontology initiated to enable systematic functional annotations of genes and proteins. These efforts were motivated by the need to integrate various biological evidences in order to achieve system-level understanding of the complex cellular processes.

Let us continue to use protein interactions as examples. There are at least two key benefits of integrating various biological evidences. Firstly, by integrating multiple data sources together, we can derive better quality data. As mentioned, the PPI data generated from the high-throughput methods are inevitably noisy and incomplete. It will be useful if we can weigh the links (i.e. the protein interactions) in the PPI networks by using appropriate confidence measures. For example, we can employ metrics such as reproducibility of the interactions from multiple experimental methods, support from such other non-interaction data as co-expression, co-localization and shared functions, as well as the conservation of the protein interactions across other genomes, etc. There are various ways to integrate biological data together. For example, in one of the works to predict protein interactions (Jansen, et al., 2003), a Bayesian network model was developed by integrating noisy experimental interaction data with the weighted genomic features that are only weakly associated with interactions, such as messenger RNA co-expression, co-essentiality, and co-localization. Another way is to use machine learning methods, such as kernel methods (Ben-Hur & Noble, 2005), to integrate different biological resources into high dimensional vector space to do classification for better prediction of protein interactions.

Secondly, integrating the PPI networks with additional biological evidences will exploit the rich information and strengths from the different sources at different biological levels (Ghazalpour, Doss, Zhang, Wang, & Plaisier, 2006; Kelley & Ideker, 2005; Mootha, et al., 2003) to provide a system-level understanding of biological processes and human diseases (Ideker & Sharan, 2008). For example, successful development of diagnostics and therapeutics that target disease-relevant protein interactions can only be attained by obtaining quantitative and dynamic PPI data across different tissue cells and integrating them with gene expression, functional, structural, and metabolic pathway data to provide a system-level picture.

Nevertheless, one should be mindful that data integration may not always result in high-quality data and results—the intrinsic noisiness of data and integration methods do matter. Computational knowledge discovery methods based on machine learning approaches, probabilistic, and statistical tools should be carefully designed to maximally exploit the knowledge from various biological evidences, while minimizing the side effects coming from the irrelevant and noisy sources.

There are two chapters in this book that cover the topic of integrating multiple data sources for system-level understanding. Chapter 5, "Clustering Genes Using Heterogeneous Data Sources," elaborates on how to integrate the incomplete data sets (e.g. from the literature) to enhance clustering algorithms. Chapter 12, as mentioned earlier, integrated known PPI network topology and transcriptomic data from 170 yeast microarray profiles (Affymertix 'GeneChips') for identifying potential protein interactions.

ORGANIZATION OF THE BOOK

This book is organized into five major sections, covering the following topics: Sequence and Function analysis (Section 1), Gene Expression Data Analysis (Section 2), Protein Interaction Analysis (Section 3), Network Mining and Systems Biology (Section 4), and Biological Data Mining for Healthcare (Section 5).

Section 1: Sequence and Function Analysis

Motif identification for DNA sequences (including single group and two group identification) has many important applications in biological studies. In Chapter 1, "Identification of Distinguishing Motifs," the authors introduce a single-group algorithm that allows indels (insertions and deletions) in the occurrences of the motif to accommodate the potential errors in the occurrences of the motif (current methods only focus on the single group and do not allow indels). An algorithm for the two-groups problem is also given along with extensive simulations evaluating algorithms.

One limitation faced by biologists when designing microarray experiments is that current probe mapping techniques only allow one mismatch between probe and genomic sequences. Chapter 2, "Mapping Affymetrix Microarray Probes to the Rat Genome via a Persistent Index," describes how a novel database implementation of a q-gram index can address this practical issue.

Chapter 3, "Improving Prediction Accuracy via Subspace Modeling in a Statistical Geometry Based Computational Protein Mutagenesis," presents a computational approach for representing a system of protein mutants, due to single residue replacements, that are experimentally determined to have either "unaffected" or "affected" activity levels relative to the native protein.

Chapter 4, "Characterization and Classification of Local Protein Surfaces Using Self-Organizing Map," describes the prediction of proteins' functions from their tertiary structures on the molecular level. The local surface patches of proteins, which are characterized by their shape and electrostatic potential, are clustered using an emergent self-organizing map approach, resulting in the discovery of 30-50 clusters of local surfaces of different characteristics.

Section 2: Gene Expression Data Analysis

Standard exploratory clustering algorithms have been used to cluster gene expression data for identifying closely related genes. The algorithms typically assumed that multiple and complete sources of data are available. In Chapter 5, "Clustering Genes Using Heterogeneous Data Sources," the authors present a new clustering algorithm to deal with multiple complete and incomplete information sources. This is important, as many incomplete sources of data are also likely to be useful in their inclusion for gene expression data analysis.

Chapter 6, "Efficient Mining Frequent Closed Discriminative Biclusters by Sample-Growth: The FDCluster Approach," proposes an algorithm FDCluster to mine frequent closed discriminative bicluster in multiple microarray datasets, which can reduce the noise influence and allow more biological biclusters to be found. The experimental results showed that FDCluster is more effectiveness than traditional methods in either a single micorarray dataset or multiple microarray datasets. The authors also tested the biological significance using GO to show that their proposed method is able to produce biologically relevant biclusters.

Chapter 7, "Wave-SOM: A Novel Wavelet-Based Clustering Algorithm for Analysis of Gene Expression Patterns," presents a wavelet-based algorithm (Wave-SOM) to help visualize and cluster oscillatory time-series data in two-dimensional gene expression microarrays. Using previously studied expression patterns of yeast cell cycle and functional genes as test data sets, the proposed algorithm outperformed five other competing programs.

Regulatory network analysis and other bioinformatics tasks require the ability to induce and represent arbitrary Boolean expressions from data sources. Chapter 8, "Mining Frequent Boolean Expressions: Application to Gene Expression and Regulatory Modeling," introduces a novel framework called BLOSOM for mining (frequent) Boolean expressions over binary-valued datasets. Application studies on gene expression and gene regulation patterns showcase the effectiveness of this approach.

Chapter 9, "Bioinformatics Methods for Studying MicroRNA and ARE-Mediated Regulation of Post-Transcriptional Gene Expression," provides a review of several miRNA target prediction tools and data sources, as well as the computational methods used for the prediction of AREs. The authors discuss the connection between miRNA and ARE-mediated post-transcriptional gene regulation. Finally, a data mining method for identifying the co-occurrences of miRNA target sites in ARE containing genes is presented.

Section 3: Protein Interaction Analysis

Chapter 10, "Prediction of Protein-Protein Interactions between Human Host and Two Mycobacterial Organisms," describes a computational method for predicting host-pathogen interactions to help biologists better understand disease processes. Given the recent re-emergence of pathogens such as *M. tubercolosis*, this work is certainly very timely and important to help defend the public against these infectious diseases.

As mentioned, it is well known that biological networks are noisy due to the limitations of biological experiments. Chapter 11, "A Transfer Learning Approach and Selective Integration of Multiple Types of Assays for Biological Network Inference," demonstrates how a transfer-learning algorithm can be used to integrate noisy interaction data obtained from multiple types of assays to result in more reliable networks.

Chapter 12, "Incorporating Network Topology Improves Prediction of Protein Interaction Networks from Transcriptomic," introduces a novel approach called Function Restricted Value Neighborhood

(FRV-N) for the identification of potential protein-protein interactions based on the integration of known PPI network topology and transcriptomic data. The proposed method was used to reconstruct PPI networks using an experimental data set consisting of 170 yeast microarray profiles. The positive results demonstrate that incorporating the interactome's topological knowledge can improve the ability of transcriptome analysis in reconstructing interaction networks with a high degree of biological relevance.

Chapter 13, "Mining Protein Interactome Networks to Measure Interaction Reliability and Select Hub Proteins," describes a k-round signal flow simulation algorithm to measure interaction reliability from the connection patterns of the interactome networks. In addition, the chapter also presents an algorithm for mining the complex interactome network structure by restructuring the network using hierarchical ordering of nodes. The structure re-formatting process is able to reveal hub proteins in the interactome networks, which are key players in the biological processes.

Section 4: Network Mining and Systems Biology

Chapter 14, "Finding Minimum Reaction Cuts of Metabolic Networks under a Boolean Model Using Integer Programming and Feedback Vertex Sets," introduces an integer programming-based method to find an optimal set of reactions to be inactivated in a large biological network, such as the *E. coli* metabolic network. One potential translational bioinformatics application for this work is the measurement of the structural robustness of metabolic networks for detecting better drug targets.

Chapter 15, "New Trends in Graph Mining: Structural and Node-Colored Network Motifs," provides a technical overview of methods for extracting motifs from biological networks. This chapter provides a survey of network motif extraction in three kinds of biological networks: gene regulatory networks, protein-protein interaction networks, and metabolic networks.

Protein classification can be performed by representing 3-D protein structures by graphs and then classifying the corresponding graphs. One effective way to classify such graphs is to use frequent subgraph patterns as features; however, the effectiveness of using subgraph patterns in graph classification is often hampered by the large search space of subgraph patterns. Chapter 16, "Discriminative Subgraph Mining for Protein Classification," presents two efficient discriminative subgraph mining algorithms: COM and GAIA. These algorithms directly search for discriminative subgraph patterns instead of frequent subgraph patterns, which can be used to generate classification rules. Experimental results show that COM and GAIA can achieve high classification accuracy and runtime efficiency.

Incongruence between species trees and gene trees has remained a challenge in molecular phylogenetics for its biological and algorithmic complexities. Chapter 17, "Infer Species Phylogenies Using Self-Organizing Maps," presents a Self-Organizing Map (SOM) based phylogeny inference method and demonstrates its superiority to the state-of-the-art gene concatenation method using the same datasets. This chapter illustrates how to cluster multispecies genes, estimate multispecies gene entropy, and visualize the species patterns through the self-organizing map mining.

Section 5: Biological Data Mining for Healthcare

Chapter 18, "$SPCC_{TDM}$: A Catalogue for Analysis of Therapeutic Drug Monitoring Related Contents in the Drug Prescription Information," describes a text mining method to analyse Therapeutic Drug Monitoring (TDM) in the prescription information of drugs (Summary of Products Characteristics, SPC), which consists of six structure-related items (dose, adverse drug reactions, drug interactions, overdose, pregnancy/breast feeding, and pharmacokinetics), four theory-guided items according to the information about ranges of plasma concentrations, and a recommendation of TDM in the SPC.

Chapter 19, "Hierarchical Density-Based Clustering of White Matter Tracts in the Human Brain," introduces a new framework for automatic white matter tract clustering using a hierarchical density-based approach. The chapter describes experiments on real data that demonstrate the proposed method can be used to effectively group these tracts into meaningful bundles on multiple scales as well as eliminating noisy fibers.

Chapter 20, "Revealing the Origin and Nature of Drug Resistance of Dynamic Tumour System," identifies the strategies that resistant subpopulations of cancer cells undertake to overcome the effect of the anticancer drug Topotecan. For the analyses of cell lineage data encoded from time-lapse microscopy, data mining tools that generate interpretable models of the data and provide the necessary statistical significance are chosen.

CONCLUSION

As computing becomes increasingly integrated into the entire pipeline of biological and medical discovery process, there is an increasingly pressing need for computer scientists, mathematicians, statisticians, and bioinformaticians to collaborate and contribute to the development of the life sciences by creating novel techniques discovering useful knowledge from large-scale real-world biomedical data. The opportunities for computational researchers to cross over from the computation domain into the biomedical domain to contribute to the meaningful scientific pursuit with the biologists and clinical scientists are abundant, as shown by the various chapters in this book.

The ultimate success of the computational knowledge discovery for bioinformatics analysis will depend on the parallel improvements both in the biological and clinical experimental techniques from the biologists and clinicians to provide rich and clean biological/clinical datasets for the knowledge discovery community, and in the computational techniques from the computer scientists, mathematicians, and statisticians to provide efficient and effective ways to exploit the data for knowledge discovery. The ultimate goal is to enable biologists and clinicians to better understand life processes. We hope this book will play a role in helping to realize what a cross-disciplinary marriage of biological science with computer science can bring.

Finally, we wish to mention that all the chapters in this book have been selected from Volume 1 of *International Journal of Knowledge Discovery in Bioinformatics* (IJKDB). As such, our heartfelt thanks go to the members of IJKDB international advisory board, the associate editors, and the members of the editorial review board in reviewing of the manuscripts and providing useful feedback. Of course, we are grateful to the authors who contributed to the exciting and important research topics of developing computational knowledge discovery approaches for applications in biology and healthcare.

Special thanks also go to the publishing team at IGI Global, for providing invaluable contributions and guidance throughout the whole process from inception of the initial idea to the final publication of this book.

Xiao-Li Li
*Institute for Infocomm Research, A*STAR, Singapore*

See-Kiong Ng
*Institute for Infocomm Research, A*STAR, Singapore*

REFERENCES

Alon, U. (2007). Network motifs: Theory and experimental approaches. *Nature Reviews. Genetics*, *8*, 450–461.

Avila, V. L. (1995). *Biology: Investigating life on earth*. Boston, MA: Jones and Bartlett.

Bader, G. D., & Hogue, C. W. (2002). Analyzing yeast protein-protein interaction data obtained from different sources. *Nature Biotechnology*, *20*(10), 991–997.

Bader, J. S., Chaudhuri, A., Rothberg, J. M., & Chant, J. (2004). Gaining confidence in high-throughput protein interaction networks. *Nature Biotechnology*, *22*(1), 78–85.

Barabasi, A. L., & Oltvai, Z. N. (2004). Network biology: Understanding the cell's functional organization. *Nature Reviews. Genetics*, *5*(2), 101–113.

Ben-Hur, A., & Noble, W. (2005). Kernel methods for predicting protein-protein interactions. *Bioinformatics (Oxford, England)*, *21*, 38–46.

Chen, J., Hsu, W., Lee, M., & Ng, S.-K. (2006). Increasing confidence of protein interactomes using network topological metrics. *Bioinformatics (Oxford, England)*, *22*(16), 1998–2004.

Fayyad, U., Gregory, P. S., & Padhraic, S. (1996). From data mining to knowledge discovery in databases. *American Association for Artificial Intelligence*. Retrieved from http://www.kdnuggets.com/gpspubs/aimag-kdd-overview-1996-Fayyad.pdf.

Fields, S., & Song, O. (1989). A novel genetic system to detect protein-protein interactions. *Nature*, *340*(6230), 245–246.

Gentleman, R., & Huber, W. (2007). Making the most of high-throughput protein-interaction data. *Genome Biology*, *8*(10), 112.

Ghazalpour, A., Doss, S., Zhang, B., Wang, S., & Plaisier, C. (2006). Integrating genetic and network analysis to characterize genes related to mouse weight. *PLOS Genetics*, *2*(8).

Han, J., & Kamber, M. (2005). *Data mining: Concepts and techniques*. New York, NY: Morgan Kaufmann.

Hart, G. T., Ramani, A. K., & Marcotte, E. M. (2006). How complete are current yeast and human protein-interaction networks? *Genome Biology*, *7*(11), 120.

Hey, A. J., Tansley, S., & Tolle, K. M. (2009). *The fourth paradigm: Data-intensive scientific discovery*. Redmond, WA: Microsoft Research.

Ideker, T., & Sharan, R. (2008). Protein networks in disease. *Genome Research*, *18*(4), 644–652.

Jansen, R., Yu, H., Greenbaum, D., Kluger, Y., Krogan, N. J., & Chung, S. (2003). A Bayesian networks approach for predicting protein-protein interactions from genomic data. *Science*, *302*(5644), 449–453.

Jeong, H., Mason, S., Barabsi, A., & Oltvai, Z. (2001). Lethality and centrality in protein networks. *Nature*, *411*(6833), 41–42.

Kelley, R., & Ideker, T. (2005). Systematic interpretation of genetic interactions using protein networks. *Nature Biotechnology*, *23*(5), 561–566.

Lalonde, S., Ehrhardt, D. W., Loque, D., Chen, J., Rhee, S. Y., & Frommer, W. B. (2008). Molecular and cellular approaches for the detection of protein-protein interactions: Latest techniques and current limitations. *The Plant Journal*, *53*(4), 610–635.

Li, X.-L., Liu, B., & Ng, S.-K. (2007). Learning to identify unexpected instances in the test set. In *Proceedings of Twentieth International Joint Conference on Artificial Intelligence*. AAAI Press.

Li, X.-L., & Ng, S.-K. (2009). *Biological data mining in protein interaction networks*. Hershey, PA: IGI Global.

Li, X.-L., Tan, S.-H., Foo, C.-S., & Ng, S.-K. N. (2005). Interaction graph mining for protein complexes using local clique merging. *Genome Informatics*, *16*(2), 260–269.

Li, X.-L., Tan, Y. C., & Ng, S.-K. (2006). Systematic gene function prediction from gene expression data by using a fuzzy nearest-cluster method. *BMC Bioinformatics*, *7*(4), 23.

Li, X.-L., Wu, M., Kwoh, C.-K., & Ng, S.-K. (2010). Computational approaches for detecting protein complexes from protein interaction networks: A survey. *BMC Genomics*, *11*(1), 3.

Mootha, V. K., Lepage, P., Miller, K., Bunkenborg, J., Reich, M., & Hjerrild, M. (2003). Identification of a gene causing human cytochrome c oxidase deficiency by integrative genomics. *Proceedings of the National Academy of Sciences of the United States of America*, *100*(2), 605–610.

Puig, O., Caspary, F., Rigaut, G., Rutz, B., Bouveret, E., & Bragado-Nilsson, E. (2001). The tandem affinity purification (TAP) method: A general procedure of protein complex purification. *Methods (San Diego, Calif.)*, *24*(3), 218–229.

Rigaut, G., Shevchenko, A., Rutz, B., Wilm, M., Mann, M., & Séraphin, B. (1999). A generic protein purification method for protein complex characterization and proteome exploration. *Nature Biotechnology*, *17*(10), 1030–1032.

Sprinzak, E., Sattath, S., & Margalit, H. (2003). How reliable are experimental protein-protein interaction data? *Journal of Molecular Biology*, *327*(5), 919–923.

Tarassov, K., Messier, V., Landry, C. R., Radinovic, S., Molina, M. M., & Shames, I. (2008). An in vivo map of the yeast protein interactome. *Science*, *320*(5882), 1465–1470.

Von Mering, C., Krause, R., Snel, B., Cornell, M., Oliver, S. G., & Fields, S. (2002). Comparative assessment of large-scale data sets of protein-protein interactions. *Nature*, *417*(6887), 399–403.

Xie, Z., Kwoh, C. K., Li, X.-L., & Wu, M. (2011). Construction of co-complex score matrix for protein complex prediction from AP-MS data. *Bioinformatics (Oxford, England)*, *27*(13), 159–166.

Yang, P., Li, X.-L., Wu, M., & Kwoh, C. K. (2011). Inferring gene-phenotype associations via global protein complex network propagation. *PLoS ONE*, *6*(7).

Section 1
Sequence and Function Analysis

Chapter 1
Identification of Distinguishing Motifs

Wangsen Feng
Peking University, China

Lusheng Wang
City University of Hong Kong, China

ABSTRACT

Motif identification for DNA sequences has many important applications in biological studies, including diagnostic probe design, locating binding sites and regulatory signals, and potential drug target identification. There are two versions—the Single Group and Two Groups. Here, the occurrences of the motif in the given sequences have errors. Currently, most of existing programs can only handle the case of single group. However, most of the programs do not allow indels (insertions and deletions) in the occurrences of the motif. In this paper, the authors propose a randomized algorithm for the one group problem that can handle indels in the occurrences of the motif. Finally, an algorithm for the two groups' problem is given along with extensive simulations evaluating algorithms.

1. INTRODUCTION

Motif identification for DNA sequences has many important applications in biological studies, e.g., diagnostic probe design, locating binding sites and regulatory signals, and potential drug target identification (Dopazo, Rodriguez, Saiz, & Sobrino, 1993; Lucas, Busch, Mossinger, & Thompson, 1991; Proutski & Holme, 1996). The most general problem is as follows:

Two Groups: Given two groups of sequences B and G, find a length-l (distinguishing) motif that appears in every sequence in B and does not appear in anywhere of the sequences in G.

The motif we want to find is called the *distinguishing* motif that can differentiate the two groups. If no error is allowed, the problem is easy.

DOI: 10.4018/978-1-4666-1785-8.ch001

However, in practice, the occurrence of the motif in each of the given sequences has errors (mutations and indels). The problem becomes extremely hard when errors appear. Many mathematic models have been proposed.

The following definition was first stated in (Lanctot, Li, Ma, Wang, & Zhang, 1999).

Distinguishing substring selection: given a set $B = \{s_1, s_2, ..., s_{n1}\}$ of n_1 (bad) strings of length at least l, and a set $G = \{g_1, g_2, ..., g_{n2}\}$ of n_2 (good) strings of length at least l, the problem is to find a string s of length l such that for each string $s_i \in B$ there exists a length-l substring t_i of s_i with $d_H(s, t_i) \leq d_b$ and for any length-l substring u_i of $g_i \in G$, $d_H(s, u_i) \geq d_g$. Here d_b and d_g are two parameters, called the radius of the two groups. We want d_b to be small and d_g to be large. $d_H(,)$ represents the Hamming distance between two strings.

When indels are allowed in the occurrences of the motif, the problem is called the *distinguishing subsequence selection* problem, where the input is the same as the previous problem and the objective here is to find a string s of length l such that for each string $s_i \in B$ there exists a substring t_i of s_i with $d(s, t_i) \leq d_b$ and for any substring u_i of $g_i \in G$, $d(s, u_i) \geq dg$. Here $d(,)$ represents the edit distance.

A special case, where only one group of sequences is involved is as follows:

Single Group: Given a group of n given sequences, find a motif that appears in each of the given sequences and those occurrences of the motif are *similar.*

There are several ways to define the similarity of the motif occurrences, e.g., the *consensus score*, *general consensus score* and SP-score (Hertz & Stormo, 1995; Li, Ma, & Wang, 1999). Here we use the following mathematical definitions.

The Closest Substring Problem: Given n sequences $\{s_1, s_2, \ldots, s_n\}$, each is of length m $(m \geq l)$, the problem is to find a center string s of length l and a substring t_i of length l in s_i such that

$$d = \max_{i=1}^{n} d_H\left(s, t_i\right)$$

is minimized, where $d_H(,)$ is the Hamming distance between the two strings of the same length.

When indels are allowed, the problem is called the *closest subsequence problem*, where the input is the same as the closest substring problem and the objective here is to find a center string s of length l and a substring t_i in s_i such that

$$d = \max_{i=1}^{n} d\left(s, t_i\right)$$

is minimized, where $d_H(,)$ is the edit distance between the two sequences.

The closest substring problem was proved to be NP-hard (Lanctot, Li, Ma, Wang, & Zhang, 1999). Thus, the more general problem, the distinguishing substring selection problem, is also NP-hard. Some polynomial time approximation schemes have been designed for both the closest substring problem and the distinguishing substring selection problem (Li, Ma, & Wang, 2002; Deng, Li, Li, Ma, & Wang, 2003). However, those polynomial time approximation schemes are too slow and do not work very well in practice.

Several programs have been developed for the single group problem based on various mathematical models. Bailey and Elkan (1995) designed MEME that uses the EM algorithm for motif identification. The algorithm allows the motif to be absent in some of the given sequences. Waterman et al introduced the extended sample-driven approach in Waterman, Arratia, and Galas (1984).

Keich and Pevzner (2002a) proposed a variant of the extended sample-driven approach in. Keich and Pevzner (2002b) also developed another motif detection program. Buhler and Tompa (2002) develop a software, PROJECTION, that combines the EM algorithm and a random projection approach. PROJECTION outperforms the programs in (Keich & Pevzner, 2002a; Keich & Pevzner, 2002b). Recently, Proce, Ranabhadran, and Pevzner (2003) designed a new algorithm that uses branching from sample strings. Their program, PatternBranching, is much faster than the previously best known program, PROJECTION. All the algorithms and programs mentioned here do not allow indels in the occurrence of the motif and can only handle the one group problem.

In this paper, we design an algorithm that allows indels in the occurrences of the motif. Our algorithm is an extension of the EM approach. We also propose an algorithm to solve the two group problem.

2. REPRESENTATIONS OF MOTIFS

In our algorithms, we use two representations of motifs, *consensus pattern*, and *profile* (Gusfield, 1997). Let $t_1, t_2, \cdots, t_n$ be n strings of length l. Each t_i is an occurrence of a motif. The *consensus pattern* of the n occurrences is obtained by choosing the letter that appears the most in each of the l columns. The *profile* of the n occurrences is a $4 \times l$ matrix W, each cell $W(i, j)$ is a number indicating the occurrence rate of letter i in column j. Figure 1 gives an example.

In this paper, we will use the two representations to design our algorithms. We find that both representations are useful and we actually use the profile representation in the early stage of the EM algorithm. The consensus pattern representation is used in the last round of the EM algorithm to improve the accuracy.

Figure 1. (a) The 5 occurrences of the motif and the consensus patterns. (b) the profile matrix.

```
                            caaccca
                            caacccc
                            catcccg
                            catccct
                            caccccca
                       --------------------
consensus pattern           caaccca
another con. pattern        catccca
                     (a)

   a   0   1   0.4   0   0   0   0.4
   c   1   0   0.2   1   1   1   0.2
   g   0   0   0.0   0   0   0   0.2
   t   0   0   0.4   0   0   0   0.2

                     (b)
```

3. COMPUTING THE CLOSEST SUBSEQUENCE

In this section, we describe the algorithm for computing the closest subsequence problem and show the simulation results.

3.1 The General Frame

From mathematical point of view, the basic ideas of the algorithm we are going to describe is from Li, Ma, and Wang (2002a) and Li, Ma, and Wang (2002b). In Li, Ma, and Wang (2002a) and Li, Ma, and Wang (2002b), a random sampling technique is used to select substrings from the given sequences to form a set of linear inequalities. Here we use the technique to directly give the starting matrix for the EM approach.

The idea is to randomly choose k positions among the l positions of the motif. Then we can guess the true motif at the k selected positions by

Figure 2. A randomized algorithm for consensus pattern

1. randomly choose k positions from the l positions of the motif;
2. try all 4^k possible (partial) strings;
3. use the partial consensus strings s_p to search all the n given strings and find a substring that is closest to the partial string from each of the n given strings;
4. reconstruct the center string s based on the n selected substrings of length l;
5. repeat 1–4 several times and choose the best result.

trying 4^k possible strings of length k. The partial motif (with k guessed letters) is used to search all the given sequences $s_1, s_2, \ldots, s_n$ to find a t_i from each s_i that is closest to the partial motif, i.e., the number of mismatches between t_i and the partial motif at those selected k positions is minimized. Let K be the set of k selected positions, and $t_i^{'}$ and s two strings of length l. We use $d(t_i', s|K)$ to denote the number of matches between $t_i^{'}$ and s at the positions in K. The algorithm is given in Figure 2.

In Step (3), when using a partial string s_p to search a given string s_i, we try to find a substring of length l such that the number of mismatches at those k selected positions is minimized. Steps (1)-(4) generate a candidate of the motif. Since this is a randomized algorithm, different executions generate different results. Thus, in the algorithm, we repeat steps (1)-(4) several times to enhance the quality of the output.

Theorem 1 *Let* $k = \left\lceil \frac{4}{\varepsilon^2} \log(nm) \right\rceil$. *With probability* $1 - 2(nm)^{-\frac{1}{3}}$, the algorithm finds a string s of length l and a length l substring t_i for each given string s_i such that for every t_i $d(s, t_i^{'}) \leq d(s, t_i) + 2\varepsilon l$. The running time of the algorithm is $O(4^k nml)$ and nm is the total number of letters in the n given sequences.

In Wang, Dong, and Fan (2004), it was shown that the algorithm can give a good approximation for the objective function $\sum_{i=1}^{n} d(s, t_i^{'})$. Theorem 1 states that the algorithm can also give a good approximation for the objective function $\max_{i=1}^{n} d(s, t_i^{'})$. However, in practice, k has to be a relatively big number in order to get satisfactory results. The speed is far below that of PROJECTION. PROJECTION uses a random projection method to find seeds and uses an EM method to do local search. In Wang, Dong, and Fan (2004), a combined approach was proposed. It uses the EM algorithm to replace Steps (3) and (4) in Figure 2.

3.2 EM Method

Lawrence and Reilly (1990) were the first to introduce the Expectation Maximization (EM) algorithm in motif finding problems. Bailey and Elkan (1995) used it in multiple motif finding. Buhler and Tompa (2002) adopted the EM method in their PROJECTION algorithm in the motif refining step. The following description of the EM algorithm is based on Bailey and Elkan (1995).

In the EM algorithm, a motif of length-l is represented by a $4 \times l$ matrix. Let a $4 \times l$ matrix W be the initial guess of the motif. $s_i(j)$ denotes the j-th letter in sequence s_i. Here is the standard EM algorithm to refine the motif:

For each position j in each sequence s_i, $s_{ij} = s_i(j)s_i(j+1)\ldots s_i(j+l-1)$ denotes the l-mer (substring of length l) starting at $s_i(j)$ and ending at $s_i(j+l-1)$. Calculate the likelihood that s_{ij} is the occurrence of the motif as follows:

$$P(i,j) = \prod_{x=1}^{l} W(s_{ij}(x), x), 1 \le i \le n, 1 \le j \le m-l+1$$

where $s_{ij}(x) = s_i(j+x-1)$ is the x-th base in l-mer s_{ij}. In order to avoid zero weights, a fixed small number δ (we use $\delta = 0.1$) is added to every element of W before calculating the likelihoods.

For each l-mer s_{ij}, we get a normalized probability from the likelihood.

$$P'(i,j) = \frac{P(i,j)}{\sum_{j=1}^{m-l+1} P(i,j)}.$$

Replace $P(i,j)$ with $P'(i,j)$. (The normalization guarantees that $\sum_{j=1}^{m-l+1} P'(i,j) = 1$, reflecting the fact that there is exactly one motif occurrence in each sequence.)

Re-estimate the (motif) matrix W from all the l-mers as follows:

$$W = \sum_{i=1}^{n} \sum_{j=1}^{m-l+1} W^{ij},$$

where W^{ij} is also a $4 \times l$ matrix, constructed from s_{ij}:

$$W^{ij}(b,x) = \begin{cases} P(i,j) & : \textit{if } b = s_{ij}(x) \\ 0 : & \textit{otherwise.} \end{cases}$$

A normalization is applied to W to ensure that the sum of each column in W is 1, i.e.,

$$W'(b,x) = \frac{W(b,x)}{\sum_{b=A,C,G,T} W(b,x)}$$

Replace W with W'.

Steps 1–4 is called a *cycle*. Let W_{q-1} and W_q be the two consecutive matrices produced in cycles $q-1$ and q. If

$$\max |W_q(b,x) - W_{q-1}(b,x)| < \varepsilon(\qquad (1)$$

then EM stops. Otherwise, goto step 1 and start next cycle.

In step 5, ε is a parameter given by the user. We use a relatively large value $\varepsilon = 0.05$ such that on average the EM algorithm stops within very few cycles.

Improved EM Algorithm

The algorithm in Section 3.1 directly finds a substring that is closest to the guessed motif, whereas the standard EM algorithm in Section 3.2 considers every substring when constructing W. In Wang, Dong, and Fan (2004), it was proposed that one can use $\frac{1}{m-l+1}$, the average $P(i,j)$ for $j = 1, 2, \ldots, m-l+1$, as the threshold when constructing W. All terms with value less than $\frac{1}{m-l+1}$ are ignored when constructing W in Step 3. Another improvement is the shifting technique. For the occurrences of motifs with high score, we try to shift the occurrences of the motifs to the left and right by one or two positions. It was shown that for the improved EM algorithm, both speed and accuracy are improved significantly (Wang, Dong, & Fan, 2004). The software MotifDetector in Wang, Dong, and Fan (2004) can outperform PROJECTION for long motifs.

However, PatternBranching in Price, Ramabhadran, and Pevzner (2003) is much faster than both PROJECTION and MotifDetector in all cases. All the methods mentioned in this subsection cannot handle indels in the occurrences of the motif.

The main reason that we still work on the EM approach is that we have ways to extend the EM algorithms to handle indels and to solve the two groups case.

3.3 Incorporating Indels

In this section, we introduce a method that extends the EM algorithms to handle insertions and deletions.

Recall that the EM algorithm needs to compute W^{ij} for every substring of length l when no indel is allowed. Let k be the maximum total number of insertions and deletions allowed in an occurrence of a motif. When indels are allowed, the matrix for EM becomes $5 \times l$, where we add the space as a letter. We have to consider all length $l + h$ substrings in all the given sequences, where $h = 0, 1, -1, 2, -2, ...k, -k$ is the number of insertions (positive number) and deletions (negative number). For each length $l + h$ substring, we will align it with the $5 \times l$ matrix.

Aligning a Length *l* + *h* String with a 5 × *l* Matrix

Let $d[i, j]$ be the cost of aligning the first i columns in W with the first j letters in the string. $d[i, j]$ can be computed as follows:

$$d[i,j] = \max\{d[i-1,j-1) \times w[x,i], d[i-1,j] \times w[\Delta,i], d[[i,j-1] \times \delta\} \quad (2)$$

where x is the j letter in the string, Δ represents the space (a new letter), and $\delta = 0.1$ is the number that is used to avoid 0-weight in the EM algorithm.

We can compute all the $d[i, j]$'s in a bottom up order. A standard backtracking procedure will give an optimal alignment.

The new EM algorithm allowing indels. The new EM algorithm also contains the 5 steps in Section 3.2. We consider all length-$l+h$ ($h = 0, 1, -1, 2, -2, \ldots, k, -k$) substrings in the given sequences. For a fixed starting position, say, position q, there are $2k+1$ substrings starting at position q and their lengths are $l, l+1, l-1, l+2, l-2, \ldots, l+k$ and $l-k$. For each starting position q, we align the $2k+1$ substrings with the matrix W and use the substring with the biggest score to do Steps 1-5 for the EM algorithm described in Section 3.2. We also apply the threshold and shifting techniques as in the improved EM algorithm here.

Note that, in equation (2), we use the score (product) for EM approach. In our algorithm, we use the profile representation (matrix) for the EM iterations. In the last round, we convert the matrix into a string (by choosing the letter with the biggest number in each column) and use Hamming distance to choose the substring (from a given sequence) that is closest to the computed motif. The results are better than that of always using the EM score. A refinement of the Hamming distance is the score $\sum_{i=1}^{l} W(s_j, i)$, where $s_1 s_2, \ldots s_l$ is the string and W is the matrix for the motif. $\sum_{i=1}^{l} W(s_j, i)$ is the *distance* between the matrix W and the string. We found that the results using refinement score is better than that of Hamming distance.

3.4 Experiment Results

In this section, we do simulations to illustrate the quality of our algorithm.

Figure 3. The results for one deletion

Problem	Maxtrials	Accuracy	Time(s)
(11,2)	1	100%	12.4
(13,3)	4	90%	66
(13,3)	6	95%	101
(13,3)	8	95%	155
(15,4)	8	90%	216
(14,4)	1	100%	5
(17,4)	1	100%	11.6
(20,5)	1	100%	22

Figure 4. The mixed case, where the motifs in 5 sequences have one insertion and the motifs in the other 5 sequences have one deletion

Problem	Maxtrials	Accuracy	Time(s)
(11,2)	1	95%	44
(11,2)	2	95%	79
(11,2)	3	100%	55
(13,3)	1	65%	165
(13,3)	3	75%	267
(15,3)	1	100%	51
(14,4)	1	100%	17
(17,4)	1	90%	150
(17,4)	2	100%	98
(20,5)	1	100%	138

Evaluation Method

To evaluate the programs, Pevzner and Sze (2000) proposed a challenge problem, which has been studied by Keich and Pevzner (2002a, 2002b). We randomly generate n ($n = 20$) sequences of length m(m = 600). Given a center string s of length l, for each of the n random sequences, we randomly choose d positions for s, randomly mutate the d letters from s and implant the mutated copy of s into the random sequence. The problem here is to find the implanted pattern. The pattern thus implanted is called an *(l, d)-pattern*. We will use this model to do simulations and test our algorithms. To incorporate indels, d here is simply the total number of insertions, deletions, and mutations.

Simulation Results

Figure 3 shows the results, where the motifs for 15 sequences has no insertion and deletion, and the motifs of the other 5 sequences have one deletion and $d - 1$ mutations. We can see that the accuracy for each case is very close to that of the algorithm in Wang, Dong, and Fan (2004), where no insertion and deletion is involved.

Figure 4 shows the results, where the motifs for 10 sequences has no insertion and deletion, the motifs of 5 sequences have one deletion and $d - 1$ mutations, and the motifs of the other 5 sequences have one insertion and $d - 1$ mutations. We can see that the running time increases significantly and accuracy in many cases is slightly worse than that in Figure 3. In particular, the case (15, 4) has very poor results. Thus, we show the accuracy for (15, 3) instead.

Figure 5 shows the results for the case, where the motifs in 5 sequences have one deletion, the motifs in the other 5 sequences have two deletions and the motifs in the remaining 10 sequences have no indel. Figure 6 shows the results for the case, where the motifs in 5 sequences have one insertion, the motifs in the other 5 sequences have two insertions and the motifs in the remaining 10 sequences have no indel. The results in Figure 6 are slightly better than that in Figure 5. The reason might be that (2) is not symmetric, i.e., $w[\Delta, i]$ and δ are not the same. The case in Figure 6 needs to insert two columns in the matrix for the motif, whereas the case in Figure 5 needs to insert two spaces in the motif sequences.

Figure 5. The motifs in 5 sequences have one deletion. The motifs in the other 5 sequences have two deletions.

Problem	Maxtrials	Accuracy	Time(s)
(11,2)	1	75%	96
(11,2)	2	90%	115
(11,2)	3	90%	232
(15,3)	1	90%	87
(15,3)	2	100%	69
(14,4)	1	100%	27.6
(17,4)	1	80%	192
(17,4)	2	90%	222
(17,4)	3	90%	314
(20,5)	1	95%	54
(20,5)	2	100%	87

4. COMPUTING THE DISTINGUISHING SUBSTRING SELECTION

In this section, we extend the EM Approach to work for the distinguishing substring selection problem, where there are two groups of sequences.

4.1 The Extended EM Algorithm for Two Groups

Let $B = \{s_1, s_2, \ldots, s_{n1}\}$ and $G = \{g_1, g_2, \ldots, g_{n2}\}$ be two groups of sequences. The problem here is that we want to find a motif s of length l such that for each string $s_i \in B$ there exists a length-l substring t_i of s_i with $d(s, t_i) \le d_b$ and for any length-l substring u_i of $g_i \in G$, $d_H(s, u_i) \ge d_g$. As a first strike, we do not consider indels.

We still follow the 5 basic Steps of the EM algorithm in Section 3.2.

For each length-l substring of s_i (g_k), we compute the likelihood $P(i, j)$ ($P(k, j)$) as in Step 1 of Section 3.2 and compute the normalized likelihood $P'(i, j)$ ($P'(k, j)$) as in Step 2 of Section 3.2. In Step 3, the $4 \times l$ matrix W_{ij} is computed as usual.

Figure 6.The motifs in 5 sequences have one insertion; the motifs in the other 5 sequences have two insertions

Problem	Maxtrials	Accuracy	Time(s)
(11,2)	1	100%	26
(13,3)	2	80%	297
(15,4)	1	40%	245
(15,4)	2	60%	417
(14,4)	1	0%	282
(17,4)	1	70%	163
(17,4)	2	100%	50
(20,5)	1	90%	101
(20,5)	2	100%	102

$$W^{ij}(b,x) = \begin{cases} P(i,j) : if\ b = s_{ij}(x) \\ 0 \ : otherwise. \end{cases}$$

The main difference is that we re-estimate the (motif) matrix W from all the l-mers in both groups in Figure 6.

$$= \sum_{s_i \in B} \sum_{j=1}^{m-l+1} W^{ij} - \rho \sum_{g_i \in G} \sum_{j\ s.t.P'(i,j) > ave} W^{ij}, \quad (3)$$

where every l-mer from B contributes positively, every l-mer from G contributes negatively, ρ is a parameter to control the intensity of the negative contributions and *ave* is the average $P'(i, j)$ over all l-mers from group B. In fact, we do not consider all the l-mers from group G. We only use those $W^{i,j}$ with $P'(i, j) > ave$ for every $g_i \in G$.

The main idea is that when the motif represented by the matrix W is too close to some l-mer from group G ($P'(i,j) > ave$), we scoop the pattern from the matrix by subtracting the corresponding matrix W^{ij}.

Steps 4 and 5 of the EM algorithm remain the same.

Figure 7. The results for the two groups, where c_2 is generated by randomly choosing 200 positions (allowing repeats) from c_1 and randomly setting the letters at those positions to one of the four letters. The average Hamming distance between c_1 and c_2 is about 128.

AveDisG1	AveDisG2	AveDisG12	l	accuracy
215	215	285	9	17%
215	215	285	10	73%
215	215	285	11	71%
215	215	285	12	95%
215	215	285	13	98%
215	215	285	14	99%
215	215	285	15	100%
215	215	285	20	100%

The Choice of ρ

The choice of ρ has influence on the performance of the program. We carry experiments to determine its value. We find that when ρ is between 0.1 and 0.4, the program has better performance. So we set ρ equal to 0.2 in our program.

The Choice of *ave*

ave here is set to be the average $P'(i, j)$ over all l-mers from group B. This setting allows us to make $d_g - d_b$ large. Using $2ave$ or $3ave$ makes the program faster. Here the users can try to play this parameter to control $d_g - d_b$ and the running time.

4.2 Experiment Results

We have done experiments based on both simulation data and real data (Figure 7).

4.3 Simulation Data

We first arbitrarily choose a sequence c_1 of length 600 as the center for group B. To get the center c_2 for group G, we randomly choose 200 positions (allowing repeats) from c_1 and randomly set the letters at those positions to one of the four letters. (Thus, it is possible that some of the letters at those selected positions may remain the same.) After obtaining the two centers c_1 and c_2, we generate two groups of sequences. Each group contains 10 sequences. The sequences in each group are generated as follows: randomly choose 200 positions (allowing repeats) from the center and randomly set the letters at those positions to be one of the four letters. The two centers are not included in the two groups of sequences. AveDisG1 is the average Hamming distance between two sequences in group B. AveDisG2 is the average Hamming distance between two sequences in group G. AveDisG12 is the average Hamming distance between any pair of sequences in different groups. l is the length of the motif. For two groups of sequences, the case is successful if our program finds a motif such that for each string $s_i \in B$ there exists a length-l substring t_i of s_i with $d_H(s, t_i) \leq d$ and for any length-l substring u_i of $g_i \in G$, $d_H(s, u_i) > d$. The accuracy is defined as the percentage of successful cases. (We run 500 examples for each case). From Figure 6, we can see that it is easy to find a motif that can distinguish the two groups when l is large.

Intuitively, it is easy to find a motif that can distinguish the two groups when the distance between the two groups (or the distance between the

Figure 8. The results for the two groups, where c_2 is generated by randomly choosing 300 positions (allowing repeats) from c_1 and randomly setting the letters at those positions to be one of the four letters. The average Hamming distance between c_1 and c_2 is about 175.

AveDisG1	AveDisG2	AveDisG12	l	accuracy
215	215	310	9	28%
215	215	310	10	87%
215	215	310	11	85%
215	215	310	12	98%
215	215	310	13	100%
215	215	310	14	100%
215	215	310	15	100%
215	215	310	20	100%

Figure 9. Experiment on real data. C, U, Ha, HE, and C-F represent Chlamydia, Uncultured bacterium, Halothiobacillus, HEDGEHOG and C-FOS, respectively.

Groups	# of sequences	l	d_b	d_g	Motif
C and U	10+5	12	0	3	GGCGGAAAGGAA
C and Ha	10+6	12	0	3	AAGAGAAAGTCT
U and Ha	5+6	12	1	3	TGCCGCAGTTAA
C and HE	10+6	12	0	4	GTACAGAAGGTA
HE and C-F	6+5	12	3	4	ACAGCAGCGAAG
C and U	10+5	15	0	6	TTTTCTAATTTTTCC
C and Ha	10+6	15	0	6	TGGCGTGTCAGCCAT
U and Ha	5+6	15	1	4	GAAAGGGGGACTAAT
C and HE	10+6	15	0	7	GATGAAGTCGTAACA
HE and C-F	6+5	15	3	5	ACTCCGATGTGTTCT
C and U	10+5	18	0	7	AAAGCACTTTCGCCTGGG
C and Ha	10+6	18	0	8	TCTCAACCCCATCCGTGT
U and Ha	5+6	18	2	7	ATAATCCCTCGAAAGGGG
C and HE	10+6	18	0	9	GGGCTTGACATGTTTTCC
HE and C-F	6+5	18	4	7	CCAAACTCCGATGTGTTC
C and U	10+5	20	0	8	ACCTCAAGTCGGCATCTAAT
C and Ha	10+6	20	0	9	TCCAATTGATTTGAGCGTAC
U and Ha	5+6	20	2	8	AATCCCTCGAAAGGGGGACT
C and HE	10+6	20	0	10	GTCGTTGACTCAACCTATTT
HE and C-F	6+5	20	5	8	TCAGCACCTGTCTTGCTTGG
C and U	10+5	24	0	10	GGAGTTGGTTTCACCTTAAGTCGT
C and Ha	10+6	24	0	11	ATTGATTTGAGCGTACCAGGTAAA
U and Ha	5+6	24	2	8	TAGTGGGGTATAATCCCTCGAAAG
C and HE	10+6	24	0	12	CTTAAAGCTATCCTCAGTTCGGAT
HE and C-F	6+5	24	6	11	CGTTTCAGCACCTGCCTGGGTAGG

two centers) is large. Figure 8 is the results, where 300 random positions are selected to generate c_2.

4.4 Real Data

We have carried some experiments on real data. We collect three groups of bacteria DNA sequences and two groups mammalia DNA sequences.

The first group of bacteria is Chlamydia, including 10 sequences. Their GenBank accession numbers are AB001774, AB001775, AB001776, AB001777, D85714, D85715, D85716, D85717, D88317 and E17343. The lengths of sequences are between 1508bp and 1548bp. The second group of bacteria is Uncultured bacterium, containing 5 sequences. Their GenBank accession numbers are AF323760, AF32761, AF323764, AF323764, AF323773 and AF323777 with lengths ranging from 1461bp to 1550bp. The last group of bacteria is Halothiobacillus, containing 6 sequences. Their GenBank accession numbers are AF170419, AF173169, AF226850, M90662, U58020 and X97534 with lengths ranging from 1456bp to 1504bp.

We also take two groups of mammalia sequences from ACUTS database. The first group is "HEDGEHOG MORPHOGEN PROTEIN 5'UTR", including 6 sequences with lengths ranging from 200bp to 319bp. The other group is "C-FOS ONCOGENE, FIRST INTRON", containing 5 sequences with lengths ranging from 376bp to 754bp.

The result is in Figure 9. The "Groups" column shows the names of two groups, B and G. l is the length of the motif, d_b is the radius for group B and d_g is the radius for group G.

ACKNOWLEDGMENT

The paper is supported by a grant from the Research Grants Council of the Hong Kong Special Administrative Region, China [Project No. CityU 121207].

REFERENCES

Bailey, T., & Elkan, C. (1994). Fitting a mixture model by expectation maximization to discover motifs in biopolymers. In *Proceedings of the Second International Conference on Intelligent Systems for Molecular Biology (ISMB-94)* (pp. 28-36). Menlo Park, CA: AAAI Press.

Bailey, T., & Elkan, C. (1995). Unsupervised learning of multiple motifs in biopolymers using expectation maximization. *Machine Learning, 21*, 51–80. doi:10.1007/BF00993379

Blanchette, M. (2001). Algorithms for phylogenetic footprinting. In *Proceedings of the Fifth Annual International Conference on Computational Molecular Biology (RECOMB 01)* (pp. 49-58).

Buhler, J., & Tompa, M. (2002). Finding motifs using random projections. *Journal of Computational Biology, 9*, 225–242. doi:10.1089/10665270252935430

Deng, X., Li, G., Li, Z., Ma, B., & Wang, L. (2003). Generic Drug Design without Side Effect. *SIAM Journal on Computing, 32*(4), 1073–1090. doi:10.1137/S0097539701397825

Dopazo, J., Rodr'ıguez, A., S'aiz, J. C., & Sobrino, F. (1993). Design of primers for PCR amplification of highly variable genomes. *CABIOS, 9*, 123–125.

Duret, L., & Bucher, P. (1997). Searching for regulatory elements in human noncoding sequences. *Current Opinion in Structural Biology, 7*, 399–406. doi:10.1016/S0959-440X(97)80058-9

Duret, L., Dorkeld, F., & Gautier, C. (1993). Strong conservation of non-coding sequences during vertebrates evolution: potential involvement in post-transcriptional regulation of gene expression. *Nucleic Acids Research, 21*, 2315–2322. doi:10.1093/nar/21.10.2315

Gusfield, D. (1997). *Algorithms on Strings, Trees, and Sequences: Computer Science and Computational Biology*. Cambridge, UK: Cambridge University Press. doi:10.1017/CBO9780511574931

Hertz, G., & Stormo, G. (1995). Identification of consensus patterns in unaligned DNA and protein sequences: a large-deviation statistical basis for penalizing gaps. In *Proceedings of the 3rd Int'l Conf. Bioinformatics and Genome Research* (pp. 201-216).

Keich, U., & Pevzner, P. (2002a). Finding motifs in the twilight zone. *Bioinformatics (Oxford, England)*, *18*, 1374–1381. doi:10.1093/bioinformatics/18.10.1374

Keich, U., & Pevzner, P. (2002b). Subtle motifs: defining the limits of motif finding algorithms. *Bioinformatics (Oxford, England)*, *18*, 1382–1390. doi:10.1093/bioinformatics/18.10.1382

Keller, G. H., & Manak, M. M. (1989). *DNA Probes* (p. 12). New York: Stockton Press.

Lanctot, K., Li, M., Ma, B., Wang, S., & Zhang, L. (1999). Distinguishing string selection problems. In *Proceedings of the 10th ACM-SIAM Symp. on Discrete Algorithms* (pp. 633-642).

Lawrence, C., & Reilly, A. (1990). An expectation maximization (EM) algorithm for the identification and characterization of common sites in unaligned biopolymer sequences. *Proteins*, *7*, 41–51. doi:10.1002/prot.340070105

Li, M., Ma, B., & Wang, L. (1999). Finding Similar Regions in Many Strings. In *Proceedings of the Thirty-first Annual ACM Symposium on Theory of Computing,* Atlanta (pp. 473-482).

Li, M., Ma, B., & Wang, L. (2002a). Finding Similar Regions in Many Sequences. *Journal of Computer and System Sciences*, *65*, 73–96. doi:10.1006/jcss.2002.1823

Li, M., Ma, B., & Wang, L. (2002b). On the closest string and substring problems. *Journal of the ACM*, *49*(2), 157–171. doi:10.1145/506147.506150

Lucas, K., Busch, M., M¨ossinger, S., & Thompson, J. A. (1991). An improved microcom-puter program for finding gene- or gene family-specific oligonucleotides suitable as primers for polymerase chain reactions or as probes. *CABIOS*, *7*, 525–529.

McPearson, M. J., Quirke, P., & Taylor, G. R. (1991). *PCR A Practical Approach* (p. 8). New York: Oxford University Press.

Pevzner, P., & Sze, S. (2000). Combinatorial approaches to finding subtle signals in DNA sequences. In *Proceedings of the 8th International Conference on Intelligent Sys-tems for Molecular Biology* (pp. 269-278).

Price, A., Ramabhadran, S., & Pevzner, P. (2003). Finding Subtle Motifs by Branching from Sample Strings. *Bioinformatics (Oxford, England)*, *19*, 149–155. doi:10.1093/bioinformatics/btg1072

Proutski, V., & Holme, E. C. (1996). Primer Master: a new program for the design and analysis of PCR primers. *CABIOS*, *12*, 253–255.

Stormo, G. (1990). Consensus patterns in DNA. In Doolittle, R. F. (Ed.), *Molecular evolution: computer analysis of protein and nucleic acid sequences* (*Vol. 183*, pp. 211–221). Methods in Enzymology.

Wang, L., Dong, L., & Fan, H. (2004). Randomized Algorithms for Motif Detection. In *Proceedings of the 15th Annual International Symposium on Algorithms and Computation (ISAAC'04)* (pp. 884-895).

Waterman, M., Arratia, R., & Galas, E. (1984). Pattern recognition in several sequences: consenus and alignment. *Bulletin of Mathematical Biology*, *46*, 515–527.

APPENDIX: PROOF OF THEOREM 1

We need the following lemma that was first proved in Li, Ma, and Wang (2002).

Lemma 2 *Let $X_1, X_2, \ldots, X_n$ be n independent random 0-1 variables, where X_i takes 1 with probability p_i, $0 < p_i < 1$. Let* $X = \sum_{i=1}^{n} X_i$ and $\mu = E[X]$. Then for any $0 < \varepsilon \leq 1$,

1. $\Pr(X > \mu + \varepsilon n) < \exp(-\frac{1}{3} n\varepsilon^2)$,

2. $\Pr(X < \mu - \varepsilon n) \leq \exp(-\frac{1}{2} n\varepsilon^2)$.

Let t_i be the true occurrence of the motif in the given string s_i and $t_i^{'}$ be a substring of length l in s_i. s denotes the consensus pattern. Assume that $t_i^{'}$ is quite different from t_i, say,

$$d(s, t_i^{'}) \geq d(s, t_i) + 2\varepsilon l \; 4 \qquad (4)$$

for a small number ε that represents the error rate. Pr $(d(t_i^{'}, s \mid K) \leq d(s, t_i \mid K))$ denotes the probability that $d(t_i^{'}, s \mid K) \leq d(s, t_i \mid K)$.We will estimate Pr $(d(t_i^{'}, s \mid K) \leq d(s, t_i \mid K))$, the probability that $t_i^{'}$ is chosen to be the closest substring to the partial consensus pattern s|K. Note that from (4), $d(t_i^{'}, s \mid K) \leq d(s, t_i \mid K)$ implies either $d(s, t_i^{'} \mid K) \leq (d(s, t_i^{'}) - \varepsilon l) \times \frac{k}{l}$ or $d(s, t_i^{'} \mid K) \leq (d(s, t_i^{'}) + \varepsilon l) \times \frac{k}{l}$.

Thus, we have

$$Pr(d(t_i^{'}, s \mid K) \leq d(s, t_i \mid K)) \leq Pr(d(s, t_i^{'} \mid K) \leq (d(s, t_i^{'}) - \varepsilon l) \times \frac{k}{l})$$

$$+Pr(d(s, t_i^{'} \mid K) \leq (d(s, t_i^{'}) + \varepsilon l) \times \frac{k}{l}) \qquad (5).$$

Note that $d(t_i^{'}, s \mid K)$ is the sum of k independent random 0-1 variables $\sum_{i=1}^{k} X_i$, where $X_i = 1$ indicating a mismatch between s and $t_i^{'}$ at the *i-th* position in K. Thus, from Lemma 2,

$$Pr(d(t_i^{'}, s \mid K) \leq d(s, t_i^{'} \mid K) \leq (d(s, t_i^{'}) - \varepsilon l) \times \frac{k}{l}) \leq \exp(-\frac{1}{2}\varepsilon^2 k)$$

If $k = \left\lceil \frac{4}{\varepsilon^2} \log(nm) \right\rceil$, then

$$Pr(d(s,t_i^{'} \mid K) \leq (d(s,t_i^{'}) - \varepsilon l) \times \frac{k}{l}) \leq (nm)^{-2}.$$

Similarly, it can be proved that if $k = \left\lceil \frac{4}{\varepsilon^2} \log(nm) \right\rceil$,

$$Pr(d(s,t_i \mid K) \leq (d(s,t_i) + \varepsilon l) \times \frac{k}{l}) \leq (nm)^{-\frac{4}{3}}$$

Therefore,

$$Pr(d(t_i^{'},s \mid K) \leq (d(s,t_i \mid K) + \varepsilon l)) \leq 2(nm)^{-\frac{4}{3}}$$

Consider all the $n(m-l+1)$ substrings of length l in the n given strings, we know that with probability $1-2(nm)^{-\frac{1}{3}}$, $d(s,t_i^{'}) \leq (d(s,t_i) + 2\varepsilon l)$ for all i = 1, 2, . . ., n.

This work was previously published in the International Journal of Knowledge Discovery in Bioinformatics, Volume 1, Issue 3, edited by Xiao-Li Li and See-Kiong Ng, pp. 53-67, copyright 2010 by IGI Publishing (an imprint of IGI Global).

Chapter 2
Mapping Affymetrix Microarray Probes to the Rat Genome via a Persistent Index

Susan Fairley
University of Glasgow, UK

Rob Irving
University of Glasgow, UK

John D. McClure
University of Glasgow, UK

Martin W. McBride
University of Glasgow, UK

Neil Hanlon
University of Glasgow, UK

Anna F. Dominiczak
University of Glasgow UK

Ela Hunt
University of Strathclyde, UK

ABSTRACT

A probe mapping technique using a novel implementation of a persistent q-gram index was developed. It guarantees to find all matches that meet certain definitions. These include exact matching of the central 19 bases of 25 base probes, matching the central 19 bases with at most one or three mismatches and exact matching of any 16 bases. In comparison with BLAST and BLAT, the new methods were either significantly faster or identified matches missed by the heuristics. The 16 bp method was used to map the 342,410 perfect match probes from the Affymetrix GeneChip Rat Genome 230 2.0 Array to the genome. When compared with the mapping from Ensembl, the new mapping included over seven million novel matches, providing additional evidence for researchers wishing to further investigate the sources of signals measured in microarray experiments. The results demonstrate the practicality of the index, which could support other q-gram based algorithms.

DOI: 10.4018/978-1-4666-1785-8.ch002

Table 1. Summary of a selection of probe mapping methods.

Group	Approach
Affymetrix	Map target sequence to genome using BLAT. Keep only best alignment.
Ensembl (Hubbard *et al.*, 2007)	Step 1 – Map probes to genome with Exonerate, allowing at most one mismatch. Step 2 – Map probesets to transcripts, requiring at least half the probes in a set to map to a transcript.
Harbig *et al.* (2005)	Step 1 - BLAST target sequences against RefSeqs. Step 2 - Map probes to the identified subset of RefSeq. Step 3 – Apply filters to link probesets to transcripts.
Gautier *et al.* (2004)	Step 1 – Map probes to RefSeq mRNAs. Step 2 – Re-group probes, creating "novel" probesets.
Dai *et al.* (2005)	Map probes to UniGenes and other sequences. Perfect matches only. One mapping to genome. Also, must match mRNA/EST sequence and genome perfectly. "Novel" probesets.
Mecham *et al.*(2004a, 2004b), Carter *et al.* (2005)	Mapping probes to UniGenes and to the cDNA sequences on cDNA probes for cross-platform comparison.
Kim *et al.* (2005)	Mapping probes to UniGenes using a q-gram supported batch implementation of BLAST.

INTRODUCTION

In this paper two things are addressed. The first is the provision of a novel implementation of a persistent q-gram index using a relational database management system (RDBMS). This implementation allows a greater value of q than has previously been reported. Further, it demonstrates the use of the RDBMS as a means of accessing the information stored on disk, the traditional bottleneck in the use of persistent indices.

Secondly, the index is applied to the task of mapping Affymetrix microarray probes, thereby demonstrating the practical use of the index implementation and providing a data set of use to those using the RAE 230 microarray. The advantage of our mapping approach is that it provides computational guarantees about the returned matches, making it possible to be certain that all matches meeting the chosen definition have been returned. Further, it is demonstrated that this can be achieved in reasonable time. The problems with the mappings provided by Affymetrix and the benefits of re-mapping are described below.

Probe Mapping Background

Probe mapping is essential to the interpretation of microarray experiments. A mapping should be current, accurate, and allow for some genetic variation, as in rat genomics various strains diverge from the published *Rattus norvegicus* sequence (Rat Genome Sequencing Consortium [RGSC], 2004). Ideally, a mapping should utilise criteria sufficient for a hybridisation signal to arise. These requirements have led to several investigations of the available mappings and to the creation of alternative mappings, summarised in Table 1.

Re-mapping has changed associations of probes to transcripts (Harbig, Sprinkle, & Enkemann, 2005; Leong, Yates, Wilson, & Miller, 2005; Stalteri & Harrison, 2007) and altered experimental interpretation (Dai et al., 2005; Lu & Zhang, 2006; Gautier, Mooller, Friis-Hansen, & Knudsen, 2004), with sequence-verification also improving measurement accuracy (Mecham et al., 2004b) and concordance between array platforms (Mecham et al., 2004a; Carter, Eklund, Mecham, Kohane & Szallasi, 2005; Severgnini et al., 2006; Barnes, Freudenberg, Thompson, Aronow, & Pavlidis, 2005) and generations (Elo et al., 2005). In addition, temporal inconsistencies (Perez-Iratxeta & Andrade, 2005) in NetAffx (Liu et al., 2003) have been highlighted. In NetAffx information centres on the representative sequence of each probeset, although problems with 'Representative Public IDs' have been identified (Dai *et al.*, 2005), which makes it difficult to link accurately to current annotation. The Affymetrix mapping of probesets to genome sequence aligns a probeset target sequence to the genome using BLAT (Kent, 2002), and chooses the best alignment to map the probes to the genome. Other alignments detected by BLAT are ignored (see www.affymetrix.com, Frequently Asked Questions, "What procedure is used to generate the genome alignment data?").

Ensembl (Hubbard et al., 2007) (www.ensembl.org) provides an alternative large-scale mapping. The mapping was generated in two steps, first mapping probes to the genome, allowing at most one base mismatch between probe and genome, and then applying filtering criteria to map probesets to transcripts. The published re-mappings (Harbig et al., 2005; Leong et al., 2005; Dai et al., 2005; Gautier et al., 2004; Mecham et al., 2004b; Chalifa-Caspi et al., 2004), along with the Affymetrix and Ensembl mappings, the known cross-platform comparisons (Mecham et al., 2004a; Carter et al., 2005; Severgnini et al., 2006), and the use of BLAST (Kim, Boyd, Athey, & Patel, 2005), all take different approaches. While Affymetrix map probesets via their target sequences, the others map the individual probe sequences, occasionally mapping mismatch (MM) probes as well as perfect match (PM) probes (Mecham et al., 2004b). With the exception of Affymetrix and Ensembl, probes are generally mapped directly to some form of transcript, although, in some cases, the genome is used (Dai et al., 2005).

Harbig et al. (2005) use probeset target sequences in BLAST queries to identify a set of RefSeqs (Pruitt, Tatusova, & Maglott, 2005). This subset of RefSeqs is used as a target for probe mapping, after which filters are used to link probesets to transcripts. In contrast, Gautier et al. (2004) map probes directly to RefSeq mRNA sequences. They also re-group the probes, effectively creating novel probesets, a tactic supported by Harbig et al. (2005) and also employed by others (Dai et al., 2005; Carter et al., 2005; Sandberg & Larsson, 2007). Such an approach means that probes capable of hybridising, or cross-hybridising, with a transcript can be grouped with other probes that can measure a signal from the same transcript.

The likelihood of a probe hybridising with a sequence depends on a number of factors, including sequence similarity. In almost all cases the mappings use exact matching, or allow at most one base mismatch. An exception is where the sequence with the greatest number of base matches to the probes in a set is chosen (Harbig et al., 2005). Default BLAST is also used to map probes (Kim *et al.*, 2005) and BLAT is used to map target sequences to the genome. In these two cases it is unclear how much variation from the probe sequences is tolerated, although Affymetrix advocate exact matching of all 25 bp.

When compared with Affymetrix mappings, novel mappings show a number of differences. Dai et al. (2005) found a 30-50% discrepancy in genes identified as being differentially expressed, while no array had greater than 19% of probes mapping to the expected RefSeq in the study by Mecham et al. (2004b). Gautier et al. (2004) concluded that in 36% of cases the probes matching a RefSeq were consistent with Affymetrix sets, while in the other cases there were differences, and Harbig et al. (2005) redefined 37% of probes. The magnitude of these differences shows the impact of re-mapping and illustrates limitations of the data available from Affymetrix.

All these mappings use stringent match criteria and require exact matching, allow at most one of the 25 bases to mismatch, or use only the best available alignment. However, it is known that probes hybridise to sequences that are not perfect matches and that such sequences can contribute to the measured signal. It has been noted that some MM probes regularly hybridise to the target more strongly than their PM partner (Naef, Lim, Patil, & Magnasco, 2002). Distinguishing between cross-hybridisation and target hybridisation is impossible without investigation and knowledge of what sequences could cross-hybridise, either as a result of perfectly matching the probe or simply binding strongly enough to contribute to the signal.

Calculations for determining melting temperature based on nearest-neighbour thermodynamics or GC content are well known. As a starting point for extending the variation from the probe sequence permitted in a mapping, however, we favoured simple rules based on certain observations from the literature. As sequences including mismatches bind, in some cases more strongly than perfect matches (Naef et al., 2002), mismatches were considered. It is also the case that mismatches are more likely to be tolerated at the ends of a probe sequence (Li & Stormo, 2001). Further, the inclusion of mismatches allows for some variation between published sequences and experimental strains. In addition, Ji *et al.* (2004) reported in a study of cross-species expression analysis that: "It appears that a relatively long stretch of perfect-matched oligonucleotides, $\geq$ 16 bp long, is essential for stable hybridisation and signal generation." Finally, Affymetrix documentation describing the design of the GeneChip Human Genome U133 Set, referred to by the equivalent document for rat, considers the identification of cross-hybridising probes. The rule resulting from their experimental work was that "a probe is not expected to cross- hybridise if it does not have at least two 8-mer perfect matches, including at least 12 consecutive matching bases" (Array Design For The Human Genome U133 Set, available from www.affymetrix.com). This supports the idea that sequences sharing 16 bp could hybridise.

Our work uses these observations to derive definitions of sequences that a probe could hybridise to. We then set out to find all occurrences of such sequences within the genome, which can then be used to link to current annotation, in a manner similar to Ensembl. The searches use a novel implementation of a q-gram index, benefiting from a database management system (DBMS), and supporting values of q greater than those previously reported. This enables matches to be found by generating all sequences meeting the match definition and locating them directly in the index.

The methods are benchmarked against BLAST (Altschul et al., 1997), BLAT (Kent, 2002) and the probe to genome mapping created by Ensembl. As the index-based method guarantees to find all matches meeting a definition, it also allows results from BLAST and Ensembl to be compared against that definition. By doing this, we provide the data necessary to make an informed choice between alternative mapping methods. Finally, one of our methods was selected to map all PM probes from the GeneChip Rat Genome 230 2.0 Array to the genome, locating over seven million new matches of length 16.

MATERIALS AND METHODS

Experimental Configuration

The indexing and search algorithms were implemented in Java version 1.4 and executed on a PC running Windows Server 2003 Enterprise Edition, and an Oracle 10g database, with a 2.4 GHz Intel Pentium 4 processor, 2 GB of RAM and two disks of 214 GB and 233 GB. Database table spaces were created to hold the partitioned q-gram index and the associated database index.

The Genome Index

Version 3.4 of the *R. norvegicus* genome was downloaded from Baylor College of Medicine (www.hgsc.bcm.tmc.edu/projects/rat). The data files use the alphabet S={A, C, G, T, N}. The indexing algorithm slides a window of length q along the text and indexes all such windows, so that for each chromosome of length n, $n-q+1$ windows are indexed. As each new character is encountered, an integer representation of the current window is calculated. For each q-gram, its chromosome (or sequence ID), location, and integer representation (code) are written to load files and loaded into the DBMS. $q = 19$ was chosen to directly locate words of length q or less.

The index data was stored in a single table with three columns: code, number and location. The table has a row for every q-gram in the genome, totaling 2,794,703,390 rows, and was range partitioned, by code, into 72 partitions. SQL*Loader used direct path loading. Code was chosen for the partitioning as it is most frequently used when querying. The space requirement of the table was 70.9 GB. A local index on code was built on the other disk, requiring 61.2 GB. Processing the genome, loading the data and constructing the index on code took 13.7 hours (hrs), composed of 3.6 hrs processing, 3.15 hrs loading and 6.95 hrs indexing.

In principle, the index is similar to q-gram indices described previously (Myers, 1994; Hyyrö & Navarro, 2003; Hunt, 2004). However, rather than using a memory-resident bitmap or header file to track the presence or absence of q-grams, it takes advantage of DBMS indexing (B-tree), similarly to Hunt (2007). This removes a traditional limitation on the size of q. When a q-gram index is constructed for a sequence with alphabet size $|S|$, there are $|S|^q$ possible q-grams, which results in a rapid growth in the number of possible q-grams with increasing q. For indices that rely on keeping a bit map or some other index for every possible q-gram in memory, there is a limit on the largest desirable q. The DBMS removes the limitation, and allows for greater flexibility in the choice of q.

A further benefit of using a DBMS is that the index structure is very simple. It is only necessary to compute an integer representation for each q-gram, in the range 0 to $|S|^q - 1$ inclusive. This saves some space, as storing a q-gram as a string would slightly increase space usage.

Example 1. Alphabet S={a, c, g, t}, |S|=4, and q=2. Assume the following encoding: enc(a)=0, enc(c)=1, enc(g)=2 and enc(t)=3. Then code(aa)=0, code(ac)=1, code(ag)=2, code(at)=3, code(ca)=4, .., and code(tt)=15. If genome='acacac', chromosome=5, and q=2, the relation **dna** will encode the q-grams 'ac', 'ca', 'ac', 'ca', and 'ac' as tuples (1,1,5), (4,2,5), (1,3,5), (4,4,5) and (1,5,5).

q-grams are encoded in $O(n)$ time, during a linear scan of the sequence, where each newcharacter is seen once. Given a q-gram w = $a_1 a_2 .. a_q$, its code, code(w)= $\sum_{i=1}^{q} enc(a_i) * |S|^{q-1}$.

Affymetrix Probes

The files RAE230A probe fasta.zip, RAE230B probe fasta.zip, RG-U34A probe fasta.zip, RG-U34B probe fasta.zip and RG-U34C probe fasta.zip were downloaded from the Affymetrix website. These were processed into load files, loaded into the database and placed in a single relation **Probe** (chipID, probeset, IP, 25bp, 19bp, 19bprc), which for each probe stores its interrogation position, 25 bp as given, a code of its central 19 bps and a code of the central 19bps reverse complement. Attributes were of type varchar2, except the codes and IP, which are numbers.

Match Definitions

Figure 1 illustrates the four match definitions used in relation to the 25 base probe sequences. The first three scenarios focus on the central 19 bases, allowing the three bases at either end of the probe to mismatch. In the first definition, 19ex, the central 19 bases are matched exactly. In 19k1 we allow at most one mismatch in the central 19 bases, and in 19k3 we allow at most three mismatches. The fourth definition, 16ex, differs from the others, and requires that 16 contiguous bases match exactly anywhere within the 25 bp probe. Multiple 16ex matches within one 25 bp probe are treated as one match. BLAST matches to the probe can be either contained in the probe, and

Figure 1. Match definitions

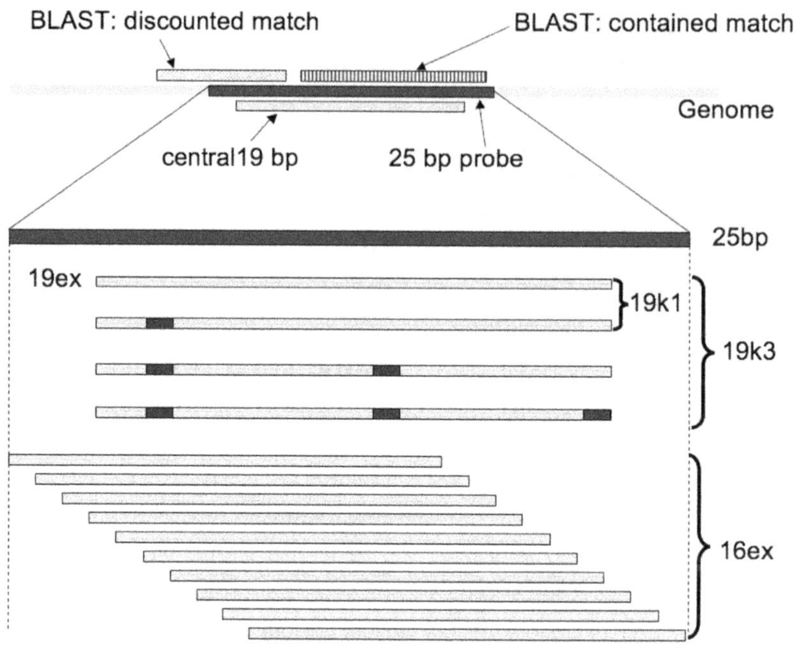

called a 'contained match', or partially overlap the probe, constituting a 'discounted match'.

Search Algorithms

In 19ex, see Figure 1, a join is performed between **dna.code** and **probe.19bp**, and **dna.code** and **probe.19bprc**. This gives genomic locations matching the central 19 bases on both strands.

Matching for 19k1 and 19k3 follows the specification of the *k*-mismatch problem (Gusfield, 1997), where at most *k* mismatches are permitted between the sequences being compared. When both the alphabet size and *k* are small, generating all possible *k*-mismatches and searching for exact matches can be efficient. For both definitions, Java stored functions were created. These accept the code for the central 19 bases (19 bp and 19 bprc) and return all possible *k*-mismatches, occurrences of which are then identified. For $k = 1$ there are 58 codes on each strand and 27,760 for $k = 3$.

In matching for 16ex, the 16 bp are treated as the prefix of a range of consecutive indexed 19-grams. As codes can be ordered lexicographically, all 19-grams with the same 16 bp prefix, say x, are adjacent, and conceptually form a block of codes from xaaa to xttt. We determine the first numerical code in the range with the given prefix x, and the last such code, i.e. code(xaaa) and code(xttt). The **dna** table is searched for codes that fall between the start and end of the range.

BLAST Benchmark

The *q*-gram methods were benchmarked against BLAST, in terms of both speed and quality of matches. The benchmark was chosen due to BLAST's popularity, combined with the fact that NCBI (www.ncbi.nlm.nih.gov) provide settings for 'short near-exact matching', suggested for cases where sequences hybridise. Two probesets from each of the five arrays were randomly selected. The resulting test set comprised 140 probes.

The executable blast-2.2.11-ia32-Win32.exe was downloaded from the NCBI and installed on the same machine as the database. Formatdb was executed on the FASTA file containing the rat genome, with -p set to false, and -o set to true, to create an index (Korf, Yandell, & Bedell, 2003).

In each of the following test settings both gapped and ungapped BLAST were executed. The settings were: (SNE) NCBI's 'short near-exact match'; (DEF) the default values, although in the context of BLAST probe mapping is atypical; (E.1 and E.01) e-value reduced to 0.1 and 0.01, with the other parameters unchanged. This produced eight sets of BLAST options, as all SNE, DEF, E.1 and E.01 were executed with both gapped (G) and ungapped (UG) settings.

The matches returned by the four exhaustive methods and the eight combinations of BLAST settings were recorded along with timings. Minitab was used for statistical analysis. A one-way ANOVA test, with Tukey's HSD pairwise comparisons, was applied to compare run times.

BLAT Benchmark

BLAT (Kent, 2002) standalone version 33.5 was run as another benchmark. For the random probes, two settings used tile size 16: (16.1) "-tileSize=16 -oneOff=1 -minMatch=1", and (16.0) "-tileSize=16 -minMatch=1". Two other settings used the default tile size of 11: (def.1) "-oneOff=1 -minMatch=1", and (def.0) "-minMatch=1". In 16.1 BLAT is to find a matching tile of length 16, with one mismatch, and then extend this to match the whole query. Similarly, in 16.0, an exact match of length 16 is sought, and then extended. In def.1 the window size is not specified (default is 11), but one mismatch is allowed, while in def.0 no mismatches in the tile are allowed. The expectation is that a window with one mismatch requires more computation and delivers more matches, and that a longer window with one mismatch will produce fewer matches than a shorter window with one mismatch, but might take longer to compute.

BLAT could not handle the entire genome, as it first builds a memory-resident index which did not fit within the available RAM (2 GB). For the timing test with the random 140 probes, data was split into three parts of about 900 MB each, with the times totaled and later averaged, over eight executions.

Ensembl and Whole Genome Index Comparison

Probe data from Ensembl, based on version 3.4 of the genome, was used in the large scale comparison. The files oligo_array.txt.table, oligo_feature.txt.table, oligo_ probe.txt.table and seq_region.txt.table were downloaded from Ensembl release 39. The Ensembl mapping was compared with the mapping based on the DBMS *q*-gram index. For all probes on the RAE230A and B chips, equivalent to the Rat Genome 230 2.0 array, the 16ex method was applied.

Next, duplicates of probes, present on both chips, and their matches were removed from the data set. This mapping for the 342,410 probes in the RAE 230 2.0 array was then compared with Ensembl. Multiple matches of the same probe to the same location, where more than one stretch of 16 bases matched, were filtered out.

Comparison of Matches

A database match and BLAST or Ensembl match are considered as the same if they are both 'contained' within the 25 bp corresponding to the definition, see Figure 1. If two matches are contained, hitting the same probe, even if different base to base alignments exist, they are classified as the same. In the BLAST comparison, all data for the 140 probes, from all methods, including Ensembl, was used. In the genome scale comparison, data generated by the 16ex method was compared to Ensembl mappings.

*Table 2. Query times (average in s) and total number of matches in a test using a batch of 140 probes. Row one shows the number of probes mapped in Ensembl. For the query times, the Tukey's HSD pairwise comparisons are significantly different from each other at the 5% cut-off, except for those indicated by the symbols *, †, ‡ and $*

Method	Mean ± SD time (s)	No of matches
Ensembl	Not known	117
19ex	4.87 ± 0.96	120
16ex	85.39 ± 2.15	2509
19k1	99.77 ± 4.48	358
BLAT 16.1	270.3 ± 14.33 *	84
BLAT 16.0	271.2 ± 16.96 *	78
BLAT def.0	344.6 ± 22.12	134
G E.01 BLAST	781.7 ± 7.378 †	121
UG E.01 BLAST	782.4 ± 21.55 †	121
UG E.1 BLAST	790.5 ± 25.44 †	125
G E.1 BLAST	821.8 ± 7.435 †	153
BLAT def.1	973.6 ± 28.92	143
G DEF BLAST	2 286 ± 25.80 ‡	2678
UG DEF BLAST	2 358 ± 11.59 ‡	2678
G SNE BLAST	5 080 ± 45.79 $	174367
UG SNE BLAST	5 317 ± 18.87 $	175704
19k3	7 971 ± 13.08	98180

RESULTS

Timing and Matches

For each of the four index-based algorithms, the eight combinations of BLAST, and four combinations of BLAT, the total time for 140 probe sequences was measured eight times. The mean and SD of time in seconds and the number of hits are shown in Table 2. ANOVA was conducted to compare the average times of the different methods using a log transformation. This found strong evidence that average times were different ($F=15907$, $p<1\times10^{-16}$). Further, all the Tukey's HSD post-hoc pairwise comparisons were significant except for those noted in Table 2.

Figure 2, which uses a log scale on both axes illustrates those results, showing that the performance of 19ex was best at under 10s/query, and 19k3 was the slowest, at around 135min/query. Further, the performance of 16ex was similar to 19k1 (less than 100s/query). Three BLAT methods (16.0, 16.1 and def.0) were slower than the three fast database methods (19ex, 16ex, 19k1) and faster than BLAST, however, they returned fewer matches than 16ex and 19k1, while taking longer to compute. BLAST with E=.01 or E=.1 required around 13 minutes, while default BLAST (DEF) was slower (around 38min), but faster than SNE BLAST (85min). BLAT with def.1 setting (window size 11, one mismatch per window) was slower than BLAST with E.1 and E.01 and returned fewer matches than BLAST G E.1. In terms of the number of matches found, the results form the following clusters: 98 to 175 matches are found by all versions of BLAT, Ensembl, 19ex, E.1 and E.01; over 300 matches are found by 19k1; around 2500 by 16ex and DEF; and large

Figure 2. A comparison of average batch execution time (seconds) over eight repetitions, each batch mapping 140 probes, and total number of matches to the 140 probes.

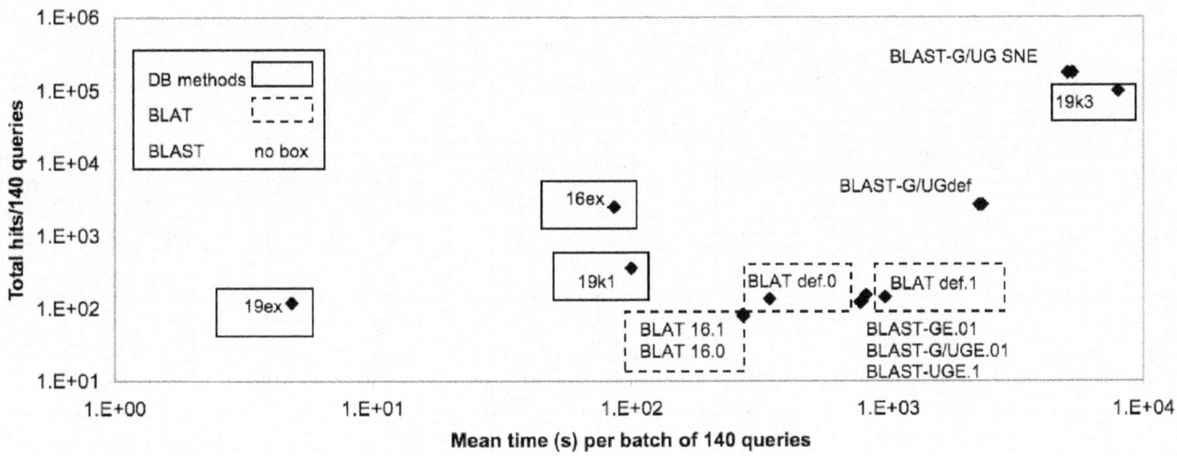

numbers, 100-200 thousand by SNE and 19k3. These raw numbers do not, with the exception of the *q*-gram methods, which are exhaustive, indicate the quality of the returned matches, or whether the same matches are being returned by all methods. Because BLAT seemed to return a lower number of matches than the database methods and was slower than them, it was decided not to include it in the detailed quality comparison described in the next section.

A Detailed Quality Comparison

Table 3 shows the matches returned by all methods in detail, using our containment criterion. The heuristics are assessed with reference to our definitions. Although each pair of methods in this comparison often find similar numbers of matches, they do not necessarily find the same matches. E.g. 19ex and G E.01, which find 120 and 121 matches, share 116 matches, but each method also finds some matches that the other method did not report. 19k3 does not miss any matches reported by E.01 or E.1, either gapped or ungapped, while DEF and SNE (G or UG) do not miss any matches reported by 16ex. The results contained in Table 3 are shown in Figure 3, which compares each of the database methods with BLAST and Ensembl. This shows that 19ex returns results most similar to Ensembl, G/UG E.01, and UG E.1, as the shared matches dominate the result set. For 19k1, large numbers of hits not found by most heuristics are returned, but 19k1 does not find hits returned by G/UG SNE. Both 19k3 and 16ex find predominantly matches not found by heuristics, however G/UG DEF share most matches with 16ex.

Table 4 shows the sensitivity of the heuristics measured against 16ex, 19ex, 19k1 and 19k3. Sensitivity values S_q, definition (1), are the percentage of the matches returned by the *q*-gram algorithm that are also found by BLAST or Ensembl, where M_q is the set of *q*-gram matches, and M_h is the set of heuristics-derived matches, and two matches are identical as defined by containment, see Figure 1. Met Definition (MD), definition (1), is the percentage of the matches returned by heuristics that had corresponding *q*-gram matches. Combined, S_q and MD show how close a method comes to returning the matches, and only those matches that correspond to locations meeting the chosen definition.

Table 3. Quality comparison: Database methods are shown as rows, and BLAST/Ensembl as columns. Bold indicates matches found by both methods. Above the shared matches, the number of matches found only by BLAST or Ensembl is given. Below, the number of matches identified only by the q-gram.

		G/UG **E.01**	G **E.1**	G/UG **DEF**	G **SNE**	UG **E.1**	UG **SNE**	Ens
	Total	121	153	2678	174367	125	175704	117
		5	33	2558	174247	8	175584	2
19ex	120	**116**	**120**	**120**	**120**	**117**	**120**	**115**
		4	0	0	0	3	0	5
		0	15	2421	174009	1	175346	0
19k1	358	**121**	**138**	**257**	**358**	**124**	**358**	**117**
		237	220	101	0	234	0	241
		0	0	1143	156784	0	158118	0
19k3	98180	**121**	**153**	**1535**	**17583**	**125**	**17586**	**117**
		98059	98027	96645	80597	98055	80594	98063
		1	1	169	171858	1	173195	1
16ex	2509	**120**	**152**	**2509**	**2509**	**124**	**2509**	**116**
		2389	2357	0	0	2385	0	2393

$$S_q = |M_q \cap M_h| / |M_q|$$

$$MD = |M_q \cap M_h| / |M_h| \qquad (1)$$

Table 4 shows that G SNE finds all matches as defined by 19ex, $S_q = 100\%$, but those BLAST matches comprise only MD= 0.1% of the matches returned by BLAST, and all other G SNE matches do not correspond to 19ex. When sensitivity is low, such as in the comparison of G/UG E.01 against 19k3, where $S_q = 0.1\%$, MD is high (100%). In a few cases there is greater concordance between our definitions and BLAST. G/UG DEF have $S_q = 100\%$ for 16ex and a high proportion of matches, MD= 93.7%, are contained matches.

Of the methods with 100% sensitivity for 19ex, G E.1 achieves the best trade-off between sensitivity and MD, with 78.4% of its matches being the same as 19ex. This is considerably higher than the other 100% sensitivity methods, with the next best, G/UG DEF, achieving 4.5%.

For 19k1 only G/UG SNE find all matches. These methods, however, return large numbers of matches that do not meet the definition, with only 0.2% of matches mapping to the same locations as 19k1. 71.8% of the matches are found by the G/UG DEF but also large numbers of additional matches are returned, and only 9.6% of matches meet the definition. A greater degree of specificity is attained by the remaining methods, which have a minimum MD of 90%. These, however, have greatly reduced sensitivity. The best, G E.1, has a sensitivity of 38.5%, while other settings vary between 32.7% and 34.6%.

For 19k3, all of the tested methods have sensitivities of less than 20%. Both SNE settings record sensitivities of 17.9%, and MD around 10%. UG/G DEF has sensitivities of 1.6% combined with MD of 57.3%. The remaining BLAST settings and Ensembl all have very low sensitivities, ranging from 0.1% to 0.2%. For each of these methods, Ensembl and the reduced e-value BLASTs, all of the matches that they return fall within the

Figure 3. Shared and disjoint matches for four database methods, in comparison with heuristics; Top-left shows 19ex, top-right 19k1, bottom left is 19k3, and bottom right 16ex.

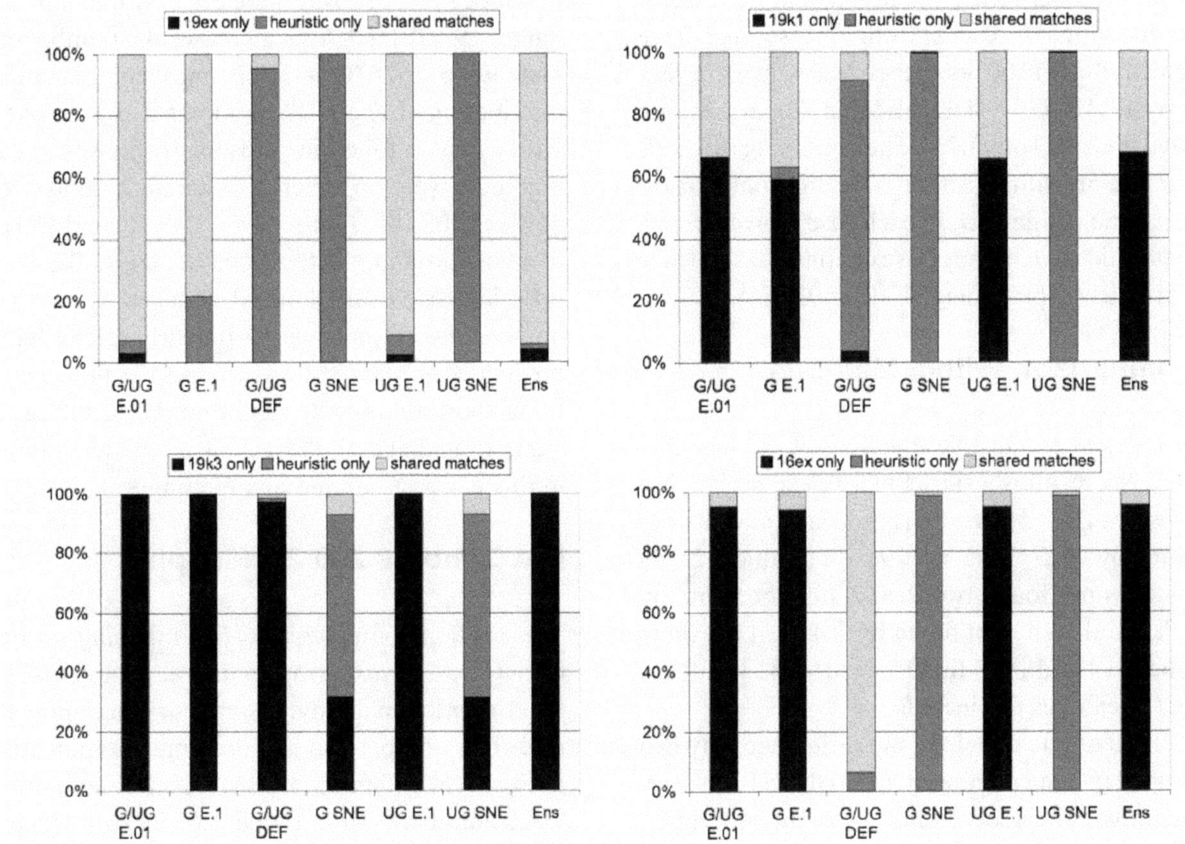

Table 4. Heuristics compared to q-grams, using S_q and MD measures.

		G/UG E.01	G E.1	G/UG DEF	G SNE	UG E.1	UG SNE	Ensembl
19ex	Sq	96.7	100	100	100	97.5	100	95.8
	MD	95.9	78.4	4.5	0.1	93.6	0.1	98.3
19k1	Sq	33.8	38.5	71.8	100	34.6	100	32.7
	MD	100	90.2	9.6	0.2	99.2	0.2	100
19k3	Sq	0.1	0.2	1.6	17.9	0.1	17.9	0.1
	MD	100	100	57.3	10.1	100	10	100
16ex	Sq	4.8	6.1	100	100	4.9	100	4.6
	MD	99.2	99.3	93.7	1.4	99.2	1.4	99.1

definition, but none of the methods coincides closely with 19k3.

In the 16ex setting the tested methods deliver a wide range of results. Both SNE settings have sensitivities of 100%, but their MD values are very low, at 1.4%. G/UG DEF also find all matches, and have high MD of 93.7%. The remaining methods are similar, with sensitivity ranging from 4.6% to 6.1% and a high MD. In each case, however, one match did not have a 16 base continuous sequence that matched, producing MD from 93.7% to 99.3%.

Comparison Within Methods

In indexed exhaustive matching it is clear that the 19ex result is a subset of 19k1, and 19k1 is a subset of 19k3. However, 16ex locates matches not found by 19k3, 19k1 and 19ex. It produces 2,389 matches not found by 19ex, 2,309 not found by 19k1, and 1,143 not found by 19k3. At the same time 19k1 and 19k3 find 158 and 96,814 matches that 16ex does not include.

In heuristic matching the difference between gapped (G) and ungapped (UG) BLAST was very small, with Pearson product moment correlation coefficient between the two result sets being 1.0. DEF and E.01 BLAST produced identical results in their G and UG forms. At E.1 G BLAST returned 28 alignments in addition to those located by UG E.1. For SNE, however, there is a divergence between G and UG BLAST, and the two methods share 174,214 matches, with an additional 153 unique to G BLAST and 1,490 found only by the UG version.

BLAST Alignments

BLAST matches can be assessed with regard to their hybridisation potential, based on alignment length, see Figure 4. For G/UG E.01 121 matches were returned. The most frequent alignment length was 25 (115 matches). This dips to zero alignments of length 23 and a few of lengths 22 and 21. A similar distribution is seen for G/UG E.1, with 125 alignments (UG) and 153 (G). A small number of alignments of 20 bp appear and in gapped BLAST (G) there are new matches of length 19. DEF BLAST increases the number of matches to 2,678 (G and UG), and changes match distribution. Length 16 dominates the results, with 17 bases following closely. It was observed that there were no matches of length 23, and two new matches of length 25. In SNE (G and UG), there are around 175 thousand matches. There is little increase in the longest matches, while the number of short matches, particularly of length 13 bp, increases massively, to around 123 thousand. Such short alignments might not be of interest, suggesting that default BLAST is a better choice in this context than the SNE settings.

Rat Genome 230 2.0 Mapping

This section compares the 16ex mapping of the GeneChip 230 2.0 array to the rat genome with the Ensembl mapping, using our containment rule. This comparison is represented graphically in Figure 5, which shows that matches unique to 16ex dominate. 16ex found 8,055,726 matches and Ensembl reported 461,366 probe locations, with 434,133 probe mappings shared by 16ex and Ensembl.

Only 5.4% of 16ex matches are in the Ensembl mapping, although they might potentially lead to hybridisation. As Ensembl allows a single mismatch in the 25 bases, the Ensembl matches with a mismatch in the middle seven bases are not found by 16ex, totalling 5.9% of the matches in the Ensembl mapping not detected by 16ex.

DISCUSSION

The DBMS *q*-Gram Index

In sequence analysis, *q*-gram indices have been investigated previously (Burkhardt et al., 1999; Cao, Li & Tung, 2005; Hunt, 2004; Hyyrö & Na-

Figure 4. BLAST: number of matches against match length for all settings.

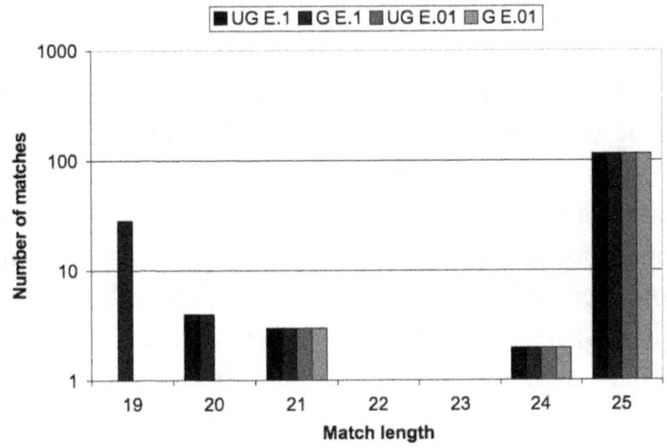

a

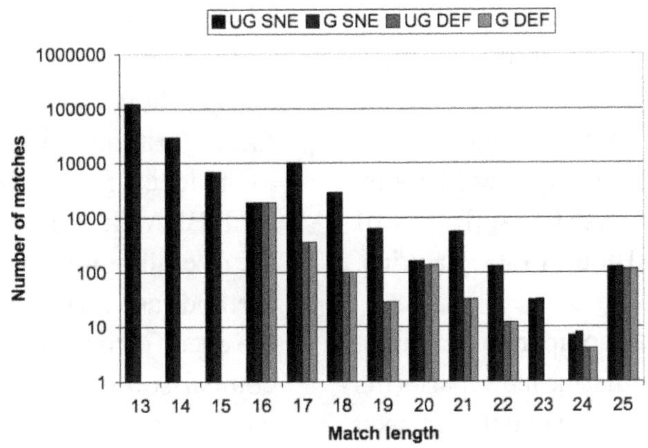

b

varro, 2003; Kent, 2002; Kim et al., 2005; Myers, 1994; Navarro & Baeza-Yates, 2000; Ning, Cox & Mullikin, 2001; Zheng, Close, Jiang, & Lonardi, 2003). While some of these are persistent, Hyyrö and Navarro (2003), for example, keep the header file in memory, limiting the desirable value of q. The largest reported q we are aware of is 12 (Hyyrö & Navarro, 2003). Searching for strings of length greater than q involves subdividing the query and then filtering the matches for those sequences that are adjacent to each other, as illustrated by miBLAST (Kim et al., 2005). Myers (1994) suggests an optimal value of q, leading to $q = 16$ for the human genome (Hyyrö & Navarro, 2003). As q increases, the header file becomes too

Figure 5. All matches delivered by 16ex and Ensembl.

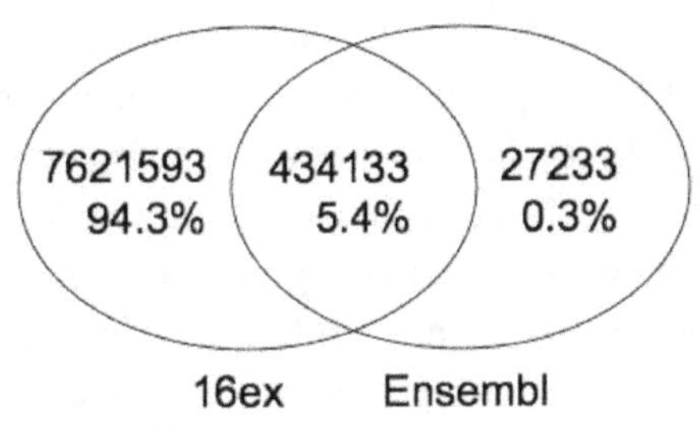

large to fit in memory. This problem is overcome by using a DBMS where a B-tree index mediates access to q-grams.

The DBMS index is a novel means of implementing q-grams for a large data set. q equals 19 in our implementation, which is larger than previously reported. An additional advantage of using the DBMS is the relative ease of subsequent data analysis and integration, even for large results sets, as shown in similar contexts (Stephens, Chen, Davidson, Thomas, & Trute, 2005; Eckman & Kaufmann, 2004).

The space requirement of the index is approximately 132.1 GB for the 2.8 Gbp rat genome, average 47.2 B/bp. Ours is not an especially space efficient implementation, however, the sizes and prices of disks favour this approach, as disk is much cheaper than main memory. The time of index construction is acceptable at under 14 hours, while the construction of a persistent suffix tree of the human genome took 30 hours (Tata, Hankins, & Patel, 2004).

Both timing and quality comparisons against BLAST and Ensembl show that the index performs well, as our methods either return matches missed by all other methods or perform significantly faster. Instead of using heuristics, we return all matches specified by a given definition. Further, the mapping of the Rat Genome 230 2.0 array shows that the method is robust and copes well with real data.

q-gram indices can support a variety of sequence analysis tasks, and are used by QUASAR (Burkhardt et al., 1999), SSAHA (Ning et al., 2001) and BLAT (Kent, 2002). None of those use a DBMS, except for Hunt (2007), which indexes 1.4 GB of protein sequence using a small window of size seven. Our experiments demonstrate clearly that a DBMS can also support DNA sequence analysis on a genomic scale in a robust and efficient manner. This indicates that the method could be adjusted to other scenarios.

Probe Mapping Performance

In all cases, other than 19k3, index-supported methods are significantly faster than BLAST and BLAT (Table 2). Unsurprisingly, the central 19 base definitions take longer to execute the more variation they allow. Within the database methods, however, execution time is not directly related to the number of matches returned. 16ex finds more matches than 19k1, by an order of magnitude, yet is significantly faster.

If time is the major consideration, either 16ex, 19ex, or 19k1 should be used for mapping, with the final selection depending on the desired level of sequence variation. Where a greater degree of variation is of importance, 19k3 may be appropriate. As shown in Table 3 and Figure 3, the three fast algorithms find numbers of matches similar to the levels reported by the reduced e-value and DEF BLAST, or Ensembl. However, even when the overall numbers of matches returned by two methods are very similar, there is a divergence in the specific matches returned. This is seen in the comparison of 19ex and gapped BLAST E.01, where 116 matches are shared, BLAST finds five new matches and 19ex another four. The same occurs on a greater scale in the comparison of SNE and 19k3. Comparing 19k3 with BLAST, in both gapped and ungapped SNE, shows that the number of shared matches is far less than those found by either of the methods separately. Indeed, BLAST identifies less than 18% of the matches found by 19k3 and only about 10% of the BLAST matches meet the 19k3 definition. The evidence that we provide enables researchers to choose the balance most appropriate to their needs.

Ensembl Comparison

This comparison illustrates that mapping results are very sensitive to the defined matching criteria. While Ensembl allows the smallest number of base variations between probe and genome, one in 25, it does list matches not detected by 19ex or

16ex. As mismatch probes on occasion hybridise more strongly than perfect match probes (Naef et al., 2002), there seems to be little question that they might need to be included in a probe mapping. This provides an argument for combining a number of criteria in mapping.

As 16ex can be sufficient to generate a signal, information on such matches should also be available to experimentalists. As the algorithm for this definition performed well, it was used in mapping the Rat Genome 230 2.0 array. This choice aimed to balance execution time with a desire to find all matches that could contribute to hybridisation. Of the tested methods, this seemed attractive to someone looking for possible matches that are not currently publicly available, and complements existing knowledge.

As over seven million matches are unique to our method (16ex), 27 thousand matches are found only by Ensembl, and 434 thousand are shared by both methods, using 16ex in addition to Ensembl in the evaluation of a microarray experiment could significantly change its interpretation. It should be noted that Ensembl employ filters during their probe to genome mapping and that this contributes to the difference in the number of matches. An avenue for future work is analysing a microarray data set with the new mapping, constructing novel probe sets, and assessing the impact.

CONCLUSION

A novel database implementation of a q-gram index was developed, allowing the use of a larger q than previously reported. The successful application of the index to a large-scale sequence analysis task was demonstrated, creating a probe mapping for the Rat Genome 230 2.0 array to the rat genome. Four matching criteria were tested: exact match to the central 19 bases of each probe (19ex), matching with at most one mismatch (19k1) and with at most three mismatches (19k3), as well as an exact match to any 16 bases within the probe (16ex). The timing and outcomes have been benchmarked against eight settings of BLAST and four settings of BLAT, on 140 probes, and the time and quality comparison for the benchmark presented. This shows that the database methods are competitive in terms of both execution time and quality of matching.

It was found that the index supported methods were either faster or found matches that were missed by BLAST and BLAT. In addition, the index guarantees to find all matches meeting the given definition.

Of the available options, the index supported 16ex algorithm was selected for mapping the probes of the Rat Genome 230 2.0 array, as it is fast and the results complement existing data from Ensembl. A comparison of 16ex and Ensembl was carried out for the whole genome, showing that the two mappings are complementary. An alternative to 16ex would be DEF BLAST, which gives similar results but is significantly slower.

Our large-scale analysis shows that using a DBMS to index and integrate data is simple, and can be done with modest equipment and programming means. We have demonstrated a successful application of the index in microarray probe mapping. Our new mapping can be used to re-interpret a microarray experiment, as it is possible to filter the data using a variety of criteria. Once the mapping is loaded into a DBMS, it can be visualised and integrated with quantitative trait loci and gene expression data. Further, the software can be extended to cater for longer or shorter indexing windows and matching or filtering criteria.

AVAILABILITY

The 16ex mapping of the probes to the Rat Genome 230 2.0 array can be downloaded from http://compbio.dcs.gla.ac.uk/sf/index.html. This URL also contains the software used in index construction and the search algorithms. A GUI driving the software package, which automates data indexing and searching, will be available at http://www.cis.strath.ac.uk/~ela.

FUNDING

Medical Research Council UK (G90/51 to E.H., G74/58 to E.H. and A.F.D.); European Union (MEIF-CT-2006-022986 to E.H, LSHG CT 2005-019015 to A.F.D.); British Heart Foundation (CH98001 and PG04/101 to A.F.D.); Wellcome Trust (066780/Z/01/Z to A.F.D.)

ACKNOWLEDGMENT

We thank B. Wiseman of the Computing Science Department at the University of Glasgow for database support.

REFERENCES

Altschul, S., Madden, T., Schaffer, A., Zhang, J., Zhang, Z., Miller, W., & Lipman, D. (1997). Gapped BLAST and PSI-BLAST: A new generation of protein database search programs. *Nucleic Acids Research*, *25*, 3389–3402. doi:10.1093/nar/25.17.3389

Barnes, M., Freudenberg, J., Thompson, S., Aronow, B., & Pavlidis, P. (2005). Experimental comparison and cross-validation of the Affymetrix and Illumina gene expression analysis platforms. *Nucleic Acids Research*, *33*, 5914–5923. doi:10.1093/nar/gki890

Burkhardt, S., Crauser, A., Ferragina, P., Lenhof, H.-P., Rivals, E., & Vingron, M. (1999). *q*-gram based database searching using a suffix array. In *RECOMB'99* (pp. 77–83).

Cao, X., Li, S. C., & Tung, A. T. (2005). Indexing DNA sequences using *q*-grams. In *CPM'03, Lecture Notes in Computer Science*, *3453*, 4–16.

Carter, S. L., Eklund, A. C., Mecham, B. H., Kohane, I. S., & Szallasi, Z. (2005). Re- definition of Affymetrix probe sets by sequence overlap with cDNA microarray probes reduces cross-platform inconsistencies in cancer-associated gene expression measurements. *BMC Bioinformatics*, *6*, 107. doi:10.1186/1471-2105-6-107

Chalifa-Caspi, V., Yanai, I., Ophir, R., Rosen, N., Shmoish, M., & Benjamin-Rodrig, H. (2004). GeneAnnot: comprehensive two-way linking between oligonuceotide array probesets and GeneCards genes. *Bioinformatics (Oxford, England)*, *20*, 1457–1458. doi:10.1093/bioinformatics/bth081

Dai, M., Wang, P., Boyd, A. D., Kostov, G., Athey, B., & Jones, E. G. (2005). Evolving gene/transcript definitions significantly alter the interpretation of GeneChip data. *Nucleic Acids Research*, *33*, e175. doi:10.1093/nar/gni179

Eckman, B. A., & Kaufmann, A. (2004). Querying BLAST within a data federation. *Bulletin of the IEEE Computer Society Technical Committee on Data Engineering*, *27*, 12–19.

Elo, L. L., Lahti, L., Skottman, H., Kyl¨aniemi, M., Lahesmaa, R., & Aittokallio, T. (2005). Integrating probe-level expression changes across generations of Affymetrix arrays. *Nucleic Acids Research*, *33*, e193. doi:10.1093/nar/gni193

Gautier, L., Mooller, M., Friis-Hansen, L., & Knudsen, S. (2004). Alternative mapping of probes to genes for Affymetrix chips. *BMC Bioinformatics*, *5*, 111. doi:10.1186/1471-2105-5-111

Gusfield, D. (1997). *Algorithms on strings, trees and sequences: computer science and computational biology.* Cambridge, UK: Cambridge University Press.

Harbig, J., Sprinkle, R., & Enkemann, S. A. (2005). A sequence-based identification of the genes detected by probesets on the Affymetrix U133 plus 2.0 array. *Nucleic Acids Research, 33*, e31. doi:10.1093/nar/gni027

Hubbard, T. J. P., Aken, B. L., Beal, K., Ballester, B., Caccamo, M., Chen, Y., Clarke, L., Coates, G., Cunningham, F., Cutts, T., et al. (2007). Ensembl 2007. *Nucleic Acids Research, 35(Database-Issue)*, 610–617.

Hunt, E. (2004). Indexed searching on proteins using a suffix sequoia. *Bulletin of the IEEE Computer Society Technical Committee on Data Engineering, 27*, 24–31.

Hunt, E. (2007). Exhaustive peptide searching using relations. In *BNCOD'07. Lecture Notes in Computer Science, 4587*, 13–24. doi:10.1007/978-3-540-73390-4_3

Hyyrö, H., & Navarro, G. (2003). A practical index for genome searching. In *SPIRE'03. Lecture Notes in Computer Science, 2857*, 341–349.

Ji, W., Zhou, W., Gregg, K., Yu, N., Davis, S., & Davis, S. (2004). A method for cross-species gene expression analysis with high-density oligonucleotide arrays. *Nucleic Acids Research, 32*, e93. doi:10.1093/nar/gnh084

Kent, W. (2002). BLAT - The BLAST-like alignment tool. *Genome Research, 12*, 656–664.

Kim, Y. J., Boyd, A., Athey, B. D., & Patel, J. M. (2005). miBLAST: scalable evaluation of a batch of nucleotide sequence queries with BLAST. *Nucleic Acids Research, 33*, 4335–4344. doi:10.1093/nar/gki739

Korf, I., Yandell, M., & Bedell, J. (2003). *BLAST.* Sebastopol, CA, USA: O'Reilly & Associates, Inc.

Leong, H. S., Yates, T., Wilson, C., & Miller, C. J. (2005). ADAPT: a database of Affymetrix probesets and transcripts. *Bioinformatics (Oxford, England), 21*, 2552–2553. doi:10.1093/bioinformatics/bti359

Li, F., & Stormo, G. D. (2001). Selection of optimal DNA oligos for gene expression arrays. *Bioinformatics (Oxford, England), 17*, 1067–1076. doi:10.1093/bioinformatics/17.11.1067

Liu, G., Loraine, A. E., Shigeta, R., Cline, M., Cheng, J., & Valmeekam, V. (2003). NetAffx: Affymetrix probesets and annotations. *Nucleic Acids Research, 31*, 82–86. doi:10.1093/nar/gkg121

Lu, X., & Zhang, X. (2006). The effect of GeneChip gene definitions on the microarray study of cancers. *BioEssays, 28*, 739–746. doi:10.1002/bies.20433

Mecham, B. H., Klus, G. T., Strovel, J., Augustus, M., Byrne, D., & Bozso, P. (2004a). Sequence-matched probes produce increased cross-platform consistency and more reproducible biological results in microarray-based gene expression measurements. *Nucleic Acids Research, 32*, e74. doi:10.1093/nar/gnh071

Mecham, B. H., Wetmore, D. Z., Szallasi, Z., Sadovsky, Y., Kohane, I., & Mariani, T. J. (2004b). Increased measurement accuracy for sequence-verified microarray probes. *Physiological Genomics, 18*, 308–315. doi:10.1152/physiolgenomics.00066.2004

Myers, E. W. (1994). A sublinear algorithm for approximate keyword searching. *Algorithmica, 12*, 345–374. doi:10.1007/BF01185432

Naef, F., Lim, D. A., Patil, N., & Magnasco, M. (2002). DNA hybridization to mismatched templates: A chip study. *Physical Review E: Statistical, Nonlinear, and Soft Matter Physics, 65*, 040902. doi:10.1103/PhysRevE.65.040902

Navarro, G., & Baeza-Yates, R. (2000). A hybrid indexing method for approximate string matching. *Journal of Discrete Algorithms*, *1*, 205–239.

Ning, Z., Cox, A., & Mullikin, J. (2001). SSAHA: A fast search method for large DNA databases. *Genome Research*, *11*, 1725–1729. doi:10.1101/gr.194201

Perez-Iratxeta, C., & Andrade, M. (2005). Inconsistencies over time in 5% of NetAffx probe-to-gene annotations. *BMC Bioinformatics*, *6*(1), 183. doi:10.1186/1471-2105-6-183

Pruitt, K. D., Tatusova, T., & Maglott, D. R. (2005). NCBI Reference Sequence (Ref-Seq): a curated non-redundant sequence database of genomes, transcripts and proteins. *Nucleic Acids Research*, *33*, D501–D504. doi:10.1093/nar/gki025

Rat Genome Sequencing Project Consortium. (2004). Genome sequence of the brown Norway rat yields insights into mammalian evolution. *Nature*, *428*, 493–521. doi:10.1038/nature02426

Sandberg, R., & Larsson, O. (2007). Improved precision and accuracy for microarrays using updated probe set definitions. *BMC Bioinformatics*, *8*, 48. doi:10.1186/1471-2105-8-48

Severgnini, M., Bicciato, S., Mangano, E., Scarlatti, F., Mezzelani, A., & Mattioli, M. (2006). Strategies for comparing gene expression profiles from different microarray platforms: application to a case-control experiment. *Analytical Biochemistry*, *353*, 43–56. doi:10.1016/j.ab.2006.03.023

Stalteri, M. A., & Harrison, A. P. (2007). Interpretation of multiple probe sets mapping to the same gene in Affymetrix GeneChips. *BMC Bioinformatics*, *8*, 13. doi:10.1186/1471-2105-8-13

Stephens, S. M., Chen, J. Y., Davidson, M. G., Thomas, S., & Trute, B. M. (2005). Oracle database 10g: a platform for BLAST search and regular expression pattern matching in life sciences. *Nucleic Acids Research*, *33*, D675–D679. doi:10.1093/nar/gki114

Tata, S., Hankins, R. A., & Patel, J. M. (2004). Practical suffix tree construction. In *VLDB'04* (pp. 36–47).

Zheng, J., Close, T. J., Jiang, T., & Lonardi, S. (2003). Efficient selection of unique and popular oligos for large EST databases. *Lecture Notes in Computer Science*, *2676*, 384–401. doi:10.1007/3-540-44888-8_28

This work was previously published in the International Journal of Knowledge Discovery in Bioinformatics, Volume 1, Issue 1, edited by Xiao-Li Li and See-Kiong Ng, pp. 48-65, copyright 2010 by IGI Publishing (an imprint of IGI Global).

Chapter 3
Improving Prediction Accuracy via Subspace Modeling in a Statistical Geometry Based Computational Protein Mutagenesis

Majid Masso
George Mason University, USA

ABSTRACT

A computational mutagenesis is detailed whereby each single residue substitution in a protein chain of primary sequence length N is represented as a sparse N-dimensional feature vector, whose M << N nonzero components locally quantify environmental perturbations occurring at the mutated position and its neighbors in the protein structure. The methodology makes use of both the Delaunay tessellation algorithm for representing protein structures, as well as a four-body, knowledge based, statistical contact potential. Feature vectors for each subset of mutants due to all possible residue substitutions at a particular position cohabit the same M-dimensional subspace, where the value of M and the identities of the M nonzero components are similarly position dependent. The approach is used to characterize a large experimental dataset of single residue substitutions in bacteriophage T4 lysozyme, each categorized as either unaffected or affected based on the measured level of mutant activity relative to that of the native protein. Performance of a single classifier trained with the collective set of mutants in N-space is compared to that of an ensemble of position-specific classifiers trained using disjoint mutant subsets residing in significantly smaller subspaces. Results suggest that significant improvements can be achieved through subspace modeling.

DOI: 10.4018/978-1-4666-1785-8.ch003

INTRODUCTION

Protein engineering experiments involving the synthesis and analysis of new proteins, each differing from the wild type via a single amino acid replacement introduced into the native sequence, yield important insights into protein folding and activity by characterizing the structural and/or functional roles of constituent residue positions (Lehmann, Pasamontes, Lassen, & Wyss, 2000; Vieille & Zeikus, 2001; Yang, Wang, & Fitzgerald, 2004). However, the prohibitively time consuming and expensive nature of conducting comprehensive single residue mutagenesis studies leads researchers to prioritize these experiments based on information gleaned from a variety of sources, including predictions obtained from *in silico* models. Such models may utilize an evolutionary scoring function (Kumar, Henikoff, & Ng, 2009), apply physical (Kollman et al., 2000; Pitera & Kollman, 2000), statistical (Kwasigroch, Gilis, Dehouck, & Rooman, 2002; Parthiban, Gromiha, Hoppe, & Schomburg, 2007; Zhou & Zhou, 2002), or empirical (Bordner & Abagyan, 2004; Guerois, Nielsen, & Serrano, 2002) potentials, or implement machine learning tools (Capriotti, Fariselli, & Casadio, 2004; Cheng, Randall, & Baldi, 2006; Huang, Gromiha, & Ho, 2007; Verzilli, Whittaker, Stallard, & Chasman, 2005). In each case the model is designed to predict a specific outcome of single residue substitutions in a protein, for example any evidence of an effect on protein activity, the relative change in protein stability, or more broadly, any pathological consequence to the organism.

With machine learning models, the single residue protein mutants populating training sets are each typically represented as a feature vector consisting of attributes that characterize the sequence or structure, frequently combined with components that reflect evolutionary information. Recently, studies have begun to focus on combining potential functions with machine learning methods by using several of the energy terms associated with the mutant proteins as attributes in feature vectors (Dehouck et al., 2009; Lise, Archambeau, Pontil, & Jones, 2009; Masso & Vaisman, 2007, 2008). One such approach involves a computational mutagenesis procedure that utilizes a four-body, knowledge based, statistical contact potential and yields, for any mutation due to a single amino acid replacement in a protein structure, an N-dimensional vector of ensuing environmental perturbations occurring at each of the N constituent residue positions (Masso & Vaisman, 2007). Since the average sized polypeptide chain in a protein structure consists of $N \sim 200$ amino acids, protein mutants are represented by vectors in high-dimensional Euclidean space. For each mutant, the methodology captures only local effects at the $M << N$ residue positions that are structurally closest to the mutated residue (including the mutated position itself) identified via Delaunay tessellation of the protein structure, a classical computational geometry technique, where the value of M and the identities of the M nonzero vector components vary according to the residue position being mutated (Masso & Vaisman, 2007). Hence the vectors are sparse, with those representing the collective set of 19 mutants, obtained by introducing all possible alternative amino acid substitutions at one particular position in the protein, residing in the same particular M-dimensional subspace.

All possible single residue replacements in a protein of size N theoretically lead to $19N$ mutants; however, experimental comprehensive mutagenesis studies generally involve the synthesis and analysis of far fewer mutants, which subsequently can be classified based on observed changes relative to the native protein (e.g., the activity of each mutant is either "unaffected" or "affected" by the respective amino acid substitution). Such an experimental dataset for a protein, with each mutant represented as an N-dimensional feature vector of inputs along with a corresponding categorical output attribute identifying the experimental mutational consequence, can be used to train and evaluate the performance of models obtained via the implementation of supervised classification algorithms. These predictive models may subsequently be used to classify the complement of single residue substitutions not in the training set that have yet to be investigated experimentally. However given a mutant dataset of limited size, the curse of dimensionality may complicate the ability of some machine learners to train an accurate model since the number of mutants needed to adequately sample a space increases exponentially with its dimensionality N (Bellman, 1961). Subspace modeling is a commonly used approach for dealing with this situation (Cai, He, & Han, 2007; De la Torre & Black, 2003; Yan et al. 2006), and the *in silico* mutagenesis procedure briefly outlined above naturally identifies N distinct M-dimensional subspaces.

In this applied data mining exercise, we begin by detailing the computational mutagenesis methodology whose underpinnings include a four-body statistical potential and the Delaunay tessellation algorithm applied to protein structures. Next, we use the approach to characterize a relatively large experimental mutant dataset generated by introducing into the bacteriophage T4 lysozyme protein (N = 164) either 12 or 13 amino acid substitutions at all but the first residue position. The researchers experimentally determined a qualitative level of activity for each mutant relative to the native T4 lysozyme, leading to two distinct categories of synthesized mutants. We use the entire dataset for training classifiers and evaluating their performance, both by representing the mutants as high-dimensional feature vectors, as well as by defining a position-dependent dimensionality reduction approach that extracts a relevant subset of only seven of these components as attributes. Finally, we perform subspace modeling by separately classifying mutant subsets that each consist of only residue substitutions at a particular position, which yields an ensemble of 163 classification tasks, and we aggregate the results. The relative impact of dimensionality reduction and subspace modeling to overall performance is discussed.

MATERIALS AND METHODS

Delaunay Tessellation of Protein Structure

The Delaunay tessellation of a set of points $P = \{x_1, x_2, x_3, \ldots, x_N\}$ in 3-dimensional (3D) Euclidean space yields a convex hull of space-filling, non-overlapping, irregular tetrahedra whose combined vertices consist of precisely all elements of P. Two adjacent tetrahedral simplices in a tessellation may share a triangular face (three out of four points in common), a linear edge (two points in common), or a single vertex. Provided that no three points of P are collinear, no four points are coplanar or on the same circle, and no five points are on the same sphere, there exists a unique Delaunay tessellation of P (de Berg, Cheong, van Kreveld, & Overmars, 2008). The technique is applied to a protein structure by initially abstracting to a point each constituent amino acid, which for this study is selected to be the center of mass of residue side chain atoms, to yield a coarse-grained representation of the protein (Figure 1 b). Each of the simplices in the ensuing protein structure tessellation objectively identifies at the vertices a quadruplet of structurally nearest neighbor amino acid residues. To ensure only biochemically feasible quadruplet interactions, all protein structure tessellations are modified by the removal of edges longer than 12 angstroms (Figure 1 c). Delaunay tessellation implementations and visualizations, and all subsequent data analyses are performed using a combination of Qhull (Barber, Dobkin, & Huhdanpaa, 1996), Matlab (Version 7.0.1.24704 (R14) Service Pack 1), and an ad hoc suite of Java and Perl codes.

Figure 1. (A) Ribbon diagram and (B) center of mass (CM) trace for bacteriophage T4 lysozyme based on the PDB coordinate file 3lzm. (C) Delaunay tessellation of T4 lysozyme subject to a 12 angstrom cutoff on allowable edge lengths. (D) Subset of all ten tetrahedral simplices from the tessellation that share the CM coordinates of residue N163 as a vertex. There are nine additional residue positions represented at the vertices of these simplices by their respective CM coordinates that collectively define the tessellation based neighborhood of residue position N163. (E) EC scores associated with the N163L mutant. The ten positions with nonzero EC scores correspond precisely to those whose CM coordinates are vertices for the simplices in (D). An alternative 7-dimensional feature vector of nonzero components for the N163L mutant is obtained from the residual profile by extracting the EC score at the mutated position, followed by the ordered EC scores at six of the closest neighbors determined by the lengths of simplex edges in (D).

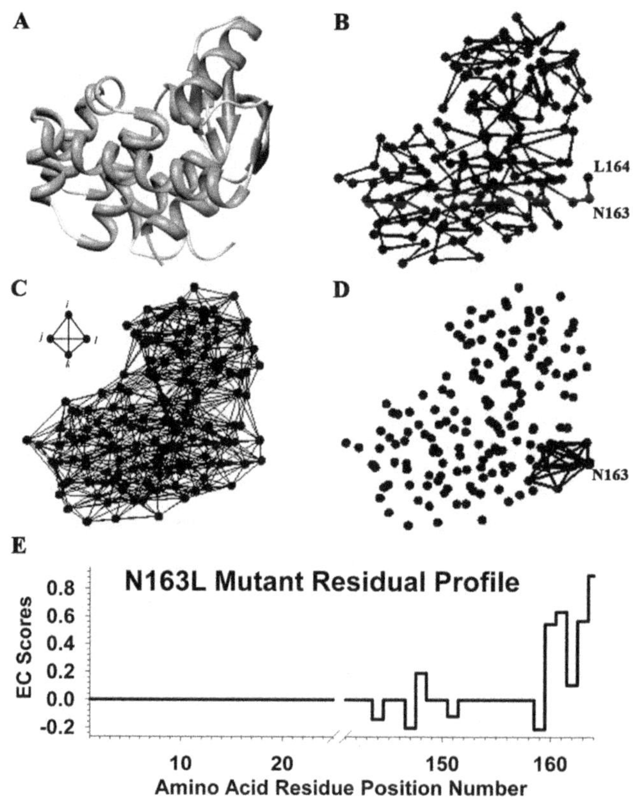

Four-Body Statistical Potential

The amino acid building blocks of proteins form a $K = 20$ letter alphabet A. The total number of $r = 4$ letter subsets (quadruplets) of A that can be enumerated, assuming that permutations of the four letters in a quadruplet do not constitute distinct subsets while allowing a quadruplet to contain repeats of the same letter, is given by

$$\binom{K + r - 1}{r} = \binom{20 + 4 - 1}{4} = 8855.$$

These conditions reflect the facts that ordering is not taken into account when four amino acids are identified at the vertices of the simplices in protein structure tessellations, and that the same amino acid may appear multiple times in a protein sequence and in close proximity structurally to form the vertices of a tetrahedral simplex. Since on average only a few hundred simplices and their respective quadruplets are encountered when a single protein is tessellated, a diverse dataset of 1375 high-resolution protein structures with low sequence and structure similarity was selected for tessellation using the PISCES server (Wang & Dunbrack, 2003) in order to reliably calculate simplicial nearest neighbor relative frequencies of occurrence f_{ijkl} for all 8855 possible quadruplets (i, j, k, l) in protein structure space. A rate expected by chance for the quadruplets is obtained with the use of a multinomial reference distribution given by

$$p_{ijkl} = \frac{4!}{\prod_{n=1}^{20}(t_n!)} \prod_{n=1}^{20} a_n^{t_n},$$

$$\text{where } \sum_{n=1}^{20} a_n = 1 \text{ and } \sum_{n=1}^{20} t_n = 4.$$

In the above formula, a_n represents the proportion of all amino acids in the 1375 tessellated protein structures that are of type n, and t_n represents the number of occurrences of amino acid n in the quadruplet. Through an application of the inverse Boltzmann principle from statistical mechanics (Sippl, 1993), a knowledge based statistical potential of amino acid quadruplet interaction is provided by the log-likelihood score s_{ijkl} = log (f_{ijkl} / p_{ijkl}), and the collection of 8855 residue quadruplet types together with their respective scores defines the four-body statistical potential (Vaisman, Tropsha, & Zheng, 1998).

Computational Mutagenesis

Given the Delaunay tessellation of any protein structure, the four-body statistical potential can be used to assign a score to each of the constituent tetrahedral simplices equivalent to that of the amino acid quadruplet identified at its four vertices. Since each amino acid vertex is generally shared by a number of adjacent tetrahedral simplices, the residue participates in multiple nearest neighbor quadruplets, and all amino acids represented at the other vertices of these simplices collectively form the *neighborhood* of that shared residue (Figure 1 d). Although the neighborhood amino acid positions are all structurally near their shared residue with respect to 3D Euclidean distance between the vertices, the neighbors are frequently distant from the shared residue in the linear (primary) sequence. For an amino acid at primary sequence position i in the protein, the sum of the scores of all tetrahedral simplices that share its vertex is referred to as the *residue environment score* q_i (Masso & Vaisman, 2007). The environment scores of all amino acids at positions $i = 1, 2, 3, \ldots, N$ in the protein can be arranged to form a *3D-1D potential profile* vector $\mathbf{Q}_{\mathbf{wt}} = \langle q_1, q_2, q_3, \ldots, q_N \rangle$, where the subscript wt (wild type) refers to the native protein (Bowie, Luthy, & Eisenberg, 1991). A similar profile $\mathbf{Q}_{\mathbf{mut}}$ can be obtained for any mutant of the protein due to a single residue substitution at some position j, by first replacing the identity of the amino acid accordingly at the vertex representing position j in the original tessellation (i.e., that of the native protein structure used for calculating $\mathbf{Q}_{\mathbf{wt}}$) and then recalculating all the residue environment scores (Masso & Vaisman, 2007).

The difference

$$\mathbf{R} = \mathbf{Q}_{\mathbf{mut}} - \mathbf{Q}_{\mathbf{wt}} = \langle EC_1, EC_2, EC_3, \ldots, EC_N \rangle$$

is referred to as the mutant *residual profile* vector (Fig. 1E), where $EC_i = q_{i,\mathrm{mut}} - q_{i,\mathrm{wt}}$ quantifies the relative *environmental change* or perturbation at each residue position i in the protein due to the mutation (Masso & Vaisman, 2007). In particular, if the mutation is due to a residue substitution at position j, then the single component EC_j in the residual profile vector **R** at the mutational epicenter is referred to as the mutant *residual score*, a critical value that empirically quantifies the overall change in sequence-structure compatibility of the mutant relative to the native protein (Masso, Lu, & Vaisman, 2006; Masso & Vaisman, 2007). Additionally, it can be shown that the only $M << N$ nonzero EC components of **R** (i.e., those amino acid positions that undergo a perturbation) correspond precisely to the mutated position j as well as those positions that form its neighborhood (i.e., those participating with it to form tetrahedral simplices in the tessellation). Hence, the approach effectively models important local effects of a mutation but does not account for any long-range consequences at structurally distant protein positions. The 19 mutants reflecting every substitution at position j each have a different set of M nonzero elements that happen to occur at the identical component locations in their respective vectors (i.e., the neighborhood of mutated position j is fixed), whose values are dependent on the particular type of residue replacement. These 19 sparse vectors consequently cohabit the same M-dimensional subspace, where the value of M and the locations of the M nonzero components of **R** vary according to the position that is mutated. Finally, an overall *comprehensive mutational profile* (**CMP**) can be generated for the protein, where each CMP_j score is calculated to be the mean of the residual scores for all 19 mutants at position j and

$$\mathbf{CMP} = \langle CMP_1, CMP_2, CMP_3, \ldots, CMP_N \rangle.$$

Residual profiles **R** sparsely represent mutant proteins in a high-dimensional feature space. On the other hand, since each position in a protein generally has a neighborhood that contains at least six amino acids, mutants can alternately be characterized by feature vectors consisting of only seven ordered nonzero components: EC value at the mutated position (i.e., the residual score), followed by EC values at the six closest positions in the neighborhood, ordered based on Euclidean distance proximity of the six neighbors to the mutated position as measured by tessellation edge lengths. Subspace modeling, which in the context of this manuscript refers to individually classifying then pooling the results for each subset of mutants due to alternative residue substitutions at a single position, can be applied to both the original high-dimensional mutant feature vectors as well as those obtained based on the dimensionality reduction approach described above (i.e., the 7-dimensional vectors of nonzero components).

Mutant Dataset Description

The methods outlined in this study are applied to an experimental set of mutants for the protein bacteriophage T4 lysozyme, an enzyme with a solved 3D structure in the Protein Data Bank (Berman et al., 2000) (PDB ID: 3lzm (Matsumura, Wozniak, Sun, & Matthews, 1989)) consisting of a single polypeptide chain with $N = 164$ amino acid positions (Figure 1 a). Bacteriophage T4 is a virus capable of infecting and replicating in *E. coli* bacterial cells, and its lysozyme protein is responsible for liberating newly formed virions by hydrolyzing bacterial cell wall peptidoglycan (Poteete & Hardy, 1994). Researchers systematically inserted the same set of 13 amino acids as substitutions at all positions except the first, resulting in the synthesis of 2015 single residue mutants as well as 104 controls at positions where the native amino acid is identical to one in the substitution set (Rennell, Bouvier, Hardy, & Poteete, 1991). Thus, the dataset consists of either 12 or 13 substitutions (out of 19 possible) at each of positions 2 – 164. With each of the 2015 mutants, the effect of residue substitution on protein activity was subsequently determined by visualizing the sizes of any *E. coli* plaque forma-

Figure 2. (A) 3D-1D potential profile and (B) comprehensive mutational profile for the T4 lysozyme protein. (C) A CMP – potential profile correlation plot for T4 lysozyme effectively segregates residues by polarity.

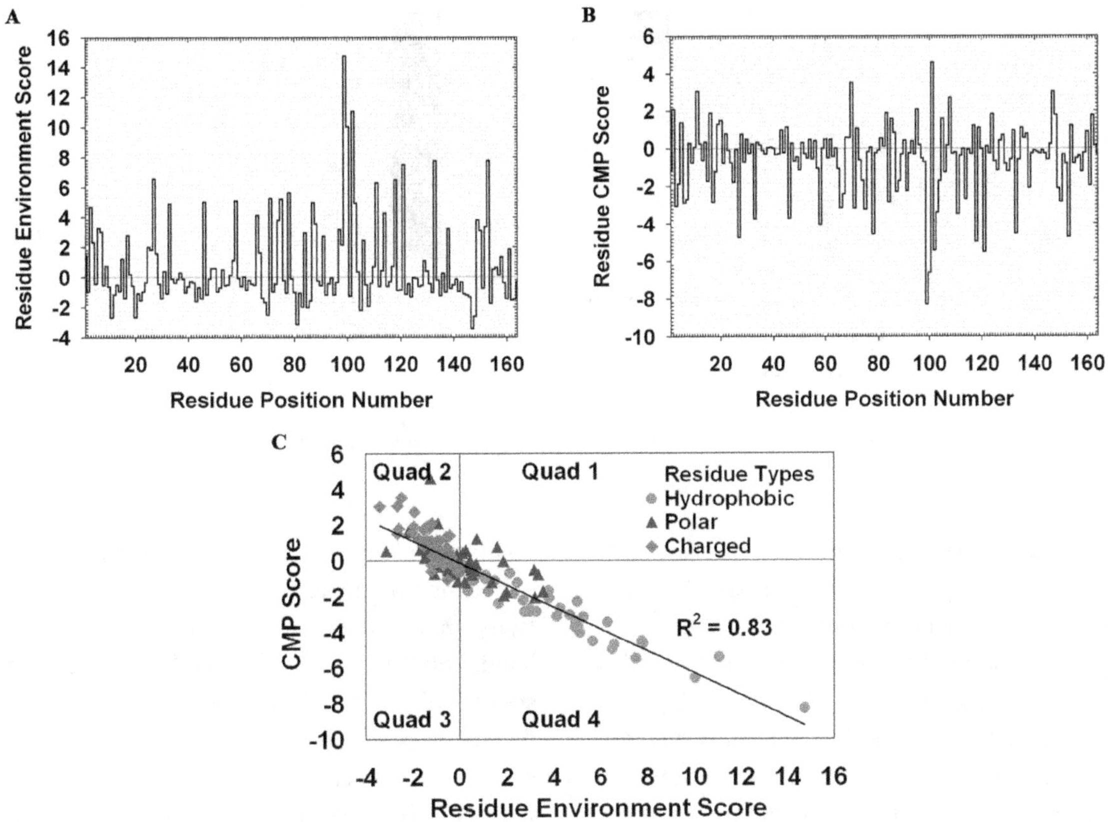

tions on Agar plates to which the lysozyme mutant was added (Rennell et al., 1991). Mutants were functionally categorized based on four size classes: large (i.e., full activity), medium, small, and no plaque (i.e., inactive). For this study, we classify as "unaffected" the 1377 mutants forming only large plaques, while the remaining 638 mutants that belong to the other three size categories are collectively classified as being "affected" by their respective residue substitutions. Though there are "mutational hotspots" at which any substitution inactivates the protein, as well as non-essential positions that do not affect activity, substitutions at a majority of positions yield mutants from both classes.

RESULTS AND DISCUSSION

Classification of T4 Lysozyme Residue Positions

The T4 lysozyme 3D-1D potential profile $\mathbf{Q}_{wt}$ (Figure 2 a) was computed as described in the Materials and Methods, based on the Delaunay tessellation of the residue side chain center of mass (CM) coordinates calculated from the T4 lysozyme PDB structure file 3lzm and by using the four-body statistical potential to calculate residue environment scores for all 164 positions. Residual profile vectors **R** were also computed for all 19 $\times 164 = 3116$ possible single residue mutations in

Figure 3. Residual scores for the T4 lysozyme experimental mutants elucidate a structure – function relationship. C / NC correspond to conservative / non-conservative amino acid substitutions, and the numbers in parentheses refer to the sizes of the respective mutant subsets.

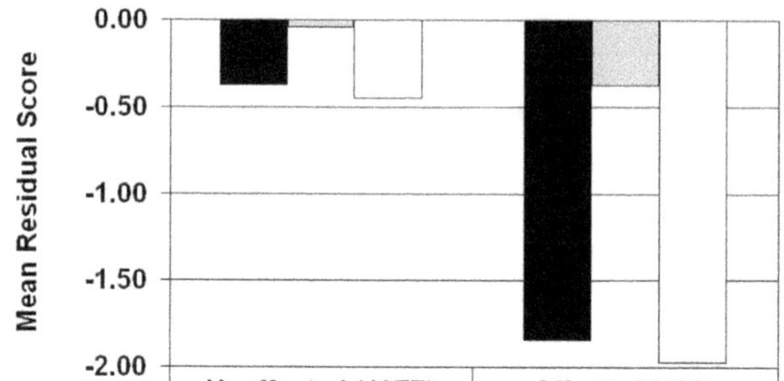

	Unaffected (1377)	Affected (638)
■ All (1377:638)	-0.37	-1.84
□ C (248:52)	-0.04	-0.37
□ NC (1129:586)	-0.45	-1.97

T4 Lysozyme Activity Upon Mutation

T4 lysozyme, from which their respective residual scores (i.e., the EC component corresponding to the mutated position number) were extracted. The T4 lysozyme comprehensive mutational profile **CMP** (Figure 2 b) was obtained by averaging the residual scores for the 19 amino acid substitutions at each position. A strong inverse correlation ($R^2 = 0.83$) exists between the CMP and 3D-1D potential profiles (Fig. 2C). The 164 residues in T4 lysozyme appear to cluster by polarity in Figure 2 c, with hydrophobic amino acids primarily residing in quadrant 4, the majority of charged residues occupying quadrant 2, and polar amino acids scattered within relatively close distance from the origin. Fisher's exact test applied to a 4×3 table that distributes all 164 amino acid counts by quadrant (rows) and residue type (columns) leads to rejection of the null hypothesis that no association exists between location and polarity ($p < 0.0001$).

Structure – Function Relationship in T4 Lysozyme

The plot in Figure 2 c is a purely computational construct that incorporates all possible single residue replacements in T4 lysozyme, and the experimental subset of mutants with known activity change data was not utilized. On the other hand, a clear trend was observed when the residual scores of the 1377 "unaffected" and 638 "affected" experimental T4 lysozyme mutants were respectively averaged (Figure 3, All category), whereby increasingly detrimental effects on structure (i.e., decreasing mean residual scores) are associated with higher levels of functional impairment (i.e., diminished activity). A *t*-test applied to these two experimental mutant activity classes revealed that a statistically significant difference exits between their mean residual scores ($p < 0.0001$). The mutants in each class of Figure 3 were further categorized based on whether the residue replacements were conservative (C) or non-conservative (NC) relative to the native amino acids, and mean residual scores were computed for each of these subgroups. By clustering the 20 naturally occurring amino acids into six subsets as {(A,S,T,G,P), (V,L,I,M), (D,E,N,Q), (R,K,H), (F,Y,W), (C)} based on similarities in physicochemical characteristics, intraclass substitutions are considered conservative while interclass residue replacements are non-conservative (Dayhoff,

Schwartz, & Orcut, 1978). Figure 3 reflects that the overall trend is driven by the NC mutations, since C substitutions minimally impact sequence-structure compatibility (i.e., residual scores of C mutations are generally relatively small in magnitude) regardless of their impact on activity.

Prediction of T4 Lysozyme Activity Changes

A variety of supervised classification algorithms were utilized for this study: pruned decision tree (DT), neural network (NN) with one hidden layer, support vector machine (SVM) with a radial basis function (RBF) kernel, and a random forest (RF) of ten trees. The relative consistency of observed trends across all four learners adds weight to any conclusions drawn, while deviations may highlight advantages or shortcomings of any particular approach. The algorithms were implemented under otherwise default classifier parameters with the WEKA suite of machine learning tools (Version 3.5.6) on a PC platform running Windows XP (Frank, Hall, Trigg, Holmes, & Witten, 2004). No data modifications were undertaken through filtering, normalization, or standardization, and cost-sensitive learning was not applied despite class imbalances.

Classifier performance was evaluated here based on leave-one-out cross-validation (LOOCV) for comparing original dataset results with those obtained via dimensionality reduction and/or subspace modeling, and tenfold cross-validation (10-fold CV) in the next subsection for establishing the statistical significance of predictions. Given TP (TN) = total number of correctly predicted "unaffected" or "U" ("affected" or "A") mutants and FN (FP) = total number of respectively misclassified mutants, overall accuracy is defined as Q = (TP + TN) / (TP + TN + FP + FN). For the "unaffected" class, S(U) = sensitivity = TP / (TP + FN) and P(U) = precision = TP / (TP + FP), while for the "affected" class, S(A) = TN / (TN + FP) and P(A) = TN / (TN + FN). Finally, the balanced error rate (BER) is defined as

$$\mathrm{BER} = \frac{1}{2} \times \left(\frac{\mathrm{FN}}{\mathrm{FN} + \mathrm{TP}} + \frac{\mathrm{FP}}{\mathrm{FP} + \mathrm{TN}} \right),$$

and Matthew's correlation coefficient (MCC) is given by

$$\mathrm{MCC} = \frac{\mathrm{TP} \times \mathrm{TN} - \mathrm{FP} \times \mathrm{FN}}{\sqrt{(\mathrm{TP} + \mathrm{FN})(\mathrm{TP} + \mathrm{FP})(\mathrm{TN} + \mathrm{FN})(\mathrm{TN} + \mathrm{FP})}}.$$

Six sets of LOOCV performance measures were obtained using each of the four supervised learning methods. First, predictions were obtained by using the full set of 2015 experimental T4 lysozyme mutants represented by their respective 164-dimensional (164D) vectors. Next, in an effort to explicitly group together substitutions at the same position, the full set was modified with the addition of a 165th attribute to each feature vector reflecting the primary sequence position number of the mutant. Third, subspace modeling (SM) was performed using the 164D vectors, whereby separate LOOCV tasks were applied to 163 disjoint subsets of the full set, each containing either 12 or 13 mutants representing only the amino acid substitutions at a particular position (2 – 164, no mutants at position 1), and the ensemble of prediction results were aggregated. The remaining three sets of results follow a similar pattern beginning with the full set of mutants represented in 7D feature space. Each nonzero vector consists of the residual score (i.e., EC score at the mutated position) followed by the ordered EC scores of the six structurally closest positions in the neighborhood, all extracted from the 164D feature vector.

A common observation among all four classifiers is the dramatic improvement to overall performance (higher Q and MCC; lower BER) that results from the application of SM relative to

Table 1. DT LOOCV classifier performance

N	Method	Q	S(U)	P(U)	S(A)	P(A)	BER	MCC
164	FS	0.77	0.85	0.82	0.60	0.65	0.27	0.46
(165)	FS + PA	0.78	0.86	0.82	0.60	0.67	0.27	0.48
164	SM	0.79	0.86	0.84	0.64	0.68	0.25	0.51
7	FS	0.73	0.88	0.77	0.43	0.62	0.35	0.34
(8)	FS + PA	0.74	0.87	0.78	0.46	0.62	0.34	0.36
7	SM	0.80	0.86	0.84	0.64	0.69	0.25	0.52

FS = Full Set; FS + PA = Full Set with an added Position Attribute; SM = Subspace Modeling.

Table 2. NN LOOCV classifier performance

N	Method	Q	S(U)	P(U)	S(A)	P(A)	BER	MCC
164	FS	0.75	0.89	0.78	0.45	0.66	0.33	0.39
(165)	FS + PA	0.68	0.99	0.69	0.02	0.56	0.50	0.05
164	SM	0.76	0.84	0.82	0.59	0.64	0.28	0.45
7	FS	0.71	0.85	0.76	0.41	0.56	0.37	0.29
(8)	FS + PA	0.68	0.99	0.68	0.01	0.29	0.50	0.00
7	SM	0.77	0.85	0.83	0.62	0.65	0.27	0.47

FS = Full Set; FS + PA = Full Set with an added Position Attribute; SM = Subspace Modeling.

the use of the full set, regardless of whether the 164D or 7D mutant feature vector representations are utilized (Tables 1, Table 2, Table 3, and Table 4). In all cases, the main driving force behind the trend is a considerable rise in S(A) that occurs with SM, which is directly due to an increase in the number of correct minority ("affected") class predictions. Additionally, performance on the full set in 164D surpasses that obtained with the 7D vectors, suggesting a modest loss of implicit structural information (via the locations of all nonzero EC components within the 164D vectors) accompanying dimensionality reduction based on EC scores at the mutated position and only its six closest neighbors. However, SM results based on both the 164D and 7D vectors are nearly identical and outperform those obtained with the full set, causing the degree of improvement of SM over use of the full set in 7D to appear greater than what is observed in 164D. Note that SM applied to the higher-dimensional vectors amounts to dimensionality reduction that takes into consideration the entire neighborhood of each position rather than only the six closest neighbors, since the associated subset of 12 or 13 mutants are represented by sparse vectors whose nonzero components are in the identical locations.

Within the framework of the full set, in order to explicitly identify and group each subset of 12 or 13 mutants associated with substitutions at the same primary sequence position, an additional attribute consisting of the position number was added to the feature vectors. In all cases with the exception of NN, results obtained with these 165D and 8D vectors fall squarely between those obtained with SM and those based on the full set of mutant vectors without the additional attribute (Tables 1, Table 2, Table 3, and Table 4). The position number attribute takes on integer values in the range of [2, 164], while the vast majority of the remaining EC score attributes take on real number values in the range of (-1, 1), and the data

Table 3. SVM LOOCV classifier performance

N	Method	Q	S(U)	P(U)	S(A)	P(A)	BER	MCC
164	FS	0.76	0.94	0.77	0.38	0.75	0.34	0.41
(165)	FS + PA	0.77	0.92	0.79	0.46	0.73	0.31	0.44
164	SM	0.77	0.85	0.82	0.60	0.64	0.28	0.45
7	FS	0.72	0.95	0.73	0.24	0.69	0.41	0.28
(8)	FS + PA	0.75	0.92	0.77	0.40	0.70	0.34	0.38
7	SM	0.78	0.85	0.83	0.62	0.65	0.27	0.48

FS = Full Set; FS + PA = Full Set with an added Position Attribute; SM = Subspace Modeling.

Table 4. RF LOOCV classifier performance

N	Method	Q	S(U)	P(U)	S(A)	P(A)	BER	MCC
164	FS	0.78	0.90	0.81	0.53	0.72	0.29	0.47
(165)	FS + PA	0.79	0.90	0.81	0.55	0.72	0.28	0.49
164	SM	0.80	0.89	0.83	0.60	0.72	0.26	0.52
7	FS	0.72	0.89	0.75	0.36	0.60	0.38	0.30
(8)	FS + PA	0.74	0.90	0.77	0.40	0.66	0.35	0.36
7	SM	0.78	0.86	0.83	0.62	0.67	0.26	0.49

FS = Full Set; FS + PA = Full Set with an added Position Attribute; SM = Subspace Modeling.

was not subsequently rescaled in any way. While DT (Table 1), SVM (Table 3), and RF (Table 4) methods can accommodate the new attribute, NN (Table 2) exhibits complications whereby S(A) is obliterated (i.e., most of the minority class is incorrectly predicted) and the MCC and BER values are consistent with a random classifier.

Classification difficulties arising from the curse of dimensionality are exasperated with increasing class imbalance (Evangelista, Embrechts, & Szymanski, 2006). The fact that S(U) >> S(A) for all methods evaluated here, based on the use of the full set in 164D mutant feature space, is a consequence of the 1377/638 ratio of "unaffected" to "affected" mutants. In particular, SVM suffers the most in this respect even though it is designed to withstand the consequences of working in a high-dimensional feature space, but it recovers on par with the other methods with the application of SM (Table 3). On the other hand, the three sets of RF results based on the use of the higher-dimensional vectors (Table 4) are highly consistent, a consequence of its implementation of bagging as well as the random selection of a small, fixed number of attributes at each tree node from which to select for splitting. Surprisingly, the corresponding DT results (Table 1) display a similar consistency and are close in value to those of RF. This is due to the fact that RF was implemented with only ten trees, and though a ten- or hundred-fold increase in the number of trees would likely lead to a significant increase in RF performance, the computing time would be prohibitively long. For the four classification methods, cpu time required to implement each LOOCV simulation on the full set in 164D or 165D feature space ranged from hours to days, while use of the 7D or 8D vectors took minutes to hours. On the other hand, application of LOOCV to each of the 163 SM subsets collectively ranged from seconds to minutes, regardless of 164D or 7D feature vector size. The algorithms can gener-

Figure 4. Distributions of 10-fold CV RF prediction performance (BER and MCC) values over 1000 permutations (random shuffles) of the activity class labels among the experimental mutants of T4 lysozyme represented as (A) 164-dimensional and (B) 7-dimensional feature vect

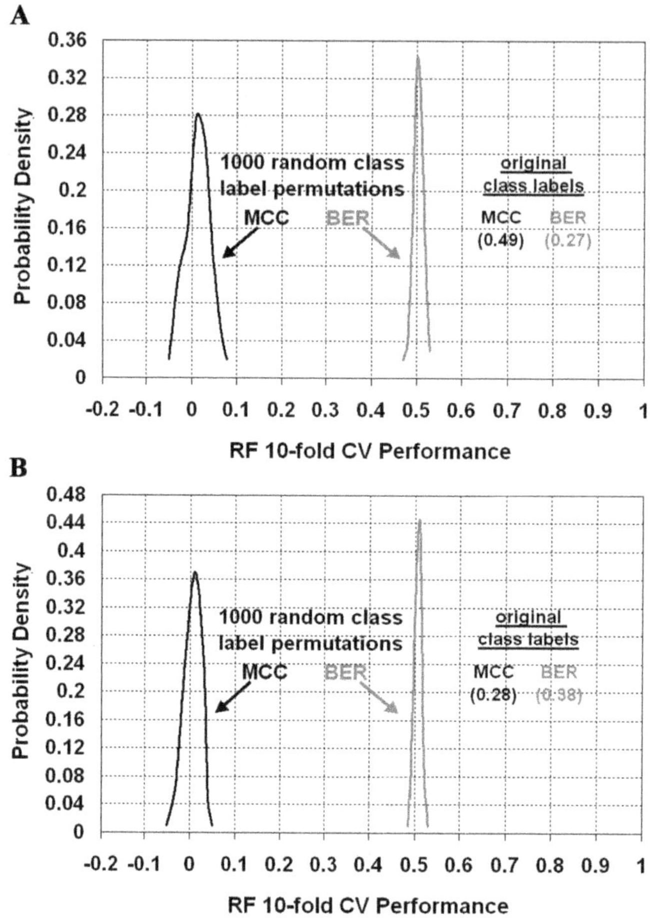

ally be ranked as SVM > NN > RF > DT based on cpu time required for running the simulations.

Internal and External Baseline Comparisons

Permutation tests were performed in order to evaluate the statistical significance of predictions obtained by representing the 2015 experimental T4 lysozyme mutants as either 164D or 7D feature vectors. Implementation of the RF algorithm prior to class shuffling (i.e., original mutant activity class labels) yielded 10-fold CV performance values of MCC = 0.49 and BER = 0.27 by using the 164D mutant feature vectors, and MCC = 0.28 and BER = 0.38 by using their 7D counterparts. Next, for each of one thousand activity class label permutations (random shuffles) generated independently for the 164D and 7D feature vector datasets, the performance of the RF algorithm was evaluated based on 10-fold CV. The resulting distributions of MCC and BER accuracy measurements based on the 164D mutant feature vectors (Figure 4 a) are narrowly centered around zero and 0.5, respectively (MCC = 0.00 ± 0.03, BER = 0.50 ± 0.01), with no permutation accuracy near what was

Table 5. Comparisons with related methods

Method	Q	S(U)	P(U)	S(A)	P(A)	BER	MCC
SIFT	0.63	0.59	0.82	0.72	0.45	0.34	0.29
MAPP	0.73	0.70	0.87	0.79	0.56	0.26	0.46
Pmut	0.52	0.42	0.77	0.74	0.37	0.42	0.15

obtained using the original class label arrangement. These results were repeated with the use of the 7D feature vectors (Figure 4 b), with MCC = 0.00 ± 0.02 and BER = 0.50 ± 0.01. Hence, in each of these cases the *p*-value for predictive power is less than 0.001. Finally, implementation of the DT, NN, and SVM algorithms generated similar results (data not shown).

In order to perform external comparisons with related methods, it was necessary to narrow the focus and utilize state-of-the-art approaches that are specifically designed to predict activity changes upon single residue substitutions. In particular, there exist a number of online servers for predicting protein stability changes upon mutation; however, they would not be useful in the present context because with the exception of relatively large changes in stability in either direction, which would likely (but not always) have a negative impact on protein activity, the effect on activity of mutants that display more modest stability changes definitely cannot be inferred. To this end, three popular online servers were identified for appropriately predicting activity changes in the 2015 experimental T4 lysozyme mutants: SIFT (http://sift.jcvi.org/) (Ng & Henikoff, 2006), MAPP (http://mendel.stanford.edu/SidowLab/downloads/MAPP/index.html) (Stone & Sidow, 2005), and Pmut (http://mmb2.pcb.ub.es:8080/PMut/) (Ferrer-Costa et al., 2005). Based on summary measures of the predictions obtained using these methods (Table 5), the approaches that we have detailed in this manuscript are highly competitive and perform at least as well as those developed for the other well-known servers.

CONCLUSION

This applied data mining study focuses on a computational approach for representing a system of protein mutants, due to single residue replacements, that are experimentally determined to have either "unaffected" or "affected" activity levels relative to the native protein. Application of the *in silico* mutagenesis procedure to an experimental dataset of bacteriophage T4 lysozyme mutants yields a collection of 164-dimensional sparse feature vectors with novel characteristics that are further analyzed via the implementation of well-known supervised classification algorithms. Efforts to improve classification include position-specific feature selection of only seven of the nonzero components; however, loss of some structural information results in minor reductions to performance. Alternatively, we apply a subspace modeling approach that involves aggregating an ensemble of position-specific classifiers, which significantly improves performance and leads to similar results in both 164- and 7-dimensional feature spaces.

REFERENCES

Barber, C. B., Dobkin, D. P., & Huhdanpaa, H. T. (1996). The quickhull algorithm for convex hulls. *ACM Transactions on Mathematical Software*, *22*(4), 469–483. doi:10.1145/235815.235821

Bellman, R. E. (1961). *Adaptive Control Processes*. Princeton, NJ: Princeton University Press.

Berman, H. M., Westbrook, J., Feng, Z., Gilliland, G., Bhat, T. N., & Weissig, H. (2000). The Protein Data Bank. *Nucleic Acids Research, 28*(1), 235–242. doi:10.1093/nar/28.1.235

Bordner, A. J., & Abagyan, R. A. (2004). Large-scale prediction of protein geometry and stability changes for arbitrary single point mutations. *Proteins, 57*(2), 400–413. doi:10.1002/prot.20185

Bowie, J. U., Luthy, R., & Eisenberg, D. (1991). A method to identify protein sequences that fold into a known three-dimensional structure. *Science, 253*(5016), 164–170. doi:10.1126/science.1853201

Cai, D., He, X., & Han, J. (2007, December). *Spectral regression: a unified approach for sparse subspace learning.* Paper presented at the IEEE International Conference on Data Mining, Omaha, NE.

Capriotti, E., Fariselli, P., & Casadio, R. (2004). A neural-network-based method for predicting protein stability changes upon single point mutations. *Bioinformatics (Oxford, England), 20*(1), I63–I68. doi:10.1093/bioinformatics/bth928

Cheng, J., Randall, A., & Baldi, P. (2006). Prediction of protein stability changes for single-site mutations using support vector machines. *Proteins, 62*(4), 1125–1132. doi:10.1002/prot.20810

Dayhoff, M. O., Schwartz, R. M., & Orcut, B. C. (1978). A model for evolutionary change in proteins. In Dayhoff, M. O. (Ed.), *Atlas of Protein Sequence and Structure* (*Vol. 5*, pp. 345–352). Washington, DC: National Biomedical Research Foundation.

de Berg, M., Cheong, O., van Kreveld, M., & Overmars, M. (2008). *Computational Geometry: Algorithms and Applications*. Berlin: Springer Verlag.

De la Torre, F., & Black, M. J. (2003). A framework for robust subspace learning. *International Journal of Computer Vision, 54*(1-3), 117–142. doi:10.1023/A:1023709501986

Dehouck, Y., Grosfils, A., Folch, B., Gilis, D., Bogaerts, P., & Rooman, M. (2009). Fast and accurate predictions of protein stability changes upon mutations using statistical potentials and neural networks: PoPMuSiC-2.0. *Bioinformatics (Oxford, England), 25*(19), 2537–2543. doi:10.1093/bioinformatics/btp445

Evangelista, P. F., Embrechts, M. J., & Szymanski, B. K. (2006). Taming the curse of dimensionality in kernels and novelty detection. In Abraham, A., Baets, B., Koppen, M., & Nickolay, B. (Eds.), *Applied Soft Computing Technologies: The Challenge of Complexity* (pp. 425–440). Berlin: Springer Verlag. doi:10.1007/3-540-31662-0_33

Ferrer-Costa, C., Gelpi, J. L., Zamakola, L., Parraga, I., de la Cruz, X., & Orozco, M. (2005). PMUT: a web-based tool for the annotation of pathological mutations on proteins. *Bioinformatics (Oxford, England), 21*(14), 3176–3178. doi:10.1093/bioinformatics/bti486

Frank, E., Hall, M., Trigg, L., Holmes, G., & Witten, I. H. (2004). Data mining in bioinformatics using Weka. *Bioinformatics (Oxford, England), 20*(15), 2479–2481. doi:10.1093/bioinformatics/bth261

Guerois, R., Nielsen, J. E., & Serrano, L. (2002). Predicting changes in the stability of proteins and protein complexes: a study of more than 1000 mutations. *Journal of Molecular Biology, 320*(2), 369–387. doi:10.1016/S0022-2836(02)00442-4

Huang, L. T., Gromiha, M. M., & Ho, S. Y. (2007). iPTREE-STAB: interpretable decision tree based method for predicting protein stability changes upon mutations. *Bioinformatics (Oxford, England), 23*(10), 1292–1293. doi:10.1093/bioinformatics/btm100

Kollman, P. A., Massova, I., Reyes, C., Kuhn, B., Huo, S., & Chong, L. (2000). Calculating structures and free energies of complex molecules: combining molecular mechanics and continuum models. *Accounts of Chemical Research*, *33*(12), 889–897. doi:10.1021/ar000033j

Kumar, P., Henikoff, S., & Ng, P. C. (2009). Predicting the effects of coding non-synonymous variants on protein function using the SIFT algorithm. *Nature Protocols*, *4*(7), 1073–1081. doi:10.1038/nprot.2009.86

Kwasigroch, J. M., Gilis, D., Dehouck, Y., & Rooman, M. (2002). PoPMuSiC, rationally designing point mutations in protein structures. *Bioinformatics (Oxford, England)*, *18*(12), 1701–1702. doi:10.1093/bioinformatics/18.12.1701

Lehmann, M., Pasamontes, L., Lassen, S. F., & Wyss, M. (2000). The consensus concept for thermostability engineering of proteins. *Biochimica et Biophysica Acta*, *1543*(2), 408–415.

Lise, S., Archambeau, C., Pontil, M., & Jones, D. T. (2009). Prediction of hot spot residues at protein-protein interfaces by combining machine learning and energy-based methods. *BMC Bioinformatics*, *10*, 365. doi:10.1186/1471-2105-10-365

Masso, M., Lu, Z., & Vaisman, I. I. (2006). Computational mutagenesis studies of protein structure-function correlations. *Proteins*, *64*, 234–245. doi:10.1002/prot.20968

Masso, M., & Vaisman, I. I. (2007). Accurate prediction of enzyme mutant activity based on a multibody statistical potential. *Bioinformatics (Oxford, England)*, *23*, 3155–3161. doi:10.1093/bioinformatics/btm509

Masso, M., & Vaisman, I. I. (2008). Accurate prediction of stability changes in protein mutants by combining machine learning with structure based computational mutagenesis. *Bioinformatics (Oxford, England)*, *24*(18), 2002–2009. doi:10.1093/bioinformatics/btn353

Matsumura, M., Wozniak, J. A., Sun, D. P., & Matthews, B. W. (1989). Structural studies of mutants of T4 lysozyme that alter hydrophobic stabilization. *The Journal of Biological Chemistry*, *264*(27), 16059–16066.

Ng, P. C., & Henikoff, S. (2006). Predicting the effects of amino acid substitutions on protein function. *Annual Review of Genomics and Human Genetics*, *7*, 61–80. doi:10.1146/annurev.genom.7.080505.115630

Parthiban, V., Gromiha, M. M., Hoppe, C., & Schomburg, D. (2007). Structural analysis and prediction of protein mutant stability using distance and torsion potentials: role of secondary structure and solvent accessibility. *Proteins*, *66*(1), 41–52. doi:10.1002/prot.21115

Pitera, J. W., & Kollman, P. A. (2000). Exhaustive mutagenesis in silico: multicoordinate free energy calculations on proteins and peptides. *Proteins*, *41*(3), 385–397. doi:10.1002/1097-0134(20001115)41:3<385::AID-PROT100>3.0.CO;2-R

Poteete, A. R., & Hardy, L. W. (1994). Genetic analysis of bacteriophage T4 lysozyme structure and function. *Journal of Bacteriology*, *176*(22), 6783–6788.

Rennell, D., Bouvier, S. E., Hardy, L. W., & Poteete, A. R. (1991). Systematic mutation of bacteriophage T4 lysozyme. *Journal of Molecular Biology*, *222*(1), 67–88. doi:10.1016/0022-2836(91)90738-R

Sippl, M. J. (1993). Boltzmann's principle, knowledge-based mean fields and protein folding. An approach to the computational determination of protein structures. *Journal of Computer-Aided Molecular Design*, *7*(4), 473–501. doi:10.1007/BF02337562

Stone, E. A., & Sidow, A. (2005). Physicochemical constraint violation by missense substitutions mediates impairment of protein function and disease severity. *Genome Research, 15*(7), 978–986. doi:10.1101/gr.3804205

Vaisman, I. I., Tropsha, A., & Zheng, W. (1998, May). *Compositional preferences in quadruplets of nearest neighbor residues in protein structures: statistical geometry analysis.* Paper presented at the IEEE Symposium on Intelligence and Systems, Washington, DC.

Verzilli, C. J., Whittaker, J. C., Stallard, N., & Chasman, D. (2005). A hierarchical Bayesian model for predicting the functional consequences of amino acid polymorphisms. *Journal of the Royal Statistical Society. Series C, Applied Statistics, 54*(1), 191–206. doi:10.1111/j.1467-9876.2005.00478.x

Vieille, C., & Zeikus, G. J. (2001). Hyperthermophilic enzymes: sources, uses, and molecular mechanisms for thermostability. *Microbiology and Molecular Biology Reviews, 65*(1), 1–43. doi:10.1128/MMBR.65.1.1-43.2001

Wang, G., & Dunbrack, R. L. Jr. (2003). PISCES: a protein sequence culling server. *Bioinformatics (Oxford, England), 19*(12), 1589–1591. doi:10.1093/bioinformatics/btg224

Yan, J., Liu, N., Zhang, B., Yang, Q., Yan, S., & Chen, Z. (2006, December). *A novel scalable algorithm for supervised subspace learning.* Paper presented at the IEEE International Conference on Data Mining, Hong Kong, China.

Yang, X., Wang, M., & Fitzgerald, M. C. (2004). Analysis of protein folding and function using backbone modified proteins. *Bioorganic Chemistry, 32*(5), 438–449. doi:10.1016/j.bioorg.2004.06.011

Zhou, H., & Zhou, Y. (2002). Distance-scaled, finite ideal-gas reference state improves structure-derived potentials of mean force for structure selection and stability prediction. *Protein Science, 11*(11), 2714–2726. doi:10.1110/ps.0217002

This work was previously published in the International Journal of Knowledge Discovery in Bioinformatics, Volume 1, Issue 4, edited by Xiao-Li Li and See-Kiong Ng, pp. 54-68, copyright 2010 by IGI Publishing (an imprint of IGI Global).

Chapter 4
Characterization and Classification of Local Protein Surfaces Using Self-Organizing Map

Lee Sael
Purdue University, USA

Daisuke Kihara
Purdue University, USA

ABSTRACT

Annotating protein structures is an urgent task as increasing number of protein structures of unknown function is being solved. To achieve this goal, it is critical to establish computational methods for characterizing and classifying protein local structures. The authors analyzed the similarity of local surface patches from 609 representative proteins considering shape and the electrostatic potential, which are represented by the 3D Zernike descriptors. Classification of local patches is done with the emergent self-organizing map (ESOM). They mapped patches at ligand binding-sites to investigate how they distribute and cluster among the ESOM map. They obtained 30-50 clusters of local surfaces of different characteristics, which will be useful for annotating surface of proteins.

DOI: 10.4018/978-1-4666-1785-8.ch004

INTRODUCTION

The importance and the impact of computational work in biology is ever increasing since large scale biological data of various types were accumulated in the past decade, which include genome sequences, protein-protein interaction networks, metabolomes, genome-scale transcriptions, and gene expression patterns. Those data contain key information for understanding orchestrated behavior of molecules in biological systems that is essential to sustain life. It is expected that bioinformatics play significant roles in analyzing such data, as computational techniques, *e.g.* clustering, feature characterization, data mining, and modeling, are indispensable in the analyses.

Of a particularly important and interesting problem is function prediction of proteins from the tertiary structures, as the structural genomics projects (Burley, 2000; Westbrook et al., 2003; Zhang & Kim, 2003) have been solving an increasing number of protein structures of unknown function. Indeed, as of May 2009, there are over 2800 proteins of unknown function in the Protein Structure Databank (PDB) (Berman et al., 2000). These proteins are remained as unknown function because so far no one has yet conducted experiments to characterize their function, and moreover, conventional sequence comparison based methods (Hawkins & Kihara, 2007), *e.g.* homology searches (Altschul et al., 1990; Altschul et al., 1997; Pearson & Lipman, 1988), functional motif (Hulo et al., 2006), and domain searches (Coggill et al., 2008), did not find significant similarity against protein sequences of known function. Ongoing efforts for better function prediction include development of sequence-based methods which are more sensitive and accurate than the conventional methods (Chitale et al., 2009a; Hawkins et al., 2008). For example, our group has recently developed two sequence-based methods, named the automated Protein Function Prediction (PFP) method (Hawkins et al., 2006; Hawkins et al., 2009) and the Extended Similarity Group (ESG) method (Chitale et al., 2009b), which efficiently and accurately mine function information from PSI-BLAST searches.

Alternatively, one can use the tertiary structure information for capturing similarity to proteins with known function that are stored in PDB (Thornton et al., 2000). Potential advantages of using structure information are two folds: Firstly, evolutionarily more distantly related proteins to a query protein could be identified because the global structure is better conserved than the primary sequence (Chothia & Lesk, 1986; Kihara & Skolnick, 2004). Secondly, physical features of functional local sites of proteins can be directly compared where interactions with ligand molecules or other proteins take place (Laskowski et al., 2005). A number of methods have been proposed which use local structure as a key feature for predicting function of proteins. Since small ligand molecules usually bind to a protein at its surface pocket regions, simply identifying pockets in the protein surface can identify active sites of enzyme in most of the cases (Li et al., 2007; Laskowski et al., 1996; Kawabata & Go, 2007). Programs which identify pockets include Visgrid (Li et al., 2007), POCKET (Levitt & Banaszak, 1992), LIGSITE (Hendlich et al., 1997; Huang & Schroeder, 2006), SURFNET (Laskowski, 1995), and PocketDepth (Kalidas & Chandra, 2008). An identified pockets can be further compared with known ligand binding pockets in a database to make prediction of the type of ligand that binds to it (Kahraman et al., 2007; Tseng et al., 2009; Kihara et al., 2009; Chikhi & Kihara, 2009; Binkowski & Joachimiak, 2008; Kalidas & Chandra, 2008; Yeturu & Chandra, 2008). In these methods, pockets are characterized by geometrical shapes, amino acid residues at pockets, and physicochemical properties such as the electrostatic potentials and hydrophobicity.

A fundamental limitation of these methods is that they only deal with pocket regions, *i.e.* prediction of ligand binding at pockets in protein surface. Although binding small ligand molecules

Figure 1. Flowchart of the methods

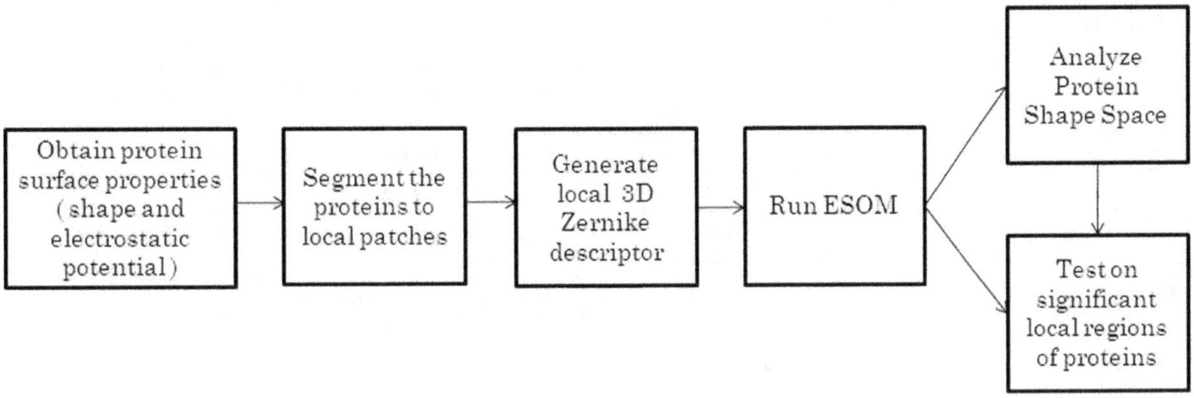

is a very important class of protein function, it only occurs in a subset of proteins, mainly enzymes and proteins that use co-factors. Moreover, on average pocket regions in an enzyme share only about 5% of the entire surface of the protein (Li et al., 2007). Hence, many of proteins in a genome and most of surface region of proteins are left out from applicability of these methods. Ideally, methods should be more generalized so that they can describe and compare any protein surface regions and annotate local regions with its potential functions.

Toward this goal, here, we classified local surface regions of proteins to see how diverse local surface patches are and how many different types of surface patches exist. We characterized a surface patch by two features, geometric shape and the electrostatic potential. Among shape descriptors that have been studied (Sael & Kihara, 2009; Tangelder & Veltkamp, 2008), we have chosen the 3D Zernike descriptors, a projection-based approach, because they have already been shown to be very effective in describing the shape and other physicochemical properties of proteins (Kihara et al., 2009; Sael et al., 2008b; Sael et al., 2008a).

Classification of surface patches is done by the emergent self-organizing map (ESOM), which is a variant of self-organizing map with a large size of neurons. Application of the ESOM to 118,003 surface patches taken from 609 proteins yields 30-50 clusters of different characteristics. To demonstrate practical usefulness of this classification, we examined classes of local surface patches occurring at binding sites of six ligand molecules and found that the combination reflects chemical similarity of the ligands. The manageable size of the clusters will be convenient to label local patches of proteins, and thus will be useful for annotating surface regions of proteins and also for defining signatures of functional sites of proteins.

MATERIALS AND METHODS

Figure 1. illustrates the procedure we conducted in this study. First, for a set of proteins, the surface is defined and the surface electrostatic potential is computed. Then, surface regions are segmented into local patches, using a sphere of a fixed size. Each local surface patches is characterized by the 3D Zernike descriptors of the surface shape and the electrostatic potential. Next, we use the ESOM to classify surface patches in the dataset. Three ESOM maps are constructed; one for classifying surface shape of patches, another one for classifying the electrostatic potential of patches, and the third one for classifying the combination of the surface shape and the electrostatic potential.

Table 1. Ligands and ligand binding proteins used in this study

Binding ligand	Number of proteins	Ligand binding proteins (Pdb ID and chain ID)
ATP	14	1a0i- 1a49A 1aylA 1b8aA 1dv2A 1dy3A 1e2qA 1e8xA 1esqA 1gn8B 1kvkA 1o9tA 1rdqE 1tidA
AMP	9	12asA 1amuA 1c0aA 1ct9A 1jp4A 1khtB 1qb8A 1tb7B 8gpbA
FMN	6	1dnlA 1f5vA 1jalA 1mvlA 1p4cA 1p4mA
HEM	16	1d0cA 1d7cA 1dk0A 1eqgA 1ew0A 1gweA 1iqcA 1nazA 1np4A 1po5A 1pp9C 1qhuA 1qlaC 1qpaB 1soxA 2cpo-
FAD	10	1cqxA 1e8gB 1eviB 1h69A 1hskA 1jqiA 1jr8B 1k87A 1poxA 3grsA

The resulting ESOM maps are further clustered to identify distinct groups of patches with different characteristics. Also, we compared binding pockets of six different ligands in terms of types of the clusters occurring at the binding pockets. Below we describe each step more in details.

Protein Dataset

A data set of representative proteins used in this study consists of 609 protein chains, each taken from a different family defined in the SCOP database (Andreeva et al., 2008). These families belong to one among 72 SCOP superfamilies. We selected the 72 superfamilies, each of which has at least three and no more than twenty families. These structures have a crystallographic resolution of 3.0 Å or better, have no more than 10 missing residues in the structure solved, have all heavy atom positions solved, are longer than 100 residues, and the structure similarity of each pair is less than a Z-score of 3.8 by the Combinatorial Extension (CE) program (Shindyalov & Bourne, 1998). The protein classification database by CE available at ftp://ftp.sdsc.edu/pub/sdsc/biology/CE/db/ata_3_8.txt is used for comparing structures of the proteins.

The ligand binding protein data set we use later in this study is composed of proteins taken from a previous work by Kahraman *et al.* (Kahraman et al., 2007) (Table 1,Fig. 2). A binding site of a ligand is defined as a set of protein heavy atoms which are closer than 3.5Å to any ligand atoms. Conformation of ATP (adenosine-5'-triphosphate) and AMP (adenosine-5'-diphosphate) are similar in shape. FAD (flavin-adenine dinucleotide) and NAD (nicotinamide-adenine-dinucleotide) are relatively flexible and take diverse bound conformations to proteins. FMN (flavin mononucleotide) and HEM (haemoglobin) are more distinct in shapes compared to the others in the dataset.

Defining Protein Surface and Computing Electrostatic Potentials

Two properties, the surface shape and the electrostatic potential are used to characterize surface of proteins. These are computed with the Adaptive Poisson-Boltzmann Solver (APBS) program (Baker et al., 2001). APBS is software for evaluating the electrostatic properties of biomolecules by solving the Poisson-Boltzmann equation. APBS first defines the solvent accessible regions of the protein and calculates the electrostatics potential for each of the voxels (3D grid points). Surface region of a protein is computed by taking the boundary of the solvent accessible and the solvent excluded region using APBS. The electrostatic potential is mapped to the computed protein surface.

Figure 2. 3D structures of the six ligands used in this study; ATP, AMP, FMN, heme, FAD, and NAD

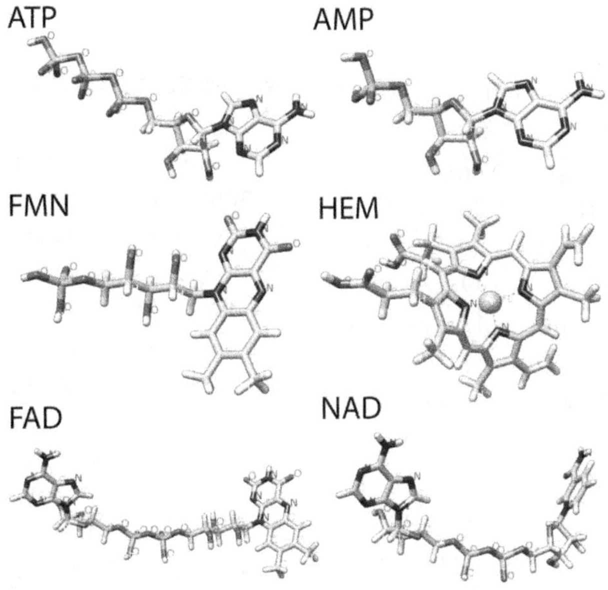

Segmenting Protein Surface to Local Patches

After obtaining the global shape and the electrostatic potential of a protein surface, the surface is segmented into local patches. First, seed points, or the centroids of the local surfaces, are selected by taking every 100th surface voxel. The average number of the seed points is approximately 194 per protein. The local patches are then extracted by taking a surface region that is within 6Å of each seed point. The resulting voxels are considered as an input 3D function, $f(x)$, which is expanded into the 3D Zernike Descriptors (3DZD) as described in the next section. 118,003 patches are obtained from the 609 proteins in the dataset.

3D Zernike Descriptors

The 3DZD are series expansion of an input 3D function, which allow rotation invariant and compact representation of a 3D object that is considered as the 3D function. The mathematical foundation of the 3DZD was laid by Canterakis (Canterakis, 1999) and later Novotni and Klein (Novotni & Klein, 2003) have applied them to 3D object retrieval. Below is brief mathematical derivation of the 3DZD. For detailed derivations and discussions, refer to the two papers (Canterakis, 1999; Novotni & Klein, 2003).

The first step in computing the 3DZD is derivation of the 3D Zernike moments. The 3D Zernike moments are series expansion of an input 3D function, $f(x)$, into the 3D Zernike polynomial. The input function, $f(x)$, in this work is a 3D grid where each grid cell (voxel) is assigned 1 if it is on the protein surface and 0 otherwise to represent a protein shape. In the case of the surface electrostatic potential, surface voxels are assigned with the electrostatic potential value of that position instead of 1. For the electrostatic potential, the voxels are classified into two groups, ones which have positive values and the other ones for negative values. The 3DZD are computed separately for the two groups of voxels and later combined (Sael et al., 2008a).

3D Zernike polynomials defined on degree l, order n, and repetition m, are given by

$$Z_{nl}^{m}(r,\vartheta,\phi)=R_{nl}(r)Y_{l}^{m}(\vartheta,\phi), \quad (1)$$

where $-l<m<l,\ 0\leq l\leq n,$ and *(n-l)* even. Spherical harmonics, $Y_l^m(\vartheta,\phi)$, are functions of a set of a polar angle, θ, and an azimuthal angle, φ. They are the angular portions of the solution to the Laplace's equations. The radial function defined by Canterakis(Canterakis, 1999), $R_{nl}(r)$, directly incorporates radius information into the basis function and are constructed so that $Z_{nl}^{m}(r,\vartheta,\phi)$ are polynomials when written in the Cartesian coordinates $\mathbf{x}$ = (x, y, z). Using all three spherical coordinate parameters (radius: *r*, polar angle: ϑ, and zimuthal angle: ϕ), the 3D Zernike polynomials are able to be formulated against any Cartesian coordinates.

The 3D Zernike moments of *f(x)* are defined as the coefficients of the expansion in this orthonormal basis:

$$\Omega_{nl}^{m}=\frac{3}{4\pi}\int_{|\mathbf{x}|\leq 1} f(\mathbf{x})\overline{Z}_{nl}^{m}(\mathbf{x})d\mathbf{x}. \quad (2)$$

Finally, rotation invariance is obtained by taking L_2 norm of 3D Zernike moments as the descriptor.

The parameter *n* is called the order of the 3DZD. The order determines the resolution, *i.e.* the number of terms, thus coefficients, of the descriptors. As mentioned earlier, *n* defines the range of *l*, and 3DZD are series of invariants for each pair of *n* and *l*, where n ranges from 0 to the order specified. In this work we set the order n=15, which yields 72 values for shape and 144 values for the electrostatic potential, to describe the local patches of proteins. The 3DZD of the electrostatic potential is twice longer because those for voxels with positive values and those with negative values are separately computed and later combined (Sael et al., 2008a). Since the ESOM assumes that data have a distribution that is close to the Gaussian distribution, we further performed log normalization to the resulting descriptors.

SOM and ESOM

The self-organizing map (SOM) is a data classification technique which reduces the dimensions of data through the use of self-organizing neural networks (Rojas & Feldman, 1996). The output of the SOM is a lower dimension map, usually two, of neurons which group similar data items together on the map. Each neuron on the map has a vector of weights, which are trained based on a set of input training data. The weights assigned to each neuron are composed of two parts. The first part represents the feature of the neuron, which has the same dimension as the input vectors. The second part represents its location in the 2D map. SOM tries to order the neurons such that they represent the distribution of the similarity of the input vectors. The SOM algorithm first initializes the weight vector map. Among several methods available for the initialization, we used sampling from the Gaussian distribution. Then, the feature weights of neurons and their location in the 2D map are updated iteratively. The process consists of three steps: selecting an input feature vector, finding the neuron that has a closest feature weights to the input, updating the weights of the closest neuron and its neighboring neurons, and updating the position of neighboring neurons so that similar neurons locate closer to each other (Rojas & Feldman, 1996).

Emergent SOM (ESOM) is a variation of SOM, which handles a larger number of neurons (at least 4000) and uses boundless maps (Ultsch, 2005; Ultsch, 2003). It embeds the maps to a finite boundless space such as sphere or toroid. Two visualization methods of the ESOM maps are used, namely, the P-matrix and the U-matrix. The P-matrix visualizes the density in the input data space using the Pareto density estimation. In general, it is suitable for dealing with slowly changing densities and overlapping clusters.

The U-matrix visualizes neurons on an ESOM map by a color coding that represents the sum of distances to all immediate neighbors normalized by the largest value in the neighboring neurons. Generally, the U-matrix is appropriate for handling data points which are clearly separated from each other. The ESOM program is available at http://databionic-esom.sourceforge.net/. The original paper by Ultsch describes the general ESOM training procedure in details.

The advantage of SOM/ESOM is that it is able to provide an intuitive visualization of the similarity of input data. On the other hand, a potential drawback is that often SOM/ESOMs trained on the same data set do not converge well to a similar map. In our work, to examine the convergence of the ESOM map we obtained, we further clustered neurons in the resulting map by using the affinity propagation clustering method (see the next section). As will be shown in Results, in our case we think that the ESOM performed satisfactorily for the purpose of the classification of protein surface patches, because neurons locating close to each other are well clustered.

The training data in this work are feature vectors of the 118,003 local surface patches in the 609 proteins. A feature vector used for local surface patches is the 3DZD describing the local shape, the electrostatic potential, or the combination of the two. As explained in the previous section, the 3DZD of the local shape have 72 values, those of the electrostatic potential have 144 values, and the combination of the shape and the electrostatic potential have 216 (*i.e.* 72+144) values. The training data set is inputted to ESOM program and trained using 1000 iterations and 4100 neurons.

Affinity Propagation Clustering Method

The affinity propagation clustering method clusters data by employing an idea of passing messages between them (Frey & Dueck, 2007). It was shown to have a low error rate and fast as compared to other common clustering methods. The number of clusters is influenced by the so-called preference parameter. Setting the preference parameter to the median of input distances results in a moderate number of clusters and setting them to the minimum of input distances results in a smaller number of clusters.

RESULTS

Reconstruction of Local Surface Shape from the 3D Zernike Descriptors

To begin with, we examine how well the 3DZD capture the shape of local surface patches by reconstructing the original shape back from the 3DZD. In Figure 3, three local patches of protein 1kvkA (PDB ID) are reconstructed using four different resolution levels, the order n of 5, 10, 15, and 20. As seen in the first row of Figure 3, the three local patches are significantly different in shape: The local patch 1 has a wrap-like shape, the local patch 2 has a pocket shape, and the local patch 3 is saddle-like. The 3DZD capture this difference well as shown in the clear distinction in the invariant in Figure 4. Figure 4 shows invariant values of the order n=20 (121 invariants). An invariant of a smaller order is a subset of a higher order invariant. For example, an invariant that corresponds to the order n=15 is the first 72 values of the 121 invariants. Figure 3 shows that the reconstruction results of the 3DZD with the order n=15 is as good as those of n=20. In our previous works, we used the order of 20 for describing the global protein surface shape (Sael et al., 2008b; Sael et al., 2008a). On the other hand, the order of 15 would be sufficient for local surface patches because the local surface is less complex than that of global surface. Thus, we use the order n = 15 in what follows.

Figure 3. Local surface shape reconstruction from the 3D Zernike descriptors. Three local patches from protein 1kvkA are reconstructed using four different resolutions, the order of 5, 10, 15, and 20.

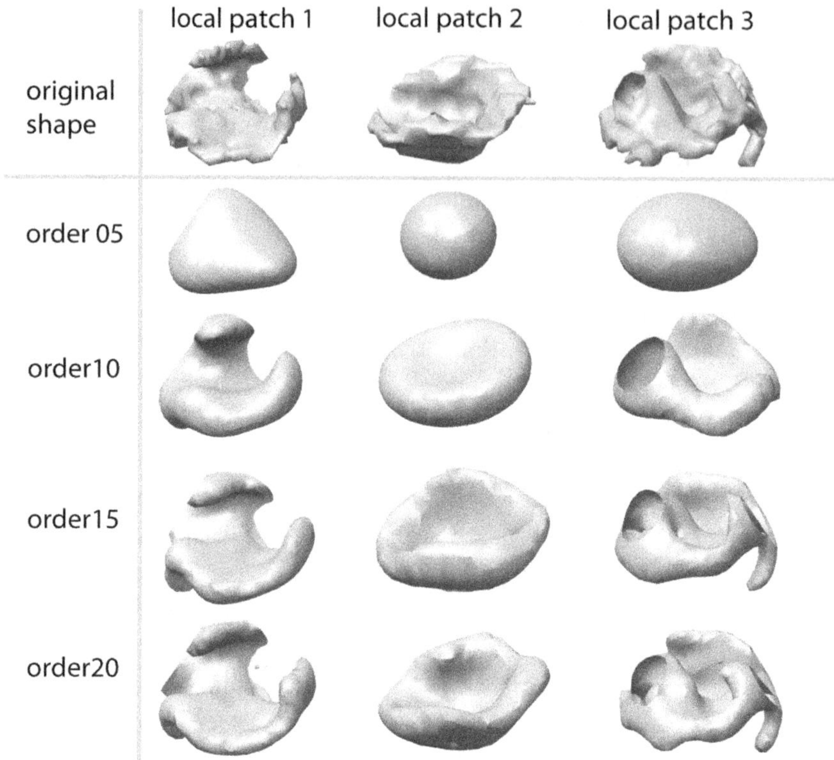

ESOM Maps

118,003 local patches from the 609 proteins represented by the 3DZD are analyzed with the ESOM with 4100 neurons. The resulting ESOM maps are boundless toroid, *i.e.* the right edge of the map is continuously connected to the left edge and the upper edge is connected to the lower edge. Three types of input feature vectors are used to characterize the surface patches: the 3DZD of 1) the surface shape, 2) the electrostatic potential, and 3) a combination of the surface shape and the electrostatic potential (Fig. 5). The maps are shown in two color-coding schemes, the P-matrix and the U-matrix. Note that these two matrices visualize the same data, although their appearance is different. Five examples of the local patches that corresponds to the labeled positions in each ESOM map are also shown in Figure 5.

The U-matrix maps show the mutual distance of neighboring neurons. In the U-matrix maps of all three types of input feature vctors, a large blue area and a couple of smaller green areas are observed. The blue area indicates that there are a large region of neurons that have a small distance to each other. These regions are mostly composed of patches of more or less flat shape. Pocket-like shape patches locate at the regions that have a higher distance (green regions). This is reasonable because it indicates that pocket shapes have more diversity in shape compared to flat regions.

The P-matrix maps show the density of the data set that are assigned to each neuron and are colored from dark to light color as the density decreases. Compared to the U-marix, more distinct

Figure 4. Invariants of local 3D Zernike descriptors of the three local patches of protein 1kvkA. The x axis is the indices of the invariants and the y axis is the value of each invariants.

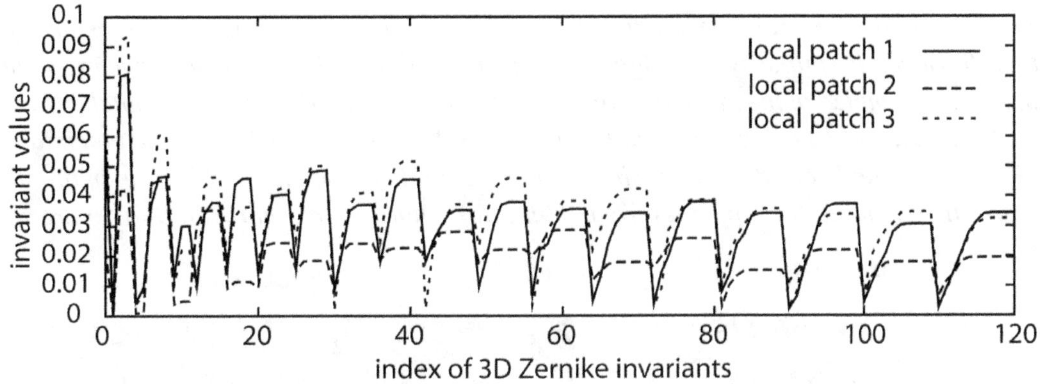

clusters are identified. Local patches of dense regions on the P-matrix (dark colored regions) are relatively flat and have mixture of the positive and negative electrostatic potential that are near 0.0. In the maps for the electrostatic potential (the middle row), clear separation of the patches is not observed except for some cases (*e.g.* a cluster with a patch from 1aquB at (e) in the U-matrix of the electrostatic potential, which is strongly positively charged). This is because the 3DZD of the electrostatic potential intrisically convolute information of the electostatic potential and the shape, as the potential is mapped on the surface. The convolution of the two information was also observed in our previous work (Sael et al., 2008a).

Figure 5 shows the overall landscape of the similarity of the surface patches. To obtain more distinct clusters, we clustered the 4100 neurons of the ESOM maps using the affinity propagation clustering method (Fig. 6). We used two ESOM maps, that of the surface shape (Fig. 6A & B) and another map of the combination of the shape and the electrostatic potential (Fig. 6C & D). Setting the preference parameter to the median of input distances results in 369 clusters for the surface ESOM map (Fig. 6A) and 215 clusters for the surface and the electrostatics combination (Fig. 6C). When the minimum of the input distances is used as the parameter, the number of clusters is reduced to 48 in the surface ESOM map (Fig. 6B) and 27 for the surface and the electrostatics combination (Fig. 6D). The small number of resulting clusters of neurons indicates that the neighboring neurons have similar weights, *i.e.* the ESOM was well trained. These distinct clusters are convenient for labeling a protein surface. As each local patch in the given surface can be assigned to one of the clusters, the surface can be represented by a set of the indices of clusters to which local patches of the surface belong.

Local Patch Types of Ligand Binding Surfaces

Next, using the distinct clusters of surface patches (Fig. 6), we classified local patches of the ligand binding sites into the clusters. Figure 7 shows histograms of surface patch clusters observed at ligand binding surfaces of the proteins listed in Table 1. The cluster compositions of all the ligand binding surfaces share some similarities that comes from the fact the binding sites have many pocket-like local shapes. In all the binding sites, the surface patch cluster 16 and 34 and the surface electrostatic potential cluster 6 and 22 are observed. The molecular shape and the binding sites of FMN are more different from the oth-

Figure 5. ESOM maps of protein local patches. Three types of feature vectors of the patches are used as input data: the 3DZD of surface shape, those of the electrostatic potential, and the combination of the two. Resulting ESOM maps are visualized in two ways, the P-matrix and the U-matrix. The color code of the P-matrix shows the density of the assigned input data to neurons, while that of the U-matrix indicates the mutual distance (dissimilarity) of neighboring neurons. For each ESOM map, five examples of the local patches are sampled at the labeled positions. In the surface ESOM map, local patches from 1a2kC (a, b, c), 1b7eA (d), and 1af6C (e) are sampled. In the ESOM map of the electrostatic potential, patches from 1a2kA (a, b, c) and 1aquB (e, f) are shown. In the ESOM map of the combination of the shape and the electrostatic potential, local patches from 1b9mB (a), 1aurB (b, c), and 1a8uB (d, e) are shown.

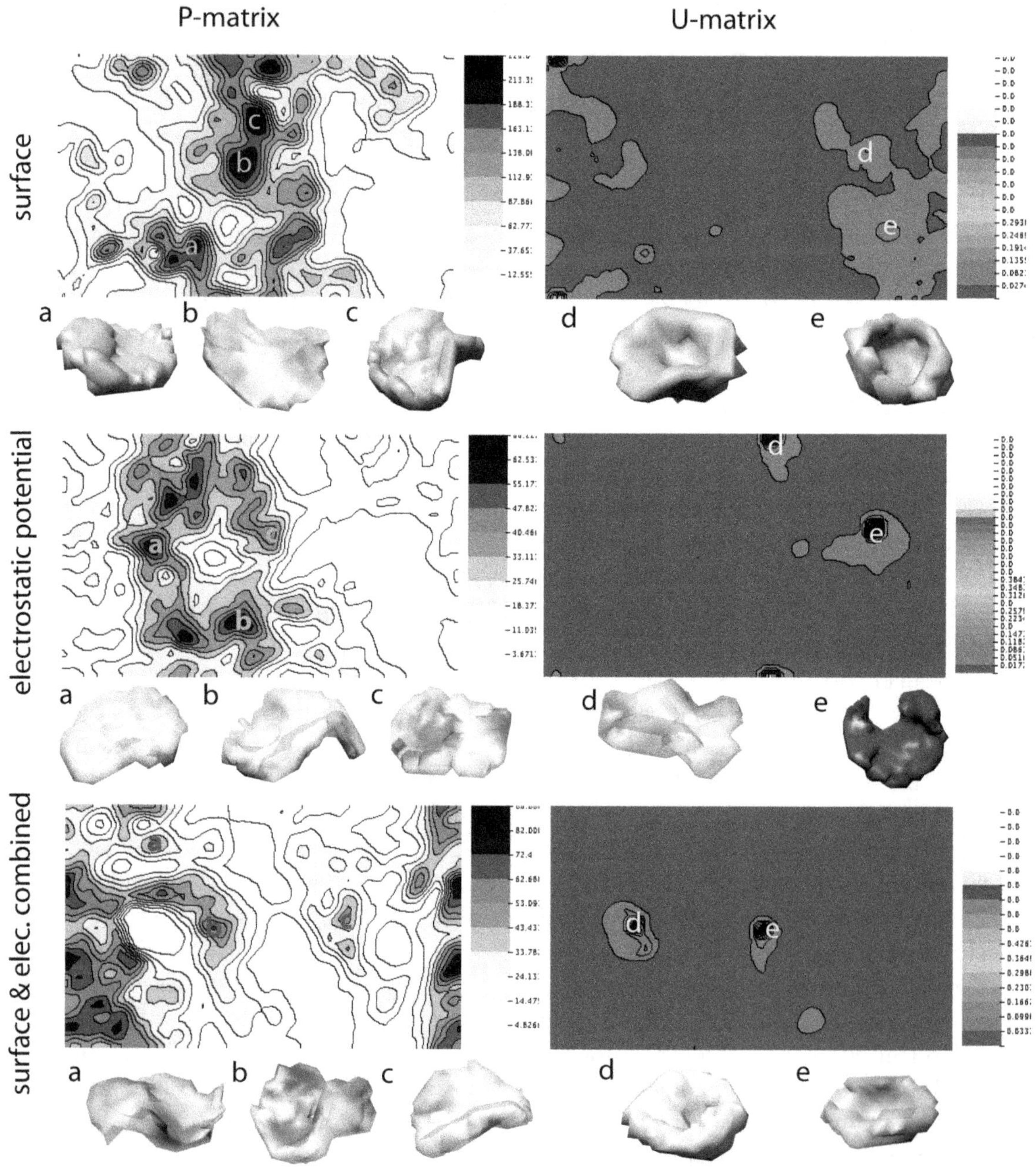

Figure 6. Affinity propagation clustering of ESOM neurons. A and B are clustering results of ESOM neurons using the surface shape, while C and D are results using the surface shape and the electrostatic potential combination as the feature of local surface patches. In A and C, the median of distance is used as the preference parameter p. B and D use the minimum distance as the preference parameter. There are 369 clusters in A, 48 clusters in B, 215 clusters in C, and 27 clusters in D. Note that the colors are used just to separate neighboring clusters; clusters in the same color at different locations do not mean similarity.

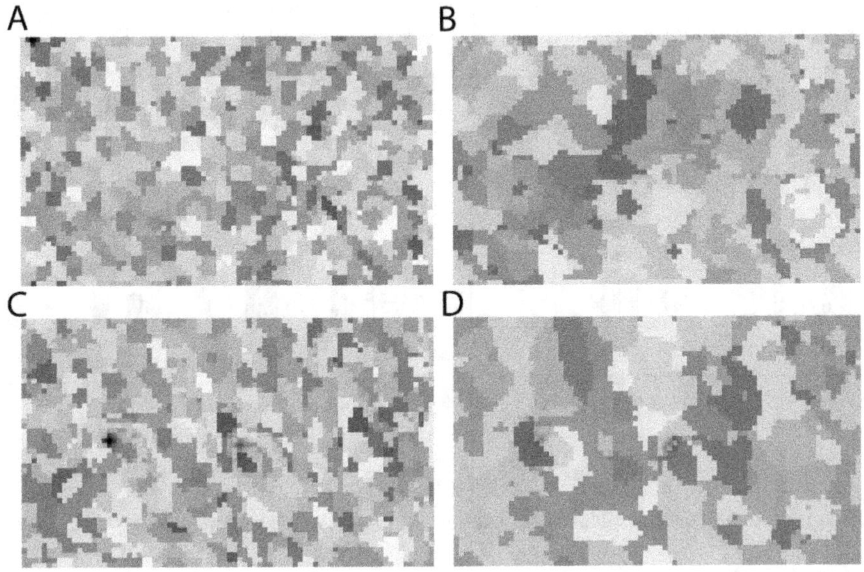

ers, which are reflected to the relatively distinct histogram.

Figure 8 shows concrete examples of local patches of ligand binding sites. They are mapped on the ESOM map of the combination of the shape and the electrostatic potential. The ATP binding site of 1dv2A consists of surface patch clusters 6(c), 13, 16(d), 21(b), and 24(a). In the parentheses, labels shown in Figure 8 are indicated. The AMP binding site of 1c0aA is composed of the clusters 6(a), 12(c), and 23(b). The FMN binding site of 1mvlA is composed of the clusters 1(a), 5(c), 10, and 21(b). The heme binding site of 1np4A consists of the clusters 2(c), 6(d), 7(b), 10, and 17(a). The FAD binding site of 1hskA consists of the clusters 6(c), 14(d), 16(b), 20, 23(a), 24, and 27. The NAD binding site of 2npxA is composed of the clusters 1, 2, 7(b), 12(a), 16(d), 22(c), and 23. The patches found at ligand binding sites of the other proteins listed in Table 1 are also included in Figure 8 with the colored dots. It turned out that binding sites of similar ligands do not always have the same set of surface patches. For example, the adenosine binding site is assigned to the patch 24 in 1dv2A, an ATP binding protein, and to the patch 12 in 1c0aA, an AMP binding protein. This is partly due to the conformational changes of the ligands (Kahraman et al., 2007) and also due to the different distribution of the seed points on a protein surface. However, of course there are cases where binding sites of the same chemical group of ligands are assigned to the same surface patch cluster, such as the phosphate binding region in 1dv2A and 1c0aA, both of which are assigned to the cluster 6. This overall similarity in consisting surface patch clusters can also be found in Figure 7, where both ATP

Figure 7. Histograms of ESOM neuron clusters to which surface patches of binding sites of the six ligand molecules (Table 1) belong. The clusters are obtained by the affinity propagation clustering method using the minimum distance as the preference parameter, i.e. the clusters for the surface shape is shown in Fig 6B and those for when the combination of the shape and the electrostatic potential are shown in Figure 6D. The x-axis is the index of the cluster and the y-axis is the density of each cluster.

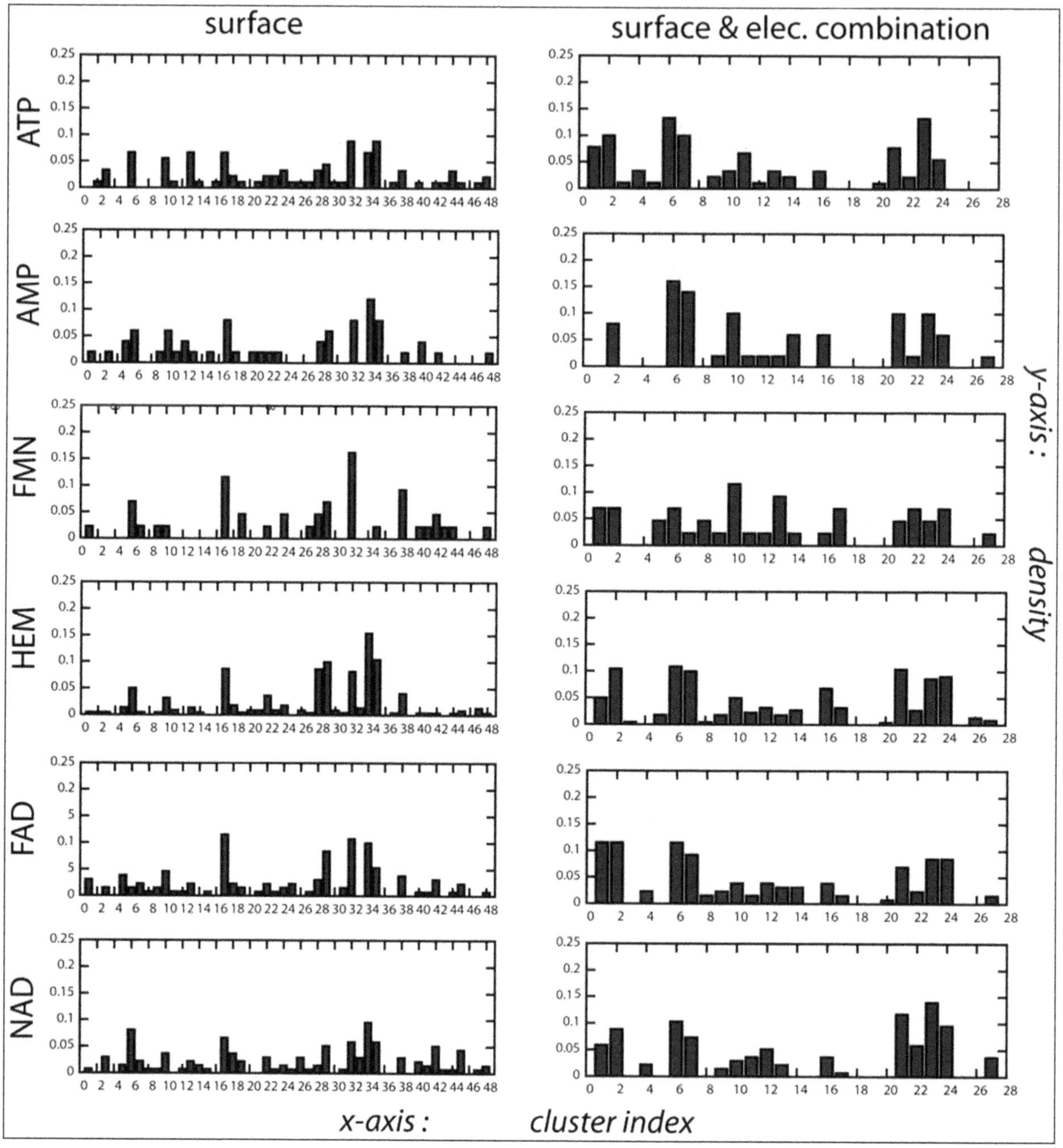

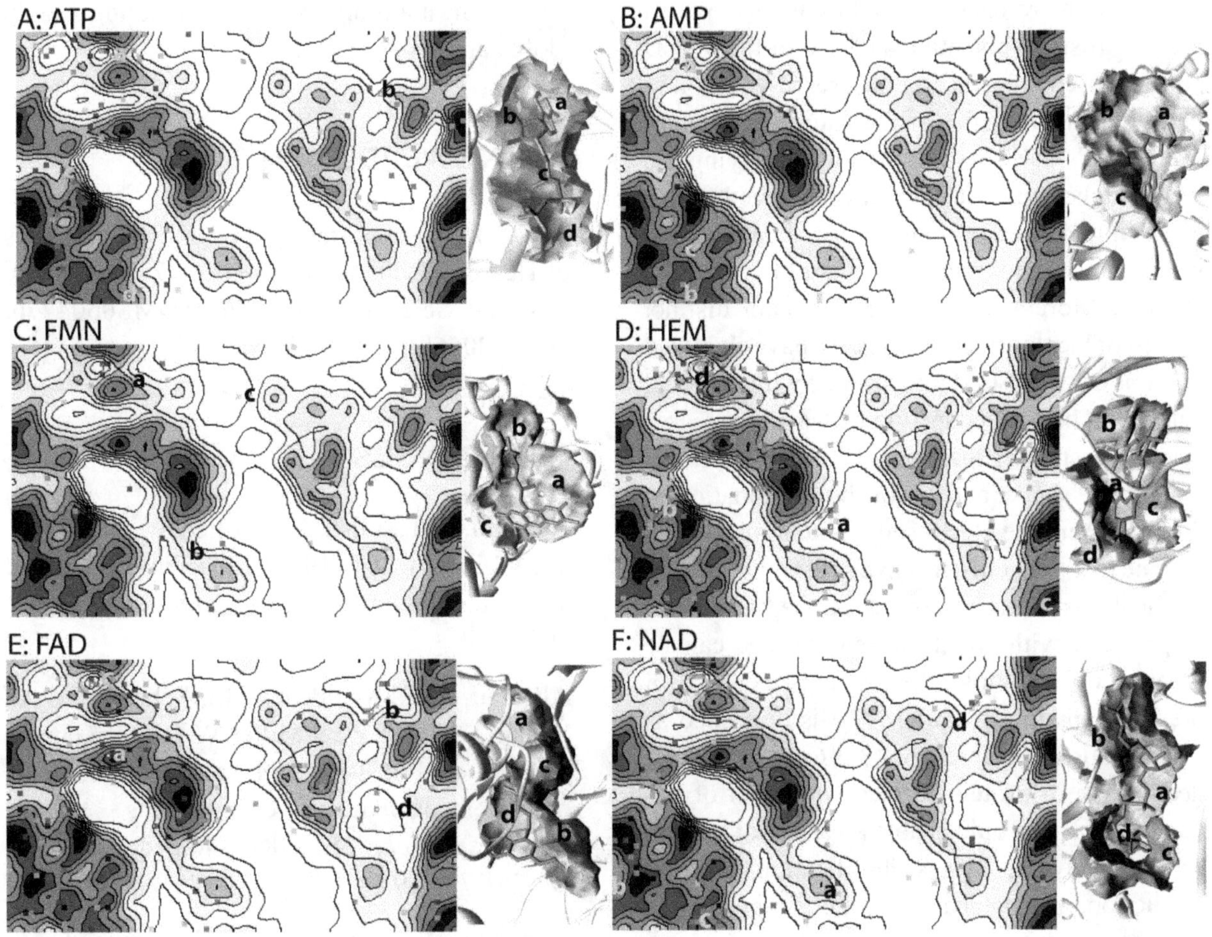

Figure 8. Distribution of ligand binding surface on the ESOM map. The ESOM map of the combination of the shape and the electrostatic potential is used (i.e. Fig. 5, bottom row). The colored dots specify the most similar neurons to the ligand binding local patches. The color of the dots indicates the source protein. For each type of ligand, one example of the binding sites is shown on the right side of each ESOM map. The proteins used are as follows: 1dv2A for ATP binding; 1c0aA for AMP; 1mvlA for FMN; 1np4A for HEM; 1hskA for FAD; and 2npxA for NAD. The labels (a, b, c, d, e) indicate the position of the corresponding local patches on the ESOM map.

and AMP have high peaks at the clusters 6, 7, 23, and 24 in their histograms.

Computation Time

The time taken to generate the descriptors for each protein depends on the size of the proteins. It takes around two minutes to generate the surface geometry and the surface electrostatic information for each protein and on an average of 37.12 seconds to compute local 3D Zernike descriptors for patches in a whole protein. The training of an ESOM using the surface patch data with 1000 iterations took approximately two weeks. However, note that the training is needed only once. Once the ESOM is trained, assignment of patches in a protein to neurons can be done in less than a second. The analysis was performed on Linux machine equipped with Intel Core TM2 CPU 6400 at 2.13GHz.

DISCUSSION

We have classified local surface patches of proteins, which are characterized by the shape and the electrostatic potential. The use of the 3DZD provided a convenient compact representation of the two characteristics of the patches as feature vectors. We used the ESOM for classifying the local patches since it is an effective method for observing landscape of similarity of data points, whose similarity is rather continuous but not very distinct. Moreover, in order to obtain distinct groups of local surface patches, we have clustered neurons of the ESOM map trained with the local patches. The procedure yielded a manageable size of clusters of local patches; 48 clusters for the surface shape, and 27 for the combination of the shape and the electrostatic potentials.

The resulting clusters have several interesting applications. The clusters can be used as surface "alphabet", with which protein surface can be labeled. For example, the previous example of the AMP binding site of 1dv2A is described as a set of surface classes (6, 13, 16, 21, 24). This description of protein surface with a set of letters (numbers) would enable a variety of protein surface analyses, such as classification, function prediction, and database searches, in analogous ways to protein sequence analyses. In particular, a strong advantage of the surface alphabet is that it allows partial matching of protein surfaces, which may be necessary for considering flexibility of protein structure and a certain degree of difference in features in protein surface.

The idea of the surface alphabet proposed in this paper certainly needs more investigations and testing. For example, different parameters and alternative methods need to be tested, such as the definition of the ligand binding sites, the way to select the seed points in a protein surface, and methods for clustering similar surface patches. We are also planning to incorporate different features for describing surface patches, including sequence conservation and hydrophobicity. To conclude, as annotation of protein structures has become an urgent task in bioinformatics, we believe that the idea of the surface alphabet will pave the way for developing many approaches for handling protein global and local surfaces.

ACKNOWLEDGMENT

This work is supported by the National Institutes of Health (GM075004 and GM077905) and the National Science Foundation (DMS0604776, DMS800568).

REFERENCES

Altschul, S. F., Gish, W., Miller, W., Myers, E. W., & Lipman, D. J. (1990). Basic local alignment search tool. *Journal of Molecular Biology, 215*, 403–410.

Altschul, S. F., Madden, T. L., Schaffer, A. A., Zhang, J., Zhang, Z., & Miller, W. (1997). Gapped BLAST and PSI-BLAST: a new generation of protein database search programs. *Nucleic Acids Research, 25*, 3389–3402. doi:10.1093/nar/25.17.3389

Andreeva, A., Howorth, D., Chandonia, J. M., Brenner, S. E., Hubbard, T. J., & Chothia, C. (2008). Data growth and its impact on the SCOP database: new developments. *Nucleic Acids Research, 36*, D419–D425. doi:10.1093/nar/gkm993

Baker, N. A., Sept, D., Joseph, S., Holst, M. J., & McCammon, J. A. (2001). Electrostatics of nanosystems: application to microtubules and the ribosome. *Proceedings of the National Academy of Sciences of the United States of America, 98*, 10037–10041. doi:10.1073/pnas.181342398

Berman, H. M., Westbrook, J., Feng, Z., Gilliland, G., Bhat, T. N., & Weissig, H. (2000). The Protein Data Bank. *Nucleic Acids Research, 28*, 235–242. doi:10.1093/nar/28.1.235

Binkowski, T. A., & Joachimiak, A. (2008). Protein functional surfaces: global shape matching and local spatial alignments of ligand binding sites. *BMC Structural Biology, 8*, 45. doi:10.1186/1472-6807-8-45

Burley, S. K. (2000). An overview of structural genomics. *Nature Structural Biology, 7*(Suppl), 932–934. doi:10.1038/80697

Canterakis, N. (1999). 3D Zernike moments and Zernike affine invariants for 3D image analysis and recognition. *Proc. 11th Scandinavian Conference on Image Analysis* (pp. 85-93).

Chikhi, R. & Kihara, D. (2009). *Protein binidng ligand prediction using local surface descriptors.*

Chitale, M., Hawkins, T., & Kihara, D. (2009a). Automated prediction of protein function from sequence. In J. Bujnicki (Ed.), *Prediction of Protein Strucutre, Functions, and Interactions* (pp. 63-86). John Wiley & Sons Ltd.

Chitale, M., Hawkins, T., Park, C., & Kihara, D. (2009b). ESG: Extended similarity group method for automated protein function prediction. *Bioinformatics (Oxford, England), 25*, 1739–1745. doi:10.1093/bioinformatics/btp309

Chothia, C., & Lesk, A. M. (1986). The relation between the divergence of sequence and structure in proteins. *The EMBO Journal, 5*, 823–826.

Coggill, P., Finn, R. D., & Bateman, A. (2008). Identifying protein domains with the Pfam database. *Curr.Protoc.Bioinformatics.*

Frey, B. J., & Dueck, D. (2007). Clustering by passing messages between data points. *Science, 315*, 972–976. doi:10.1126/science.1136800

Hawkins, T., Chitale, M., & Kihara, D. (2008). New paradigm in protein function prediction for large scale omics analysis. *Molecular BioSystems, 4*, 223–231. doi:10.1039/b718229e

Hawkins, T., Chitale, M., Luban, S., & Kihara, D. (2009). PFP: Automated prediction of gene ontology functional annotations with confidence scores using protein sequence data. *Proteins, 74*, 566–582. doi:10.1002/prot.22172

Hawkins, T., & Kihara, D. (2007). Function prediction of uncharacterized proteins. *Journal of Bioinformatics and Computational Biology, 5*, 1–30. doi:10.1142/S0219720007002503

Hawkins, T., Luban, S., & Kihara, D. (2006). Enhanced automated function prediction using distantly related sequences and contextual association by PFP. *Protein Science, 15*, 1550–1556. doi:10.1110/ps.062153506

Hendlich, M., Rippmann, F., & Barnickel, G. (1997). LIGSITE: automatic and efficient detection of potential small molecule-binding sites in proteins. *Journal of Molecular Graphics & Modelling, 15*, 359–363, 389. doi:10.1016/S1093-3263(98)00002-3

Huang, B., & Schroeder, M. (2006). LIGSITEcsc: predicting ligand binding sites using the Connolly surface and degree of conservation. *BMC Structural Biology, 6*, 19. doi:10.1186/1472-6807-6-19

Hulo, N., Bairoch, A., Bulliard, V., Cerutti, L., De, C. E., & Langendijk-Genevaux, P. S. (2006). The PROSITE database. *Nucleic Acids Research, 34*, D227–D230. doi:10.1093/nar/gkj063

Kahraman, A., Morris, R. J., Laskowski, R. A., & Thornton, J. M. (2007). Shape variation in protein binding pockets and their ligands. *Journal of Molecular Biology, 368*, 283–301. doi:10.1016/j.jmb.2007.01.086

Kalidas, Y., & Chandra, N. (2008). PocketDepth: a new depth based algorithm for identification of ligand binding sites in proteins. *Journal of Structural Biology, 161*, 31–42. doi:10.1016/j.jsb.2007.09.005

Kawabata, T., & Go, N. (2007). Detection of pockets on protein surfaces using small and large probe spheres to find putative ligand binding sites. *Proteins, 68*, 516–529. doi:10.1002/prot.21283

Kihara, D., Sael, L., & Chikhi, R. (2009). Local surface shape-based protein function prediction using Zernike descriptors. *Biophysical Journal, 96*, 650a. doi:10.1016/j.bpj.2008.12.3435

Kihara, D., & Skolnick, J. (2004). Microbial Genomes have over 72% structure assignment by the threading algorithm PROSPECTOR_Q. *Proteins, 55*, 464–473. doi:10.1002/prot.20044

Laskowski, R. A. (1995). SURFNET: a program for visualizing molecular surfaces, cavities, and intermolecular interactions. *Journal of Molecular Graphics, 13*, 323–328. doi:10.1016/0263-7855(95)00073-9

Laskowski, R. A., Luscombe, N. M., Swindells, M. B., & Thornton, J. M. (1996). Protein clefts in molecular recognition and function. *Protein Science, 5*, 2438–2452.

Laskowski, R. A., Watson, J. D., & Thornton, J. M. (2005). Protein function prediction using local 3D templates. *Journal of Molecular Biology, 351*, 614–626. doi:10.1016/j.jmb.2005.05.067

Levitt, D. G., & Banaszak, L. J. (1992). POCKET: a computer graphics method for identifying and displaying protein cavities and their surrounding amino acids. *Journal of Molecular Graphics, 10*, 229–234. doi:10.1016/0263-7855(92)80074-N

Li, B., Turuvekere, S., Agrawal, M., La, D., Ramani, K., & Kihara, D. (2007). Characterization of local geometry of protein surfaces with the visibility criterion. *Proteins, 71*, 670–683. doi:10.1002/prot.21732

Novotni, M., & Klein, R. (2003). 3D Zernike descriptors for content based shape retrieval. *ACM Symposium on Solid and Physical Modeling, Proceedings of the eighth ACM symposium on Solid modeling and applications* (pp. 216-225).

Pearson, W. R., & Lipman, D. J. (1988). Improved tools for biological sequence comparison. *Proceedings of the National Academy of Sciences of the United States of America, 85*, 2444–2448. doi:10.1073/pnas.85.8.2444

Rojas, R., & Feldman, J. (1996). *Neural networks: A systematic introduction*. Berlin, New York: Springer-Verlag.

Sael, L., & Kihara, D. (2009). Protein surface representation and comparison: New approaches in structural proteomics. In J. Chen & S. Lonardi (Eds.), *Biological data mining*. Boca Raton, Florida, USA: Chapman & Hall/CRC Press.

Sael, L., La, D., Li, B., Rustamov, R., & Kihara, D. (2008a). Rapid comparison of properties on protein surface. *Proteins, 73*, 1–10. doi:10.1002/prot.22141

Sael, L., Li, B., La, D., Fang, Y., Ramani, K., & Rustamov, R. (2008b). Fast protein tertiary structure retrieval based on global surface shape similarity. *Proteins, 72*, 1259–1273. doi:10.1002/prot.22030

Shindyalov, I. N., & Bourne, P. E. (1998). Protein structure alignment by incremental combinatorial extension (CE) of the optimal path. *Protein Engineering, 11*, 739–747. doi:10.1093/protein/11.9.739

Tangelder, J. H., & Veltkamp, R. C. (2008). A survey of content based 3D shape retrieval methods. *Multimedia Tools and Applications*, *39*, 441–471. doi:10.1007/s11042-007-0181-0

Thornton, J. M., Todd, A. E., Milburn, D., Borkakoti, N., & Orengo, C. A. (2000). From structure to function: approaches and limitations. *Nature Structural Biology*, *7*(Suppl), 991–994. doi:10.1038/80784

Tseng, Y. Y., Dundas, J., & Liang, J. (2009). Predicting Protein Function and Binding Profile via Matching of Local Evolutionary and Geometric Surface Patterns. *J. Mol. Biol.*

Ultsch, A. (2003). Maps for the visualization of high-dimensional data spaces. *Proc. WSOM'03*, 225-230.

Ultsch, A. (2005). ESOM-Maps: tools for clustering, visualization, and classification with Emergent SOM. *Technical report. Dept. of Mathematics and Computer Sciece, Univeristy of Marburg.*

Westbrook, J., Feng, Z., Chen, L., Yang, H., & Berman, H. M. (2003). The Protein Data Bank and structural genomics. *Nucleic Acids Research*, *31*, 489–491. doi:10.1093/nar/gkg068

Yeturu, K., & Chandra, N. (2008). PocketMatch: a new algorithm to compare binding sites in protein structures. *BMC Bioinformatics*, *9*, 543. doi:10.1186/1471-2105-9-543

Zhang, C., & Kim, S. H. (2003). Overview of structural genomics: from structure to function. *Current Opinion in Chemical Biology*, *7*, 28–32. doi:10.1016/S1367-5931(02)00015-7

This work was previously published in the International Journal of Knowledge Discovery in Bioinformatics, Volume 1, Issue 1, edited by Xiao-Li Li and See-Kiong Ng, pp. 32-47, copyright 2010 by IGI Publishing (an imprint of IGI Global).

Section 2
Gene Expression Data Analysis

Chapter 5
Clustering Genes Using Heterogeneous Data Sources

Erliang Zeng
University of Notre Dame, USA

Chengyong Yang
Life Technologies Inc., USA

Tao Li
Florida International University, USA

Giri Narasimhan
Florida International University, USA

ABSTRACT

Clustering of gene expression data is a standard exploratory technique used to identify closely related genes. Many other sources of data are also likely to be of great assistance in the analysis of gene expression data. This data provides a mean to begin elucidating the large-scale modular organization of the cell. The authors consider the challenging task of developing exploratory analytical techniques to deal with multiple complete and incomplete information sources. The Multi-Source Clustering (MSC) algorithm developed performs clustering with multiple, but complete, sources of data. To deal with incomplete data sources, the authors adopted the MPCK-means clustering algorithms to perform exploratory analysis on one complete source and other potentially incomplete sources provided in the form of constraints. This paper presents a new clustering algorithm MSC to perform exploratory analysis using two or more diverse but complete data sources, studies the effectiveness of constraints sets and robustness of the constrained clustering algorithm using multiple sources of incomplete biological data, and incorporates such incomplete data into constrained clustering algorithm in form of constraints sets.

DOI: 10.4018/978-1-4666-1785-8.ch005

1. INTRODUCTION

Large scale microarray experiments that have been performed under a variety of conditions or at various stages during a biological process have resulted in huge amounts of gene expression data, and have presented big challenges for the field of data mining (de Souto et al., 2008; Kerr et al., 2008). Challenges include rapidly analyzing and interpreting data on thousands of genes measured with hundreds of different conditions, and assessing the biological significance of the results. Clustering is the exploratory, unsupervised process of partitioning the expression data into groups (or clusters) of genes sharing similar expression patterns (Yeung et al., 2003; Kerr et al., 2008). However, the quality of clusters can vary greatly, as can their ability to lead to biologically meaningful conclusions.

On a different note, the biological and medical literature databases are information warehouses with a vast store of useful knowledge. In fact, text analysis has been successfully applied in bioinformatics for various purposes such as identifying relevant literature for genes and proteins, connecting genes with diseases, and reconstructing gene networks (Yandell & Majoros, 2002). Hence, including the literature in the analysis of gene expression data offers an opportunity to incorporate additional functional information about the genes when defining expression clusters. In more general terms, with the availability of multiple information sources, it is a challenging problem to conduct integrated exploratory analyses with the aim of extracting more information than what is possible from only a single source.

The basic problem of learning from multiple information sources has been extensively studied by the machine learning community. In computer vision this problem is referred to as multi-modal learning. In general, there are two approaches to multi-modal learning: feature level integration and semantic integration (Wu et al., 1999). Methods that use feature level integration combine the information at the feature level and then perform the analysis in the joint feature space (Glenisson et al., 2003). On the other hand, the semantic level integration methods first build individual models based on separate information sources and then combine these models via techniques such as mutual information maximization (Becker, 1996).

Microarray experiments usually provide gene expression data on all the genes in a genome. Hence they are inherently "complete". A major challenge using other sources of data to assist the analysis of gene expression data is that they may not always be complete, i.e., do not provide information on all the genes in the genome.

Recent work from the machine learning community has focused on the use of background information in the form of instance-level constraints. Two types of pair-wise constraints have been proposed: positive constraints that specify that two instances must remain in the same cluster, and negative constraints that specify that two instances must not be placed in the same cluster. Recent examples of work include methods that ensured that constraints were satisfied at each iteration (Wagsta et al., 2001), algorithms that used constraints as initial conditions (Basu et al., 2002), algorithms that learned a distance metric trained by a shortest-path algorithm (Klein et al., 2002), a convex optimization method using Mahalanobis distances (Xing et al., 2002), and semi-supervised clustering that incorporated both metric learning and the use of pair-wise constraints in a principled manner (Bilenko et al., 2004).

While great efforts have been made to develop efficient constrained clustering algorithm variants, the role of constraint sets in constrained clustering algorithm has not been fully studied yet. Recently, Wagstaff et al. (2006) and Davidson et al. (2006) attempted to link the quality of constraint sets with clustering algorithm performance (Davidson et al., 2006; Wagsta et al., 2006). Two properties of constraint set – inconsistency and incoherence – were shown to be strongly negative correlated with clustering algorithm performance.

Table 1. Notation

n	Number of genes
m	Number of data sources
$M_1, M_2, \ldots, M_m$	Dimensions for each data source
K	Number of clusters
$\lambda_i^{(l)}$	Model parameters for i-th cluster on data source l
$\lambda_i = (\lambda_i^{(1)}, \lambda_i^{(2)}, \ldots, \lambda_i^{(m)})$	Model parameters for i-th cluster
$O = \{(o_1^{(1)}, o_1^{(2)}, \ldots, o_1^{(m)}), \ldots, (o_n^{(1)}, o_n^{(2)}, \ldots, o_n^{(m)})\}$	The dataset
$\Lambda = \{(\lambda_1^{(1)}, \lambda_1^{(2)}, \ldots, \lambda_1^{(m)}), \ldots, (\lambda_k^{(1)}, \lambda_k^{(2)}, \ldots, \lambda_k^{(m)})\}$	The cluster model
$Y = \{y_1, \ldots, y_n\}, y_i \in \{1, \ldots, k\}$	Cluster assignment vector
P	Vector denoting the weights for each data source.

This paper makes several important contributions. First, we investigate the problem of integrating two or more heterogeneous sources of data to produce biologically significant clusters. To address this problem, we have proposed an EM-based clustering algorithm called Multi-Source Clustering (MSC) to perform clustering using multiple, but complete, data sources. For our experiments with MSC, we focused on algorithms using gene expression data and biological text literature as the two main sources, although the techniques are general enough to be applicable to other data sources. Then, we consider the non-trivial task of proposing evaluation schemes to evaluate the results of clustering techniques by utilizing independent sources of data. This is performed by considering functional annotation data and transcription factor binding site data as the independent data sources. Next, we tackle the data integration problem, when the data sources provide incomplete data sets. Most data sources (such as biological text literature) are incomplete since data is not available on all the genes in the genome. The problem of incomplete data sources is addressed by adopting MPCK-Means algorithm. Finally, we tackle the case when the data sources provide inconsistent and conflicting pieces of information or are unreliable or inaccurate to varying degrees.

2. METHODS

In this section we describe our new approaches. We first introduce our new MSC clustering algorithm and then present the adaptive MPCK-means clustering algorithm.

2.1 Multi-Source Clustering

In this section we describe MSC algorithm in detail. The problem of clustering data arises in many disciplines and has a wide range of applications. Intuitively, clustering is the problem of partitioning a finite set of points in a multi-dimensional space into classes (called clusters) so that (i) the points belonging to the same class are similar and (ii) the points belonging to different classes are dissimilar (Jain & Dubes, 1988). In this paper, our goal is to identify clusters of related genes using multiple available datasets. The notation used in the paper is listed in Table 1.

Table 2. The MSC Algorithm

Input: Data samples O, model structure Λ, and weight vector P.
Output: Trained models Λ and a partition of the data samples given by the cluster identity vector Y
1. **Initialization**: initialize the cluster identity vector Y
2. **while** stopping criterion is not met a. Randomly pick a data source j according to P b. Model re-estimation for source j: for each cluster i, let $O_i^{(j)} = \{o_s^{(j)} \mid y_s = i\}$. The parameters of the model for cluster i, $»_i^{(j)}$, are re-estimated as $\lambda_i^{(j)} = \arg\max_{\lambda} \sum_{o^{(j)} \in O_i^{(j)}} \log P(o^{(j)} \mid \lambda^{(j)})$ c. Sample re-assignment: for each data sample s, set $y_s = \arg\max_{i} \log P(o_s^{(j)} \mid \lambda_i^{(j)})$
3. **Return** Λ and Y.

2.1.1 The MSC Algorithm

The method extends the model-based K-means clustering algorithm to allow for combined learning of different data sources. The algorithm assumes that there are m parameterized models, one for each cluster. The set of parameters for the i-th model is denoted by λ_i. Typically, all the models are assumed to be from the same family, e.g., Gaussian or multinomial distribution. In the sample re-assignment step, a data point could be assigned to a cluster using one of three possible approaches: maximum likelihood (ML), soft, or stochastic (Zhong & Ghosh, 2003).

The MSC algorithm, shown in Table 2, is a variant of the EM method (Dempster et al., 1977). It stochastically builds the models for each data source by boosting the models using the cluster assignments from the other models. Let $\lambda^{(j)}$ be the set of parameters for the models of data source j. In each iteration, we first randomly select a data source j based on the weight vector P. We then perform the following steps: (i) find the model parameters ($\lambda_1^{(j)}$, $\lambda_2^{(j)}$,..., $\lambda_m^{(j)}$) that maximize the likelihood of the data given the current cluster assignment (shown in step *2b* of the algorithm); (ii) assign the data points to the cluster that maximizes the posterior probability (shown in step *2c* of the algorithm). Our experiments show that the MSC algorithm implicitly learns the correlation structure among the multiple data sources.

2.1.2 Cluster Assignment

In order to obtain the final clustering, we develop a new approach to combine the clustering results from each data source. Note that in each iteration, one data source is randomly picked and every data point (i.e., gene) is reassigned to one of the k clusters based on information from that data source and on its previous assignment. After all the iterations are completed, each data point has to be given a final assignment to one of the k different clusters based on some criteria that depends on its cluster assignment for each of the m data sources. Note that the cluster assignments for each of the data sources may be different.

One approach is to assign each data point to the maximum probability cluster, as suggested in (Kasturi & Acharya, 2004). This approach has the underlying assumption that cluster assignments for the m data sources are correlated. However, this need not be true. Another approach is to compute a consensus mean of the cluster assignments obtained from the m data sources. This approach may not always be successful especially when the number, variability and reliability of the data sources are large (Bickel & Tobias, 2004).

We introduce a new method of assigning cluster membership to data points by taking into account the cluster assignment obtained from each data source. The cluster assignment for each point, for each data source, can be thought as a k- dimensional vector in which only one entry (corresponding to the assigned cluster) is equal to *one* and all the others are zero. By combining the results obtained from the m data sources, the cluster assignment for each data point now constitutes a km-dimensional vector. Thus we obtain a $n \times km$ matrix whose entries are as follows:

$$\mathrm{C}_{\mathrm{i}((\mathrm{j}-1)\mathrm{k}+\mathrm{s}-1)} = \mathrm{p}_{\mathrm{j}} \cdot {}'(\mathrm{C}_{\mathrm{s}}^{(\mathrm{j})},\mathrm{i}), \tag{1}$$

where p_j is the prior probability of data type j, $\mathrm{C}_{\mathrm{s}}^{(\mathrm{j})}$ is a cluster s in data type j and

$$\delta(C_s^{(j)}, i) = \begin{cases} 1 & \text{if gene i} \in C_s^{(j)} \\ 0 & \text{otherwise} \end{cases}$$

The above matrix is used to cluster using one of standard clustering algorithms, such as K-means. Clearly, genes with similar cluster assignments across all data sources will be assigned to the same cluster.

2.2 Constrained Clustering Algorithm

The original MPCK-means is a constrained clustering algorithm (Bilenko et al., 2004). In this section we will introduce it briefly.

2.2.1 The MPCK-means Algorithm

We assume that the input may consist of two types of pair-wise constraints: (1) positive constraints, which specify that two genes must lie in the same cluster, and (2) negative constraints, which specify that two genes must not be placed in the same cluster. Furthermore, all constraints may be provided with a confidence measure. Positive constraints define a transitive binary relation over the instances; a transitive closure over the input constraints is computed and then presented to our modified algorithm. In general, constraints may be derived from any given data source. In Section 3, we discuss from what data sources these constraints can be generated and how.

The original MPCK-means integrates constraints and metric learning. This semi-supervised clustering algorithm was shown to have better performance than constraint-based learning methods and metric-based learning methods (Bilenko et al., 2004). MPCK-means has the following objective function.

$$\mathrm{GCVQE} = \frac{1}{2}\sum_{j=1}^{k} VQE_{A_j} + PM + PC \tag{2}$$

where

$$\mathrm{PM} = \sum_{x_i,\, x_j \in M} p_{ij}^{m} \neg\Delta(y(x_i), y(x_j)), \tag{3}$$

and

$$\mathrm{PC} = \sum_{x_i,\, x_j \in C} p_{ij}^{c} \Delta(y(x_i), y(x_j)). \tag{4}$$

Here VQE_{A_j} is the Euclidean distance function parameterized by a symmetric positive-definite matrix A_j that accommodates the constraints. p_{ij}^{m} and p_{ij}^{c} are penalty parameters for the positive and negative constraints respectively, and $y(x_i)$ returns the index of the cluster to which x_i belongs. Δ is the Kronecker delta function defined by: $\Delta(x,y) = 1$, if $x = y$, and 0 otherwise, and $\neg\Delta$ denotes its negation.

The MPCK-means algorithm is an iterative algorithm. As in K-means, the first order derivative of GCVQE (Eq. (2)) is set to zero and solved. The result of the derivative is dependent on the choice of penalty functions (Bilenko et al., 2004).

3. DATA SOURCES AND THEIR REPRESENTATIONS

As discussed in this section, the goal is to convert the information in each data source into numeric feature vectors. Once the vector representations are obtained, the clustering algorithms have several choices for integrating the information. Integration can either happen before the analysis or after the analysis. Feature-level integration simply concatenates the feature vectors (Glenisson et al., 2003). An alternative and more sophisticated approach is that of semantic-level integration, as described in the previous section.

3.1 Gene Expression Data

The main source of data for algorithms that cluster genes is gene expression data, typically arising from microarray experiments. The advantage of this data source is that it is often complete, in the sense that it covers all the genes in the genome. Since this data is numeric, representation of this information is now standard, i.e., it is represented as a point in d-dimensional space, where d is the number of different conditions under which the data for each gene is collected.

3.2 Text Literature Data

In order to efficiently extract and use this information in a homogeneous way with other numeric data, textual domain knowledge needs to be transformed into numeric data. In this paper, we adopt the TF-IDF (Term Frequency Inverse Document Frequency) model often used in information retrieval and text mining, as described below.

Each document from the data source can be represented by a vector in which each component of the vector corresponds to a single term from the entire set of terms, i.e., the vocabulary (Raghavan & Wong, 1986). The value of each component can be calculated using the term weight indexing as follows:

$$w_{ij}=tf_{ij} \times idf_i = tf_{ij} \log\left(\frac{N}{df_j}\right) \tag{5}$$

where term frequency tf_{ij} measures the occurrences of a term j in a document i, and idf is inverse document frequency, which is equal to the logarithm of the ratio of the total number of documents (N) divided by the number of documents containing term j in the collection (df_j).

To integrate with the gene expression data, text data can be represented as a gene-term matrix, which can be obtained by combining the document-term matrix with the gene-document matrix. The textual profile of a gene i, a vector of terms j, can thus be obtained by taking the average over the N_i documents containing gene i:

$$g_i=\{g_i\}_j=\{\frac{1}{N_i}\sum_{k=1}^{N_i}\frac{w_{ki}}{N_{kg}}\}_j \tag{6}$$

Here N_{kg} denotes the number of genes linking to document k. Note that this factor was not considered in the previous research (Glenisson et al., 2003). It accounts for the distribution of the number of genes associated with each document. Our proposal is that the weight indexing be first averaged over the number of genes in a specific document before being averaged over all documents linking gene i.

3.3 Constraints Generation

To experiment with the MPCK-means algorithm, instead of generating gene-term matrix from literature text data, the key issue is how to drive constraint sets from literature text data.

Constraints can be generated from the data with known labels (such as functional categories) or from prior knowledge about the data from other knowledge domain (such as literature text).

For labeled datasets, i.e., datasets where the clusters are known beforehand, constraints were generated based on the known labels. For each randomly chosen gene pair, if they had the same label, then a positive constraint was generated. A negative constraint was generated otherwise.

For unlabeled data sets, constraints were generated from literature text data. A simple approach is to generate positive constraint pairs from genes associated with the same document. After merging all positive constraint pairs into connected components, neighborhood sets could be produced and genes belonging to different neighborhood sets could be used to build the negative constraint set. A cautionary note is that this approach ignores contextual information and natural language issues and could potentially generate erroneous constraints. Although this approach is efficient, we used the gene-term matrix as described above.

We explored two ways to generate gene pair constraints from the gene-term matrix. We refer to the two approaches as cluster-based and similarity-based constraint generation methods. In the first approach, cosine similarity was calculated between all pairs of genes and then sorted. Positive constraint pairs were generated by picking gene pairs having high similarity. Negative constraint pairs may be generated by picking gene pairs with low similarity values. However, for the text literature data source, this is not appropriate since it is difficult to differentiate between negative information and lack of information. In the second approach, the spherical K-means algorithm, which is a K-means algorithm using cosine-based distance, was applied to the gene-term matrix T times (we chose $T = 100$ here). Frequencies of gene pairs appearing in the same cluster were counted and sorted into $T + 1$ bins using the bucket sort algorithm. Gene pairs with high frequency were listed as positive constraint pairs. In other words, the constraints can be generated from the consensus matrix M, where M_{ij} gives the probability that gene i and j belongs to the same cluster. Two sets of constraint pairs were generated: constraint pairs from text similarity, and constraint pairs from text clustering.

4. CLUSTERING ASSESSMENT

Different metrics were tried to evaluate clustering results. For the clustering resulting from the data set without known labels, the clusters were assessed using the knowledge from the Gene Ontology (GO) database (http://www.geneontology.org). The SGD gene association file was downloaded directly from the Gene Ontology website (Revision: 1.1230). A table of 6470 genes and 4032 GO terms was produced in which a *1* in position (i, j) meant that gene i is associated with GO term j, and a *0* indicates a lack of knowledge about their association. Each GO term is seen as an attribute and the gene-attribute table is the knowledge base used to assess the quality of gene clusters. The GO-based metric called **CMI** (corrected mutual information) was proposed. The definition of the CMI measure is shown below.

$$CMI = MI_{cluster} - \overline{MI}_{random} \tag{7}$$

where

$$\overline{MI}_{random} = \sum_{j=1}^{m} MI_{random(j)}/m, \tag{8}$$

and

$$MI_{cluster} = \sum_i MI(C, A_i) = N_A H(C) + \sum_i H(A_i, C). \tag{9}$$

Here $MI_{cluster}$ corresponds to the mutual information between the cluster assignment produced by a clustering algorithm and the GO database;

MI_{random} refers to the mutual information between a random cluster assignment and the GO database. *H(.)* refers to entropy function. The mutual information between a cluster assignment and all attributes (GO terms) is defined as the sum of the mutual information between the cluster assignment and the genes associated with each individual attribute (Gibbons & Roth, 2002), using the standard definition of mutual information. Thus, higher corrected CMI values suggest that the clustering results are more significantly related to gene function. The GO-based CMI measure is similar to the z-score measure (Gibbons & Roth, 2002).

Evaluation was performed using the Rand Index (Rand, 1971) for the clustering resulting from the data set with known labels. Rand Index allows for a measure of agreement between the clustering results and their true labels. Let n be the size of data set *D*. The clustering result is viewed as a collection of $n \times (n-1)/2$ pairwise decisions. For each pair of items d_i and d_j in *D*, the algorithm either assigns them to the same cluster or to different clusters. Let p_1 (p_2, respectively) be the number of decisions where the algorithm correctly assigns the same (different, respectively) label to items d_i and d_j. Then

$$RandIndex = \frac{p_1 + p_2}{n \times (n-1)/2}. \tag{10}$$

Metrics CMI is gene annotation based method, and is used for the clustering resulting from the data set without known labels. Rand Index is suitable for the clustering resulting from the data set with known labels. In this paper, the CMI was used to compare the performance of the four clustering methods: K-means clustering of expression data, K-means clustering of text data, K-means clustering of the feature-level integrated expression and text data, and the MSC algorithm applied to expression and text data. The CMI was also used to compare the performance of MSC algorithm and constrained clustering algorithm. Rand Index was used to evaluate the effectiveness of constraints sets in clustering algorithm.

5. EXPERIMENTAL RESULTS

Five algorithms were implemented in Java. These included model-based K-means algorithm on the individual data sources (text and gene expression), model-based K-means algorithm on feature-level integration of the multi-source data, the MSC algorithm, and the MPCK-means algorithm. We also include an meta-clustering algorithm developed by Zare et al. (2006) for comparison. We first discuss the data sets used for our experiments.

5.1 Expression and Literature Data Sets

For text information, a literature index for yeast genes was constructed from 31924 yeast-related MEDLINE abstracts, which were downloaded using Entrez/Pubmed search engine based on text matching in an entry's fields (Roberts, 2001). The abstract-gene relation information was constructed from the Saccharomyces Genome Database (SGD, http://www.yeastgenome.org/). It is also available by ftp from the following URL: (ftp://genome-ftp.stanford.edu/pub/yeast/data_download/literature_curation/). Gene expression data set was generated from cultures synchronized in cell cycle by three independent methods and consisted of measurements of 6206 genes over 77 experimental conditions (Spellman et al., 1998). After removing those having no literature references, the remaining 5473 genes were retained for further analysis.

Two other gene expression data sets were used in this research. One is the yeast galactose dataset originally from the study of Ideker et al. (2001). Yeung et al. (2003) compiled a subset of 205 genes from four functional categories. Genes from each functional category would be expected to cluster together. In the original experiment, microarrays were used to measure the mRNA expression profiles of yeast growing under 20 differ perturbations to the GAL pathway. Four replicates were performed for each condition.

All MEDLINE abstracts referred to in SGD's literature database were considered as acceptable, noise-free, domain-specific source of information for the yeast genes being considered (Stephens et al., 2001). A restricted vocabulary is suggested in several recent papers (Stephens et al., 2001; Chiang & Yu, 2003; Glenisson et al., 2003). Often these restricted vocabularies involve terms from the GO database. In this work, we chose to eliminate such constraints and have resorted to generic text mining methods to extract the terms. Our reasoning was as follows. Since GO terms were used in the validation of the clusters, it would be inappropriate to bias the text mining part of the process by allowing only terms that would validate positively.

5.2 Results of MSC Algorithm

5.2.1 Comparison Results

As discussed above, we compared the performance of the five clustering methods: K-means clustering of expression data, K-means clustering of text data, K-means clustering of the feature-level integrated expression and text data, meta-clustering algorithm, and the MSC algorithm applied to expression and text data. Equal weights were used for the expression and text data in all the three multi-source algorithms, although the weights could be specified using expert knowledge to specify the importance of each data source. The expression data consisted of 5473 genes under 77 experimental conditions and the text data consisted of 5473 genes and 250 terms. The CMI were plotted against the number of clusters, k, for all values of k from 2 to 100. The results are shown in Figure 1.

Using CMI as a criterion, the results from the multi-source data clustering exhibited the best performance for over 80% of values of k and were better than the method that only used text data. Surprisingly, the results from the feature-level integration were worse than the methods that used only a single data source. It suggests that a simple combination of features and distance functions may not be the best approach to improve the quality of clustering. Figure 1 shows that CMI decays as k grows over 70, indicating clustering with k greater than 70 is not appropriate because CMI decrease means clustering results are less significantly related to gene function. In the next two subsections, we explore other ways to evaluate the quality of the resulting clusters.

5.2.2 Functional Enrichment

To assess the classification capability of the clustering algorithms, gene ontology information was used to evaluate whether the clusters have significant enrichment of one or more terms from the gene ontology (GO) database; this was done using FuncAssociate (Berriz et al., 2003), a program that takes a list of genes as input and produces a ranked (by P-values) list of the GO attributes for which the input gene list is enriched. Each query gene set is composed of the genes from each cluster in a clustering, and the output gives the terms significantly enriched in each cluster among all genes (in this case, the number of all genes is 5473 which is the total number of genes for clustering). Table 3 shows details of 5 typical clusters with enriched functional groups.

Function enrichment analysis also reveals that within a given cluster, often the enriched functions are closely related. For instance, in cluster 2, which has 277 genes, 101 ontology categories are enriched. These categories include, but not limited to, DNA replication and chromosome cycle, cell proliferation, helicase activity, mitotic recombination, and so on. All the 101 enriched ontology categories in this cluster are involved in cell

Figure 1. Clustering results from expression, text, expression-text feature level integration, meta-clustering, and multi-source clustering. The horizontal axis shows the number of clusters desired, and the vertical axis shows corrected mutual information

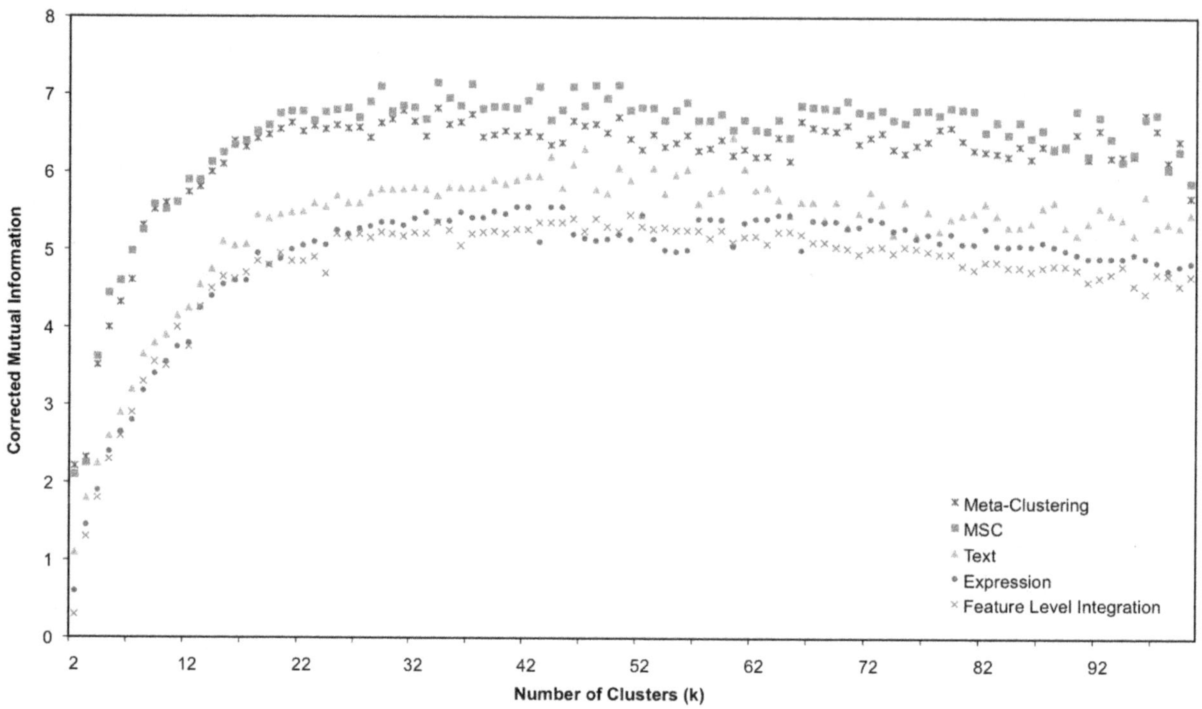

proliferation and DNA replication, which is biologically meaningful because when cell occurs, DNA replication and regulation must also occur. Similar observations can be made about other clusters. For example, the genes in cluster 4 are predominantly involved in regulation; genes in cluster 5 perform cell proliferation and cell division and so on.

5.2.3 Transcription Factor Binding Motifs

Next we decided to perform further exploratory analysis of some of the clusters obtained by the MSC algorithm. For the purpose of comparison, we examined the clusters obtained using the MSC algorithm and those obtained using only gene expression data. Upon further examination, we found that cluster 8 (with a total of 53 genes) from the MSC clustering shared 41 of the 43 genes in cluster 14 from the expression clustering. The genes contained in these two clusters are shown in Table 4, with the common ones not shown in bold font. Gene function enrichment tests showed that the significant categories were metabolism, cell growth, cell division, and DNA synthesis (as cluster 8 shown in Table 4). Most genes in cluster 8 that were not in cluster 14 also belonged to these same categories, implying that text data can enrich genes with similar functions. Our new approach takes into account both expression and function, giving it increased ability to capture more biologically meaningful features. Further exploratory analysis was performed from the point of view of shared motifs. One underlying assumption in clustering is that genes in a cluster are functionally

Table 3. Function enrichment of clusters generated from Multi-Data clustering. The first column (C) refers to the cluster number. The second column (N) refers to the number of genes in that cluster. The fourth column (NC) corresponds to the number of genes from that category in that cluster. Finally, the last column (P) gives the log of the p-value ($-\log_{10}$(P-value)) of the enrichment

C	N	Enriched Functional Category (Total genes in category)	NC	P
1	70	Sterol biosynthesis(29) sterol metabolism(34)	4 4	5 5
2	277	DNA helicase activity(27) base-excision repair(10) postreplication repair(10) spindle pole body(49) DNA replication and chromosome cycle(238)	9 5 5 9 45	9 6 6 6 28
3	39	cytosolic ribosome(167) eukaryotic 48S initiation complex(63)	36 19	65 34
4	624	regulation of physiological process(376) regulation of metabolism(376) regulation of biological process/regulation(426) regulation of transcription, DNA-dependent(297) protein modification(398)	57 57 61 49 59	13 13 13 13 13
5	53	cell proliferation(588) acid phosphatase activity(5) cell cycle/cell-division cycle(516)	21 4 15	9 8 5

related, implying that there is a strong possibility that many of them are also co-regulated, and co-regulated genes share transcription factor binding motifs (i.e., regulatory elements) in their upstream sequences. Motif detection is often performed on clusters obtained by clustering gene expression data. Thus clustering schemes can be evaluated by looking for the presence of motifs in gene clusters.

Toward this end, we applied the motif discovery tool, AlignACE (Roth et al., 1998) to find shared motifs in each of the two clusters. Results revealed a motif in cluster 8, GGCACTCACACGTGGG, located in the upstream sequence of YBR093C, which, according to TRANSFAC (Matys et al., 2003), is known to be the binding site for the transcription factor PHO4 and has been reported previously in the literature (Vogel et al., 1989). Genes that shared this motif are YBR093C, YAR018C, YDR055W, YHR215W, YML034W, and YOR313C. In particular, YBR093C and YHR215W are two of three repressible acid phosphatases (SGD). However, this motif was not found in cluster 14. Thus, clustering obtained by integrating information from the literature databases, as performed by the MSC algorithm was able to better detect motifs.

5.3 Results of Constrained Clustering Algorithm

As discussed above, the MSC algorithm performed significantly better than the feature-level integration approach. However, a major disadvantage of the MSC algorithm is that it requires all the data sources to be complete (i.e., data available on all genes to be analyzed). Constrained clustering algorithm, on the other hand, should overcome this weakness since constraints can be generated from incomplete data sources.

The motivation for designing the constrained clustering algorithm is based on the assumption that constraints based on prior correct knowledge can assist a clustering algorithm to improve its performance. Our experimental results confirm the hypothesis, but also show situations where it is not true.

Figure 2. Performance of constrained clustering algorithm when using different constraints set size and different types of constraints sets. The experiments involved computed the average Rand index on 10 runs for each constraints set size.

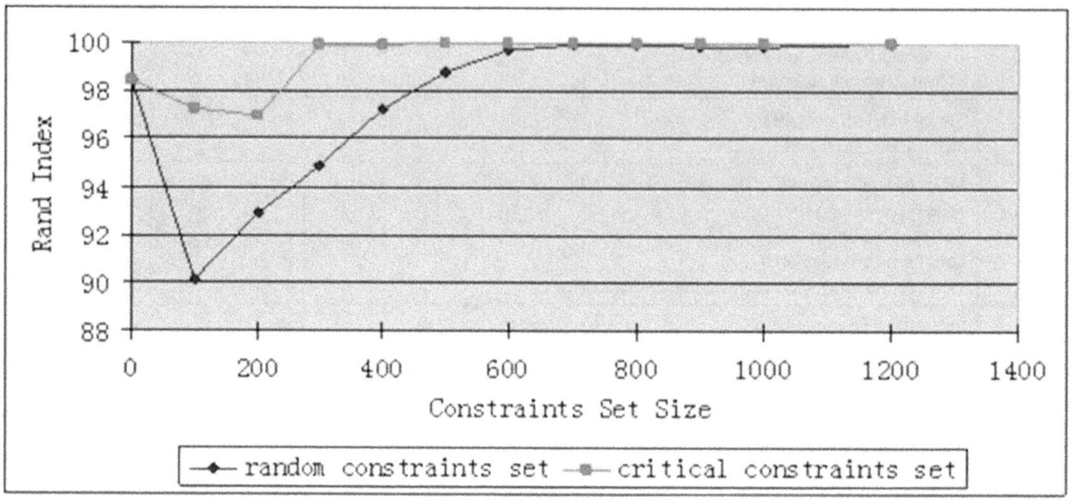

5.3.1 Effectiveness of Constraints Set

We used the yeast galactose data set (with all replicates) to investigate the effectiveness of constraints sets. The first observation was the performance of clustering with randomly generated constraints can deteriorate even if the constraints conformed to the known functional categories (i.e., labels). This was observed earlier in a different context (Davidson et al., 2006). In fact, many of the errors in constrained clustering had been correctly classified when no constraints were provided. As shown in Figure 2, constraints sets with small size can be detrimental to the performance of the algorithm. However, when the size of the constraints set is sufficiently large, then accurate clustering is achieved with both types of constraints sets. Therefore, we investigated the alternative option of generating constraints from relationships that the clustering algorithm could not learn when no constraints were provided. A second constraints set was generated by the following procedure. First a clustering was obtained by running standard clustering algorithm (without any constraint); then constraint pairs were chosen at random from all pairs of genes that included at least one misclassified gene. The pairs were then placed in the positive constraints set or negative constraints set using the known labels. We refer to the resulting set of constraints as the critical constraints set.

Figure 2 shows that accurate clustering can be achieved with much smaller critical constraints sets (roughly, 300 is enough in comparison to the roughly 700 that is needed in the case of random constraints sets). In practice, it is difficult to estimate the size of the critical constraints set required to achieve accurate clustering.

5.3.2 Results with Inconsistent Constraints

The results shown in Section 5.3.1 showed that small constraints set might hurt the ability to infer accurate clusters. However, all constraints used in Section 5.3.1 were known to be true and were based on known functional categories. Next we investigated how robust the algorithm was, i.e., how does it perform when a small number of constraints are incorrect or inconsistent with known functional categories. Once again, we consider two ways of generating constraints sets. As before, pairs were either picked at random or picked from among pairs with at least one misclassified gene. A small number of the generated constraints were then made inconsistent (i.e., put into positive constraints when they should have been put into negative constraints, vice versa).

Figure 3a and Figure 3b show that a small number of inconsistent constraints do not affect clustering algorithm performance a lot when both consistent and inconsistent pairs of constraints were generated from the critical constraints set. Otherwise, the performance of the algorithm decreases dramatically. However, Figure 3c shows that algorithm performance decreases considerably if the fraction of inconsistent constraints rises above some threshold even when the consistent and inconsistent pairs of constraints were generated from critical constraints set. Overall, the results confirm the hypothesis that inconsistent constraints will diminish clustering performance and should be avoided.

5.3.3 Results with Constraints from Text Data

For the yeast cell cycle data set, since the functional categories are not available, the constraints set cannot be generated as described before. But the constraints set can be generated from other knowledge-bases such as biological text literature. We discussed two approaches to generate constraints sets from biological text literature: cluster-based and similarity-based. In this subsection, we compare the effectiveness of constraint generation by the two methods. 1000 pairs of constraints were randomly picked from the positive constraint set. These were then provided as inputs to the MPCK-means algorithm. The expression data consisted of 5473 yeast genes under 77 experimental conditions.

The experiments were performed in 10 runs for each parameter setting (same k, same constraint type, if any) and the averaged CMI was plotted against the number of clusters, k, for all values of k from 2 to 100.

As mentioned earlier, no negative constraints were used in this set of experiments. Figure 4 shows the CMI measures resulting using three approaches. All three used gene expression data and differed in the constraints generation method used. The first performed clustering using no constraints. The second performed clustering using constraints generated by the cluster-based method while the third used constraints generated by the similarity-based approach. The version with constraints performed better than the one with no constraints, with a better performance resulting from the version with cluster-based constraints. The constraints generated from the clustering-based method performed better than those from the similarity-based approach.

5.3.4 Results Comparison with MSC

Finally we compare the results of our experiments with MPCK-means to that of the MSC algorithm. According to the results in section 5.3.3, k was set to 50. In all the previous experiments, only 1000 constraints were used. Here different constraint sizes were tried to investigate the effect of the size of the constraint on the overall clustering performance. Since the total size of constraint pool was 20,000, constraint set sizes were selected starting from 2000 with a step of 2000, along with a constraint set of size of 1000, resulting in 11 different constraint set sizes for each type of constraint set.

Figure 4 presents the CMI measures results for constrained clusters obtained by cluster-based both, cluster-based positive, and similarity-based positive constraints and MSC results in dashed line. As constraint set size increases, the CMI measures for each of three types of constraints steadily improved and flattened out after 16000. MSC performed better than the cluster-based MPCK-means version when constraint set size

Figure 3. Performance of constrained clustering algorithm when using inconsistent constraints. $N(m_1,m_2)$ means that experiment was performed for constraint set size N generated by approach m_1, and the inconsistent constraints pair were from set m_2. $m_1 = 1$ means that constraint set was randomly generated, $m_1 = 2$ means that constraint set was critical constraint set. $m_2 = 1$ means inconsistent constraints pair were from randomly generated constraint set, $m_2 = 2$ means inconsistent constraints pair were from critical constraint set.

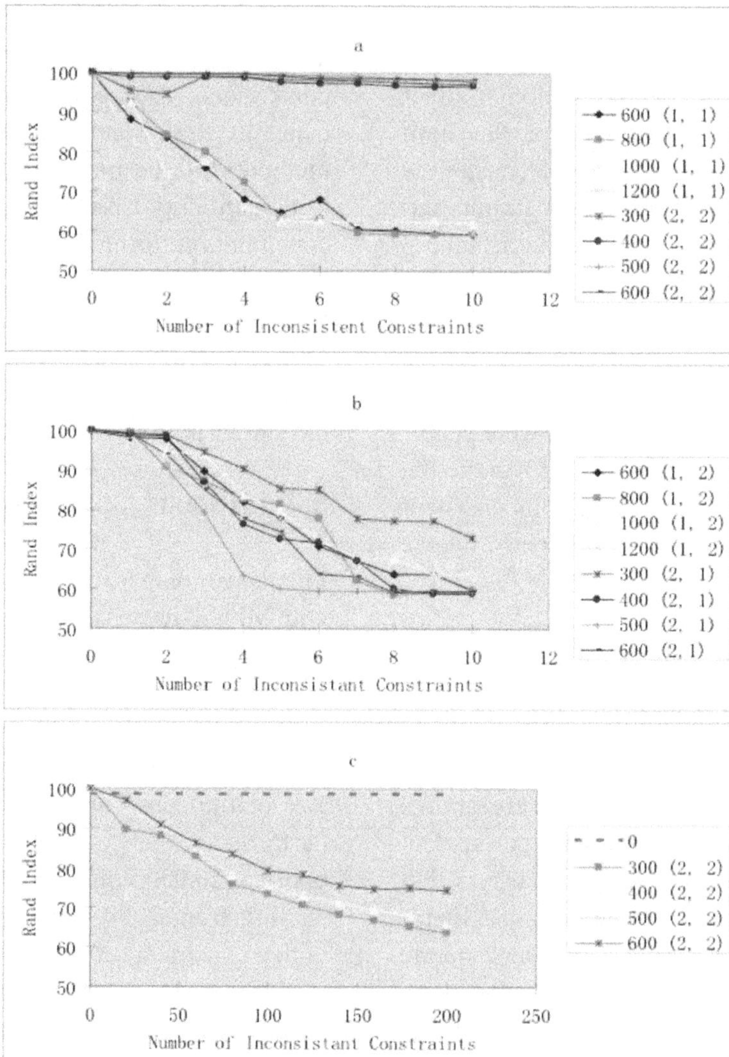

was less than 10000, and was outperformed for larger set sizes. However, MSC performed significantly better when constraints were generated by text similarity.

6. DISCUSSION AND CONCLUSION

Clusters obtained from gene expression data analysis must correlate to the existing knowledge. Mining on gene expression alone may not be able

Figure 4. Comparison of clustering generation methods as a function of constraint set size

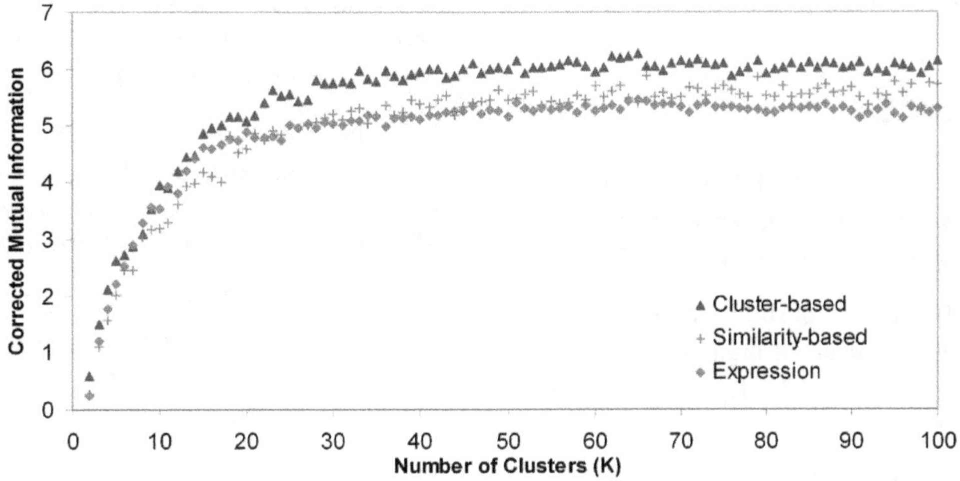

to reveal the biological information related to the gene expressions. Most biologists focus their research on a small select set of genes, which they know to be functionally related. Consequently, their publications focus on these genes. Therefore publications stored in medical literature databases, such as PubMed, can provide valuable additional information. In this paper, we use text literature as a guide for microarray data analysis. In particular, we want to identify subgroups of genes with commonalities in gene expression and in biological function.

We developed two clustering algorithms for clustering genes incorporating information from literature text data. These two algorithms were applied to gene expression data for the same purpose, but differ in the way using literature text data. When the information from literature text data is complete (i.e., data available on all genes to be analyzed), the MSC algorithm performed significantly better than the feature-level integration approach. Also, the clusters from the MSC algorithm shared regulatory elements that were not found using gene expression data alone.

For the situation that complete information from literature text data is not available, constrained clustering algorithm can be applied to gene expression data with constraints derived from a subset of the literature data. We showed that constraints set may not improve the performance of a clustering algorithm if it is not carefully chosen or not large enough. We concluded that inconsistent constraints should be avoided since small number of them will hurt the clustering. We explored several ways of deriving constraints from prior knowledge about the data and from external biological text literature. Our results suggested that clustering using constraints generated from a clustering-based approach outperformed the case when constraints were generated using a similarity-based approach. Comparisons with MSC algorithm showed that constraint-based methods performed better assuming that the constraints were of sufficiently high quality. Based on our experience, negative constraints are not appropriate for text literature data source. However, we conjecture that it is appropriate for other data sources such as protein-protein interactions, and to a limited extent, regulatory element information.

Clustering using heterogeneous data sources is a rapidly growing area of research. Future plans include incorporating more data sources for clustering and comparative studies on the clustering methods proposed in this paper and other methods, such as ensemble clustering, which involves finding a single consensus solution by combining a set of clusterings obtained from different data sources.

REFERENCES

Basu, S., Banerjee, A., et al. (2002). *Semi-supervised clustering by seeding*. Paper presented at the International Conference on Machine Learning.

Becker, S. (1996). Mutual information maximization: Models of cortical self-organization. *Network (Bristol, England)*, *7*(1), 7–31. doi:10.1088/0954-898X/7/1/003

Berriz, G. F., & King, O. D. (2003). Characterizing gene sets with FuncAssociate. *Bioinformatics (Oxford, England)*, *19*(18), 2502–2504. doi:10.1093/bioinformatics/btg363

Bickel, S., & Tobias, S. (2004). Multi-View Clustering. In *Proceedings of the Fourth IEEE International Conference on Data Mining (ICDM'04)*.

Bilenko, M., Basu, S., et al. (2004). Integrating constraints and metric learning in semi-supervised clustering. In *Proceedings of the International Conference on Machine Learning (ICML '04)*.

Chiang, J. H., & Yu, H. C. (2003). MeKE: discovering the functions of gene products from biomedical literature via sentence alignment. *Bioinformatics (Oxford, England)*, *19*(11), 1417–1422. doi:10.1093/bioinformatics/btg160

Davidson, I., & Wagstaff, K. L. (2006). Measuring constraint-set utility for partitional clustering algorithms. In *Proceedings of Knowledge Discovery in Databases* (*Vol. 4213*, pp. 115–126). Pkdd. doi:10.1007/11871637_15

de Souto, M. C., & Costa, I. G. (2008). Clustering cancer gene expression data: a comparative study. *BMC Bioinformatics*, *9*, 497. doi:10.1186/1471-2105-9-497

Dempster, A. P., & Laird, N. M. (1977). Maximum likelihood from incomplete data via the em algorithm. *Journal of the Royal Statistical Society. Series A (General)*, *39*, 1–38.

Gibbons, F. D., & Roth, F. P. (2002). Judging the quality of gene expression-based clustering methods using gene annotation. *Genome Research*, *12*(10), 1574–1581. doi:10.1101/gr.397002

Glenisson, P., & Antal, P. (2003). Evaluation of the vector space representation in text-based gene clustering. *Pacific Symposium on Biocomputing. Pacific Symposium on Biocomputing*, 391–402.

Glenisson, P., & Mathys, J. (2003). Meta-clustering of gene expression data and literature-based information. *SIGKDD Explor. Newsl.*, *5*(2), 101–112. doi:10.1145/980972.980985

Ideker, T., & Thorsson, V. (2001). Integrated genomic and proteomic analyses of a systematically perturbed metabolic network. *Science*, *292*(5518), 929–934. doi:10.1126/science.292.5518.929

Jain, A. K., & Dubes, R. C. (1988). *Algorithms for clustering data*. Upper Saddle River, NJ: Prentice Hall.

Kasturi, J., & Acharya, R. (2004). Clustering of diverse genomic data using information fusion. In *Proceedings of the 2004 ACM symposium on Applied computing,* Nicosia, Cyprus (pp. 116-120). New York: ACM Press.

Kerr, G., & Ruskin, H. J. (2008). Techniques for clustering gene expression data. *Computers in Biology and Medicine*, *38*(3), 283–293. doi:10.1016/j.compbiomed.2007.11.001

Klein, D., Kamvar, S. D., et al. (2002). *From instance-level constraints to space- level constraints: Making the most of prior knowledge in data clustering*. Paper presented at the International Conference on Machine Learning (ICML'02).

Matys, V., & Fricke, E. (2003). TRANSFAC(R): transcriptional regulation, from patterns to profiles. *Nucleic Acids Research, 31*(1), 374–378. doi:10.1093/nar/gkg108

Raghavan, V. V., & Wong, S. K. M. (1986). A critical analysis of vector space model for information retrieval. *Journal of the American Society for Information Science American Society for Information Science, 37*(5), 279–287.

Rand, W. M. (1971). Objective Criteria for Evaluation of Clustering Methods. *Journal of the American Statistical Association, 66*(336), 846. doi:10.2307/2284239

Roberts, R. J. (2001). PubMed Central: The GenBank of the published literature. *Proceedings of the National Academy of Sciences of the United States of America, 98*(2), 381–382. doi:10.1073/pnas.98.2.381

Roth, F. P., & Hughes, J. D. (1998). Finding DNA regulatory motifs within unaligned noncoding sequences clustered by whole-genome mRNA quantitation. *Nature Biotechnology, 16*(10), 939–945. doi:10.1038/nbt1098-939

Spellman, P. T., & Sherlock, G. (1998). Identification of cell cycle regulated genes in yeast by DNA microarray hybridization. *Molecular Biology of the Cell, 9*, 371a–371a.

Stephens, M., Palakal, M., et al. (2001). Detecting gene relations from MEDLINE abstracts. In *Proceedings of the sixth Ann Pac Symp Biocomp (PSB 2001)*.

Vogel, K., & Horz, W. (1989). The two positively acting regulatory proteins PHO2 and PHO4 physically interact with PHO5 upstream activation regions. *Molecular and Cellular Biology, 9*(5), 2050–2057.

Wagsta, K., Basu, S., et al. (2006). When is constrained clustering beneficial, and why? In *Proceedings of the AAAI.*

Wagsta, K., Cardie, C., et al. (2001). *Constrained k-means clustering with background knowledge.* Paper presented at the 18th International Conference on Machine Learning (ICML-01).

Wu, L., & Oviatt, S. L. (1999). Multimodal integration - a statistical view. *IEEE Transactions on Multimedia, 1*(4), 334–341. doi:10.1109/6046.807953

Xing, E. P., & Ng, A. Y. (2002). Distance metric learning, with application to clustering with side-information. In *Proceedings of Advances in Neural Information Processing Systems, 15*, 505–512.

Yandell, M. D., & Majoros, W. H. (2002). Genomics and natural language processing. *Nature Reviews. Genetics, 3*(8), 601–610.

Yeung, K. Y., & Medvedovic, M. (2003). Clustering gene-expression data with repeated measurements. *Genome Biology, 4*(5), R34. doi:10.1186/gb-2003-4-5-r34

Zare, H., & Khodursky, A. B. (2006). Gene clustering and gene function prediction using multiple sources of data. In *Proceedings of the IEEE Genomic Signal Processing and Statistics, 2006*, 113–114.

Zhong, S., & Ghosh, J. (2003). A comparative study of generative models for document clustering. In *Proceedings of the workshop on Clustering High Dimensional Data and Its Applications in SIAM Data Mining Conference*.

This work was previously published in the International Journal of Knowledge Discovery in Bioinformatics, Volume 1, Issue 2, edited by Xiao-Li Li and See-Kiong Ng, pp. 12-28, copyright 2010 by IGI Publishing (an imprint of IGI Global).

Chapter 6
Efficient Mining Frequent Closed Discriminative Biclusters by Sample-Growth:
The FDCluster Approach

Miao Wang
Northwestern Polytechnical University, China

Xuequn Shang
Northwestern Polytechnical University, China

Shaohua Zhang
Northwestern Polytechnical University, China

Zhanhuai Li
Northwestern Polytechnical University, China

ABSTRACT

DNA microarray technology has generated a large number of gene expression data. Biclustering is a methodology allowing for condition set and gene set points clustering simultaneously. It finds clusters of genes possessing similar characteristics together with biological conditions creating these similarities. Almost all the current biclustering algorithms find bicluster in one microarray dataset. In order to reduce the noise influence and find more biological biclusters, the authors propose the FDCluster algorithm in order to mine frequent closed discriminative bicluster in multiple microarray datasets. FDCluster uses Apriori property and several novel techniques for pruning to mine biclusters efficiently. To increase the space usage, FDCluster also utilizes several techniques to generate frequent closed bicluster without candidate maintenance in memory. The experimental results show that FDCluster is more effective than traditional methods in either single micorarray dataset or multiple microarray datasets. This paper tests the biological significance using GO to show the proposed method is able to produce biologically relevant biclusters.

DOI: 10.4018/978-1-4666-1785-8.ch006

INTRODUCTION

Nowadays, in the post-genomic era, there have many bioinformatics data sets available. Due to the lack of accurate machine learning or intelligent tools in the bioinformatics community, the information embedded in most of these data has not yet completely exploited. Recently, DNA microarray technology has generated a large number of gene expression data, which is typically represented by a matrix where each cell represents the gene expression level of a gene under an experimental condition. How to use these data to reveal the function and biological process of genes poses a great challenge of analysis algorithms. Various data mining techniques have been employed to infer useful biological information from the huge and rapid growing microarray data set.

One widely used method to infer relationship among genes in microarray data set is frequent pattern mining. Based on the characteristic of microarray data, (Pan et al., 2004; Cong et al., 2004) proposed to use condition enumeration method to exploit the gene patterns. However, both of above algorithms need to maintain the candidate patterns in memory, which limits the scalability. Association rules mining method is another way to analyze the gene expression data (Becquet et al., 2003; Creighton & Hanash, 2003; McIntosh & Chawla, 2007; Cong et al., 2004), which can discover the relationship among genes. However, it only can identify genes whose expression levels correlated across some conditions, it can not reveal the regulatory relations among genes. Using association rule to exploit regulatory modules has its limitations (Yeung et al., 2004).

How to identify genes with similar behavior with respect to different samples? Biclustering (Cheng & Church, 2000) is a methodology allowing for condition set and gene set points clustering simultaneously. It finds clusters of genes possessing similar characteristics together with biological conditions creating these similarities. The main advantage of biclustering is the simultaneous mining module on genes and experimental condition, another advantage is its applicability on original data instead of discretized data (Zhao & Zaki, 2005). However, mining microarray data for biclusters presents the following four challenges. First, the computing of biclustering method is NP-hard (Cheng & Church, 2000). Second, biclustering method deals with original data, it should adapt to the noise-sensitive character of microarray dataset. Third, the biclustering method should allow overlapping biclusters which share some genes or conditions, which would increase the complex of biclustering algorithm. Finally, the biclustering method should be flexible enough to handle different types of biclusters. (Madeira & Oliverira, 2004) classified biclusters into four categories: (i) constant value biclusters, (ii) constant row or column biclusters, (iii) biclusters with coherent values, where each row and column is obtained by addition or multiplication of the previous row and column by a constant value and (iv) biclusters with coherent evolutions, where the direction of change of values is important rather than the coherence of the values (Pandey et al., 2009).

Facing with the former three challenges above of biclustering method, some algorithms proposed to use greedy or heuristics approach for mining biclusters. In (Cheng & Church, 2000), Cheng and Church employed a greedy node deletion algorithm in their search based on using a low mean squared residue. Once a bicluster is created, its entries are replaced by random numbers and the procedure is repeated iteratively. Since then, there have been many greedy algorithms (Ben et al., 2003; Yang et al., 2003; Liu & Wang, 2007; Cheng et al., 2008; Teng & Chan, 2008; Dharan & Nair, 2009). A recent review of biclustering algorithms for biological data analysis can be found in (Madeira & Oliveira, 2004). Although these algorithms may improve the result, yet the efficiency is not very well. Another biclustering algorithm, MicroCluster (Zhao & Zaki, 2005), used weighted directed range multigraph to generate deterministic bicluster. The experimental results show this algorithm is very efficient. It also used some deletion or merging method to reduce the influence of noise. However, the main drawback of MicroCluster is that all the candidate biclusters need to be maintained in memory, which would increase computing complex and reduce space usage.

How to handle the last drawback of biclustering method is very intractable. Since the clustering formulation cannot be suitable to exploit some or all types of biclusters. For instance, SAMBA (Subramanian et al., 2005) is designed to find constant value biclusters. And Cheng and Church's method (Cheng & Church, 2000) can find both constant value and constant row or column biclusters. Biclusters with coherent trends of up or down regulation can be found by OPSM (Ben et al., 2003). While xMotifs (Murali & Kasif, 2003a) is designed to infer biclusters with constant columns and MicroCluster (Zhao & Zaki, 2005) uses a weighted multigraph to mine scaling biclusters. One way to increase the scalability of biclustering algorithm is to discretize the original data, which can reduce the data sensitivity. But it may lose some biological information. Therefore, there is a tradeoff whether using original data or descredized data. In order to improve the scalability of biclustering method, this paper uses discretized data to mine bicluster.

Recently, discriminative patterns have been shown to be useful for classification analysis (Cheng et al., 2007; Cheng et al., 2008; Deshpande et al., 2005; Fan et al., 2008; Lo et al., 2009; Vliet et al., 2007). Mining discriminative patterns for microarray data is very interesting, which has the potential to find groups of genes that are individually not informative but are highly associated with a phenotype when considered as a group (Gowtham et al., 2009). Take the cancer microarray data set for example; discriminative patterns denote the patterns are co-expressed in healthy person but unexpressed in cancer patients. Mining discriminative patterns can infer the reasonable genes for cancer. (Gang et al., 2010) is a novel method to mine subspace differential coexpression pattern. (Gang et al., 2010) proposes a new definition of support, which has property called anti-monotonicity that guarantees the method can be used in an efficient pattern mining framework for finding all the differential co-expression patterns. Although (Gang et al., 2010) can find differential patterns using Apriori property, this approach still have some limitations. (1) Due to the limitation of Aprioir framework, the computing efficiency is not very well. (2) In order to prune non-interesting patterns, all the candidate patterns should be maintained in memory for checking, which increase the computing complex and reduce the space usage. (3) The scalability is poor. (4) Based on the definition of support, mining differential patterns need to be double checking. One is to find the patterns existing in Class A but Class B, the other is to find the patterns existing in Class B but Class A. Therefore, using the method proposed by (Gang et al., 2010) to mine discriminative pattern is not very efficiently.

Discriminative co-expression patterns can be considered as biclusters that exist mostly in one class but not in the other. Some studies (Murali & Kasif, 2003 b; Tanay et al., 2004; Zhao & Zaki, 2005) use two-step approach to discovery discriminative co-expression patterns. A set of biclusters are discovered in the first step and then the ones that unique to a single class are selected in the second step. Such two-step approach is very time-consuming. The approach proposed in (Gang et al., 2010) can also be used to generate discriminative biclusters. However, according to the discussion in (Gang et al., 2010), the double checking procedure is less efficient. In this paper, we propose to mine discriminative co-expression biclusters in differential data set, which is shown in Figure 1. Using differential data set to analyze has a premise, which is the sample must be a unique one. It may not be suitable to mine cancer microarray data set, since we cannot get the expression value of a sample of health and cancerous. However, the AGEMAP

Figure 1. The overview of FDCluster approach

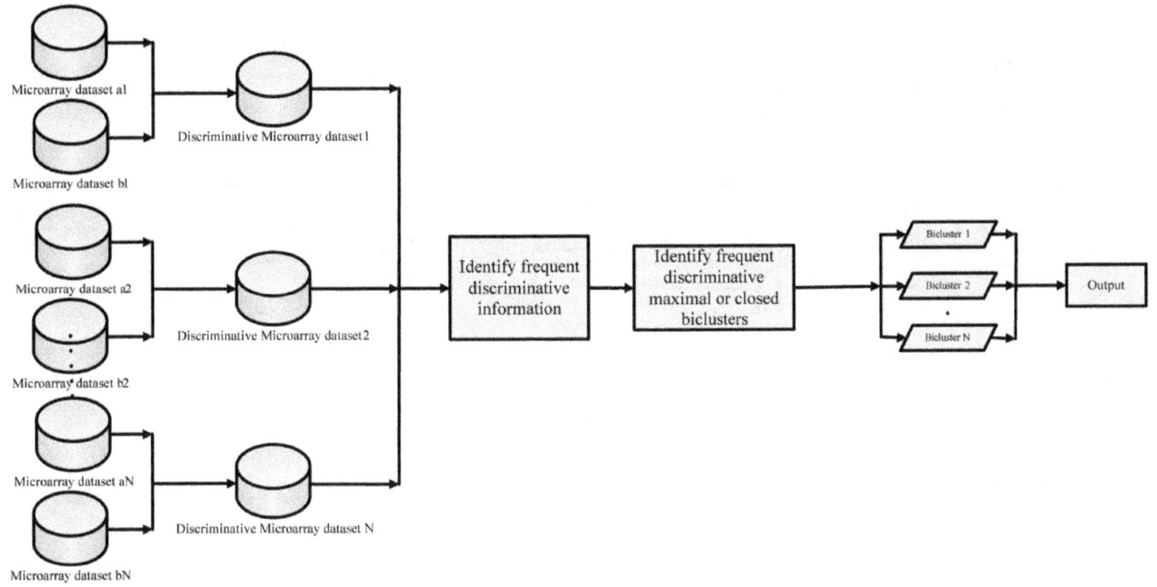

(Zahn et al., 2007) data set provides a unique opportunity to create difference matrix as the data set from the 8-month old mice is matched to the data set from the 16-month old mice. Since the mice sample (five males, five females) is unique one during aging, so the difference matrix does not lose information comparing to original data sets. The detail of AGEMAP data set can be found in experimental result section.

According to above analysis and in hopes of overcome the limitations of biclustering methods, we propose a method that utilizes multiple microarray datasets to identify frequent discriminative biclusters. The reason to mine multiply microarray datasets is to reduce the influence of noise and get more accurate results. A straightforward approach towards this aim is to mine biclusters in each microarray dataset, then count the number of times. Based on the support threshold defined by user, the frequent biclusters is output. The overview of this naïve method is shown in Figure 2. A major drawback in this approach is that the entire set of samples needs to be checked and the support pruning technique is not used, so the computing is very time-consuming. Another drawback is that all the biclusters need to be maintained in memory, which influences the scalability and the space usage.

The computational and space usage cost raised by naïve framework motivates us to investigate an alternative approach: In stead of generating all the biclusters in each microarray dataset, directly mine frequent highly discriminative biclusters. It leads to our approach of a frequent closed discriminative bicluster mining method without candidate maintenance, FDCluster. It uses Apriori property and several novel techniques for pruning to mine biclusters efficiently. In order to increase the space usage, FDCluster also utilizes several techniques to generate frequent closed bicluster without candidate maintenance in memory. FDCluster considers both positive and negative relations among genes in bicluster, which can deliver more valuable potential regulation information of inferring the function of gene and gene network construction. The overview of our approach is illustrated in Figure 1.

The contributions of this paper can be summarized as follows:

Figure 2. The overview of naïve method for mining frequent bicluster

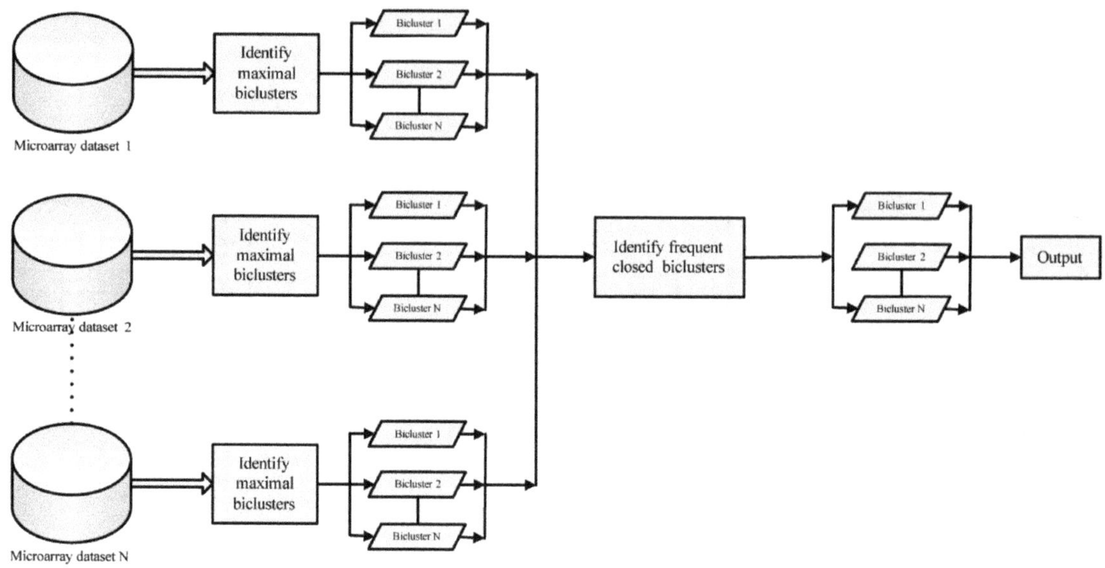

- We address the necessity of trading off using the real value of microarray dataset with discretization data. Using discretization data may lose some information, but can improve the scalability of algorithm and handle different type of biclusters.
- We propose to mine bicluster in multiple microarray datasets, which can reduce the influence of noise. In stead of using naïve method, FDCluster is proposed by using Apriori property to prune non-frequent bicluster in time.
- In stead of mining discriminative bicluster in both original datasets, we mine discriminative bicluster in differential dataset, which is the difference between original datasets.
- FDCluster can find frequent closed discriminative biclusters without candidate maintenance.

PRELIMINARY CONCEPTS AND DEFINITIONS

The microarray dataset is defined as a gene expression matrix in real expression numbers, denoted as $D = C \times G$, where the column set $C = \{c_1, c_2, \cdots, c_i\}$ represents the different experimental conditions or samples, the row set $G = \{g_1, g_2, \cdots, g_i\}$ represents genes, and the element value of D_{ij} is a real number that represents the expression level of gene i under experimental condition j. Each row corresponds to the expression levels of a gene over all the experimental conditions, and each column corresponds to the expression levels of genes in one experimental condition.

A bicluster B is defined as a submatrix of D, where $B=X \times Y$ with $X \subseteq C$, $Y \subseteq G$ and the genes in Y show similar behavior (coexpression) under the experimental conditions X. Given M be the set of all biclusters in D, which satisfy the given coexpression condition, then $N = P \times Q \subseteq M$ is a maxical bicluster if and only if there does not exist a bicluster $E = J \times K$ such that $P \subseteq J$ and $K \subseteq Q$. A bicluster can be denoted as

Table 1. Converted Microarray Dataset

	SB_{0B}	SB_{1B}	SB_{2B}	SB_{3B}	SB_{4B}
A	1	-1	1	-1	1
B	1	-1	1	-1	1
C	0	-1	1	0	0
D	1	0	0	-1	0
E	0	-1	0	-1	1

Samples(Genes), which represents the samples of the bicluster and its co-expressed genes.

In order to improve the scalability of algorithm and handle different type of biclusters, the original microarray data would be converted each gene expression number into one of the three numbers: 1, -1 and 0, which denotes expressed, depressed and non-expressed, respectively, as shown in Table 1. The discretization method of our algorithm is shown in experimental result section. Therefore, three types of relations between genes X and Y can be respectively defined as follows:

1. X and Y is positive expressed denoted as (XY) if $X=1$ and $Y=1$, or $X=-1$ and $Y=-1$.
2. X and Y is negative expressed denoted as $(X\bar{Y})$ if $X=1$ and $Y=-1$, or $X=-1$ and $Y=1$.
3. X and Y is unexpressed if X is if $X=1$ and $Y=0$, or $X=-1$ and $Y=0$, or $X=0$ and $Y=1$, or $X=0$ and $Y=-1$.

In our method, we mine bicluster by using undirected weighted sample relational graph, which is similar to the weighted directed range multigraph in (Zhao & Zaki, 2005). The definition of undirected weighted sample relational graph is shown as following:

Definition 1. The weighted undirected sample relational graph $G=\{E, V, W\}$, each vertex V_i in the graph represent an unique sample, there exist an edge E_i between a pair of samples only if the two samples has co-expressed genes and weighted item set W_i is the gene patterns which are co-expressed in both vertexes samples.

Figure 3. The weighted undirected sample relational graph of Table I

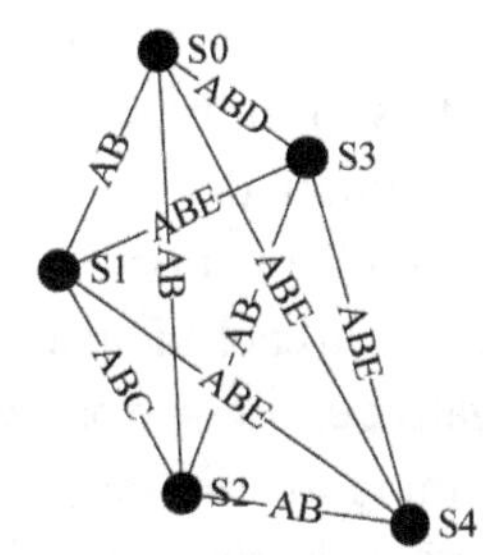

Figure 3 shows the weighted undirected sample relational graph constructed from Table 1. The weighted item set of each sample pair can be got using above genes relation definition, we only focus on the positive and negative expressed pattern. The detail of how to construct the weighted undirected sample relational graph can be found in next section.

Definition 2. A weighted undirected sample relational graph set can be defined as $S = \{D_1, D_2, \cdots, D_i\}$, where D_i is the weighted undirected sample relational graph of one microarray dataset.

Definition 3. Given a weighted undirected sample relational graph set S, the support of a bicluster T is the number of graphs in D where T is a subset, denoted as $Sup(T)$. T

is claimed to be frequent bicluster if *Sup(T)* is greater than a minimum user defined support threshold. *T.Transaction* denotes the id set of each graph dataset which contains *T*.

Definition 4. Given a weighted undirected sample relational graph set *S*, a bicluster *T* is claimed to be frequent closed bicluster if there exists no proper superset of *T* with the same support, that is, there exists no a bicluster M such that $T \subseteq M$ and *Sup(M)=Sup(T)*.

Our goal is to mine all the frequent closed discriminative biclusters in multiple differential microarray datasets, which denoted by using weighted undirected sample relational graph. And the relations among genes in each bicluster include both positive and negative expressed patterns. Unlike the traditional mining discriminative pattern method, our method mine discriminative bicluster in differential datasets. The differential method can be found in experimental result section.

THE FDCluster ALGORITHM

This section describes the FDCluster algorithm in detail for mining frequent closed discriminative bicluster in multiple differential microarray datasets. There is a hypothesis in this section that the input dataset of microarray is the differential microarray dataset. The differential method would be shown in next section.

A. Construct the Weighted Undirected Sample Relational Graph

The weighted undirected sample relational graph can be got by using the discretization microarray dataset, which can escape of designing computing function which is shown in (Zhao & Zaki, 2005). As discussed in Section 1, using discretization data can increase the scalability of algorithm and can also handle more types of biclusters. The key to construct this relational graph is to mine co-expressed patterns in two samples. According to the definition of co-expressed pattern, we are only interested in both positive co-expressed pattern and negative co-expressed pattern. Based on the definition of positive and negative co-expressed patterns, it is easy to generate co-expressed pattern in each pair of samples according to the following lemma.

Lemma 1. Given one gene *G*, if it is co-expressed with an existing co-expressed pattern, it exists in either positive co-expressed or negative co-expressed pattern.

Proof. According to the definition of both co-expressed patterns, if *G* is positive co-expressed with the existing pattern *P* in sample S_1 and S_2, the value of *G* in S_1 and S_2 should be same (1 or -1). If g is negative co-expressed with the existing pattern *Q* in S_1 and S_2, the value of G in both samples should be negative (the value in S_1 is 1 and the value in S_2 is -1, or the value in S_1 is -1 and the value in S_2 is 1). However, it does not be existed one gene which has two values in one sample. Therefore, one gene can only exist in one type of co-expressed pattern.

Algorithm 1 outlines the main routine of the WURGraph algorithm implementation to construct the weighted undirected relational graph from the discretization microarray dataset. Figure 3 is an example by converting original microarray dataset show in Table I into the weighted undirected relational graph. In our FDCluster framework, all the original microarray datasets should be converted to weighted undirected relational graphs respectively, then using these graphs to mine biclusters.

Algorithm 1. WURGraph algorithm

Input: the minimum row threshold $r_{min,}$ the discretization microarray dataset *D*.
Output: the weighted undirected sample relational graph of *D*: *G*.
Initialization: *G* =∅, the positive co-expressed pattern *PSet*=∅, the negative co-expressed pattern *NSet*=∅.
Method: WURGraph ($r_{min,}$ *D*)
(1) scan *D* and find the sample set *S* and gene set *G* of *D*;
(2) **for** each pair samples S_i and S_j in *S*
(3) **for** each pair genes G_i and G_j in *G*
(4) **if** the value of G_i in S_i and S_j are not zero and the same as the value of G_j in S_i and S_j **then**
(5) $PSet = G_iG_j \cup PSet;$
(6) **else if** the value of G_i in S_i is the opposite of the value of G_j in S_i and the value of G_i in S_j is the opposite of the value of G_j in S_j
(7) $NSet = G_iG_j \cup NSet;$
(8) **endif**
(9) **endfor**
(10) **endfor**
(11) **if** the number of genes in *PSet* or *NSet* is greater than r_{min} **then**
(12) insert S_i , S_j , *PSet* and *NSet* to *G*;
(13) **endif**
(14) **endfor**
(15) **return**

B. Mining Maximal Bicluster

In this part, we will introduce how our method finding the maximal bicluster in one microarray dataset, then the extended algorithm to mine frequent bicluster in multiple microarray would be shown in next part. Before the algorithm is presented, let us discuss some approaches for mining maximal bicluster. There are two approaches for mining maximal bicluster. (1) Find all the biclusters, and only out the maximal ones based on the definition. (2) Find the candidate set and prune the biclusters which are not maximal. For each new generated bicluster, scanning the previous biclusters to check whether the new one is maximal, if it is maximal, then update the previous maintained biclusters, or else the new one is pruned. Both of above approaches has rather poor scalability in the number of biclusters, as candidate maintenance-and-test scheme would lead to memory redundancy. Since they would maintain a number of biclusters in memory. In our method, a novel method called maximum checking is used to check the generated bicluster is maximal or not. Maximum checking method can avoid maintaining all the candidate set and prune non-maximal biclusters efficiently.

Then, we will introduce how our algorithm finding maximal bicluster in one microarray dataset which is denoted as a weighted undirected sample relational graph. In order to distinguish it from the multiple microarray datasets mining algorithm, we denote the algorithm in one microarray dataset as FDCluster_One algorithm. FDCluster_One exploits a fundamental different approach from previous works. In order to prune the non-maximal bicluster in time and avoid candidate set maintenance in memory, we use several techniques to achieve efficient mining. As we known, the microarray has more genes and less samples. Therefore, FDCluster_One mines bicluster by using sample-growth strategy. The following lemmas are made to guarantee the completeness and correctness of generating all the maximal biclusters.

Definition 5. Let S be a candidate of the extended bicluster, which can be denoted as *S.Sample(S.Geneset)*, where *S.Sample* is the sample which links all the samples of S and *S.Geneset* is the gene set which is co-expressed both in *S.Sample* and each samples of S.

Lemma 2. Let P be a size-2 extended bicluster and $\{S_1, S_2, \cdots, S_n\}$ be the set of its candidate samples. The candidate set of the extending bicluster growth from P, must be the subset of $\{S_1, S_2, \cdots, S_n\}$.

Proof. Supposed there exists a super bicluster $\{S_1, S_2, \cdots, S_n\}$ which is extended by P, and the candidate sample set of PS_m has a sample S_i that does not in $\{S_1, S_2, \cdots, S_n\}$. Since S_i is the candidate sample of PS_m, so S_i must be linked to both the samples of P and S_i in the weighted undirected sample relational graph. Therefore, S_i must be the candidate sample set of P. However, this result is conflicting with above hypothesis that S_i does not exist in $\{S_1, S_2, \cdots, S_n\}$. Therefore, the candidate sample set of the extending bicluster growth from P must be the subset of the candidate sample set of the extended bicluster.

Lemma 3. Given P be the current extending bicluster, if the candidate S_i can be extended to P. PS_i is a maximal bicluster if it satisfies both of following conditions. (1) There should not exist a previous candidate of P denoted as S_j, where the genes in PS_i is the subset of the genes in PS_j; (2) There does not exist an extending bicluster PS_iS_p, where S_p is the candidate of PS_i and PS_i is the subset of PS_iS_p.

Proof. (1) Given $\{S_1, S_2, \cdots, S_{m-1}, S_{m+1}, \cdots, S_n\}$ be the candidate set of S_m, and the genes in $S_mS_i(S_i \in [S_{m+1}, S_n])$ is the subset of the genes in $S_mS_j(S_j \in [S_1, S_{m-1}])$. Since the sample-growth method is depth-first, so if the genes in S_mS_i is the subset of genes in S_mS_j, it must get $S_mS_jS_i$ which has the same genes as S_mS_i. Therefore, S_mS_i must be the subset of existing bicluster. (2) Based on the definition of maximal bicluster, the result is straightforward.

Using Lemma 3 can generate maximal biclusters without candidate maintenance in memory. However, it needs to maintain some prior candidate information in memory. If the microarray dataset is very large, it may store more candidate information, which reduces the efficiency. In order to escape of maintaining prior candidate information, we use a Hyper List (H-List) to store the weighted undirected sample relational graph. H-List is a special list storage structure, which is similar the orthogonal list. The H-List consists of sample header table, sample table and hyper node structure. To facilitate maximal checking which is show in the follow, a sample header table is built to link all the hyper nodes which have the same sample. The detail description of H-List is shown as following definition. Figure 4 shows a Hyper List constructed based.

Definition 6. A hyper list structure is a graph-based structure defined as below:

1. It consists of one sample header table which has two fields: sample-id and sample-link, where sample-id stores the unique sample information and sample-link stores address of the first hyper node which has the same sample.
2. Each entry in the sample table consists of two fields: sample-id and node-link, which store sample information and the address of hyper node respectively.
3. Each hyper node consists of four fields: sample-id, geneset-id, node-link and node-next, where sample-id stores the sample name, the co-expressed gene

Figure 4. The Hyper List of microarray dataset

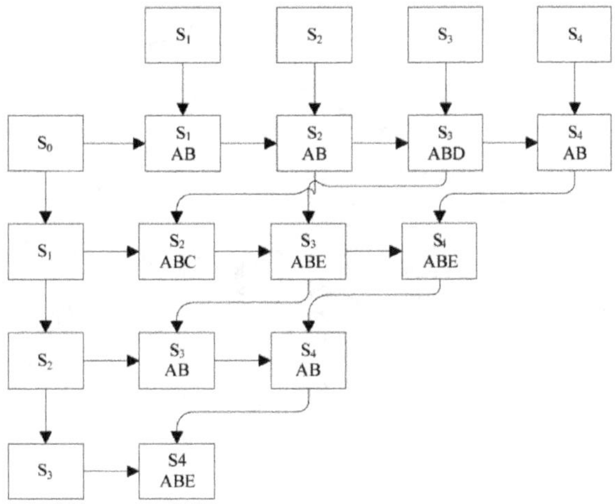

set is stored in geneset-id, node-link stores the next hyper node in this layer and the next node in the same layer is stored in node-next.

Lemma 4. If extending the candidate *S* to a bicluster *Q*, it should check the above layers of *S* in H-List. If there exists a sample, whose sample name is the same as *S* and its gene set is the superset of the candidate sample *S.Geneset*, the candidate sample *S* should not be extended and pruned.

Proof. We use disproval method to prove this lemma. Supposed *S* should not be pruned. The current extending bicluster is *Q*. There exists a previous candidate *P*, whose gene set is the superset of *S.Geneset*. Since the maximal bicluster is generated by using depth-first method and *P* is the previous candidate of *S*, whose gene set has the same subset of *P.Geneset*. Therefore, there must be existed a bicluster which contains both *P* and *S*. And *S* must be extended by *P* previously. Based on the Lemma 2, the candidate set of *QS* must be the candidate set of *QP*, which is the superset of *QS*. Therefore, *QS* should be pruned, which is conflicted to our hypothesis. So *S* should be pruned.

For example, if the current extended bicluster is S_1S_2 *(ABC)* and its candidate is S_3. There exists a sample S_3 in the first layer in the H-List shown in Fig. 3 and $S_1S_2S_3$*.Geneset (AB)* is the subset of above layer S_0S_3*.Geneset (ABD)*, so S_3 should not be extended to S_1S_2 *(ABC)*.

Based on the above lemmas, the FDCluster_One algorithm exploits several pruning techniques to achieve efficient mining.

Pruning 1. If there exists a previous candidate sample whose gene set is the superset of the current candidate sample' gene set, the current sample should be pruned.

Pruning 2. If there exists an above layer of the current candidate sample, whose gene set is the superset of the current candidate sample, the current candidate sample should be pruned.

Based on the related definitions of bicluster and above lemmas and pruning techniques, FDCluster_One algorithm is designed for mining maximal bicluster without candidate maintenance in one microarray dataset. FDCluster_One algorithm adopts the sample-growth and depth-first mining technique to generate maximal bicluster. Our algorithm is not naïve. It never searches the

Figure 5. The process of FDCluster_One dealing with an example dataset

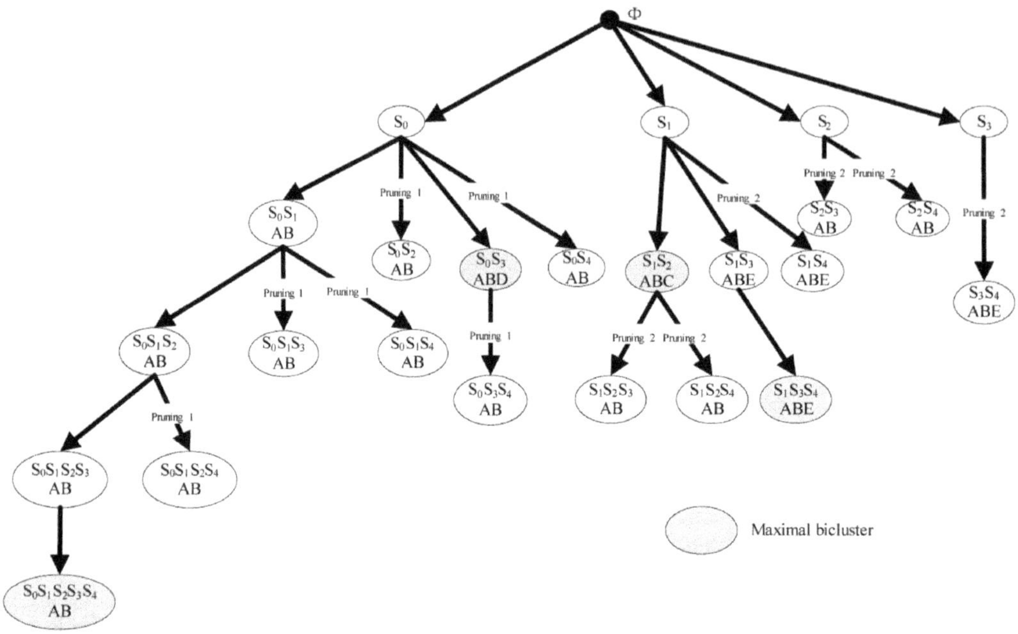

complete set of biclusters, which most bicluster mining algorithms adopt. Instead, to prune the search space based on the above lemmas. The detail of FDCluster_One algorithm is shown in Algorithm 2. The example dataset is shown in Figure 5. The minimum row threshold r_{min} and the minimum column threshold c_{min} are both set to 2.

C. Mining Frequent Closed Bicluster

In this part, we will introduce how FDCluster Algorithm finding frequent closed bicluster in multiple microarray datasets. FDCluster uses a novel scheme to generate frequent closed bicluster without candidate maintenance, which is similar to the maximum checking method of FDCluster_One algorithm. Before the FDCluster algorithm is presented, we will discuss some approaches for mining frequent closed bicluster. There are two approaches for mining frequent closed bicluster. (1) Find all the closed biclusters in each microarray datasets, then using the definition of frequent pattern to output frequent closed bicluster. This naïve method is easy to implement, but has poor scalability in the number of biclusters and microarray datasets. (2) Use Apriori property to find frequent closed bicluster by pattern-growth method. If a bicluster is frequent, the any subset of it must be frequent. Using patter-growth method can prune infrequent bicluster in time, but it may scan datasets more times and the efficiency is not very well. In our method, we use the second method to find frequent closed bicluster. In order to reduce the times of scanning datasets, the multiple microarray datasets would be converted to one dataset which contains frequent information.

In the part of FDCluster_One algorithm, we introduce a special storage structure, Hyper List, to store the weighted undirected sample relational graph of one microarray dataset. In order to facilitate closure checking, we design a storage structure denoted as Multiple Hyper List (MH-List) to store information of each microarray datasets. The detail description of MH-List is similar to H-List defined by Definition 6, the difference is that the hyper node of MH-List has

Algorithm 2. The FDCluster_One algorithm

Input: the minimum row threshold r_{min}; the minimum column threshold c_{min}; the weighted undirected sample relational graph D; the current extending bicluster P;
Output: the maximal bicluster set of D;
Initialization: $P=\varnothing$;
Method: FDCluster_One ($r_{min}, c_{min}, s_{min}, S, P$)
(1) flag=**false**;
(2) scan D and find all the candidate set S of P;
(3) **for** each candidate S_i in S
(4) **if** the biclsuter PS_i does not satisfy **Pruning 1** and **Pruning 2**, and the number of genes in PS_i is greater than r_{min}, the number of samples is greater than c_{min}, **then**
(5) *P.Sample= P.Sample +* S_i *Sample*;
(6) *P.Geneset= P.Geneset*$\cap$ S_i *Geneset*;
(7) CallFDCluster_One ($r_{min}, c_{min}, s_{min}, S, P$);
(8) **endif**
(9) **if** S_i *Geneset = P.Geneset*
(10) flag=**true**;
(11) **endif**
(12) **endfor**
(13) **if** flag=**false**
(14) **output**(P);
(15) **endif**
(16) **return**

Figure 6. The Multiple Hyper List of multiple microarray datasets

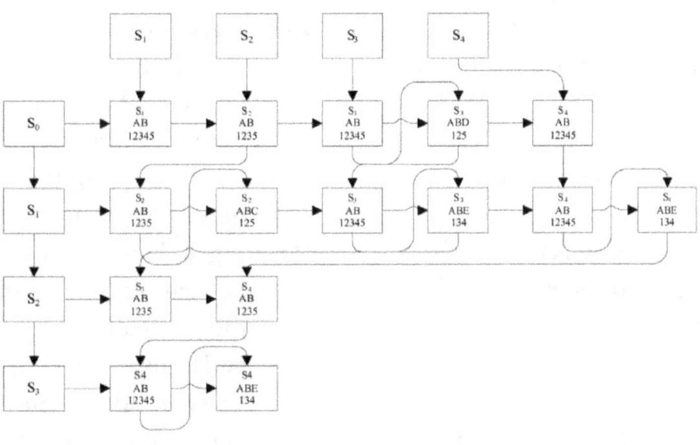

five fields: sample-id, node-link, node-next, geneset-id and trans-id. The former three fields have the same mean of H-List, which is shown in Definition 6. The geneset-id in MH-List stores the frequent co-expressed gene set between samples. The trans-id stores the graph dataset id set where the gene set is co-expressed. Figure 6 shows an example of MH-List structure. The letter S_i in each node indicates the sample name, the letter set below of sample name in hyper node indicates the co-expressed gene set, and the transaction number of above co-expressed gene set is shown in the below of geneset-id.

The key of MH-List construction is how to generate frequent co-expressed gene set between two samples. Many frequent closed pattern mining methods can be used to generate frequent co-expressed gene set. The reason why the closed pattern is used is that closed pattern has no redundant information without losing information. In this paper, we use the recent proposed frequent closed pattern mining method, WIBE (Wang et al., 2010), to generate co-expressed gene set. WIBE can mine not only positive expressed patterns, but also negative patterns. It also adopts some techniques to generate closed patterns without candidate maintenance. Since there may exist multiple co-expressed gene sets, so it exists several hyper nodes which have the same sample-id but different co-expressed gene set. MH-List structure of an example can be found in Figure 6.

MH-List contains not only the information of each microarray dataset, but also the frequent co-expressed gene set information. Using MH-List to generate frequent closed bicluster can reduce the times and scale of scanning original datasets and easy to design pruning techniques to check closed bicluster.

Then, we will introduce how FDCluster finding frequent closed bicluster in multiple microarray datasets. As shown above, there are two approaches for mining closed bicluster. However, both of them have drawbacks. FDCluster exploits a different approach from them. It adopts several techniques to achieve efficient mining without candidate maintenance. Before FDCluster is presented, the following lemma is made to guarantee the completeness and correctness for mining frequent closed bicluster.

Lemma 5. Given P be the current extending bicluster, if the candidate S_i can be extended to P. PS_i is a frequent closed bicluster if it satisfies both of following situations. (1) There should not exist a previous candidate of P, denoted as S_j, where PS_i.*Geneset* is the subset of PS_j.*Geneset* and PS_i.*Transaction* is the subset of PS_j.*Transaction*; (2) There does not exist an extending bicluster PS_iS_p, where S_p is the candidate of PS_i and PS_i is the subset of PS_iS_p, and their support is not equal.

Proof. The process of proof is similar to the proof of Lemma 3. For the simplicity reason, we omit the detailed proof here.

Lemma 6. If extending the candidate S to a bicluster Q, it should check the above layers of S in H-List. If there exists a sample, whose sample name is the same as S and its gene set is the superset of the candidate sample S.*Geneset* and its transaction set is the superset of the candidate sample S.*Transaction*, the candidate sample S should not be extended and pruned.

Proof. The process of proof is similar to the proof of Lemma 4. For the simplicity reason, we omit the detailed proof here.

Based on the above lemmas, the FDCluster algorithm exploits several pruning techniques to achieve efficient mining.

Pruning 3. If there exists a previous candidate sample whose gene set is the superset of the current candidate sample's gene set and transaction set is the superset of the current candidate sample's transaction set, the current sample should be pruned.

Pruning 4. If there exists an above layer of the current candidate sample, whose gene set is the superset of the current candidate sample and transaction set is the superset of the current candidate sample's transaction set, the current candidate sample should be pruned.

Based on above lemmas and pruning techniques, FDCluster algorithm is proposed for mining frequent closed bicluster without candidate maintenance in multiple microarray datasets, which is shown in Algorithm 3. Fig 7 shows an example to illustrate the process of FDCluster. The example dataset is shown in Figure 6. The minimum row threshold r_{min} and the minimum column threshold c_{min} are both set to 2. The support threshold is 3.

Figure 7. The process of FDCluster mining frequent closed biclusters from the example dataset

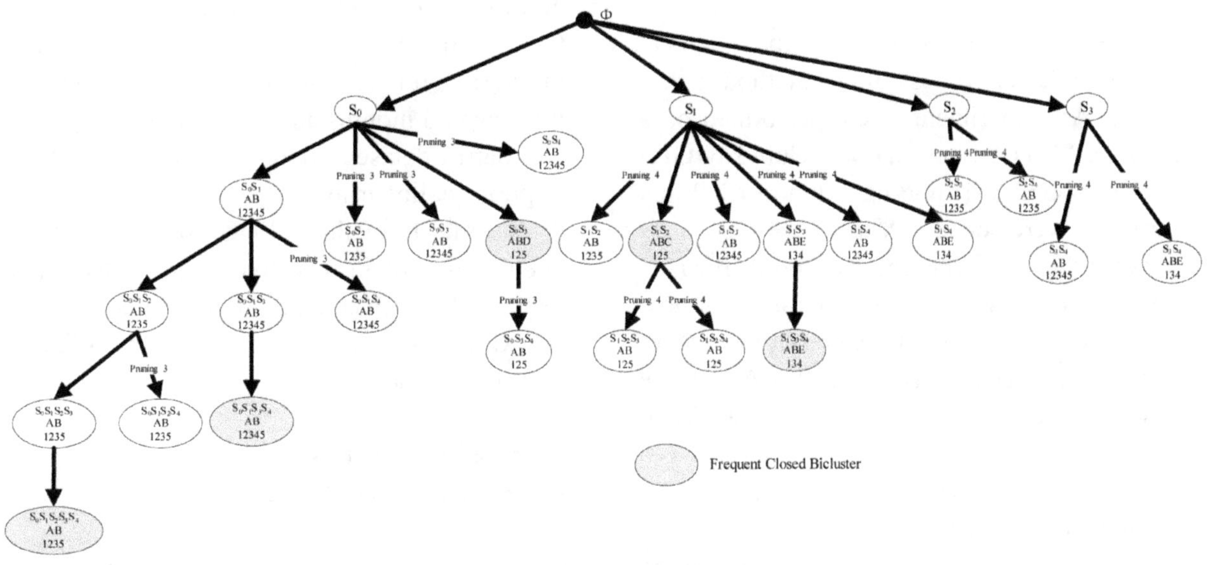

Algorithm 3. The FDCluster algorithm

Input: the minimum row threshold r_{min}; the minimum column threshold c_{min}; the support threshold s_{min}; the weighted undirected sample relational graph set S; the current extending bicluster P;
Output: the frequent closed bicluster set of F;
Initialization: P =∅; the multiple hyper list M =∅;
Method: FDCluster ($r_{min}, c_{min}, s_{min}, S, P$)
(1) **if** M =∅**then**
(2) construct M;
(3) **endif**
(4) scan M and find all the candidate set S of P;
(5) **for** each candidate S_i in S
(6) **if** the biclsuter PS_i does not satisfy **Pruning 3** and **Pruning 4**, and the number of genes in PS_i is greater than r_{min}, the number of samples is greater than c_{min}, and the support of PS_i is greater than s_{min}, **then**
(7) *P.Sample= P.Sample +* $S_i.Sample$;
(8) *P.Geneset= P.Geneset*∩ $S_i.Geneset$;
(9) *P.Transaction= P. Transaction*∩ $S_i.Transaction$
(10) CallFDCluster ($r_{min}, c_{min}, s_{min}, S, P$);
(11) **endif**
(12) **if** $S_i.Geneset = P.Geneset$
(13) flag=**true**;
(14) **endif**
(15) **endfor**
(16) **if** flag=**false**
(17) **output**(P);
(18) **endif**
(19) **return**

EXPERIMENTAL RESULTS

In this section we present several experiments to evaluate the effectiveness of our method in real datasets. All experiments were performed on a PC with a AMD Athlon(tm) Dual Core 2.61GHz CPU and 2G RAM running Windows XP and algorithms were coded in Visual C++ 6.0.

We use the microarray datasets from AGEMAP, a large microarray database which is a resource that catalogs changes in gene expression as a function of age in mice (Zahn et al., 2007). AGEMAP includes expression changes for 8,932 genes and a total of 16,896 cDNA clones in 16 tissues as a function of age (Zahn et al., 2007).For each tissues, there are five male and five female mice aged 1, 6, 16, 24 month. We only analyze three tissues in this study, which are Hippocampus, Heart and Gonads respectively. Our goal is to find potential co-expressed genes which are age-related. As discussed in above section, using discretized data to mine bicluster can improve the scalability of biclusting method. And since the AGEMAP dataset provides a unique opportunity to create difference matrix by matched the expressed value between two periods in the same sample, so the discretization method can be done with difference by using the following changed support (Wang et al., 2010).

Definition 7: Let g_1 and g_2 be the expressed value of one gene G in two adjacent ages in one sample. The changed support of G in this period is defined as follow:

$$C\_sup(G) = \frac{g_2 - g_1}{|g_1|} \quad (1)$$

The discretization can be performed by binning the changed support of the expression changed level into the three classes with bounds $\leq -n$, $\geq n$, or in between, where n is a user defined threshold. In this paper, we claim the gene is expressed if its changed support is no less than 1.0; and on the contrary, depressed; other else is non-expressed, which is represented as 1, -1 and 0, respectively. In this paper, we only focus on mining age-related co-expressed gene set between mice aged 6 months and 16 months. Other aged microarray datasets are discarded due to missing or poor-quality arrays.

In order to evaluate FDCluster to mine frequent closed biclusters, five frequent datasets is generated, which contains all the paired samples' co-expressed gene sets generated by WIBE in five sets of mice.

D. Evaluating FDCluster_One's Effectiveness

In this section, we demonstrate the effectiveness of FDCluster_One algorithm to mine maximal bicluster in one mocroarray dataset in comparison with MicroCluster. Deleting or merging clusters step is not included in our implemented MicroCluster. We evaluate the algorithms on five microarray datasets in term of runtime. The five differential microarray datasets are generated by using above discretization method on five mice sets (one male and one female), and each differential dataset has 16,896 cDNA clones and 12 samples. In order to evaluate the scalability of our algorithm, each differential dataset is set to five different scale datasets by using different row support threshold in each gene, where the number of non-zero discretezed value in each gene (row support) is not less than 8, 7, 6, 5 and 4, respectively. The number of genes in each dataset would be shown in the each related figure respectively. The number of samples of each dataset is 12. In this part, we only focus on the maximal bicluster with at least two co-expressed genes and two samples (Figure 7).

Figure 8 a through Figure 8 e shows the running time between MicroCluster and FDCluster_One in different scale microarray datasets. From these figures, we can see FDCluster_One algorithm is much faster than MicroCluster in all datasets. This is because the candidate pruning in MicroCluster

Figure 8. The running time comparison of MicroCluster and FDCluster_One in each mice set

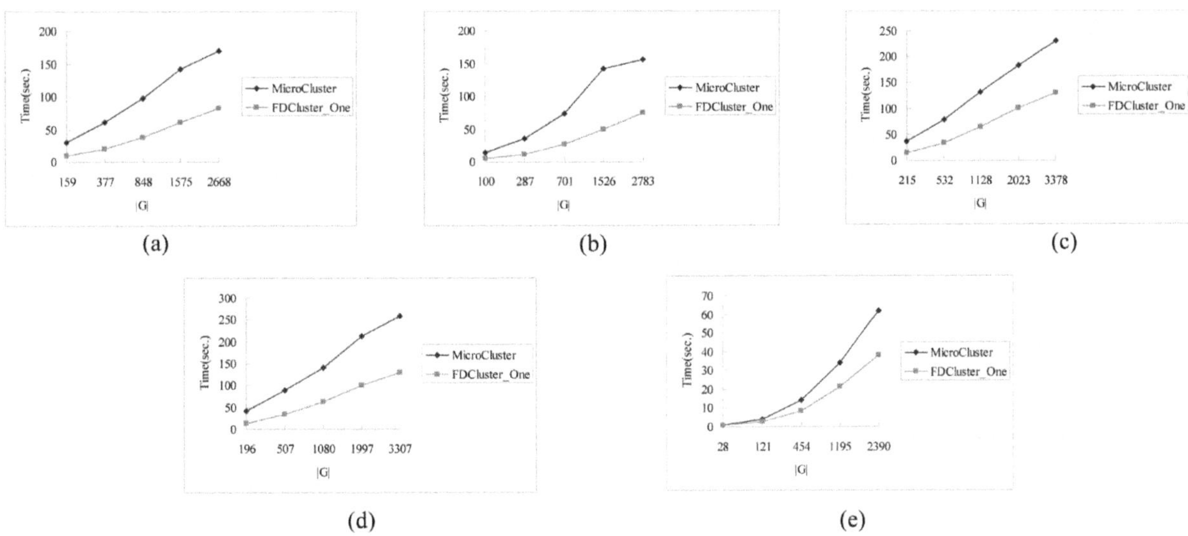

is only based on the definition of maximal bicluster, so the efficiency is poor. And it does not consider the characteristic of microarray dataset. Therefore, the efficient pruning techniques are not used. However, in our FDCluster_One algorithm, we analyze the characteristic of microarray data in detail and using proposed pruning scheme can reduce non-maximal biclusters in time and increase the efficiency of mining procedure.

E. Evaluation of FDCluster Algorithm

We demonstrate the performance of FDCluster algorithm in comparison with naïve method for mining frequent closed discriminative bicluster. The process of naïve method is that it generates maximal bicluster in each dataset by using FDCluster_One algorithm, then outputs the frequent closed bicluster based on the definition of closed bicluster. In this section, we use five datasets to evaluate the algorithms. Each dataset has mice sets (one male and one female), which is obtained by using different row support which is 8, 7, 6, 5 and 4, respectively. All the paired samples co-expressed gene sets would be generated by WIBE in each dataset. Using the frequent information FDCluster can find frequent closed bicluster efficiently. We only focus on the frequent closed bicluster with at least two co-expressed genes and two samples, and the frequent support is 2.

Figure 9 shows the two algorithms found the same biclusters; FDCluster is 20 to 200 times faster than naïve method in most cases for the unbalanced microarray datasets. The reason is that, due to the noise influence, the bicluster found in one microarray dataset may not be found in other one. The naïve method does not use frequent pruning technique, so it can not prune non-frequent bicluster in time. However, FDCluster algorithm generates frequent co-expressed gene sets between two samples, the non-frequent gene sets would be reduce which improves our efficiency.

F. Biological Analysis

In this part, we will show how the biological significance of the biclusters found by FDCluster is evaluated. To quantify the comparison, we assess the bicluster quality by determining the percentage of functionally homogeneous biclusters among all identified biclusters. We use the Gene Ontology (GO) annotation to evaluate our results. If the

Figure 9. The running time comparison

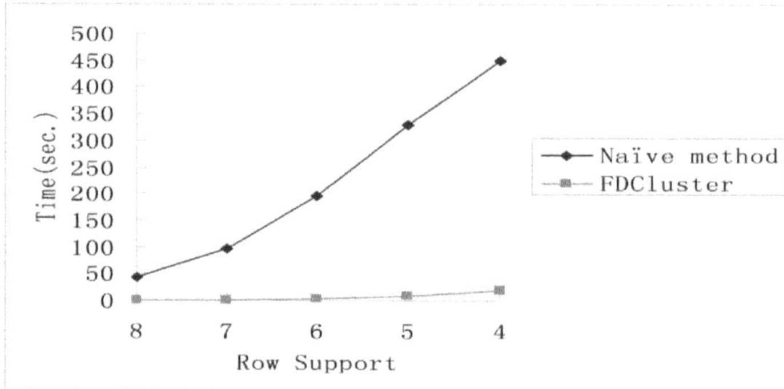

Figure 10. The number of gene sets assessed by GO and Leave-One-Out

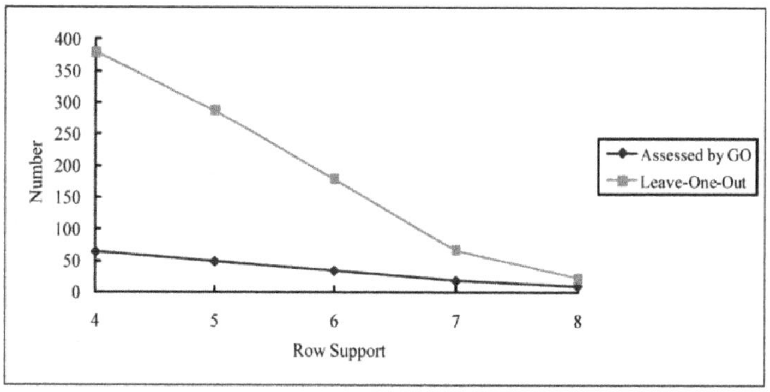

ratio of cluster members having the same known annotations which belong to a specific GO functional category is greater than the threshold, the bicluster is claimed as meaningful one. In this section, the homogeneous threshold is set to be 0.6. In order to assess the prediction accuracy of our method, we also employee the 'leave-one-out' method by masking a known gene to be unknown and assign its function based on the remaining known co-expressed in the bicluster. In our study, we only assess the biclusters annotated by GO. If the function of mask gene is the same as the real function, it is considered the prediction is correct.

Due to the limitation of paper space, we only assess the result generated by FDCluster. Figure 10 shows the number of gene sets evaluated by GO and 'leave-one-out' approach. Due to the limitation of the confirmed function of many mouse genes, it is not an easy task. So most of our results may not be confirmed by GO, but it does not imply it is incorrect. Our goal is to give some potential results for biologist research. Table II shows some GO-annotated biclusters found by FDCluster. For example, based on original GO database, BG077520 can be annotated to the first layer of GO structure. According to the results of FDCluster mined, BG077520 can be annotated to GO:0016020 which are the forth layer of GO structure (Table 2).

CONCLUSION

In this paper, we propose two algorithms to mine bicluster. FDCluster_One algorithm can find maximal bicluster in one dataset without candi-

Table 2. Groups of Co-Expressed Genes Annotated by Go

Bicluster number	Co-expressed genes in bicluster	GO ID
1	BG078982,BG077352,AW555339	GO:0016020
2	BG077352,BG077520,AW555339	GO:0016020
3	C76810,BG079074,BG072042	GO:0005576
4	BG088889,AW556222,BG086892	GO:0005515
5	BG087892,BG087037,BG088029	GO:0005840,GO:0030529,GO:0003735,GO:0005622,GO:0006412

date maintenance. The other algorithm FDCluster can generate frequent closed bicluster in multiple microarray datasets, which can reduce the influence of noise. In stead of using original value, our algorithms use discritized value to generate bicluster, which can increase the scalability of method. We also use difference matrix to generate discriminative bicluster. However, there remain lots of further investigations. As future work, we will plan to mine bicluster in multiple uncertain microarray datasets.

ACKNOWLEDGMENT

The work is supported by the National Natural Science Foundation of China under Grant No.60703105 and the Natural Science Foundation of Shaanxi Province under Grant No.2007F27. It is in part supported by the Research Foundation at Northwestern Polytechnical University of China under Grant No.JC201042.

REFERENCES

Becquet, C., Blachon, S., Jeudy, B., Boulicaut, J. F., & Gandrillon, O. (2003). Strong-association-rule mining for large-scale gene-expression data analysis: a case study o human SAGE data. *Genome Biology*, *12*, 1–16.

Ben-Dor, A., Chor, B., Karp, R., & Yakhini, Z. (2003). Discovering local structure in gene expression data: the order-preserving submatrix problem. *Journal of Computational Biology*, *10*, 373–384. doi:10.1089/10665270360688075

Cheng, H., Yan, X., Han, J., & Hsu, C.-W. (2007). Discriminative frequent pattern analysis for effective classification. In *Proceedings of International Conference on Data Engineering* (pp. 716-725).

Cheng, H., Yan, X., Han, J., & Yu, P. (2008). Direct discriminative pattern mining for effective classification. In *Proceedings of International Conference on Data Engineering* (pp. 169-178).

Cheng, K., Law, N., Siu, W., & Liew, A. (2008). Identification of coherent patterns in gene expression data using an efficient biclustering algorithm and parallel coordinate visualization. *BMC Bioinformatics*, *9*, 210. doi:10.1186/1471-2105-9-210

Cheng, Y., & Church, G. M. (2000). Biclustering of Expression Data. In *Proceedings of the 8th Int'l Conf. Intelligent Systems for Molecular Biology (ISMB 00)* (pp. 93-103). New York: ACM Press.

Cong, G., Tan, K., Tung, A., & Feng, P. (2004). Mining Frequent Closed Patterns in Microarray Data. In *ICDM'04* (pp. 363–366). Washington, DC: IEEE Press.

Cong, G., Tung, A., Xu, X., Pan, F., & Yang, J. (2004). FARMER: Finding Interesting Rule Groups in Microarray Datasets. In *Proceedings of the ACM SIGMOD Int'l Conf. Management of Data* (pp. 143-154).

Creighton, C., & Hanash, S. (2003). Mining gene expression databases for association rules. *Bioinformatics (Oxford, England)*, *19*, 79–86. doi:10.1093/bioinformatics/19.1.79

Deshpande, M., Kuramochi, M., Wale, N., & Karypis, G. (2005). Frequent sub-structure based approaches for classifying chemical compounds. *IEEE Transactions on Knowledge and Data Engineering*, *17*(8), 1036–1050. doi:10.1109/TKDE.2005.127

Dharan, A., & Nair, A. S. (2009). Biclustering of gene expression data using reactive greedy randomized adaptive search procedure. *BMC Bioinformatics*, *10*(1), S27. doi:10.1186/1471-2105-10-S1-S27

Fan, W., Zhang, K., Cheng, H., Gao, J., Yan, X., Han, J., et al. (2008). Direct mining of discriminative and essential graphical and itemset features via model-based search tree. In *Proceedings of the ACM SIGKDD international conference on knowledge discovery in databases* (pp. 230-238).

Fang, G., Kuang, R., Pandey, G., Steinbach, M., Myers, C. L., & Kumar, V. (2010). Subspace Differential Coexpression Analysis: Problem Definition and A General Approach. In *Proceedings of the 15th Pacific Symposium on Biocomputing (PSB)* (Vol. 15, pp. 145-156).

Gowtham, A., Gupta, R., Fang, G., Pandey, G., Steinbach, M., & Kumar, V. (2009). Association Analysis Techniques for Bioinformatics Problems. In *Proceedings of the 1st International Conference on Bioinformatics and Computational Biology (BICoB)* (pp. 1-13).

Liu, X., & Wang, L. (2007). Computing the maxim um similarity bi-clusters of gene expression data. *Bioinformatics (Oxford, England)*, *23*(1), 50–56. doi:10.1093/bioinformatics/btl560

Lo, D., Cheng, H., Han, J., Khoo, S., & Sun, C. (2009). Classification of software behaviors for failure detection: a discriminative pattern mining approach. In *Proceedings of the 15th ACM SIGKDD international conference on Knowledge discovery and data mining* (pp. 557-566).

Madeira, S. C., & Oliveira, A. L. (2004). Biclustering algorithms for biological data analysis: a survey. *IEEE/ACM TCBB*, *1*(1), 24-45.

McIntosh, T., & Chawla, S. (2007). High confidence rule mining for microarray analysis. *IEEE/ACM TCBB*, *4*(4), 611-623.

Murali, T., & Kasif, S. (2003a). Extracting conserved gene expression motifs from gene expression data. In *Proceedings of the Pac Symp Biocomput* (pp. 77-88).

Murali, T., & Kasif, S. (2003 b). RankGene: identification of diagnostic genes based on expression data. In *Proceedings of the Pacific Symposium on Biocomputing*, *8*, 77–88.

Pan, F., Cong, G., Tung, K., Yang, J., & Zaki, M. (2004). Carpenter: Finding closed patterns in long biological datasets. In *Proceedings of the ACM SIGKDD Intl. Conf. Knowledge Discovery and Data Mining (KDD)* (pp. 637-642).

Pandey, G., Atluri, G., Steinbach, M., Myers, C. L., & Kumar, V. (2009). An association analysis approach to biclusting. In *Proceedings of the ACM Conf. on Kownlege Discovery and Data Mining* (pp. 677-686).

Subramanian, A., Tamayo, P., Mootha, V., Mukherjee, S., Ebert, B., & Gillette, M. (2005). Gene set enrichment analysis: A knowledge-based approach for interpreting genome-wide expression profiles. *Proceedings of the National Academy of Sciences of the United States of America*, *102*(43), 15545–15550. doi:10.1073/pnas.0506580102

Tanay, A., Sharan, R., Kupiec, M., & Shamir, R. (2004). Revealing modularity and organization in the yeast molecular network by integrated analysis of highly heterogeneous genomewide data. *Proceedings of the National Academy of Sciences of the United States of America, 101*(9), 2981–2986. doi:10.1073/pnas.0308661100

Teng, L., & Chan, L. (2008). Discovering biclusters by iteratively sorting with weighted correlation coefficient in gene expression data. *J Signal Process Syst, 50*(3), 267–280. doi:10.1007/s11265-007-0121-2

van Vliet, M., Klijn, C., Wessels, L., & Reinders, M. (2007). Module-based outcome prediction using breast cancer compendia. *PLoS ONE, 2*(10), 1047. doi:10.1371/journal.pone.0001047

Wang, M., Shang, X. Q., Diao, J. N., & Li, Z. H. (2010). WIBE: Mining frequent closed patterns without candidate maintenance in microarray dataset. In *Proceedings of the 6th International Conference on Data Mining.*

Yang, J., Wang, H., Wang, W., & Yu, P. (2003). Enhanced biclustering on expression data. In *Proceedings of the Third IEEE Symposium on Bioinformatics and Bioengineering (BIBE'03)* (pp. 1-7).

Yeung, K., Medvedovic, M., & Bumgarner, R. (2004). From co-expression to co-regulation: how many microarray experiments do we need? *Genome Biology, 5*(7). doi:10.1186/gb-2004-5-7-r48

Zahn, J. M., & Poosala, S. (2007). AGEMAP: A gene expression database for aging in mice. *PLOS Genetics, 3*(11), 2326–2337. doi:10.1371/journal.pgen.0030201

Zhao, L., & Zaki, M. (2005). MicroCluster: Efficient deterministic biclustering of Microarray data. *IEEE Intelligent Systems, 20*(6), 40–49. doi:10.1109/MIS.2005.112

This work was previously published in the International Journal of Knowledge Discovery in Bioinformatics, Volume 1, Issue 4, edited by Xiao-Li Li and See-Kiong Ng, pp. 69-88, copyright 2010 by IGI Publishing (an imprint of IGI Global).

Chapter 7
Wave-SOM:
A Novel Wavelet-Based Clustering Algorithm for Analysis of Gene Expression Patterns

Andrew Blanchard
University of Arkansas, USA

Christopher Wolter
University of Arkansas, USA & University of Minnesota, USA

David S. McNabb
University of Arkansas, USA

Eitan Gross
University of Arkansas, USA

ABSTRACT

In this paper, the authors present a wavelet-based algorithm (Wave-SOM) to help visualize and cluster oscillatory time-series data in two-dimensional gene expression micro-arrays. Using various wavelet transformations, raw data are first de-noised by decomposing the time-series into low and high frequency wavelet coefficients. Following thresholding, the coefficients are fed as an input vector into a two-dimensional Self-Organizing-Map clustering algorithm. Transformed data are then clustered by minimizing the Euclidean (L^2) distance between their corresponding fluctuation patterns. A multi-resolution analysis by Wave-SOM of expression data from the yeast Saccharomyces cerevisiae, exposed to oxidative stress and glucose-limited growth, identified 29 genes with correlated expression patterns that were mapped into 5 different nodes. The ordered clustering of yeast genes by Wave-SOM illustrates that the same set of genes (encoding ribosomal proteins) can be regulated by two different environmental stresses, oxidative stress and starvation. The algorithm provides heuristic information regarding the similarity of different genes. Using previously studied expression patterns of yeast cell-cycle and functional genes as test data sets, the authors' algorithm outperformed five other competing programs.

DOI: 10.4018/978-1-4666-1785-8.ch007

INTRODUCTION

Saccharomyces cerevisiae (Bakers' yeast) is commonly used by molecular biologists as a model eukaryote for the study of basic cellular processes. In addition, studies of the response of yeast to different stresses have implications in the brewing and baking industries and for developing strains of yeast that produce lower levels of acetic acid (Mizuno et al., 2006) in response to varying ethanol concentrations, with potential for being a viable source of ethanol for the bio-fuel industry.

In the budding yeast there are over 6000 expressed genes. Until very recently, analyzing the expression patterns of these genes was confined to only a few gene products or message levels at a time, or could only be addressed by means of perturbation analysis in combination with computer simulations (Glass & Mackey, 1979). The advent of microarray techniques can now provide a means for assessing the sum total of expressed genes in a cell in a quantitative, reproducible, and internally standardized manner. A single array provides a snapshot of the transcriptional state of the cell at some point in time. When multiple snapshots are taken from a temporally coherent system such as synchronous cells, these signals yield the characteristic time signature of each of the genes in the cell. Thus, by clustering the genes according to the similarity in their expression patterns, one can learn about potential functions of the gene products that otherwise would be very difficult to determine.

Furthermore, the regulation of gene expression in the yeast is a non-linear process under control of a network of connected regulatory proteins called transcription factors (Nicholas & Prigogine, 1977) although many epigenetic factors such as RNA splicing and degradation, are known to modify the expression levels. Consequently, concentrations of a variety of cellular reagents oscillate as the cell goes through the different phases of cell cycle (Mitchison, 1971; Klevecz et al., 1984). Chemical or physical perturbations to the cell cycle can thus introduce a phase shift in the onset time of a given cell cycle event (i.e. mitosis) (Klevecz et al., 1978). If the oscillatory kinetics of expression is confined to a small number of genes, then finding these 'cell cycle regulated' genes becomes a fairly easy clustering task. However, if a large number of genes or indeed the entire genome oscillates and the fundamental harmonic of this oscillation is significantly distinguishable from the characteristic period of the cell cycle, then the different functional groups can be identified by their characteristic kinetics or oscillation frequency.

A successful gene-expression clustering program must be able to handle noisy time series data with possible uneven time interval data points. In addition the algorithm needs to be able to offset time shifts in the onset time for the expression of one or more genes in the dataset due to variability in experimental protocols, or variations in timing between different labs.

To address these issues we present here a new computer algorithm. Our clustering algorithm has two main components, a neural network based module (Self Organizing Map, or SOM, Kohonen, 2001) to cluster the gene expression patterns and a wavelet module to filter (de-noise) the data prior to clustering them. SOMs have been used previously for gene clustering (Tamayo et al, 1999). The use of wavelets for de-noising such data is innovative. Before we can cluster the genes, we need to reduce or if possible remove the noise from the data. Currently, there are several approaches used for de-noising. One approach is based on spectral analysis and includes methods such as Wiener filtering, Kalman filtering and Fourier transform (Haykin, 2002). These methods work well only with linear systems however, and the results of the analysis depend on the establishment of appropriate state space functions. Furthermore, Fourier analyses are not suited for the sparse sampling and short duration data sets frequently encountered in genomic research. While some variants of Fourier analysis, such as Maximum Entropy Spectral Analysis and discrete-time Fourier Transform, have been successful in detecting periodicity in noisy, periodic biological time series data (Tang, 2008) they are not suitable to be used with gene expression time series as the latter is often characterized by extremely non-stationary and nonlinear features which result from spatial and dynamical heterogeneities. In addition gene expression time series can also exhibit multi-temporal scale variations. Thus a de-noising method is required that will not de-localize the signal on the time axis and at the same time can remove high frequency noise "contamination". Our algorithm uses wavelet transforms to de-noise the data. Wavelets are now emerging as the approach of choice for analysis of short-duration, noisy biological time-series (Vetterli and Herley, 1992). In particular, the ability to decompose a signal into high-frequency noise and low-frequency trends, and to distinguish these effects from those that are characteristic of the signal by matching the wavelet scale to the scale of the signal, makes wavelets a promising alternative for the analysis of time series data and perturbation studies in combination with microarray techniques. A major obstacle however that still remains in the application of this versatile de-noising method is deciding which high frequency wavelet coefficients to throw away (as noise) and which ones to keep. This problem is also known as the thresholding problem (Donoho & Johnstone, 1995). To address this problem we employed a thresholding approach which is based on a discrete Hilbert transform (Bracewell, 1999). Discrete Hilbert transform is an analytic stationary wavelet transform thresholding technique that allows us to compare the size of the signal component relative to the noise level at each transform level by creating a complex-valued analytic vector from which an amplitude vector can be defined. In addition, the discrete Hilbert transform is invariant to phase or time shifts (Percival & Walden, 2000; Olhede & Walden, 2004) often observed with gene expression time series data obtained from different sources due for instance to variations in onset/application time of an external stress signal or treatment. Using previously studied expression patterns of yeast' cell-cycle and functional genes as test data sets our algorithm outperformed five other competing programs namely Cluster 3.0, MCLUST, CurveSOM, SSClust and

GENECLUSTER

Using our algorithm, we analyzed microarray gene expression data from *Saccharomyces cerevisiae,* exposed to two environmental stresses: oxidative stress and glucose-limited growth in a chemostat. To de-noise the original time-series data sets and visualize the patterns of gene expression, we decomposed the signal from each gene, with wavelet filters, and then reassembled the de-noised expression patterns at each scale to find the peaks in expression in the genome as a whole. Using our analytic stationary wavelet transform thresholding procedure, high frequency wavelet coefficients were discarded prior to reassembly. Following wavelet decomposition and de-noising of the gene expression time-series, genes were clustered according to their expression kinetics. We identified 29 genes with correlated expression patterns that were mapped into 5 different nodes.

The ordered clustering of yeast genes with our algorithm (which we call Wave-SOM) offers insight into possible functions of these genes that were previously unknown.

METHODS

Gene Expression Data

Data from a time series experiment measuring the effects of oxidative stress on the yeast *Saccharomyces cerevisiae* were from Shapira et al. (2004). In one set of experiments, yeast cells were exposed to a solution containing 0.24 mM hydrogen peroxide (H_20_2) to induce oxidative stress. Twenty one sampling times were taken between

Figure 1. Two dimensional gene expression array for (A) All 536 genes and (B) a sample of 100 genes after clustering into 6 different nodes. Each column represents a gene in the figure and the color intensity indicates the value of the wavelet coefficient.

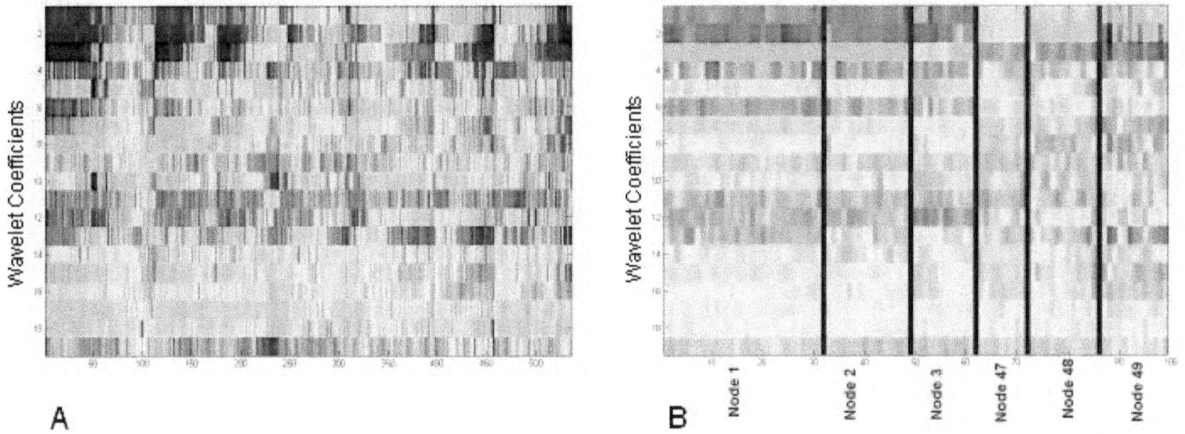

$t = 0$ to 245 min. The changes, as compared to baseline, in expression levels were color coded with a red (R) dot indicating an increase in expression levels and a green (G) dot indicating a decrease. We used the normalized ratio of $\log_2 R / G$ values as a measure of expression level for each gene.

A second analysis was done on data obtained from cells grown in a chemostat under glucose-limited conditions, using galactose as the carbon source (Brauer et al., 2005). This time series data set contained 12 sampling times during which the cells were subjected to a glucose pulse (0.2 g/l).

Both data sets are available for download from the Stanford Microarray Database at: http://genome-www5.stanford.edu/MicroArray/SMD/. Clusters of genes obtained for the two stress experiments were compared with each other to identify genes that respond to the two types of stresses, with activation patterns that exhibit temporal correlations.

Inclusion criteria: In order to have a uniform data set for the wavelet transformation, a gene needed to have available data for all 21 time series runs. This resulted in some genes being excluded since a microarray spot may have been flagged and excluded in one or more of the time series runs. The data files were also filtered to select only those spots with a pixel-to-pixel regression correlation greater than 0.6. Overall, these criteria are compatible with those employed for other clustering algorithms and were used here to facilitate the comparison of Wave-SOM with the performance of its competitors. Only 536 genes met the inclusion criteria and were included in the current study of patterns of gene expression in response to oxidative stress and glucose-limited growth. Figure 1A depicts the wavelet coefficients (in lieu of the expression levels) for the 536 genes.

ALGORITHMS

Wavelet Analysis

A wavelet function can be constructed to have certain desirable mathematical properties (Ogden 1997) and it normally grows and decays in a confined time period (Percival & Walden, 2000). The wavelet function integrates to zero while its square integrates to unity. Denoting the scale as

a and the position as ***b***, the continuous wavelet basis is given by (Daubechies, 1992):

$$\psi_{a,b}(t) = \frac{1}{\sqrt{a}}\psi\left(\frac{t-b}{a}\right) \quad (1)$$

For a discrete wavelet basis, ***a*** and ***b*** are replaced, respectively, by 2^j and $k2^j$, where ***j*** and ***k*** are the dyadic dilation and translation indexes, thus

$$\psi_{j,k}(t) = 2^{j/2}\psi\left(2^{-j}t - k\right) \quad (2)$$

In a discrete wavelet basis, a 2^N long time series is expressed as:

$$f(t) = \sum_{j=0}^{N}\sum_{k=1}^{2^{N-j}} d_{j,k}\psi_{j,k}(t) \quad (3)$$

Where the wavelet or detail coefficients $d_{j,k}$ are simply the inner products of ***f(t)*** with the corresponding basis functions. Each "mother wavelet" is associated with a "father wavelet" or scaling function which, by means of dilations and translations, produces an orthogonal basis:

$$\varphi_{j,k}(t) = 2^{-j/2}\varphi\left(2^{-j}t - k\right) \quad (4)$$

A jth order smoothed time series can be obtained by combining the set $\{\phi_{j,k}, k = 1,\ldots,2^{N-j}\}$ with the scaling coefficients $\mathbf{a}_{j,k}$ given by the inner products of ***f(t)*** with the corresponding basis scaling functions:

$$f(t)_{j^{th}smooth} = \sum_{k=1}^{2^{N-j}} a_{j,k}\varphi_{j,k}(t) \quad (5)$$

The decomposition of the signal with the scaling function performs an averaging of the original data over the scale interval, while wavelet transform coefficients are related to differences of averages of successive scales. The decision of which wavelet to use is not always trivial and it largely depends on the problem at hand. In some situations, the results are similar for most wavelets, but for some problems, the result of the analysis critically depends upon the wavelet type used. Of note is the fact that the sets $\{\psi_{j,k}\}$ and $\{\phi_{j,k}\}$ are mutually orthogonal, therefore the function can be expressed as:

$$f(t) = \sum_{k} a_{j,k}\varphi_{j,k}(t) + \sum_{j\le J}^{N}\sum_{k} d_{j,k}\psi_{j,k}(t) \quad (6)$$

where ***J*** is the index of the last decomposition level of interest where the decomposition sequence, which reaches level ***N***, is truncated. For a 2^N point vector, the indexes vary as follows: $1 \le j \le J \le N$ ***and*** $1 \le k \le 2^{N-j}$. Thus, a time series ***f(t)*** can be written as a sum of a coarse scaling series and ***J*** increasingly fine detail series.

For describing a signal that is changing over time, one needs to know the time-frequency localization of the related amplitudes. In Eqn 3 the coefficients $\mathbf{d}_{j,k}$ are related to the amplitudes at a given position ***k***, corresponding to instant in time $t = 2^j k\Delta t$ and a period $2^j\Delta t$, where Δt is the acquisition interval. With each frequency interval $[1/(2^{j+1}\Delta t), 1/(2^j\Delta t)]$ there is an associated vector $\{d_{j,k}\}$.

When used in signal processing, wavelet analysis is capable of detecting singularities, as wavelet coefficients are very sensible to discontinuities and abrupt behavior changes of the signal. The features of the signal are more visible at continuous wavelet analysis, but the discrete analysis is more feasible, because it demands less computing time and less memory space than the continuous one. Each set of detail coefficients retains the information contained within a definite frequency interval. The signal can thus be reconstructed separately for each one of these intervals. These sets $f_j(t)$, which are part of the original signal, reveal the localized time behavior of the signal at the corresponding frequency interval. However, the width of the frequency interval depends on the decomposition level, being larger for higher frequencies. To obtain narrower frequency intervals a new wavelet decomposition procedure must be applied. Treating each detail coefficient vector as a series and decomposing that series using the same wavelet and scaling functions, two series with respectively lower and upper half width frequency interval are thus obtained. This scheme, applied recursively at all levels, generates a "wavelet tree", called wavelet packet transform. Each wavelet packet transform is associated with a level $\boldsymbol{j}$, and the $\boldsymbol{j}^{th}$ level decomposes the frequency interval of the original signal $\boldsymbol{[0,f/2]}$ into $\boldsymbol{2^j}$ equal and individual intervals. The advantage of this band pass filter operation of wavelet packet decomposition is that, by reconstructing separately each frequency interval, the signal can now be presented and its behavior analyzed in a time frequency domain.

For the case of discrete wavelet analysis, we consider a vector $\boldsymbol{y}_i$ of data with $i=0,1,2,\dots N-1$ and $N=2^J$ and a family of discrete wavelet transforms, DWT(L,M), with the parameter L representing the coarsest resolution and M representing the number of vanishing moments of the wavelet (Donoho & Johnstone, 1995). For fixed values of these parameters we get a matrix $\boldsymbol{W}$ defined by (wavelet coefficients will be labeled with either upper case and/or bold face w):

$$\boldsymbol{w}=\boldsymbol{W}\boldsymbol{y} \tag{7}$$

where $\boldsymbol{w}$ is the wavelet coefficients vector. For an orthogonal transform the inverse matrix $\boldsymbol{W}^T$ satisfies:

$$y=\boldsymbol{W}^T\boldsymbol{y} \tag{8}$$

The vector w has $N=2^J$ elements:

$$\boldsymbol{w}_{j,k}: j=0,\dots,J;\ k=0,\dots,2^J-1. \tag{9}$$

We can now write $\boldsymbol{y}$ in terms of the sum of the basis matrix elements $W_{j,k}$ and the coefficients $\boldsymbol{w}_{j,k}$:

$$y_i=\Sigma_{j,k}\boldsymbol{w}_{j,k}\boldsymbol{W}_{j,k}(i) \tag{10}$$

In the special case when $L=M=0$, the transform reduces to the discrete Haar transform. For the Haar transform $W_{j,k}(i)$ is proportional to 1 for $2^{-j}k \le i/N < 2^{-j}(k+½)$ and to -1 for $2^{-j}(k+½) \le i/N < 2^{-j}(k+1)$. Thus the wavelet coefficients measure the differences of the function across various scales. The function is reconstructed from building blocks of zero-mean localized square waves. For the simulations performed in this study we used the Haar wavelet as defined below:

$$\begin{aligned} &f(\mathrm{x})=1, \text{ if } \mathrm{x}\in[0, 0.5]\\ &f(\mathrm{x})=-1, \text{ if } \mathrm{x}\in[0.5, 1]\\ &f(\mathrm{x})=0, \text{ if } \mathrm{x}\notin[0, 1]\\ &\phi(\mathrm{x})=1, \text{ if } \mathrm{x}\in[0, 1]\\ &\phi(\mathrm{x})=0, \text{ if } \mathrm{x}\notin[0, 1] \end{aligned} \tag{11}$$

Level 4 and 3 Haar wavelet decompositions were performed on the first (oxidative stress) and second (glucose-limited chemostat) data sets, respectively, using the MATLAB wavelet toolbox. The computer algorithm produces two sets of coefficients: approximation coefficients (*cA1*) and detail coefficients (*cD1*), by convolving $\boldsymbol{y}$ with low-pass and high-pass filters, respectively, as illustrated in Figure 2. The next step splits the approximation coefficients *cA1* in two parts using the same scheme, but with modified filters obtained by up sampling the filters used in the previous step and replacing $\boldsymbol{y}$ by *cA1*. High frequency, detail, coefficients that qualified as noise as determined by our thresholding protocol were discarded. The remaining wavelet coefficients were then sent as input vectors to the SOM program described below.

Figure 2. Level 3 hierarchy wavelet decomposition protocol for the microarray data. Vector X contains the data to be transformed. cA1 and cD1 represent the first level decomposition where cA1 is the approximation coefficients and cD1 is the detail coefficients. The approximation coefficients cA1 are further broken down into second and third levels of approximation with detail coefficients, cA2, cD2 and cA3, cD3, respectively. The remaining coefficients are gathered into the vector C which is the wavelet coefficients that the program returns. The method also returns a bookkeeping vector L which keeps track of the coefficients assignments.

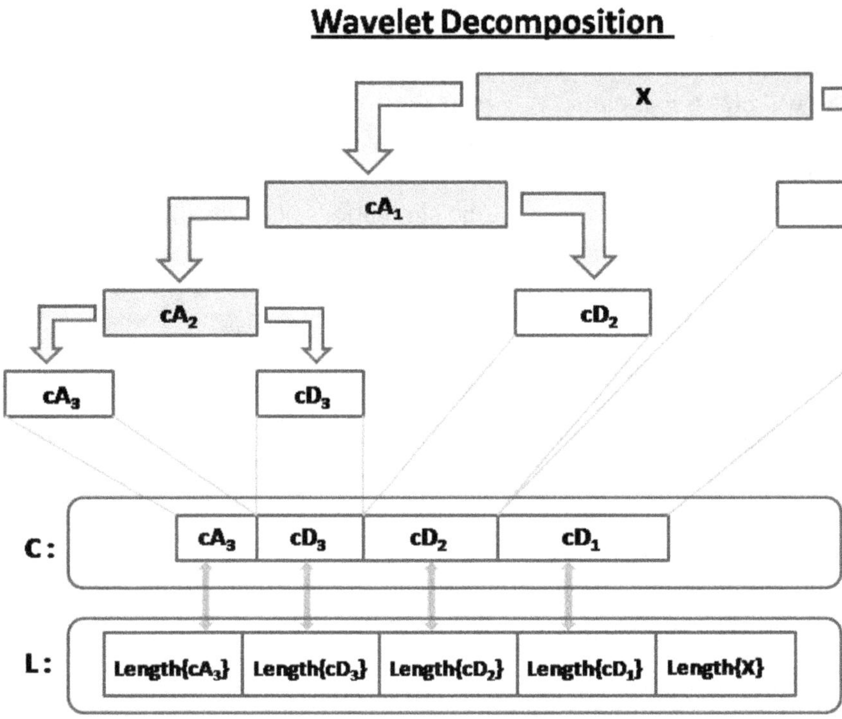

Thresholding

Small amplitudes and phase-shifts can render the signal-to-noise ratio of oscillatory signal components extremely low. We have thus employed analytic stationary wavelet transform thresholding to compare the size of the signal component relative to the noise level at each transform level. This was achieved by using the discrete Hilbert transform to create a complex-valued analytic vector from which an amplitude vector was defined (Olhede & Walden, 2004). Thresholding of the real-valued wavelet coefficients was then performed as determined by the corresponding value in the amplitude vector. The mean square error obtained with the analytic stationary wavelet transform thresholding was lower then that produced with standard stationary wavelet transform thresholding (Donoho & Johnstone, 1994). In addition, the analytic stationary wavelet transform thresholding was invariant to phase shifts in the data.

To illustrate our approach let's assume that the level ***j*** coefficients for a given signal is given by $W^{(x)}_{j,t} = a_t \cos(\phi_t)$ for $t= 1,...,100$, where the amplitude is $a_t = exp(-t/\tau)$ for $t \in (1,100)$ and zero otherwise, and $\varphi_t = -2\pi f t^2 + \pi/2$. This is an oscillatory function with a decreasing period and a decreasing amplitude (Figure 3A). We can re-write the wavelet coefficients as $W^{(x)}_{j,t} = \Re\{a_t e^{i\phi_t})$ where $\Re$ indicates the real part. The decreasing amplitude a_t and the existence of the phase shift φ_t, both produce sub-threshold $W^{(x)}_{j,t}$ values. Following thresholding, these small values of the signal are set to zero. For comparison, we consider hard thresholding the signal. With a threshold level of 0.1 for instance, the standard approach would be to set $W^{(x)}_{j,t}$ to zero if $| W^{(x)}_{j,t} |$ is less than 0.1. Otherwise $W^{(x)}_{j,t}$ is left unchanged. In the amplitude-based approach we set $W^{(x)}_{j,t}$ to zero if a_t is less than 0.1. The amplitude-based approach yields a smoother signal with no discontinuities, but in order to use this approach in practice we need to know the amplitude a_t. For that we need to define an appropriate amplitude vector by first creating a complex-valued vector $W^{(x)}_j = iHW^{(x)}_j$ where H denotes the discrete Hilbert transform matrix (see Eqn. A15 in the Appendix). We refer to this complex-valued vector as an analytic vector by analogy with the theory of continuous-time signals. Thus, the amplitudes vector corresponds to the vector of moduli. In order to obtain the corresponding analytic vector we apply $\boldsymbol{H}$ to $W^{(x)}_j$ for each level of the transform. Alternatively, we can obtain the analytic vector by noting that $\boldsymbol{H}$ and $W^{(x)}_j$ commute, so that

$$W^{(x)}_j + iHW^{(x)}_j = W_j x + iHW_j x = W_j x + iW_j Hx = W_j z = W^{(z)}_j \quad (12)$$

where $z = x + iHx = x + i\tilde{x}$, and $\tilde{x} = Hx$. Thus, to obtain the analytic vectors

$$W^{(x)}_j = iHW^{(x)}_j \quad (13)$$

for different levels $\boldsymbol{j}$ of the stationary wavelet transform, we need only calculate the analytic vector $\boldsymbol{z} = \boldsymbol{x} + \boldsymbol{iHx}$ once, with, $\boldsymbol{x} = \boldsymbol{d} + \varepsilon$, and then apply the stationary wavelet transform to it in the standard fashion. In order to avoid mixing transform and analytic vector contributions, the scaling and wavelet filters associated with W_j are real-valued. $\mathbf{z}$, the analytic vector corresponding to $\boldsymbol{x}$ is then given by

$$z = d + \varepsilon + i(\tilde{d} + \tilde{\varepsilon}) = (d + i\,\tilde{d}) + (\varepsilon + i\,\tilde{\varepsilon}) \quad (14)$$

the sum of the analytic vector of the deterministic signal and the analytic vector of the noise. In order to obtain the wavelet coefficients $W^{(x)}_{j,t}$ at level $\boldsymbol{j}$, we applied the stationary wavelet transform to $\boldsymbol{x}$. In order to obtain $\{W^{(z)}_{j,t}\}$, we applied the stationary wavelet transform to the complex-valued analytic vector z. Thresholding of $W^{(x)}_{j,t}$ was then performed according to the size of $\{W^{(z)}_{j,t}\}$. We compared the two thresholding techniques on the clustering accuracy of simulated synthetic data sets with noise added and on clustering of genes which are involved in galactose uitilization in the yeast. In both of these cases, the analytic stationary wavelet transform thresholding yielded far more accurate clustering results compared to the corresponding standard stationary thresholding (Table 1). We used the analytic stationary wavelet transform thresholding in all subsequent clustering applications.

Self Organizing Maps (SOM)

Genes where grouped by clustering the wavelet coefficients associated with the genes' expression levels. Neural networks of $\boldsymbol{n}$ neurons (nodes) arranged either as a 1D vector or a 2D grid were constructed to read and process the data. Genes were grouped into the closest node in the network

Figure 3. Simulated and real (yeast) time-series data used for evaluation of Wave-SOM clustering performance. (A), (top panel) synthetic oscillatory time-series before error is added (black, $e^{-(t/\tau)}sin(2\pi \cdot 3t)$; red, $e^{-(t/\tau)}sin(2\pi t)$; green, $e^{-(t/\tau)}sin(2\pi t + \pi/4)$; blue, $e^{-(t/\tau)}cos((2\pi t)$; dark blue, $e^{-(t/\tau)}cos(2\pi t/2)$; pink, $e^{-(t/\tau)}cos(2\pi t + \pi/4)$; τ=40. (middle panel) same as in top panel with noise added. (bottom panel) synthetic oscillatory time-series after de-noising. (B) expression patterns of 205 genes involved in galactose uitilization in the yeast, organized in 4 distinct functional groups (Ideker, 2006). (C) expression patterns of same 205 genes organized in 4 groups proposed by our Wave-SOM clustering algorithm. (D) expression patterns of 383 genes involved with cell cycle in the yeast organized in 5 groups according to the five cell-cycle phases (Spellman et al, 1998). (E) expression patterns of same 383 genes involved with cell cycle in the yeast organized in 5 groups proposed by our Wave-SOM clustering algorithm.

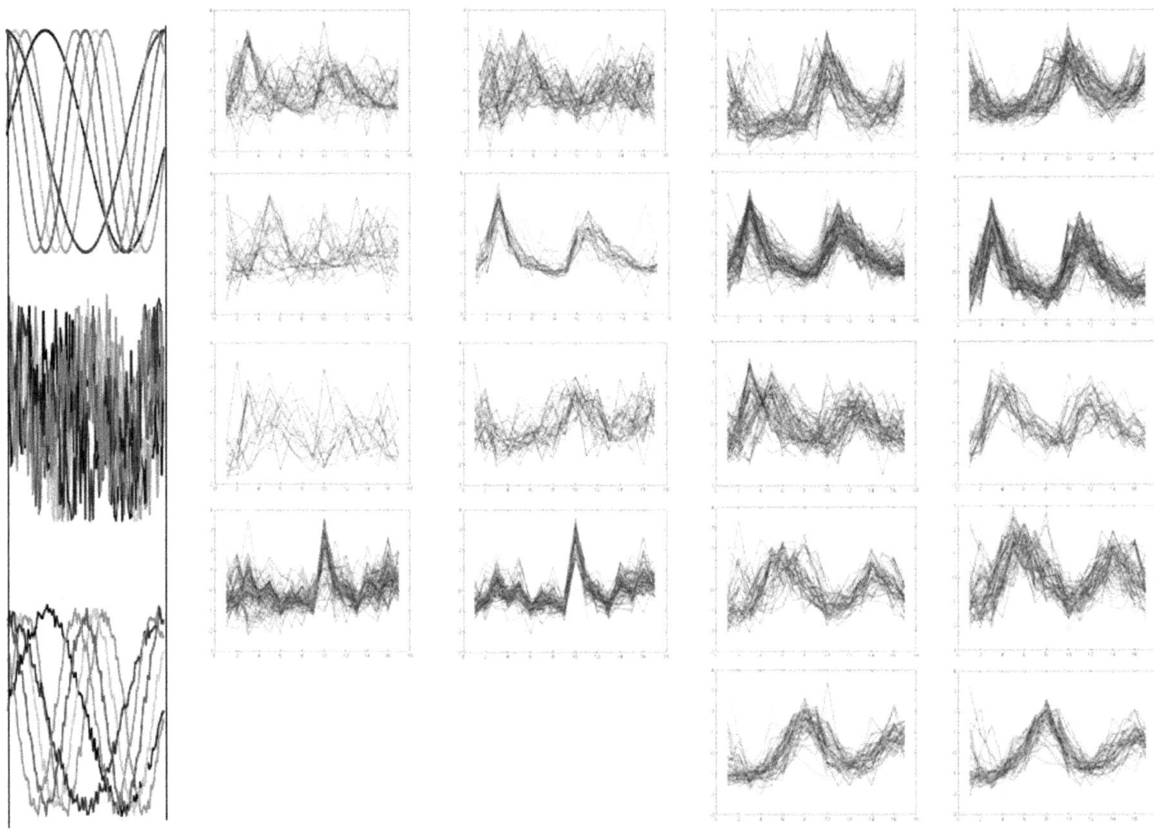

based on Euclidean (L^2) distance parameter and by employing a SOM algorithm. The algorithm performs, iteratively, the four basic steps as described below (plain lower-case w is used for SOM weight vector to distinguish it from wavelet coefficients which are labeled with either upper case and/or bold face w):

1. **Initialize:** Initialize the weight vector $\left[w_j(0)\right]_{j=1}^{l}$ by selecting values from the available set of input vector $\left[x_j\right]_{j=1}^{N}$ in a random manner.
2. **Input:** Read in the activation pattern. The activation pattern for the SOM is the input

Table 1. ARI and SW scores for clustering of noisy simulated time-series data and of yeast expression data

Algorithm[1]	ARI			SW		
	Yeast		**Simulated**	**Yeast**[1]		Simulated[5]
Wave-SOM[2]	Functional[3]	Cell Phase[4]		Functional[3]	Cell Phase[4]	
(analyt. thresh)	0.861	0.924	0.791	0.421	0.572	0.584
(stand. thresh.)	0.804***	0.825***	0.686***	0.299***	0.341***	0.412***
(no de-noising)	0.471***	0.501***	0.416***	.212***	0.308***	0.346***
Cluster 3.0	0.542***	0.413***	0.372***	0.108***	.304***	0.313***
MCLUST	0.617***	0.672***	0.489***	0.212***	0.129***	0.091***
CurveSOM	0.793***	0.811***	0.785**	0.278***	0.235***	0.292***
SSClust	0.506***	0.516***	0.592***	0.238***	0.124***	0.089***
GENECLUSTER	0.837*	0.802***	0.711*	0.202***	.0322***	0.081***

[1] Clustering results (ARI and SW scores) were varied with different distance metrics and linkage algorithms used. Scores reported here for the other programs were the highest obtained with that program.

[2] Results for Wave-SOM with analytical thresholding (analyt. thresh.) was compared to clustering results obtained with 7 other algorithms (including Wave-SOM with standard- and no-thresholding). Data obtained with the other 7 algorithms had the following p-values compared with Wave-SOM (analyt. thresh.): *** $p<0.01$, **$p<0.05$, *$p<0.1$.

[3] Gene expression data from Ideker et al (2001).

[4] Gene expression data from Spellman et al (1998).

[5] Analytical expression for the simulated data are given in the Evaluation Criteria section under Methods.

vector, $\mathbf{x}$, which contains the entire gene data set. The dimension of the vector $\mathbf{x}$ is m.

3. **Similarity matching:** find the best-matching neuron $i(\mathbf{x})$ at time step n by using the minimum-distance Euclidean criterion:

$$i(x) = \arg\min_j \left\| x(n) - w_j \right\|, j = 1, 2, ...l \quad (15)$$

4. **Update:** Adjust the synaptic weight vectors of all neurons by using the update relation:

$$w_j(n+1) = w_j(n) + \eta(n) h_{j,i(x)}(n)(x(n) - w_j(n)) \quad (16)$$

where $\eta(n)$ is the learning-rate parameter and $h_{j,i(x)}(n)$ is the neighborhood function centered around the winning neuron $i(\mathbf{x})$.

5. **Continue:** Continue with step 2 until no noticeable changes in the feature map are observed.

For the simulations in this study, a training time of 4000 epochs was employed for the network neurons. Figure 1B shows the wavelet coefficients for a representative group of 100 genes clustered in six different nodes of the network.

Evaluation Criteria

Two methods were used to evaluate the performance of our algorithm as follows:

1. *Synthetic Data:* Genes in the yeast can be classified according to their expression patterns. These include, among other groups, cell cycle genes which exhibit oscillatory expression patterns, functional or "metabolic" genes whose expression can rise or fall in response to external stimuli and housekeeping or "homeostatic" genes which exhibit a relatively flat expression pattern. To simulate these different classes of gene, we generated simulated "gene expression" patterns data using four different classes of functions: $e^{-(t/\tau)}$

$sine(2\pi jt+2\pi/k)$; $e^{-(t/\tau)} cos(2\pi jt+2\pi/k)$; $e^{(-jt/\tau,)}$ $1-e^{(-jt/\tau)}$, with $j \in [2,3]$; $\tau = 40$. While the simulated expression patterns in each class have identical patterns, the cluster structure of real data will, in general, be less distinguishable than that of these synthetic data. Hence, it is of interest to study the performance of our clustering algorithm as a function of noise level in the simulated data. Error was thus added to the simulated time-series using an experimentally derived error distribution. The error for each data point was randomly sampled from the distribution of standard deviations of log ratios of the yeast galactose expression data (Ideker at al, 2000) with the mean equal to the value of the simulated pattern, and SD equal to the sampled error. The signal-to-noise ratio of the simulated dataset is the ratio of the range of signals to the average sampled error. The signal-to-noise was adjusted by linearly scaling the error before adding it to the pattern. Figure 3A shows the "expression" profiles of the two oscillatory classes (sin and cosines) in our typical simulated datasets without (*top panel*) and with (*middle panel*) noise added, respectively. Figure 3A also shows the noisy time-series after de-noising with Wave-SOM (*lower panel*).

2. *Yeast Data*: We used two sets of gene expression data from yeast as follows:

   Set I: The first set of gene expression data were from Ideker et al. (2001) who studied galactose utilization in yeast using cDNA arrays by deleting nine genes on the galactose utilization pathway in the presence or absence of galactose and raffinose. Twenty experiments in total were carried out as follows: nine single-gene deletions and one wild-type experiment with galactose and raffinose, nine deletions and one wild-type without galactose and raffinose. Four replicate hybridizations were performed for each experiment. We analyzed the expression of a subset of 205 genes from this data set, whose expression patterns reflect four functional categories in the Gene Ontology (Ashburner et al., 2000). These functional categories include *(i)* DNA synthesis and replication, *(ii)* organization of centrosome, *(iii)* nitrogen and sulfur metabolism and *(iv)* ribosomal proteins. Figure 3B depicts the expression patterns of the genes in each of the four functional groups, while Figure 3C shows the expression patterns of the same genes as clustered by Wave-SOM. As can be seen from these two figures there is a high degree of similarity between the two sets of clusters.

   Set II: The second set are expression data of 6200 genes from a synchronized yeast cell culture (Spellman et al., 1998) each containing 17 time points with 10 min intervals. This expression patterns span two cell cycles. Accordingly, we identified a subset of 383 genes with expression patterns that exhibit peaks at five different time points corresponding to the five distinct phases in the cell cycle. The phases in sequential order are early G1, late G1, S, G2, and M. Figure 3D shows the expression patterns of the genes in each of the five phases. Figure 3E shows the expression patterns of the same genes as clustered by Wave-SOM. As was the case with the functional genes, these figures suggest a high degree of similarity between the experimental and computed sets of clusters.

To evaluate the accuracy of our clustering algorithm in a more objective manner, we also calculated the adjusted Rand index (ARI) with the five functional categories mentioned above used as our external knowledge (see below). We evaluated the performance of our algorithm by how closely our computed clusters reproduce the four functional categories and five cell-cycle phases before and after wavelet transformation.

Adjusted Rand Index: We used Adjusted Rand index (ARI) to measure the agreement between the computed clustering to the partition determined by the four synthetic function classes in the case of simulated data or the partitions of the yeast gene expression data reported in Ideker et al (2001) and in Spellman et al. (1998). ARI is a statistic designed to assess the degree of agreement between these two types of partitions (Hubert & Arabie, 1985). The Rand index (Rand, 1971) is defined as the fraction of agreement, that is, the number of pairs of objects that are either in the same groups in both partitions or in different groups in both partitions, divided by the total number of pairs of objects. The Rand index assumes values between 0 and 1. When the two partitions agree perfectly, the Rand index is 1. The adjusted Rand index adjusts the score so that its expected value in the case of random partitions is 0. Formally, the adjusted Rand index is given by:

$$ARI(T,C) = \frac{2(ad - bc)}{(a+b)(b+d)+(a+c)(c+d)} \tag{17}$$

Where T and C stand for the known (or True) and computed clustering, respectively. a is the number of gene pairs that belong to the same cluster in both T and C; b is the number of gene pairs within the same cluster in C but different clusters in T; c is the number of gene pairs within the same cluster in T but different clusters in C; and d is the number of gene pairs that belong to different clusters in both T and C.

Table 1 compares ARI scores for the clustering of the synthetic time-series data and the expression data for the functional and cell-cycle gene sets in yeast.

Another measure for validating clustering results is the silhouette width, SW (Rousseeuw, 1987; Chen, 2009). This measure, often used when the "True" clustering is not known, reflects the compactness of each cluster and the separation between the clusters. For each gene i, we denote a value a_i corresponding to the average distance to all other genes in its own cluster, and a value b_i corresponding to the minimum of the average distance of gene i to all the genes in the other clusters. SW for gene i is then defined as:

$$s_i = \frac{b_i - a_i}{\max\{a_i, b_i\}}$$

The values for s_i can be averaged for all genes to provide a measure for the performance of the clustering method. A high value of SW implies that the clusters are well separated and hence may provide meaningful results. In contrast to ARI, SW evaluates the clustering results based on information specific and within a particular clustering result, and thus cannot determine how closely a given computed result agrees with the "True" clustering.

Application

Figure 4A shows a cluster map of gene expression patterns for the oxidative stress experiment. Figure 4B shows the corresponding map for the expression patterns of the same genes after de-noising using our wavelet filtering transform. De-noising was performed prior to network training. Figure 4C provides a closer look at the group

Figure 4. Gene expression patterns of yeast under oxidative stress conditions (A) without and (B) with de-noising. The patterns are organized into 49 clusters which correspond to the different nodes of the 7x7 self-organizing-map neural network. (C) The names of the 9 genes found in node 7 cluster of the SOM map for the de-noised oxidative stress data; and (D) expression patterns for six of the genes.

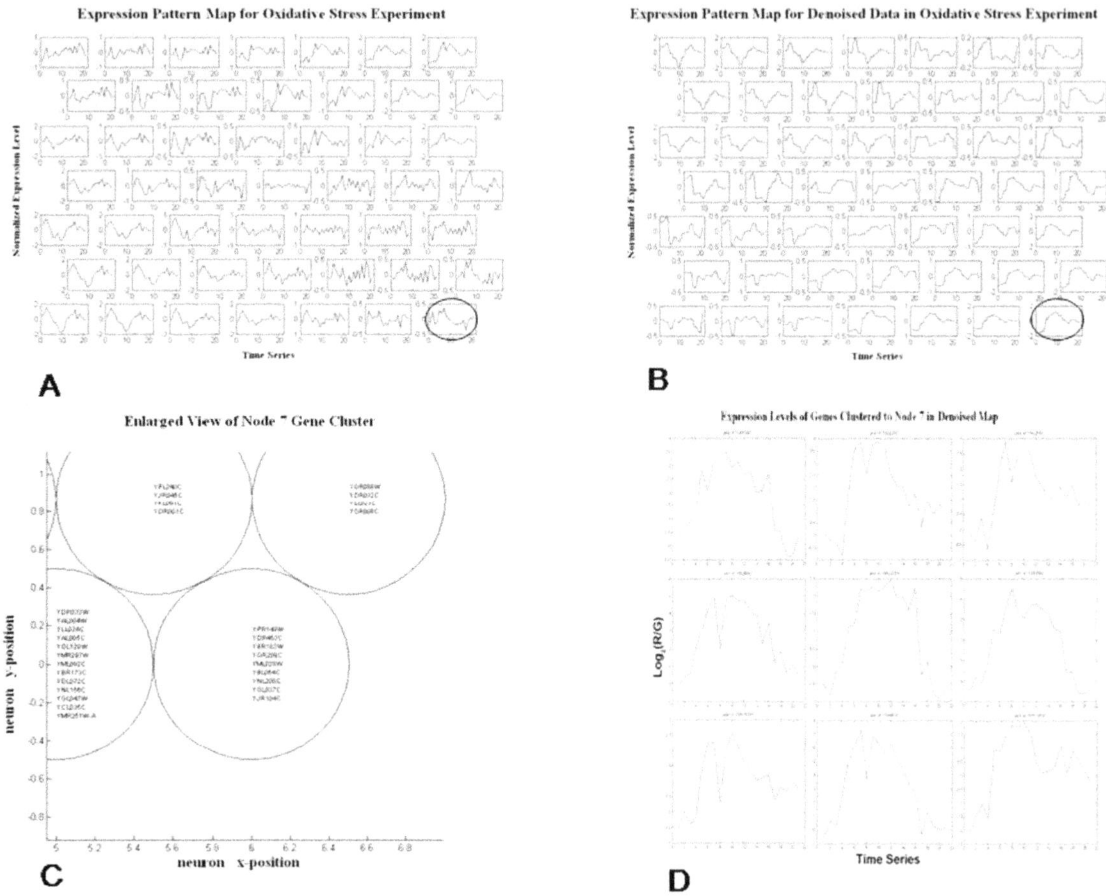

of genes clustered together in node 7 in the "de-noised" cluster map of Figure 4B (the bottom right node in Figure 4B). Figure 4D depicts the de-noised expression patterns for the 9 genes. It is evident that the $\log_2 R / G$ data shown in Figure 4A has some noise that was filtered out in the cluster map. Figure 5 depicts the corresponding gene expression patterns and the resulting cluster map for yeast cells grown on glucose-limited medium in a typical chemostat experiment.

Next, we intersected the clusters obtained from the oxidative stress data with those from the glucose-limited chemostat data. Following sorting, we identified five different groups that contain at least five genes which were co-expressed in *both* experiments and that exhibit expression patterns that are closely correlated with each other. Figure 6 lists the systemic names of the genes in these five groups and depicts the expression patterns of the genes in each experiment.

Using the Saccharomyces Genome Database (SGD), we determined the genes that were identified by our analysis and the function of the associated proteins as reported by. Table 2 lists the

Figure 5. Gene expression patterns of yeast under glucose restriction (chemostat) conditions. The patterns are organized into 49 clusters which correspond to the nodes of the 7×7 self-organizing-map neural network.

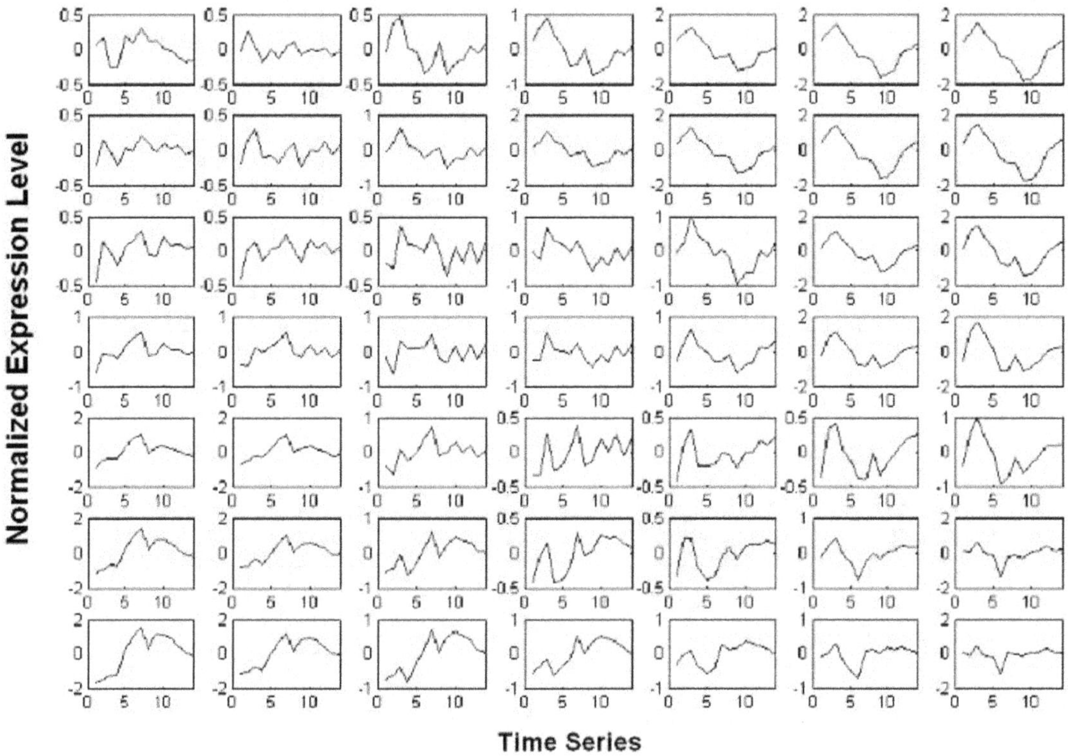

systematic and standard names for all 29 genes along with their cellular functions. As can be seen, except for a very few exceptions, all of the genes encode proteins which are part of the ribosomal complex (Lascaris et al., 1999). Therefore, they all belong to a common metabolic pathway.

DISCUSSION AND CONCLUSION

Current gene expression analysis techniques cluster genes based on their response to a variety of environmental challenges. However, in most cases the clustering algorithms are limited to handling data from one microarray per environmental stress. Clustering genes based on similarity in the *kinetics* of their activation patterns across multiple experiments could provide a unique and valuable dimension of information for gene function analysis. To take advantage of this new dimension, we presented a new algorithm capable of clustering genes that have overall different expression levels based on similarities in the kinetics of their expression.

Our algorithm employs a neural network with a Self Organizing Map (SOM) learning scheme. SOM is a powerful technique for clustering gene expression data. SOM are versatile neural network models developed by Kohonen that are currently being used in a variety of applications including speech recognition, image and textual analysis and textual data mining (Lagus et al., 2004). SOMs are particularly useful in microarray analysis as they allow the user to visualize multi dimensional data matrices and relationships between the cluster nodes. We combined our neural network algorithm

Figure 6. Genes found in both the oxidative stress and glucose chemostat experiments which exhibit expression patterns that are closely correlated with each other. Only clusters that contain 5 genes or more are listed.

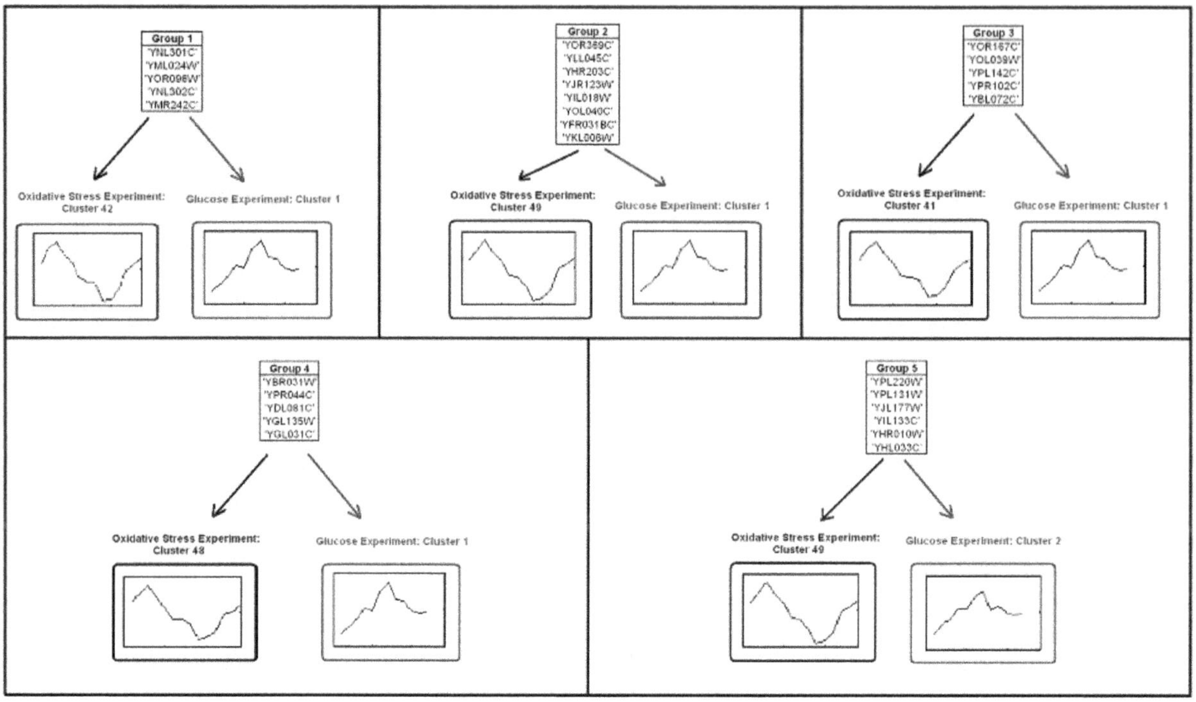

with a wavelet decomposition transform, which pre-filters the raw data to remove the noise (de-noising).

Wavelet analysis is a relatively new signal-processing method that offers some advantages over Fourier analyses. Fourier analyses are not suited for the sparse sampling and short duration data sets frequently encountered in genomic research. While some variants of Fourier analysis, such as Maximum Entropy Spectral Analysis and discrete-time Fourier Transform, have been successful in detecting periodicity in noisy, periodic biological time series data, wavelets are now emerging as the approach of choice for analysis of short-duration, noisy biological time-series. In particular, the ability to decompose a signal into high-frequency noise and low-frequency trends, and to distinguish these effects from those that are characteristic of the signal by matching the wavelet scale to the scale of the signal, makes wavelets a promising alternative for the analysis of time series data and perturbation studies in combination with microarray techniques.

Our algorithm is designed to cluster gene expression data based on the kinetics of their expression patterns and to generate maps that are robust against noise in the data. Since wavelet decomposition does not de-localize the data on the time axis, the algorithm clusters genes which exhibit similarities in their expression kinetics whether the patterns are time-shifted relative to each other or not, as long as the time shift is not larger than the wavelet's largest period used for the analysis. Filters are designed to pass only expression levels that occur during a specific time interval in the cell cycle, while simultaneously normalizing the expression patterns and reducing the signal noise. To implement this feature-

Table 2. Gene ontology information for genes listed in Fig. 6

Gene Group	Systematic Gene Name	Standard Gene Name	Gene Function	TF binding sites[5]
1	YNL301c	RPL18B	Rp[1]	
	YML024w	RPS17A	Rp	Rap1p
	YNL302c	RPS19B	Rp	Rap1p
	YMR242c	RPL20A	Rp	
	YOR096w	RPS7A	Rp	Rap1p, Abf1p
2	YOR369c	RPS12	Rp	Rap1p
	YLL045c	RPL8B	Rp	Rap1p
	YHR203c	RPS4B	Rp	Rap1p
	YJR123w	RPS5	Rp	Rap1p, Abf1p
	YIL018w	RPL2B	Rp	Rap1p
	YOL040c	RPS15	Rp	Rap1p
	YFR031c-A	RPL2A	Rp	Rap1p
	YKL066w	dubious ORF	[2]	
3	YOR167c	RPS28A	Rp	Rap1p, Abf1p
	YOL039w	RPP2A	Rp	Rap1p
	YPL142c	(RPL33A/YPL143w)	[3]	Rap1p
	YPR102c	RPL11A	Rp	Rap1p, Reb1p
	YBL072c	RPS8A	Rp	Rap1p
4	YBR031w	RPL4A	Rp	Abf1p
	YPR044c	(RPL43A/YPR043w)	[4]	Rap1p
	YDL081c	RPP1A	Rp	Abf1p
	YGL135w	RPL1B	Rp	Rap1p
	YGL031c	RPL24A	Rp	Rap1p
5	YPL220w	RPL1A	Rp	
	YPL131w	RPL5	Rp	Rap1p
	YJL177w	RPL17B	Rp	Rap1p
	YIL133c	RPL16A	Rp	Rap1p
	YHR010w	RPL27A	Rp	Rap1p
	YHL033c	RPL8A	Rp	Rap1p

[1]Rp, ribosomal protein.

[2]No match was found for YFR031bc in the yeast database. YFR031c has a match to SMC2, which is a component of the condensin complex involved in chromosome maintenance.

[3]Yeast database list YPL142c as dubious. This ORF overlaps with the gene encoding the ribosomal protein RPL33A.

[4]Yeast database list YPR044c as dubious. This ORF overlaps with the gene encoding the ribosomal protein RPL43A that is transcribed in the opposite direction.

[5]TF, transcription factor binding sites were from Lascaris et al. (1999).

oriented gene clustering, the SOM-based neural network is trained using a filter that focuses on the particular scale size of that feature. A possible application of this algorithm, which was not pursued in the current study, would be to see at what point during the cell cycle a particular set of genes are turned on and off in the specific pattern emphasized. Therefore, based on the timing and patterns of their expression, one may be able to infer causality relationships for enzymes in regulatory pathways and metabolic networks from the temporal response pattern of their encoded genes. In that regard, it is interesting to note that our intersection algorithm identified genes that encode ribosomal proteins (Figure 6). The vast majority of the known genes encoding ribosomal proteins, in the yeast, are activated by the global DNA-binding factor Rap1p; while a small subset of them is activated by Abf1p. Table 2 also lists genes that contain putative Rap1p binding sites within their promoter region for the different genes listed in Figure 6. With a few exceptions, the majority of the genes are reported to have *Rap1p* binding sites in their promoter region (Lascaris et al., 1999). These binding sites are consistent with our finding of strong temporal correlation in the expression patterns for these genes.

As illustrated in Table 1, our wavelet-based noise reduction algorithm has produced a significant improvement in ARI over the unfiltered data. As the ARI scores suggest, our wavelet-based algorithm outperformed several other commonly used clustering algorithms. Furthermore, while GENECLUSTER came in close second place to Wave-SOM on the ARI score for the "functional" genes (only less than 3% behind, with an ARI score of 0.837 compared to 0.861 for Wave-SOM, $p<0.1$), Wave-SOM had an ARI score which was more than 12% higher for clustering the cell-cycle genes (0.924 compared to 0.802, $p<0.01$). This result illustrates the decisive advantage of Wave-SOM over the other programs in clustering genes with oscillatory expression patterns. Furthermore, the quality of the individual clusters, as expressed by the SW parameter, generated with Wave-SOM were significantly higher ($p<0.01$) than all other 5 algorithms tested; for both the functional and cell cycle genes. The only drawback with our algorithm is the overhead involved in computing the discrete Hilbert transform of the data vector and computing the stationary wavelet transform of the complex analytic vector. The latter however can be carried out using a fast Fourier transform algorithm as given by Eqn A11 in the Appendix.

The kinetics based cluster analysis described in this paper may also be applied to fields outside of microarray analysis. One possible application would be a study of an organism's response to environmental challenges such as disease or infection. For example, one could count cells that participate in an immune response as well as measure blood chemistry at regular time intervals and then cluster the data by the network to better understand the response of the organism's immune system to the challenge.

REFERENCES

Ashburner, M., Ball, C. A., Blake, J. A., Botstein, D., Butler, H., & Cherry, J. M. (2000). Gene Ontology: tool for the unification of biology. *Nature Genetics*, *25*, 25–29. doi:10.1038/75556

Bracewell, R. (1999). The Hilbert Transform. In *The Fourier Transform and Its Applications* (3rd ed., pp. 267–272). New York: McGraw-Hill.

Brauer, M. J., Saldanha, A. J., Dolinski, K., & Botstein, D. (2005). Homeostatic adjustment and metabolic remodeling in glucose-limited yeast cultures. *Molecular Biology of the Cell*, *16*, 2503–2517. doi:10.1091/mbc.E04-11-0968

Chen, X. (2009). Curve-Based Clustering of Time Course Gene Expression Data Using Self-Organizing Maps. *Journal of Bioinformatics and Computational Biology*, *7*, 645–661. doi:10.1142/S0219720009004291

Daubechies, I. (1992). *Ten Lectures on Wavelets*. Philadelphia: SIAM. doi:10.1137/1.9781611970104

Donoho, D. L., & Johnstone, I. M. (1994). Ideal spatial adaptation by wavelet shrinkage. *Biometrika*, *81*, 425–455. doi:10.1093/biomet/81.3.425

Donoho, D. L., & Johnstone, I. M. (1995). Adapting to unknown smoothness via wavelet shrinkage. *Journal of the American Statistical Association*, *90*, 1200–1224. doi:10.2307/2291512

Glass, L., & Mackey, M. C. (1979). A simple model for phase locking of biological oscillators. *Journal of Mathematical Biology*, *7*, 339–352. doi:10.1007/BF00275153

Haykin, S. (2002). *Adaptive Filter Theory*. Englewood Cliffs, NJ: Prentice Hall.

Hohmann, S., & Mager, W. H. (Eds.). (2003). *Yeast Stress Responses*. Berlin: Springer. doi:10.1007/3-540-45611-2

Hubert, L., & Arabie, P. (1985). Comparing partitions. *J. Classification*, *2*, 193–218. doi:10.1007/BF01908075

Ideker, T., Thorsson, V., Ranish, J. A., Christmas, R., Buhler, J., & Eng, J. K. (2001). Integrated genomic and proteomic analyses of a systemically perturbed metabolic network. *Science*, *292*, 929–934. doi:10.1126/science.292.5518.929

Klevecz, R. R., Kauffman, S. A., & Shymko, R. M. (1984). Cellular clocks and oscillators. *International Review of Cytology*, *86*, 97–128. doi:10.1016/S0074-7696(08)60178-3

Klevecz, R. R., Kros, J., & Gross, S. D. (1978). Phase response versus positive and negative division delay in animal cells. *Experimental Cell Research*, *116*, 285–290. doi:10.1016/0014-4827(78)90450-0

Kohonen, T. (2001). *Self Oganizing Maps* (3rd ed.). Berlin: Springer.

Lagus, K., Kaski, S., & Kohonen, T. (2004). Mining massive document collections by the WEBSOM method. *Information Sciences*, *163*, 135–156. doi:10.1016/j.ins.2003.03.017

Lascaris, R. F., Mager, W. H., & Planta, R. J. (1999). DNA-binding requirements of the yeast protein Rap1p as selected in silico from ribosomal protein gene promoter sequences. *Bioinformatics (Oxford, England)*, *15*, 267–277. doi:10.1093/bioinformatics/15.4.267

Marple, S. L. Jr. (1999). Computing the discrete-time 'analytic' signal via FFT. *IEEE Transactions on Signal Processing*, *47*, 2600–2603. doi:10.1109/78.782222

Mitchison, J. M. (1971). *The Biology of the Cell Cycle*. London: Cambridge University Press.

Mizuno, A., Tabei, H., & Iwahuti, M. (2006). Characterization of low-acetic-acid-producing yeast isolated from 2-deoxyglucose-resistant mutants and its application to high-gravity brewing. *Journal of Bioscience and Bioengineering*, *101*, 31–37. doi:10.1263/jbb.101.31

Nicholas, G., & Prigogine, I. (1977). *Self-Organization in Non-Equilibrium Systems*. New York: Wiley.

Ogden, R. T. (1997). *Essential Wavelets for Statistical Applications and Data Analysis*. Boston: Birkhäuser. doi:10.1007/978-1-4612-0709-2

Olhede, S. C., & Walden, A. T. (2004). Analytic wavelet thresholding. *Biometrika*, *91*, 955–973. doi:10.1093/biomet/91.4.955

Oppenheimer, A. V., & Schafer, R. W. (1975). *Digital Signal Processing*. Englewood Cliffs, NJ: Prentice-Hall.

Papoulis, A. (1991). *Probability, Random Variables, and Stochastic Processes* (3rd ed.). New York: McGraw-Hill.

Percival, D. B., & Walden, A. T. (2000). *Wavelet Methods for Time Series Analysis*. London: Cambridge University Press.

Rand, W. M. (1971). Objective criteria for the evaluation of clustering methods. *Journal of the American Statistical Association*, *66*, 846–850.

Rousseeuw, P. (1987). Silhouettes. a graphical aid to the interpretation and validation of cluster analysis. *Journal of Computational and Applied Mathematics*, *20*, 53–65. doi:10.1016/0377-0427(87)90125-7

Shapira, M., Segal, E., & Botstein, D. (2004). Disruption of yeast forkhead-associated cell cycle transcription by oxidative stress. *Molecular Biology of the Cell*, *15*, 5659–5669. doi:10.1091/mbc.E04-04-0340

Spellman, P. T., Gavin, S., Zhang, M. Q., Iyer, V. R., Anders, A., & Eisen, M. B. (1998). Comprehensive Identification of Cell Cycle-regulated Genes of the Yeast Saccharomyces cerevisiae by Microarray Hybridization. *Molecular Biology of the Cell*, *9*, 3273–3297.

Tamayo, P., Slonim, D., Mesirov, J., Zhu, Q., Kitareewan, S., & Dmitrovsky, E. (1999). Interpreting patterns of gene expression with self-organizing maps: Methods and application to hematopietic differentiation. *Proceedings of the National Academy of Sciences of the United States of America*, *96*, 2907–2912. doi:10.1073/pnas.96.6.2907

Tang, A., Jackson, D., Hobbs, J., Chen, W., Smith, J. L., & Patel, H. (2008). A maximum entropy model applied to spatial and temporal correlations from cortical networks in vitro. *The Journal of Neuroscience*, *28*, 505–518. doi:10.1523/JNEUROSCI.3359-07.2008

Vetterli, M., & Herley, C. (1992). Wavelets and filter banks: theory and design. *IEEE Transactions on Signal Processing*, *40*, 2207–2232. doi:10.1109/78.157221

APPENDIX

The Hilbert transform:

$$\tilde{x}(t) = x(t) * \frac{1}{\pi t} = \frac{1}{\pi} \int_{-\infty}^{\infty} \frac{x(u)}{t-u} du \qquad \text{(A1)}$$

$$G(f) = -i \operatorname{sgn}(f) = \begin{cases} -i, & \text{if } f > 0 \\ 0, & \text{if } f = 0 \\ i, & \text{if } f < 0 \end{cases} \qquad \text{(A2)}$$

The convolution of $\boldsymbol{x(t)}$ with $\boldsymbol{(\pi t)^{-1}}$ is equivalent to filtering $\boldsymbol{x(t)}$. The frequency response function (sometimes referred to as the transfer function) of the filter is $\boldsymbol{G(f)}$ and is given by $\boldsymbol{G(f)=|G(f)|\ exp\ \{i\theta(f)\}}$, where $|\mathbf{G}(.)|$ and $\boldsymbol{\theta\ (.)}$ are the gain function and phase function, respectively, of the filter. It should be noted however that $\boldsymbol{G(f)}$ is imaginary, thus $\boldsymbol{|G(f)|=1}$ for all $\boldsymbol{f}$, and

$$\theta(f) = \begin{cases} -\pi / 2, & \text{if } f > 0 \\ 0, & \text{if } f = 0 \\ \pi / 2, & \text{if } f < 0 \end{cases} \qquad \text{(A3)}$$

Thus the operation of Hilbert transformation is equivalent to leaving the energy spectrum of the signal unchanged while changing the phase by $-\boldsymbol{\pi}/2$ for positive frequencies and by $\boldsymbol{\pi}/2$ for negative frequencies. Fourier transformation of the complex-valued signal $z(i) = x(t) + i\tilde{x}(t)$ gives

$$Z(f) = X(f) + i\{-\operatorname{sgn}(f)\}X(f) = \begin{cases} 2X(f), & \text{if } f > 0 \\ X(f), & \text{if } f = 0 \\ 0, & \text{if } f < 0 \end{cases} \qquad \text{(A4)}$$

Consequently, since $Z(f)$ is zero for negative frequencies,

$$z(t) = \int_0^{\infty} Z(f) e^{i2\pi ft} df \qquad \text{(A5)}$$

$$z(\tau) = \int_0^{\infty} Z(f) e^{i2\pi f\tau} df, \tau = t + iu, u > 0, \qquad \text{(A6)}$$

Discrete time/continuous frequency: The digital filter corresponding to the frequency response ***−i sgn (f)*** is given by (Oppenheim & Schafer, 1975, p. 360):

$$q_t = \int_{-1/2}^{1/2} -i\,\mathrm{sgn}(f) e^{i2\pi ft} df = \begin{cases} [2\sin 2(l\pi / 2] / (l\pi), if\ l\ odd \\ 0, \qquad\qquad\qquad if\ l\ even \end{cases} \tag{A7}$$

This is also the infinite length impulse response sequence.

Discrete time/discrete frequency: For even values of ***N*** and with reference to the Fourier frequencies $f_k = k/N$, we define the following discrete sign sequence:

$$\mathrm{sgn}\, d_k = \begin{cases} 0\ (\mathrm{K} = 0 &) \\ 0\ (\mathrm{K} = 1,...(\mathrm{N}/2)\text{-}1 &) \\ 0\ (\mathrm{K} = \mathrm{N}/2 &) \\ -1\ (\mathrm{K} = (\mathrm{N}/2){+}1,\ ...\ \mathrm{N}\text{-}1) \end{cases} \tag{A8}$$

Given the vector $x^T = (x_0, \ldots, x_{N-1})$ expressed in terms of discrete Fourier transform

$$X_k = \sum_{t=0}^{N-1} x_t e^{-i2\pi tk/N} (k = 0, \ldots, N-1) \tag{A9}$$

The equivalent form of Eqn A4 is:

$$Z_k = X_k + i(-i\,\mathrm{sgn}\, d_k) X_k = \begin{cases} X_0\ \ (\mathrm{K} = 0 &) \\ 2X_k\ \ (\mathrm{K} = 1,...(\mathrm{N}/2)\text{-}1 &) \\ X_{N/2}\ \ (\mathrm{K} = \mathrm{N}/ &) \\ 0\ (\mathrm{K} = (\mathrm{N}/2){+}1,\ ...\ \mathrm{N}\text{-}1) \end{cases} \tag{A10}$$

The elements of the analytic vector are found from the inverse discrete Fourier transform of $\{Z_k\}$:

$$Z_t = \frac{1}{N} \sum_{k=0}^{N-1} Z_k e^{-i2\pi tkt/N} (t = 0, \ldots N-1) \tag{A11}$$

(*k=0, . . ., N−1*), or

$$q^{o}_{tt} = -\frac{1}{N}\sum_{k=0}^{N-1} i \,\mathrm{sgnd}_k e^{2\pi kl/N} (l = 0, ..., N-1) \quad \text{(A12)}$$

$$q^{o}_{tt} = \sum\nolimits_{n=-\infty}^{\infty} q_{t+nN} \quad \text{(A13)}$$

Thus $x_t = R\{z_t\}$ and $\tilde{x}_t = \Im\{z_t\}$ with $\tilde{x}_t$ given by

$$\tilde{X}_t = \sum_{t=0}^{N-1} q^{o}_{l} x_{(t-l) \bmod N} (t = 0, ... N-1),$$

And the periodized filter is given by:

(2/N)cos(π/N), if l odd

$$q^{o}_{l} = \begin{cases} (2 / N)\cos(\pi / N), if \ l \ odd \\ 0, \qquad\qquad\qquad if \ l \ even \end{cases} = \{0\, if\, l\, even \quad \text{(A14)}$$

For $l = 0, ... N\text{-}1$. Therefore, we can write the discrete Hilbert transform in matrix form as $\tilde{x} = Hx$ where:

(matrix)H=

0	q^{o}_{N-1}	0	q^{o}_{N-3}	0	…	0	q^{o}_{1}
q^{o}_{1}	0	q^{o}_{N-1}	0	q^{o}_{N-3}	…	q^{o}_{3}	0
0	q^{o}_{1}	0	q^{o}_{N-1}	0	…	0	q^{o}_{3}
q^{o}_{3}	0	q^{o}_{1}	0	q^{o}_{N-1}	…	q^{o}_{5}	0
0	q^{o}_{3}	0	q^{o}_{1}	0	…	0	q^{o}_{5}
M					O	M	M
M						O	M
q^{o}_{N-1}	0					q^{o}_{1}	0

(A15)

With $\{z_t\}$ defined as in Eqn. A11. Note that the real and imaginary parts of the analytic vector z are orthogonal (Marple, 1999, p. 2601).

$$\sum_{t=0}^{N-1} x_t \tilde{x}_t = 0 \tag{A16}$$

This work was previously published in the International Journal of Knowledge Discovery in Bioinformatics, Volume 1, Issue 2, edited by Xiao-Li Li and See-Kiong Ng, pp. 50-73, copyright 2010 by IGI Publishing (an imprint of IGI Global).

Chapter 8
Mining Frequent Boolean Expressions:
Application to Gene Expression and Regulatory Modeling

Mohammed J. Zaki
Rensselaer Polytechnic Institute, USA

Naren Ramakrishnan
Virginia Tech., USA

Lizhuang Zhao
Microsoft Corp., USA

ABSTRACT

Regulatory network analysis and other bioinformatics tasks require the ability to induce and represent arbitrary boolean expressions from data sources. In this paper, the authors introduce a novel framework called BLOSOM for mining (frequent) boolean expressions over binary-valued datasets. Boolean expressions can be grouped into four categories: pure conjunctions, pure disjunctions, conjunction of disjunctions, and disjunction of conjunctions. The authors' main focus is on mining the simplest expressions (the minimal generators), but also to propose closure operators that yield closed (or unique maximal) boolean expressions. BLOSOM efficiently mines frequent boolean expressions by utilizing a number of methodical pruning techniques. Experiments showcase the behavior of BLOSOM for different input settings and parameter thresholds. Application studies on gene expression and gene regulation patterns showcase the effectiveness of this approach.

DOI: 10.4018/978-1-4666-1785-8.ch008

INTRODUCTION

A key class of problems in biological knowledge discovery pertains to modeling complex relationships between entities, such as GO category descriptions, genetic regulatory connections, and metabolic pathway networks. To effectively address these tasks, we require data mining techniques that possess the expressiveness to model *arbitrary* boolean connections. However, many state-of-the-art techniques are rather restrictive in the class of expressions they can mine.

For example, itemset mining (Agrawal et al., 1996) takes as input a binary-valued dataset and discovers patterns that are *pure conjunctions* of items. Admittedly, such techniques are useful in bioinformatics, but have limited scope. For example, given the expression levels (high or low) of genes for various cancers, we may find that cancerous tissues have high levels of genes (g_1 AND g_5 AND g_{13}) OR (g_{10} AND g_{15}). This might indicate that these groups of genes are co-regulated and somehow linked to cancer. These techniques can be generalized to include negations of descriptions and used to mine redescriptions (Ramakrishnan et al., 2004), that is, finding alternative ways of describing a group of genes. For instance, from yeast gene expression data, we may find that the gene highly expressed at time point 15 mins (*d183*), but not involved in mannose transportation (*d388*) or fructose metabolism (*d515*) or not part of external protective structure (*d460*), are the exact same genes also highly expressed at time point 20 mins (*d184*), provided they do not have unknown molecular function (*d309*). This redescription can be written as an equivalence of boolean expressions: *d183* AND (NOT *d388*) AND (NOT *d460*) AND (NOT *d515*) ⇔ *d184* AND (NOT *d309*).

We seek to generalize these approaches to mine arbitrary boolean expressions. Boolean expression mining can provide tremendous value, but there are two main challenges to contend with. The first deals with the problem of high computational complexity. With n items (or variables), there are 2^{2^n} possible distinct-valued Boolean expressions. With n items, there are 2^n distinct ways of assigning true/false values to the n terms, and for any such assignment the truth value of the expression can be set to true or false, giving the 2^{2^n} value. These are far too many to enumerate. To make the search practical we focus on only the frequent boolean expressions. Also instead of mining all frequent boolean expressions, we focus on mining a lossless subset that retains complete frequency information, namely *closed* boolean expression. The second challenge relates to comprehension of the patterns, i.e., they may be complex and difficult to understand. Here we focus on mining the simplest or *minimal* expressions (which are in fact the *minimal generators* of the closed expressions) that still from a lossless representation of all possible boolean expressions.

In this paper, we present a novel framework, called BLOSOM (an anagram of the bold letters in **BOOL**ean expression **M**ining over attribute **S**ets), the first such approach to simultaneously mine closed boolean expressions over attribute sets and their minimal generators. Our main contributions are as follows: We organize boolean expressions into four categories: (i) pure conjunctions (AND-clauses), (ii) pure disjunctions (OR-clauses), (iii) conjunctive normal form (CNF; conjunction of disjunctions), and (iv) disjunctive normal form (DNF; disjunction of conjunctions). For both CNF and DNF expressions we propose a *closure* operator, and we give a characterization of the *minimal generators*. BLOSOM employs a number of effective pruning techniques for searching over the space of boolean expressions, yielding orders of magnitude in speedup. We conducted several experiments on synthetic datasets to study the behavior of BLOSOM with respect to (w.r.t.) different input settings and parameter thresholds. We also highlight some of the patterns found using BLOSOM on real datasets from bioinformatics applications such as analysis of gene expression patterns, and boolean gene regulatory network discovery.

PRELIMINARY CONCEPTS

Lattice Theory: Let us first review a few facts from lattice theory (Davey & Priestley, 1990), which will be useful in our discussion. Let $(P, \subseteq)$ be a partially ordered set (also called a *poset*). Let $X, Y \in P$ and let $f : P \to P$ be a function on *P*. *f* is called *monotone* if $X \subseteq Y \Rightarrow f(X) \subseteq f(Y)$. We say that *f* is *idempotent* if $f(X) = f(f(X))$. *f* is called *extensive* (or expansive) if $X \subseteq f(X)$. Finally, *f* is called *intensive* (or contractive) if $f(X) \subseteq X$. A *closure operator* on *P* is a function $\mathbb{C} : P \to P$ such that $\mathbb{C}$ is monotone, idempotent, and extensive. *X* is called *closed* if $\mathbb{C}(X) = X$. On the other hand, a *kernel operator* on *P* is a function $\mathbb{K} : P \to P$, which is monotone, idempotent, and intensive. *X* is called *open* if $\mathbb{K}(X) = X$. The set of all closed and open members of *P* form the fixed-point of the closure ($\mathbb{C}$) and kernel ($\mathbb{K}$) operators, respectively. Given two posets $(P, \subseteq)$ and $(Q, \leq)$, a *monotone Galois connection* between them consists of two *order-preserving* functions, $\phi : P \to Q$ and $\psi : Q \to P$, such that for all $X \in P$ and $Y \in Q$, we have: $X \subseteq \psi(Y) \Leftrightarrow \phi(X) \leq Y$. The composite function $\psi \circ \phi : P \to P$ is a closure operator, whereas the function $\phi \circ \psi : Q \to Q$ is a kernel operator, on *P* and *Q*, respectively (Davey & Priestley, 1990). Given posets $(P, \subseteq)$ and $(Q, \leq)$, a *anti-monotone Galois connection* (Davey & Priestley, 1990) between them consists of two *order-reversing* functions, $\phi : P \to Q$ and $\psi : Q \to P$, such that for all $X \in P$ and $Y \in Q$, we have: $X \subseteq \psi(Y) \Leftrightarrow Y \leq \phi(X)$. The composite functions $\psi \circ \phi : P \to P$ and $\phi \circ \psi : Q \to Q$ are both closure operators on *P* and *Q*, respectively (Davey & Priestley, 1990).

Boolean Expressions: Let $\mathcal{I} = \{i_1, i_2, \ldots, i_m\}$ be a set of binary-valued attributes or *items*. Let AND, OR, and NOT denote the usual logical operators. We denote a negated item (NOT *i*) as $\overline{i}$. We use the symbol | to denote OR, and we simply omit the AND operator whenever there is no ambiguity. For example, (i_3 AND i_4) OR (i_1 AND (NOT i_7)) is rewritten as $(i_3 i_4) \mid (i_1 \overline{i_7})$. A *literal* is either an item *i* or its negation $\overline{i}$. A *clause* is either the logical AND or logical OR of one or more literals. An AND-clause is a clause that has only the AND operator over all its literals, and an OR-clause is one that has only the OR operator over all its literals. We assume without loss of generality that a clause has all distinct literals (since a clause is either an AND- or an OR-clause, repeated literals are logically redundant). A boolean *expression* is the logical AND or OR of one or more clauses.

A boolean expression is said to be in *negation normal form* (NNF) if all NOT operators directly precede literals (any expression can be converted to NNF by pushing all negations into the clauses). An NNF boolean expression is said to be in *conjunctive normal form* (CNF) if it is an AND of OR-clauses. An NNF expression is said to be in *disjunctive normal form* (DNF) if it is an OR of AND-clauses. For example, $(i_3 i_4) \mid (i_1 i_5 i_7)$ is in DNF, whereas $(i_2 \mid i_3)(i_0 \mid i_1 \mid \overline{i_3})$ is in CNF. Note that by definition, single OR-clauses and single AND-clauses, are in both CNF and DNF. Furthermore, when considering negated literals, we disallow a tautology like the OR-clause containing $(i \mid \overline{i})$ which is always true. Similarly, we disallow a contradiction like the AND-clause containing $(i \, \overline{i})$ since this is always false. Note that a CNF expression is a tautology if and only

Figure 1. Dataset $\mathcal{D}$ and its transpose $\mathcal{D}^T$ and complement $\overline{\mathcal{D}}$

$\mathcal{D}$

tid	set of items
1	ACD
2	BC
3	ABCD
4	ADE
5	E

$\mathcal{D}^T$

item	tidset
A	134
B	23
C	123
D	134
E	45

$\overline{\mathcal{D}}$

tid	set of items
1	$\overline{B}\,\overline{E}$
2	$\overline{A}\,\overline{D}\,\overline{E}$
3	$\overline{E}$
4	$\overline{B}\,\overline{C}$
5	$\overline{A}\,\overline{B}\,\overline{C}\,\overline{D}$

if (iff) each one of its clauses contains both an item and its negation. Likewise, a DNF expression is a contradiction iff each one of its clauses contains both a variable and its negation. Thus by disallowing the tautologies in individual clauses, we disallow tautologies in CNF expressions. Likewise, by eliminating contradictions in clauses, we eliminate

Dataset: Let $\mathcal{I} = \{i_1, i_2, \ldots, i_m\}$ be a set of items, let $\mathcal{T} = \{t_1, t_2, \ldots, t_n\}$ be a set of transaction identifiers (tids). A dataset $\mathcal{D}$ is then a subset of $\mathcal{T} \times \mathcal{I}$; in other words, the dataset $\mathcal{D}$ is a set of tuples of the form $(t, t.X)$, where $t \in \mathcal{T}$ is the tid of the transaction containing the set of items $t.X \subseteq \mathcal{I}$. Note that any categorical dataset can be transformed into this transactional form, by assigning a unique item for each attribute-value pair.

Given dataset $\mathcal{D}$, we denote by $\mathcal{D}^T$ the transposed dataset that consists of a set of tuples of the form $(i, i.Y)$, where $i \in \mathcal{I}$ and $i.Y \subseteq \mathcal{T}$ is the set of tids of transactions containing *i*. Figure 1 shows a dataset and its transpose, which we will use as a running example in this paper. It has five items $\mathcal{I} = \{A, B, C, D, E\}$ and five tids $\mathcal{T} = \{1, 2, 3, 4, 5\}$. Note that in $\mathcal{D}$, transaction t_1 contains the set of items $\{A, C, D\}$ (for convenience, we write it as *ACD*), and in $\mathcal{D}^T$, the set of tids of transactions that contain item *A* is {1,3,4} (again, for convenience we write it as *134*).

Let $\aleph(t.X) = \{\overline{i} \mid i \in \mathcal{I} \setminus t.X\}$, denote the negated items that do not appear in a transaction $t.X$. For example, for $(1, ACD)$, we have $\aleph(ACD) = \overline{B}\,\overline{E}$. Define $\overline{\mathcal{D}}$ as the complement of database $\mathcal{D}$, given as $\overline{\mathcal{D}} = \{(t, \aleph(t.X))\# \ t.X \in \mathcal{D}\}$.

Figure 1 also shows the complemented dataset $\overline{\mathcal{D}}$ for our running example.

Tidset and Support: Given a transaction $(t, t.X) \in \mathcal{D}$, with $t \in \mathcal{T}$ and $t.X \subseteq \mathcal{I}$, we say that tid *t satisfies* an item/literal $i \in \mathcal{I}$ if $i \in t.X$, and *t* satisfies the literal $\overline{i}$ if $i NOTin t.X$. For a literal *l*, the *truth value* of *l* in transaction *t*, denoted $V_t(l)$ is given as follows:

$$V_t(l) = \begin{cases} 1 & \text{if } t \text{ satisfies } l \\ 0 & \text{if } t \text{ does not satisfy } l \end{cases}$$

We say that a transaction $t \in \mathcal{T}$ *satisfies* a Boolean expression *E* if the truth-value of *E*, denoted $V_t(E)$, evaluates to true when we replace every literal *l* in *E* with $V_t(l)$. For any boolean expression *E*, $\mathrm{t}(E) = \{t \in \mathcal{T} : V_t(E) = 1\}$ denotes the set of tids (also called a *tidset*), that satisfy *E*.

The *support* of a boolean expression *E* in dataset $\mathcal{D}$ is the number of transactions which satisfy *E*, i.e., $|\ \mathrm{t}(E)\ |$. An expression is *frequent* if its support is more than or equal to a user-specified *minimum support* (*minsup*) value, i.e., if $| t(E) | \geq minsup$. For disjunctive expressions, we also impose a *maximum support* threshold (*maxsup*) to disallow any expression with too high a support. Setting *minsup*=1 and $maxsup = \infty$ allows mining all possible expressions.

Boolean Expression Mining Tasks: Given a dataset $\mathcal{D}$ and support thresholds *minsup* and *maxsup*, the task is to mine minimal and closed frequent boolean expressions, such as AND-clauses, OR-clauses, CNF and DNF, as specified by the user.

One of the goals of mining for different boolean expressions to find the minimal (and/or closed) boolean expressions that characterize a set of transactions (the tidset), i.e., the tids in the tidset satisfy only the given boolean expressions. In this view, both AND-clauses and OR-clauses, by definition, convey different kinds of information. For example, some tidsets can only be characterized by an AND-clause, and while others may only be characterized by an OR-clause. For some others, the simpler AND- and OR-clauses are not enough, and we have to consider CNF (or DNF) expressions to completely characterize them. As such, CNF and DNF are logically equivalent, in the sense that any CNF expression can be transformed into a DNF expression and vice-versa. However, sometimes such a transformation can lead to an exponential growth in the number of clauses. For example, the DNF expression with *n* clauses, $(a_1\ b_1) \mid (a_2\ b_2) \mid \cdots \mid (a_n\ b_n)$, yields a CNF expression with 2^n clauses. For instance, $(a_1\ b_1) \mid (a_2\ b_2) = (a_1 \mid a_2)(a_1 \mid b_2)(b_1 \mid a_2)(b_1 \mid b_2)$. For the sake of simplicity, we may desire to retain either the DNF or CNF expression for the analyst. Furthermore, note that for different *minsup* and *maxsup* settings, due to support based pruning, the two sets of expressions from CNF or DNF mining may be different. For these reasons, we explicitly consider both CNF and DNF expression in this paper.

Before presenting the BLOSOM framework, we will first study the structure and properties of four classes of boolean expressions; we consider each case separately -- AND-clauses, OR-clauses, CNF and DNF. For simplicity of exposition, we will restrict our examples to only positive literals, but our approach is applicable to negated literals as well. In essence, we treat each negated literal as a separate item. We eliminate simple tautologies (contradictions) in OR (AND)-clauses by eliminating any clause containing both literals i and $\overline{i}$. However, since our current approach does not check for tautologies/contradictions in more complex expressions, it can find logically redundant or contradictory patterns. Integrating full tautology/contradiction checking is part of future work.

RELATED WORK

Mining frequent itemsets (i.e., pure conjunctions) has been extensively studied within the context of itemset mining (Agrawal et al., 1996). The closure operator for itemsets (AND-clauses) was proposed in (Ganter & Wille, 1999), and the notion of minimal generators for itemsets was introduced in (Bastide et al., 2000). Many algorithms for mining closed itemsets (see Goethals & Zaki, 2003), and a few to mine minimal generators (Bastide et al., 2000; Zaki & Ramakrishnan, 2005; Dong et al., 2005) have also been proposed in the past. The work in (Dong et al., 2005) focuses on finding the succinct (or essential) minimal generators for itemsets. CHARM-L (Zaki & Ramakrishnan, 2005) finds the minimal generators for itemsets (MA), based on the differential lower shadows of closed itemsets. The task of mining closed and minimal monotone DNF expressions was proposed in (Shima et al., 2004). It gives a direct definition of the closed and minimal DNF expressions (i.e., a closed expression is one that doesn't have a superset with the same support and a minimal expression is one that doesn't have a subset with the same support). The authors further give a level-wise Apriori-style algorithm. In contrast, the novel contribution of our work is the structural characterization of the different classes of boolean expressions via the use of closure operators and minimal generators, as well as the framework for mining arbitrary expressions.

Within the association rule context, there has been previous work on mining negative rules (Savasere et al., 1998; Yuan et al., 2002; Wu et al., 2004; Antonie & Zaiane, 2004), as well as disjunctive rules (Nanavati et al., 2001). Unlike these methods we are interested in characterizing such rules within the general framework of boolean expression mining. However, as pointed out above, our current framework relies on a relatively straightforward approach to handle negated literals. Using more sophisticated methods, as suggested by some of the existing works on negative rules, can speed up the computation time. One can also use approaches that approximate the support of arbitrary boolean expressions (Calders & Goethals, 2005; Jaroszewicz & Simovici, 2002; Mannila & Toivonen, 1996) to deliver further performance gains.

Also related is the mining of optimal rules according to some constraints constraints (Bayardo & Agrawal, 1999), since the boolean expressions can be considered as constraints on the patterns. More general notions of itemsets (including negated items and disjunctions) have been considered in the context of concise representations (Calders & Goethals, 2003; Kryszkiewicz, 2001; Kryszkiewicz, 2005). Another point of comparison is w.r.t. the work in (Gunopulos et al., 2003) where the authors aim to find frequent and (maximally) interesting sentences w.r.t. a variety of criteria. Many data mining tasks, including inferring boolean functions, are instantiations of this problem. Also of relevance is the task of mining redescriptions. The CARTwheels algorithm (Ramakrishnan et al., 2004) mines redescriptions only between length-limited boolean expressions in disjunctive normal form and CHARM-L (Zaki & Ramakrishnan, 2005) is restricted to redescriptions between conjunctions. None of these algorithms can mine redescriptions between *arbitrary* boolean expressions, as done here.

The theoretical machine learning (PAC) community has focused on learning Boolean expressions (Bshouty, 1995) in the presence of membership queries and equivalence queries. Mitchell (Mitchell, 1982) proposed the concept of version spaces (which are basically a partial order over expressions) to organize the search for expressions consistent with a given set of data. However, these works conform to the classical supervised learning scenario where both positive and negative examples of the unknown function are supplied. In contrast, our work aims to find boolean expressions without explicit direction about the examples they cover. We note that a preliminary version of this paper appeared in Zhao et al. (2006).

Characterizing AND- and OR-Clauses

As already noted, mining AND-clauses, has been widely studied ever since frequent itemsets (i.e., pure conjunctions) were introduced (Agrawal et al., 1996). Mining of OR-clauses, on the other hand has not received that much attention. Note that OR and AND clauses are *duals* of each other, in the sense that $Y = (l_1 l_2 \ldots l_k)$ is an AND-clause in $\mathcal{D}$ iff $\bar{Y} = (\overline{l_1} \mid \overline{l_2} \mid \cdots \mid \overline{l_k})$ is an OR-clause in $\overline{\mathcal{D}}$. This duality between OR- and AND-clauses allows one to transform each problem into the other for the purposes of mining. Since the theory of minimal and closed AND-clauses is well developed, we include only a brief overview here, for the sake of completeness.

Characterizing AND-Clauses

Closed AND-clauses have been well studied in data mining as closed itemsets (Pasquier et al., 1999), as well as in the Formal Concept Analysis as concepts (Ganter & Wille, 1999). The notion of minimal generators for AND-clauses has also been previously proposed in (Bastide et al., 2000). Many algorithms for mining closed AND-clauses (e.g., Charm; Zaki & Hsiao, 2005) and others (Goethals & Zaki, 2003), and a few to mine minimal clauses (Bastide et al., 2000; Zaki & Ramakrishnan, 2005), have been proposed.

Given dataset $\mathcal{D}$, and thresholds *minsup* and *maxsup*, the goal here is to mine AND-clauses that occur in at least *minsup* and in at most *maxsup* transactions. Let $\mathcal{E}^{\wedge}$ be the set of all AND-clauses over the set of items $\mathcal{I}$. Given posets $(\mathcal{E}^{\wedge}, \subseteq)$ and $(2^{\mathcal{T}}, \subseteq)$, and $X \in \mathcal{E}^{\wedge}, Y \in 2^{\mathcal{T}}$, the following two mappings form an anti-monotone Galois connection (Ganter & Wille, 1999):

$$\phi = \mathbf{t} : \mathcal{E}^{\wedge} \mapsto 2^{\mathcal{T}}, \quad \mathbf{t}(X) = \{t \in \mathcal{T} \# \ \ t \text{ satisfies } X\}$$

$$\psi = \mathbf{i} : 2^{\mathcal{T}} \mapsto \mathcal{E}^{\wedge}, \quad \mathbf{i}(Y) = \{i \in \mathcal{I} \# \ \ Y \subseteq \mathbf{t}(i)\}$$

Since $(\mathbf{t}, \mathbf{i})$ forms an anti-monotone Galois connection, it follows that $\mathbb{C} = \mathbf{i}^{\circ}\mathbf{t} : \mathcal{E}^{\wedge} \to \mathcal{E}^{\wedge}$, and $\mathbf{t}^{\circ}\mathbf{i} : 2^{\mathcal{T}} \to 2^{\mathcal{T}}$ both form a closure operator for AND-clauses (Ganter & Wille, 1999). An AND-clause *X* is called closed if $\mathbb{C}(X) = X$. *Y* is called a *minimal generator* for a closed AND-clause *X* iff *Y* is a minimal subset of *X* such that $\mathbf{t}(Y) = \mathbf{t}(X)$. We use the notation $\mathbb{C}(\mathcal{E}^{\wedge})$ and $\mathcal{G}(\mathcal{E}^{\wedge})$ to denote the set of all closed and minimal generators of AND-clauses, respectively. Consider our example dataset (Figure 1) for which Figure 2 shows the set of all closed AND-clauses (in rectangles) and their minimal generators (in circles), as well as the groupings of the generators for the AND-clauses. The tidset for each clause is also shown. As in the case of OR-clauses, it is clear that the closed and minimal AND-clauses are the most specific and most general clauses relating to a group of tidsets.

In terms of the structure of minimal generators for a closed AND-clause *X*, they can be considered as the minimal hitting sets of its differential lower shadow (Pfaltz & Jamison, 2001). Furthermore, closed AND-clauses have also been characterized as those resulting from finite intersections of transactions (Mielikainen, 2003). We focus instead on a direct characterization of the closed and minimal generators for AND-clauses. We define an *intersection tidset* to be a tidset obtained by finite intersections over the set of tidsets for single items, $\{\mathbf{t}(i) \mid i \in \mathcal{I}\}$. It is easy to see that every distinct tidset *T* obtained as a finite intersection of other tidsets is closed, with an associated closed AND-clause, and its minimal generators. Table 1 lists the set of all closed (CA) AND-clauses, the minimal generators (MA), and the corresponding tidsets.

Characterizing OR-Clauses

Given dataset $\mathcal{D}$, and thresholds *minsup* and *maxsup*, the goal here is to mine OR-clauses that occur in at least *minsup* and in at most *maxsup* transactions. Let $\mathcal{E}^{\vee}$ be the set of all possible OR-clauses over the set of items $\mathcal{I}$; $\mathcal{T}$ is the set of all tids as before. For an OR-clause $X \in \mathcal{E}^{\vee}$, let $\mathcal{L}(X) = \{l \mid l \text{ is a literal in } X\}$ denote the set of its literals. Given $X, Y \in \mathcal{E}^{\vee}$, we define the relation $\subseteq$ between OR-clauses as follows: $X \subseteq Y$ iff $\mathcal{L}(X) \subseteq \mathcal{L}(Y)$. Then $\subseteq$ induces a partial order over $\mathcal{E}^{\vee}$. For example, $A \mid C \subseteq A \mid B \mid C$, since $\mathcal{L}(A \mid C) \subseteq \mathcal{L}(A \mid B \mid C)$. Let $X \in \mathcal{E}^{\vee}$ be an OR-clause, and let $l \in X$ be some literal in *X*. Then $\mathbf{t}(X) = \bigcup_{l \in X} \mathbf{t}(l)$. For example,

$$\mathbf{t}(A \mid B) = \mathbf{t}(A) \cup \mathbf{t}(B) = 134 \cup 23 = 1234.$$

Figure 2. The Lattice of AND-Clauses (minsup=1, maxsup=5)

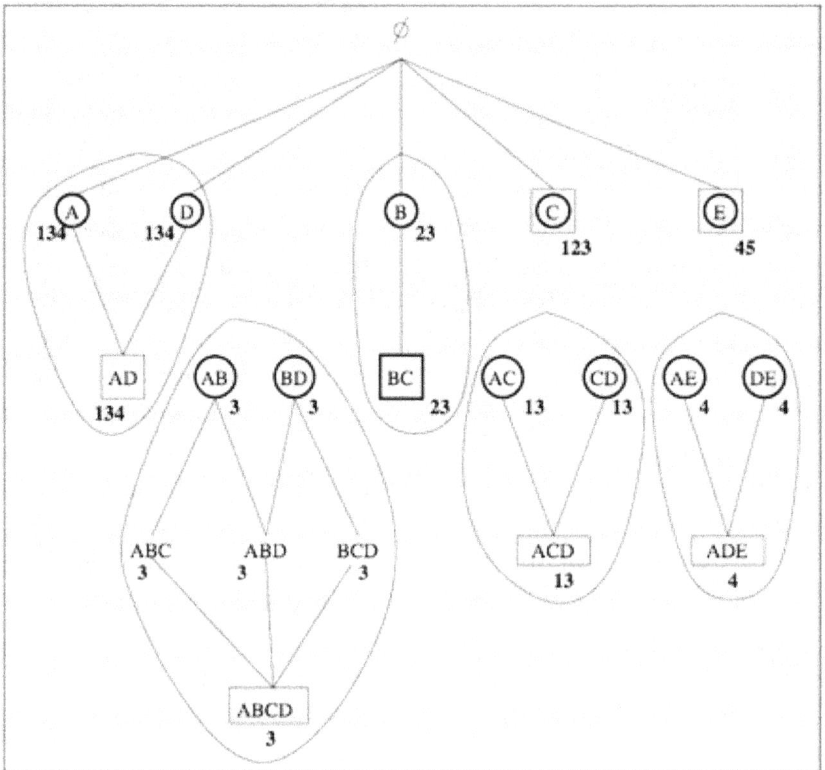

Table 1. Closed (CA) and minimal generators (MA) for AND-clauses

Tidset	Closed	Min Generators
3	ABCD	AB, BD
4	ADE	AE, DE
13	ACD	AC, CD
23	BC	B
45	E	E
123	C	C
134	AD	A, D

For convenience we write the OR-clause $l_1 \mid l_2 \mid \cdots \mid l_k$ as $\vee\{l_1 l_2 \cdots l_k\}$. Also for OR-clauses, define $\mathbf{t}(\varnothing) = \varnothing$.

Let $(\mathcal{E}^{\vee}, \subseteq)$ be the partial order over OR-clauses, and let $(2^{\mathcal{T}}, \subseteq)$ be the partial order over the tidsets under the usual subset ($\subseteq$) relationship.

Theorem 1: Given posets $(\mathcal{E}^{\vee}, \subseteq)$ and $(2^{\mathcal{T}}, \subseteq)$. Let $X \in \mathcal{E}^{\vee}$ and $Y \in 2^{\mathcal{T}}$. Then the following two mappings form a monotone Galois connection over $\mathcal{E}^{\vee}$ and $2^{\mathcal{T}}$:

$$\phi = \mathbf{t} : \mathcal{E}^{\vee} \mapsto 2^{\mathcal{T}}, \quad \mathbf{t}(X) = \{t \in \mathcal{T}\# \ \ t \text{ satisfies } X\}$$

$$\psi = \mathbf{i} : 2^{\mathcal{T}} \mapsto \mathcal{E}^{\vee}, \quad \mathbf{i}(Y) = \vee\{i \in \mathcal{I}\# \ \ \mathbf{t}(i) \subseteq Y\}$$

Proof: We have to show $X \subseteq \mathbf{i}(Y) \Leftrightarrow \mathbf{t}(X) \subseteq Y$

a) First we prove $X \subseteq \mathbf{i}(Y) \Rightarrow \mathbf{t}(X) \subseteq Y$. Observe that $X \subseteq \mathbf{i}(Y) \Rightarrow \mathbf{t}(X) \subseteq \mathbf{t}(\mathbf{i}(Y))$. However, by definition of $\mathbf{i}$, $\forall i \in \mathbf{i}(Y)$ we have $\mathbf{t}(i) \subseteq Y$. This means that $\mathbf{t}(\mathbf{i}(Y)) = \bigcup_{I \in \mathbf{i}(Y)} \mathbf{t}(i) \subseteq Y$. Thus $\mathbf{t}(X) \subseteq Y$.

b) We now prove $\mathbf{t}(X) \subseteq Y \Rightarrow X \subseteq \mathbf{i}(Y)$. Assume that $X \not\subseteq \mathbf{i}(Y)$. This implies that $\exists i \in X$, such that $I \notin \mathbf{i}(Y)$. By definition of $\mathbf{i}$, this means that $\mathbf{t}(i) \not\subseteq Y$. But, this is a contradiction since $i \in X \Rightarrow \mathbf{t}(i) \subseteq Y$. Thus $X \subseteq \mathbf{i}(Y)$.

Since $(\mathbf{t}, \mathbf{i})$ forms a monotone Galois connection, it follows immediately that $\mathbb{C} = \mathbf{i}^\circ \mathbf{t} : \mathcal{E}^\vee \rightarrow \mathcal{E}^\vee$ is a closure operator and $\mathbb{K} = \mathbf{t}^\circ \mathbf{i} : 2^T \rightarrow 2^T$ is a kernel operator for OR-clauses. For example, in our example dataset from Figure 1, $\mathbb{C}(A \mid C) = \mathbf{i}(\mathbf{t}(A \mid C)) = \mathbf{i}(1234) = A \mid B \mid C \mid D$. Thus $A|B|C|D$ is a closed OR-clause. On the other hand $\mathbb{K}(234) = \mathbf{t}(\mathbf{i}(234)) = \mathbf{t}(B) = 23$. Thus *23* is an open tidset. It is also easy to see that the corresponding tidset for a closed OR-clause is always open. Figure 3 shows all the closed OR-clauses (enclosed in boxes) and their corresponding open tidsets obtained from our example dataset. For example, the closed OR-clause $A|D$ has the open tidset *134*. We use the notation $\mathbb{C}(\mathcal{E}^\vee)$ to denote the set of all closed OR-clauses. The set of all minimal generators of closed OR-clauses in $\mathcal{E}^\vee$ is given as $\mathcal{G}(\mathcal{E}^\vee) = \bigcup_{X \in \mathbb{C}(\mathcal{E}^\vee)} \mathcal{G}(X)$. Recall that a closed OR-clause X is the *unique* maximal OR-clause that describes a given set of objects $T = \mathbf{t}(X)$. On the other hand, a minimal generator of X is the minimal or simplest OR-clause that still describes the same object set T. Figure 3 shows the groups of all generators of closed OR-clauses in our example dataset (from Figure 1). Within each group the unique maximal element is the closed clause (enclosed in a rectangle). The minimal elements of each group are the minimal generators (enclosed in dark circles). One can observe that the tidsets are the same for each member of a group, but different across groups. All the minimal generators and closed OR-clauses, and their corresponding tidsets, for our example database (from Figure 1), are summarized in Table 2.

Both closed OR-clauses (CO) and minimal OR-clauses (MO), along with their tidsets, individually serve as a lossless representation of the set of all possible (frequent) OR-clauses. We are particularly interested in minimal OR-clauses, since they represent the most general expressions that characterize the corresponding tidsets, and as such may be easier to comprehend. We would like to gain further insight into the structure of these minimal generators.

We give two separate characterizations of the minimal generators, one based on the closed clauses and the other a direct one. They are both based on the notion of hitting sets. Let $\mathbf{X} = \{X_1, X_2, \cdots, X_k\}$ be a set of subsets over some universe *U*. The set $Z \subseteq U$ is called a *hitting set* (or *transversal*) of $\mathbf{X}$ iff $Z \cap X_i \neq \varnothing$ for all $i \in [1, k]$. Let $\mathcal{H}(\mathbf{X})$ denote the set of all hitting sets of $\mathbf{X}$. *Z* is called a *minimal hitting set* if there does not exist another hitting set Z', such that $Z' \subset Z$. Let *X* be a closed OR-clause, we define the *lower-shadow* of X as the set $\mathbf{X}^\ell = \{X_i\}_{i \in [1,k]}$ where for all $i \in [1, k]$, $X_i \subset X$, X_i is closed, and there doesn't exist any other closed OR-clause *Y*, such that $X_i \subset Y \subset X$. In other words, the lower shadow of *X* is the set of closed OR-clauses that are immediate subsets of *X*. We further define the *differential lower-shadow* of X as the set $\mathbf{X}^\delta = \{X - X_i\}_{i \in [1,k]}$, where $X_i \in \mathbf{X}^\ell$.

Using the duality between AND- and OR-clauses, and the characterization of minimal AND-clauses (Pfaltz & Jamison, 2001), it is not hard to show that the minimal generators of a closed OR-clause *X* are exactly the minimal hitting sets of the differential lower shadow of *X*. For example, consider the closed clause $X = A \mid B \mid C \mid D \mid E$. From Figure 3 we obtain as its lower shadow the set of closed clauses $\mathbf{X}^\ell = \{A \mid B \mid C \mid D, A \mid D \mid E, B \mid E\}$. Thus the differential lower shadow of X is given as $\mathbf{X}^\delta = \{E, B \mid C, A \mid C \mid D\}$. The minimal hitting sets of $\mathbf{X}^\delta$ are given as $\min_{\subseteq}\{\mathcal{H}(\mathbf{X}^\delta)\} = \{C \mid E, A \mid B \mid E, B \mid D \mid E\}$.

Figure 3. The Lattice of OR-clauses (minsup=1, maxsup=5)

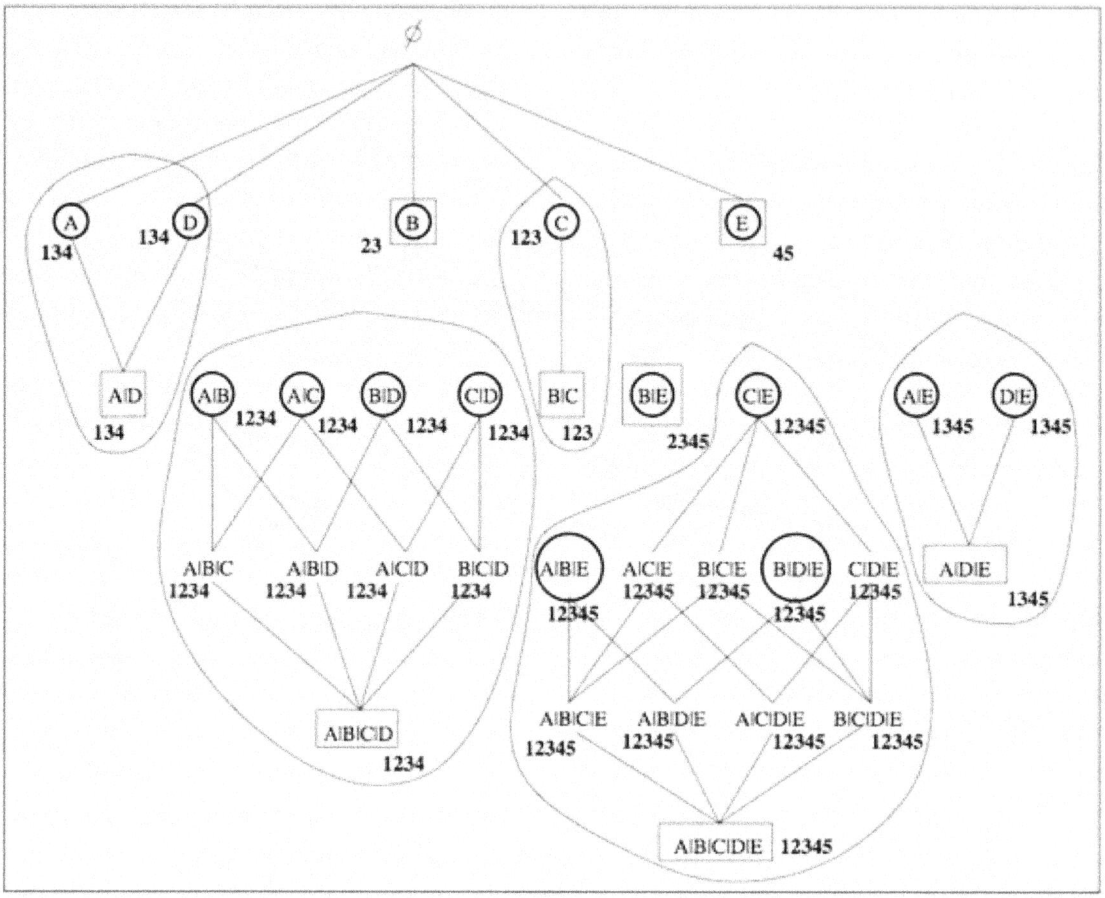

We can see from Table 2, that the minimal hitting sets are identical to the minimal generators of $A|B|C|D|E$.

The above characterization of minimal generators relies on knowing the closed sets and their lower shadows. There is in fact a direct structural description of the closed and minimal OR-clauses. We define a *union tidset* to be a tidset obtained by finite unions over the set of tidsets for single items, $\{\mathbf{t}(i) \mid i \in \mathcal{I}\}$. Let $\mathcal{U}$ be the set of all distinct union tidsets. For a union tidset $T \in \mathcal{U}$, we define the *transaction set* of *T*, as the set $\mathcal{R}(T) = \{t.X \mid t \in T, (t, t.X) \in \mathcal{D}\}$, i.e., the set of transactions (i.e., itemsets) in $\mathcal{D}$ with tids $t \in T$. One can show that every distinct tidset *T* obtained as a union of other tidsets produces minimal generators for the closed OR-clause associated with the open tidset *T*. For example, let $T = 1345 = \mathbf{t}(A) \cup \mathbf{t}(E)$ in our example database in Figure 1. Then

Table 2. Closed (CO) and Minimal (MO) OR-Clauses

Tidset	Closed	Min Generators
23	B	B
45	E	E
123	B/C	C
134	A/D	A,D
1234	A/B/C/D	A/B, A/C, B/D, C/D
1345	A/D/E	A/E, D/E
2345	B/E	B/E
12345	A/B/C/D/E	A/B/E, B/D/E, C/E

$\mathcal{R}(T) = \{ACD, ABCD, ADE, E\}$.

The hitting sets Z with $\mathbf{t}(Z) = T$ and that are minimal are given as follows $\{A \mid E, D \mid E\}$. Note that $C|E$ is a minimal hitting set of $\mathcal{R}(T)$, but $\mathbf{t}(C \mid E) = 12345 \neq T$, thus we reject it. We can see from Table 2 that these minimal hitting sets form the minimal generators of the closed OR-clause $A|D|E$ with tidset $T = 1345$.

Characterizing Normal Forms

Our approach for DNF and CNF mining builds upon the pure OR- and AND-clauses. Here we give structural characterizations for the minimal DNF and CNF expressions.

Characterizing DNF Expressions

Let $\mathcal{E}^{dnf}$ denote the set of all boolean expressions in DNF, i.e., each $X \in \mathcal{E}^{dnf}$ is an OR of AND-clauses. For convenience we denote a DNF-expression X as $X = \vee X_i$, where each X_i is an AND-clause. By definition $\mathcal{E}^{\wedge} \subseteq \mathcal{E}^{dnf}$. Also $\mathcal{E}^{\vee} \subseteq \mathcal{E}^{dnf}$, since an OR-clause is a DNF-expression over single literal (AND) clauses. We assume we have already computed the closed $(\mathbb{C}(\mathcal{E}^{\wedge}))$ and minimal $(\mathcal{G}(\mathcal{E}^{\wedge}))$ AND-clauses, and their corresponding tidsets.

Note that any DNF-expression $X = \vee X_i$ is equivalent to the DNF expression $X' = \vee \mathbb{C}(X_i)$, since any tidset that satisfies X_i must satisfy $\mathbb{C}(X_i)$ as well. Similarly X is also equivalent to the DNF expression $X'' = \vee \mathcal{G}(X_i)$, since $\mathbf{t}(X_i) = \mathbf{t}(\mathcal{G}(X_i))$. We say that $X = \vee X_i$ is a *min-DNF-expression* if for each AND-clause X_i there does not exist another X_j $(= (i \neq j)$ such that $X_i \subseteq X_j$. Note that any DNF-expression can easily be made a min-DNF-expression by simply deleting the offending clauses. For example in the DNF-expression $(AD) \mid (ADE)$, we have $AD \subseteq ADE$; thus the expression is logically equivalent to its min-DNF form (AD). For any DNF-expression X, we use the notation $\min^{dnf}(X)$ to denote its min-DNF form. Given $X, Y \in \mathcal{E}^{dnf}$, with $X = \vee X_i$ and $Y = \vee Y_i$, we say that X is more general than Y, denoted $X \subseteq Y$, if there exists an injective mapping f that maps each $X_i \in X$ to $f(X_i) = Y_j \in Y$, such that $X_i \subseteq Y_j$.

Closed DNF: We now define a closure operator for DNF expressions. First, we consider DNF expressions consisting only of closed AND-clauses. Then if we treat each $X_i \in \mathbb{C}(\mathcal{E}^{\wedge})$ as a *composite item*, we can define two monotone mappings that form a monotone Galois connection as follows: Let $X = \vee X_i$ be a DNF expression, such that $X_i \in \mathbb{C}(\mathcal{E}^{\wedge})$, and let $Y \in 2^{\mathcal{T}}$. Define $\mathbf{t}(X) = \{t \in \mathcal{T} \# \ t \text{ satisfies } X\}$, and $\mathbf{i}(Y) = \vee\{X_i \# \ X_i \in \mathbb{C}(\mathcal{E}^{\wedge}) \text{ and } \mathbf{t}(X_i) \subseteq Y\}$. This implies that $\mathbb{C} = \mathbf{i}^{\circ}\mathbf{t}$ is a closure operator. For example, consider each closed AND-clause in Table 1 as an "item". Consider $X = AD \mid E$.
$\mathbb{C}(X) = \mathbf{i}(\mathbf{t}(AD \mid E)) = \mathbf{i}(1345) = ABCD \mid ADE \mid ACD \mid E \mid AD$, which is a closed DNF expression. However it is logically redundant. What we want is the maximal min-DNF expression equivalent to $\mathbb{C}(X)$, which is $ACD|E$. That is, among all the min-DNF expressions (i.e., with no clause a subset of another) with tidset equal to $\mathbf{t}(X)$, the maximum one is defined as the closure.

Minimal DNF: Let $T \in 2^{\mathcal{T}}$ be a union tidset obtained as the finite union of tidsets of closed AND-clauses. As before the transaction set of T, denoted $\mathcal{R}(T)$, is the set of transactions in $\mathcal{D}$ with tid $t \in T$. We can characterize the minimal DNF expressions as the set

$$\min_{\subseteq}\{Z \in \mathcal{E}^{dnf} \# \ Z \in \mathcal{H}(\mathcal{R}(T)) \text{ and } \nexists T' \supset T \text{ such that } Z \in \mathcal{H}(\mathcal{R}(T'))\}$$

which is the set of all minimal hitting sets of $\mathcal{R}(T)$ having the tidset T. For example,

Table 3. Additional/changed Closed (CD) and Minimal Generators (MD)

Tidset	Closed (maximal min-DNF)	Min Generators
34	(ABCD)/(ADE)	(AB)/(AE), (AB)/(DE), (BD)/(AE), (BD)/(DE)
123	(ACD)/(BC)	C
134	(ACD)/(ADE)	A, D
234	(ADE)/(BC)	B/(AE), B/(DE)
345	(ABCD)/E	(AB)/E, (BD)/E
1234	(ACD)/(ADE)/(BC)	A/B, A/C, B/D, C/D
1345	(ACD)/E	A/E, D/E
2345	(BC)/E	B/E
12345	(ACD)/(BC)/E	A/B/E, B/D/E, C/E

consider the union tidset $T = 34$ (which is the union of $\mathbf{t}(ABCD)$ and $\mathbf{t}(ADE)$. The minimal DNF hitting sets that hit only the tidset *34* are *AB|AE, AB|DE, BD|AE, BD|DE*. In fact minimal hitting sets can be obtained directly from minimal AND-clauses.

Lemma 2: Let *T* be a union tidset, and let *X* be the closed DNF-expression with $\mathbf{t}(X) = T$. Then

$\mathcal{G}(X) = \min_{\subseteq}\{Z = \vee Z_i \mid Z_i \in \mathcal{G}(\mathcal{E}^{\wedge})\ and\ \mathbf{t}(Z) = T\}$.

Proof: Let $G = \vee_i G_i \in \mathcal{G}(X)$. Then $\mathbf{t}(G) = T$. Further, each $G_i \in \mathcal{G}(\mathcal{E}^{\wedge})$, since otherwise *G* would not be minimal. Finally minimality of *G* implies it is a minimal DNF expression satisfying the right hand side. For the opposite direction, let $Z = \vee Z_i \in \mathcal{G}(\mathcal{E}^{\wedge})$, be a minimal expression with $\mathbf{t}(Z) = T$. But this means *Z* is a minimal generator of *X*.

For example, for $T = 34$, we see in Table 1 that $\{AB, BD\}$ are the minimal generators with tidset 3, and $\{AE, DE\}$ have tidset 4. Taking the minimal OR expressions obtained from these two sets, we get all the minimal generators having tidset $T = 34$, namely *AB|AE, AB|DE, BD|AE, BD|DE*. Table 3 shows the closed and minimal DNF expressions, in addition to those shown in Table 1 and Table 2. Some entries are repeated since the closed expressions in DNF have changed. Also the new union tidsets are marked in bold. These new tidsets represent expressions that cannot be generated using pure clauses; they require the full power of a DNF expression to characterize them.

Characterizing CNF Expressions

Let $\mathcal{E}^{cnf}$ denote the set of all boolean expressions in CNF, i.e., each $X \in \mathcal{E}^{cnf}$ is an AND of OR-clauses. By definition $\mathcal{E}^{\vee} \subseteq \mathcal{E}^{cnf}$ and $\mathcal{E}^{\wedge} \subseteq \mathcal{E}^{cnf}$. For convenience we denote a CNF-expression *X* as $X = \wedge X_i$, where each X_i is an OR-clause. We say that *X* is a *min-CNF-expression* if for each OR-clause X_i there does not exist another X_j $(i \neq j)$ such that $X_i \subseteq X_j$. For a CNF expression *X*, we use the notation $\min^{cnf}(X)$ to denote its min-CNF form. Let $\mathcal{E}^{cnf}$ denote the set of all min-CNF-expressions. Given $X, Y \in \mathcal{E}^{cnf}$, with $X = \wedge X_i$ and $Y = \wedge Y_i$, we say that *X* is more general than *Y*, denoted $X \subseteq Y$, if there exists an injective *f* that maps each $X_i \in X$ to $f(X_i) = Y_j \in Y$, such that $X_i \subseteq Y_j$. Analogously to DNF expressions, we can define the closed and minimal CNF expressions directly from the set of all closed $(\mathbb{C}(\mathcal{E}^{\vee}))$ and minimal $(\mathcal{G}(\mathcal{E}^{\vee}))$ OR-clauses, and their corresponding tidsets.

Let us treat each $X_i \in \mathbb{C}(\mathcal{E}^{\vee})$ as a *composite item*; we can then define two anti-monotone mappings that form an anti-monotone Galois connection as follows: Let $X = \wedge X_i$ be a CNF expression, such that $X_i \in \mathbb{C}(\mathcal{E}^{\vee})$, and let $Y \in 2^{T}$.

Table 4. Additional/changed Closed (CC) and Minimal Generators (MC) for CNF

Tidset	Closed (maximal min-CNF)	Min Generators
3	B(A/D)	AB, BD
13	(B/C)(A/D/E)	AC, CD
34	(A/D)(B/E)	A(B/E), D(B/E)
234	(A/B/C/D)(B/E)	(A/B)(B/E), (A/C)(B/E), (B/D)(B/E), (C/D)(B/E)
345	(A/D/E)(B/E)	(A/E)(B/E), (D/E)(B/E)

Define $\mathbf{t}(X) = \{t \in \mathcal{T} \# \ t \text{ satisfies } X\}$, and $\mathbf{i}(Y) = \wedge\{X_i \# \ X_i \in \mathbb{C}(\mathcal{E}^{\vee}) \text{ and } Y \subseteq \mathbf{t}(X)\}$. This implies that $\mathbb{C} = \mathbf{i}^{\circ}\mathbf{t}$ is a closure operator. For example, consider each closed OR-clause in Table 2 as an "item". Consider $X = (A \mid D)(B \mid C)$. $\mathbb{C}(X) = \mathbf{i}(\mathbf{t}((A \mid D)(B \mid C))) = \mathbf{i}(13) = (B \mid C)(A \mid D)(A \mid B \mid C \mid D)(A \mid D \mid E)(A \mid B \mid C \mid D \mid E)$, which is closed. However it is logically redundant. What we want is the maximal min-CNF expression equivalent to $\mathbb{C}(X)$, which is *(A|D|E)(B|C)*. That is among all the min-CNF expressions with tidset equal to $\mathbf{t}(X)$ the maximum one is defined as the closure.

Lemma 3: Let $T \in 2^{\mathcal{T}}$ be an intersection tidset obtained as the finite intersection of tidsets of closed OR-clauses. Let *X* be the closed CNF-expression with $\mathbf{t}(X) = T$. Then $\mathcal{G}(X) = \min_{\subseteq}\{Z = \wedge Z_i \mid Z_i \in \mathcal{G}(\mathcal{E}^{\vee}) \text{ and } \mathbf{t}(Z) = T\}$.

Proof: Let $G = \wedge_i G_i \in \mathcal{G}(X)$. Then $\mathbf{t}(G) = T$. Further, each $G_i \in \mathcal{G}(\mathcal{E}^{\vee})$, since otherwise *G* would not be minimal. Finally minimality of *G* implies it is a minimal CNF expression satisfying the right hand side. In the other direction, let $Z = \wedge Z_i \in \mathcal{G}(\mathcal{E}^{\vee})$, be a minimal expression with $\mathbf{t}(Z) = T$. But this means *Z* is a minimal generator of *X*.

For example, let $T = 13$. We can obtain *13* as the intersection of several minimal OR-clauses' tidsets, e.g., the minimal OR-clauses *C* and $\{A \mid E, D \mid E\}$. However the minimal among all of these are *C* and $\{A, D\}$, giving the two minimal CNF expressions: *AC* and *CD*. Table 4 shows the closed CNF expressions and their minimal generators in addition to those already shown in Table 1 and Table 2, or those that have changed. The bold tidsets represent expressions that cannot be generated using pure clauses; they require the full power of a CNF expression to characterize them.

Table 5. Algorithms in the BLOSOM Framework

Algorithm	**Mining Task**
BLOSOM-MO	Minimal OR-clauses
BLOSOM-MA	Minimal AND-clauses
BLOSOM-MD	Minimal DNF expressions
BLOSOM-MC	Minimal CNF expressions
BLOSOM-CO	Closed OR-clauses
BLOSOM-CA	Closed AND-clauses
BLOSOM-CD	Closed DNF expressions
BLOSOM-CC	Closed CNF expressions

THE BLOSOM FRAMEWORK

The BLOSOM framework for mining arbitrary boolean expressions supports several different algorithms, as listed in Table 5. Our main focus in on efficiently mining the minimal boolean expressions due to their inherent simplicity. We do propose algorithms for mining closed clauses, which can easily be extended to mine closed normal forms.

BLOSOM assumes that the input dataset is $\mathcal{D}$, and it then transforms it to work with the transposed dataset $\mathcal{D}^T$. Starting with the single items (literals) and their tidsets, BLOSOM performs a depth-first search (DFS) extending an existing expression by one more "item". BLOSOM employs a number of effective pruning techniques for searching over the space of boolean expressions, yielding orders of magnitude in speedup. These include: dynamic sibling reordering, parent-child pruning, sibling merging, threshold pruning, and fast subsumption checking. Further BLOSOM utilizes a novel *extraset* data structure for fast frequency computations, and to identify the corresponding transaction set for a given arbitrary Boolean expression.

BLOSOM-MO: Minimal OR-Clauses

BLOSOM-MO mines all the minimal OR-generators, and is broadly based on CHARM (Zaki & Hsiao, 2005). However, CHARM mines only the closed AND-clauses, whereas BLOSOM-MO mines the minimal OR-clauses. BLOSOM-MO takes as input the set of parameter values *minsup*, *maxsup*, *maxitem* and a dataset $\mathcal{D}$ (we implicitly convert it to $\mathcal{D}^T$). The *maxitem* constraint is used to limit the maximum size of any boolean expression, if desired. BLOSOM-MO conceptually utilizes a DFS tree to search over the OR-clauses. Each OR-clause is stored as a set of items (the OR is implicitly assumed). Thus each node of the search tree is a pair of $(T \times I)$, where *I* is an item set denoting an OR-clause and *T* is a tidset (as shown in Figure 5). In the description below, we use MO to denote a minimal generator of OR-clauses. Before describing the pseudo-code we briefly describe each of the optimizations used in the BLOSOM-MO.

Pruning and Optimization Techniques

Threshold Pruning: BLOSOM uses the three thresholds to prune the generators, based on *minsup*, *maxsup* and *maxitem*. Furthermore, if the item set *I* of the current node equals the set of all items $\mathcal{I}$, then we stop its expansion immediately, since any of its descendants will be pruned.

Dynamic Sibling Reordering: Before expanding a group of sibling nodes in the search tree, BLOSOM-MO dynamically reorders them by their tidset size dynamically. Since smaller tidsets are more likely to be contained in longer previous tidsets, this can prune out many branches.

Relationship Pruning: BLOSOM-MO makes use of two kinds of relationship pruning: parent-child and sibling based. During the DFS expansion, if the tidset of a node is the same as any of its parents', the node and all of its descendants are pruned (based on the definition of minimal generators). If some sibling nodes of the same parent node have the same tidset, they are merged together by unioning their itemsets into a "composite node". A prefix item set data structure is utilized to deal with sibling merging. All sibling nodes of the same parent share a common prefix, containing the item sets on the path from the root to the current node. The prefix also saves memory, since all the siblings do not need to keep their separate prefix copies.

Fast Subsumption Checking: BLOSOM-MO maintains a hash table for storing the current MO; the hash key of the MO is the *tidsum* (summation of tids in *T*) of its transaction set *T*. An element of the hash table is a pair $(T \times MS)$, where *T* is a tidset and *MS* is the set of *T* 's MOs. Subsumption checking on MOs guarantees that the current generators are minimal, i.e., for some transaction set *T*, any two of its generators do not contain each other. Before adding a new generator *G* we remove any of its supersets in *MS*. Due to the nature of the enumeration process, it is not necessary to check if *G* is minimal, i.e., G cannot be a superset of any MO of entry T, since otherwise G would have been pruned previously by one of its ancestor nodes using the parent-child pruning. So we only need to do the subsumption checking in one direction.

***Extraset* Technique**: BLOSOM-MO utilizes a novel *extraset* technique to save set operation time and memory usage for tidsets. The DFS expansion involves a lot of tidset unions, and each node has to keep an intermediate set, which is a superset of the parent node's tidset. So for each node, inspired by the diffset (Zaki & Hsiao, 2005) technique, we simply retain the extra-part of its parent's tidset in a data structure we call *extraset*, which can save a lotof memory while storing the intermediate tidsets. For example, in Figure 1, we have $\mathbf{t}(A) = 134$, and $\mathbf{t}(A \mid B) = 1234$. The *extraset* of $A|B$ is given as $\mathbf{t}(A \mid B) - \mathbf{t}(A) = 2$. In effect an *extraset* keeps track of the additional items added while performing unions of tidsets; as such it is the opposite of a diffset (Zaki & Hsiao, 2005), which keeps tracks of the items that drop out when performing intersections of tidsets, while mining AND-clauses. In addition, the set union operations on the original tidset are changed to the operations on the *extraset*, which also saves a lot of time, due to the small size of *extrasets*.

Transaction-Set Optimization: Sometimes one needs only the frequency of a MO, instead of the entire tidset. If tidsets are not required, we can avoid keeping large tidsets for each generator and the corresponding costly comparisons. The additional overhead is to separate those generators that share the same hash entry (if they share the same tidsum, but not the same tidset); however, since most entries have one or only a few generators for most datasets, this overhead is minimal.

Algorithmic Description

We now describe the BLOSOM-MO algorithm in detail, based on the pseudo-code in Figure 4. It is a recursive algorithm that, at each call accepts a current prefix queue P containing the I sets on the path from the root to the current node, a node list NL containing a group of sibling nodes that share the same parent, and two parameters of the current tidset T: support ($|T|$) and summation $(\sum_{t \in T} t)$. We use these instead of the original tidset T to save memory and corresponding copying cost. The initial call is made with an empty P, the NL containing all the single items, and the support and summation are set to 0. Line 1 sorts the current sibling nodes in NL in decreasing order of support, which speeds up the convergence of the algorithm (*dynamic sibling reordering*). Lines 2-4 merge the sibling nodes with the same tidset T by unioning their item sets together (*sibling merging*). Lines 7-20 form the main loop to process each of the sibling nodes one by one. Line 6 updates the current prefix queue for further recursive calls. Lines 7-8 generate the transaction set support *sup'* and summation *sum'* by adding tids from the *extraset* $n_i.T$. Line 9 checks the *maxsup* threshold. Lines 10-13 try to add the current sibling node satisfying *minsup* and *maxsup* to the hash table M. If the node satisfies *maxsup* (line 9) and *minsup* (line 10), then the algorithm will try to add every possible queue P' item combination to the current MO hash table (lines 11-13). Each combination is generated by picking one and only one item from each $P \in P'$. At the same time, we also need to do subsumption checking and to delete any supersets of the new MO I' to be added (line 12). The new MO I' must be minimal, i.e., it should not be subsumed by any current MO in the hash table (*fast subsumption checking*). Line 14 checks the *maxitem* threshold. Lines 15-22 produce node n_i's valid children nodes NL' generated with other sibling nodes ranked after n_i (in sorted order). BLOSOM-MO tries every node pair (line 16) and generates children nodes (line 17). It only saves the *extraset* $(n_i.T \cup n_j.T - n_i.T = n_j.T - n_i.T)$ to save memory and set operation time, which proves very efficient for dense datasets. If a child node is not subsumed by its parents (line 18), then it is added to the valid children node list in lines 19-20 (*parent-child relationship pruning*). Lines 21-22 make a recursive call with node n_i's valid children. After the steps showed in Figure 4, BLOSOM-MO needs to do some further work to separate those generators that have the same support and tidsum but actually have different tidsets. Since most entries of hash table *M* have only one or a few generators, the separating stage is not too costly.

An Example: Consider how BLOSOM-MO works on our example dataset $\mathcal{D}^T$ (from Figure 1). Figure 5 shows the DFS search tree. Initially the prefix queue P is empty, and node list *NL* contains five single items. After sibling merging and reordering, they become $AD \times 134, C \times 123, B \times 23$ and $E \times 45$ as shown in Figure 5. From $P' = \{A, D\}$, we get two item combinations: *A* and *D*. We thus add *A* and *D* to the

Figure 4. BLOSOM-MO algorithm

```
Input          : min_sup, max_sup, max_item and D^T
Output         : hash table M containing all MO of D^T
Initialization: NL = {t(i) × {i}|i ∈ I}, call BLOSOM-MO(∅, NL, 0, 0)

BLOSOM-MO(P, NL, sup, sum):
1    sort NL = {n_i.T × n_i} in decreasing order of |n_i.T|
2    while ∃ n_i and n_j ∈ NL such that n_i.T = n_j.T do
3        n_i.I ← n_i.I ∪ n_j.I /* sibling merging */
4        NL ← NL − n_j
5    foreach n_i ∈ NL do
6        P' ← P + n_i.I
7        sup' = sup + |n_i.T|
8        sum' = sum + summation(n_i.T)
9        If sup' > max_sup then skip to next n_i /* goto line 6 */
10       if sup' ≥ min_sup then
11           foreach combination I' of P' do
12               delete all supersets of I' in M[sup'×sum']
13               M[sup'×sum'] ← M[sup'×sum'] + I'
14       If |P'| ≥ max_item then skip to next n_i /* goto line 6 */
15       NL' ← ∅
16       foreach n_j ∈ NL ranking after n_i do
17           n'.T ← n_j.T − n_i.T /* get extraset */
18           if |n'.T| ≠ ∅ then
19               n'.I ← n_j.I
20               NL' ← NL' + n'
21       if NL' ≠ ∅ then
22           BLOSOM-MO( P', NL', sup', sum' )
```

MOs hash table M with the entry $T' = \varnothing \cup 134 = 134$, *sup*=3, and *tidsum*=1+3+4=8, as shown in Table 6. In the table the subscript numbers represent the order in which MOs are added to the table. Then we expand node $\{A, D\} \times 134$ by combining with node $C \times 123, B \times 23$ and $E \times 45$, and save the *extraset* of the tidset union for each combination pair as follows. Since $X \cup Y - X = Y - X$, the *extrasets* for the three new candidates (children of the node $\{A, D\} \times 134$) are 123-134=2, 23-134=2, and 45-134=5, respectively. After reordering and merging, *NL'* = $\{\{B, C\} \times 2, E \times 5\}$. Along with $P = \{\{A, D\}\}$ and *T'*=134, BLOSOM is called recursively. Next step from $P' = \{\{A, D\}, \{B, C\}\}$, we get four combinations: $\{A, B\}, \{A, C\}, \{B, D\}$ and $\{C, D\}$,, and we add them to the hash table with entry: $T' = 134 \cup 2 = 1234$, *sup*=4, *tidsum*=10. Readers can continue this process until all the MOs are generated as shown in Table 6 in subscript order. Note that node $B \times \varnothing$ under the path $Root \to C \to B$ is pruned using parent-child relationship because of its empty *extraset* (i.e., generator *BC* is not minimal and generates the same tidset as one of its parents). In addition, when generator *CE* is added to entry *T*=12345 of *M*, previously added MOs *ACE* and *CDE* will be deleted (as marked by underlines) since they are supersets of generator *CE*.

Figure 5. OR-clause mining example: DFS search tree

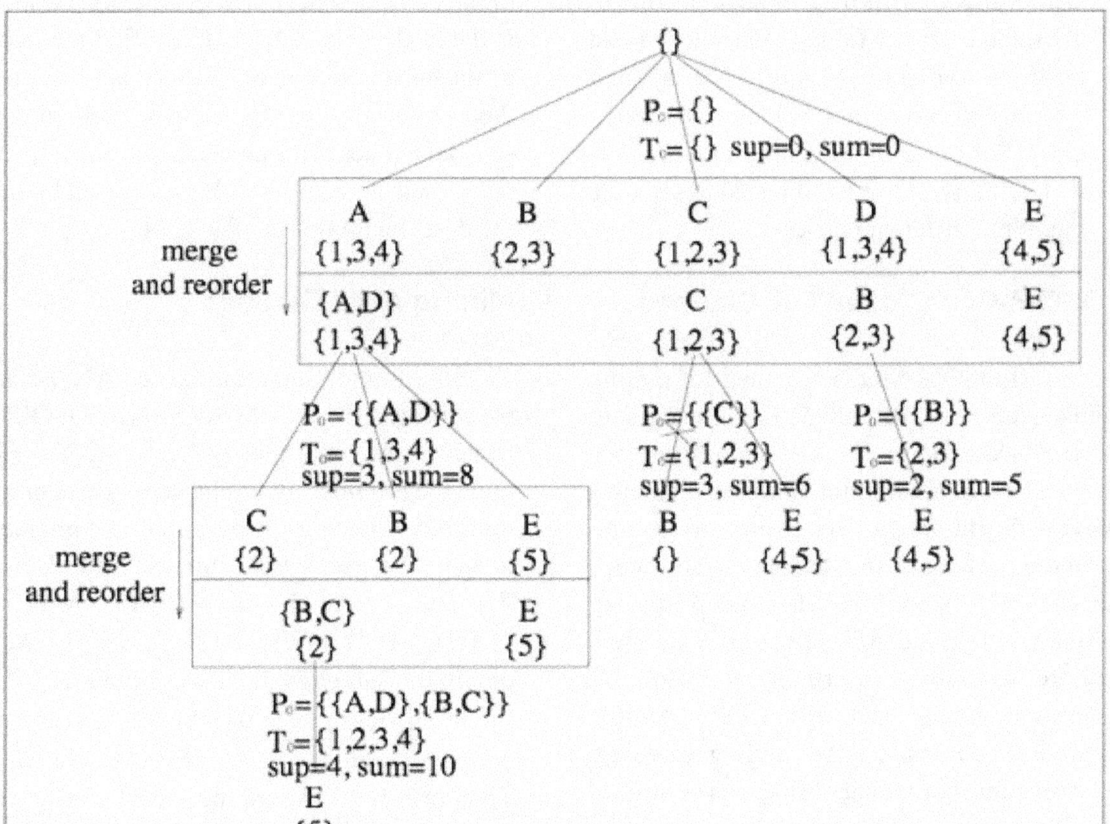

Table 6. Hash Table ${\cal M}$ for MOs

T	*sup*	*sum*	Minimal OR Generators
134	3	8	$\{A\}_1, \{D\}_1$
1234	4	10	$\{AB\}_2, \{AC\}_2, \{BD\}_2, \{CD\}_2$
12345	5	15	$\{ABE\}_3, \underline{\{ACE\}_3}, \{BDE\}_3, \underline{\{CDE\}_3}, \{CE\}_6$
1345	4	13	$\{AE\}_4, \{DE\}_4$
123	3	6	$\{C\}_5$
23	2	5	$\{B\}_7$
2345	4	14	$\{BE\}_8$
45	2	9	$\{E\}_9$

Note that in Table 6, the tidsets T are actually *not kept* during the mining process, if only support is desired. In this case we have to verify that the generators in each cell $(sup \times sum)$ of the hash table do belong to the same entry. In Table 6, this check must be done for cells $(4\times 10), (5 \times 15)$ and (4×13) to assure they belong to the same entry. For our example, all entries are correct. Otherwise, we would need to separate them into different entries.

BLOSOM-CO: Closed OR-Clauses

Whereas BLOSOM-MO is designed for mining minimal generators, BLOSOM-CO is specific to mining closed expressions. The main difference is that instead of finding the *minimal* elements, we have to find the *maximal* elements corresponding to the given tidsets. Thus the logic of subsumption checking, as well as relationship pruning (both parent-child and sibling), has to be reversed. There are two additional pruning optimizations utilized in BLOSOM-CO, which are: a) *Sibling Containment Pruning*: After sibling merging, another relationship among siblings can be utilized for CO generation, i.e., containment relationship. When expanding a node N, any sibling node ranking after N whose tidset is a subset of that of N, can be pruned in the next level of N's expansion tree, and its item set is merged with the item set of the current node. b) *Hash Pruning*: Assume M is the hash table of COs. If we find the item set of the current tree node $(T \times I)$ is subsumed by the entry of M, i.e., $I \subset M(T)$, then all the descendant nodes of the current node will be pruned, since their item sets will be subsumed by that of the descendant nodes of $T \times M(T)$.

Mining AND-Clauses

To mine the minimal and closed AND-clauses, we build upon BLOSOM-MO and BLOSOM-CO, respectively. Note that by DeMorgan's law, to mine the minimal AND-clauses, we can mine the minimal OR-clauses over the complemented tidsets. For example in our example dataset in Figure 1, with $T = 12345$, we have $\mathbf{t}(AB) = \mathbf{t}(A) \cap \mathbf{t}(B) = 134 \cap 23 = 3$. We can obtain the same results if we mine for $\overline{A} \mid \overline{B}$ in the complemented database. For example $\overline{\mathbf{t}(\overline{A}) \mid \mathbf{t}(\overline{B})} = \overline{\mathbf{t}(\overline{A}) \cup \mathbf{t}(\overline{B})} = \overline{25 \cup 145} = \overline{1245} = 3$. Thus to mine MA, we mine MO over complemented tidsets. Likewise, to mine CA, we mine

Figure 6. The BLOSOM-MD algorithm

```
Input      : min_sup ≥ min_sup', max_sup, max_item and dataset D^T
Output     : hash table M_MD containing all MD of D^T

BLOSOM-MD(min_sup, minsup', max_sup, max_item, D^T):
1   M_MA ← BLOSOM-MA(min_sup', max_sup, max_item, D^T)
2   foreach G ∈ M_MA do
3       re-generate t(G)
4   D^T_new ← CreateNewDB({G × t(G)|G ∈ M_MA})
5   M'_MD ← BLOSOM-MO(min_sup, max_sup, max_item, D^T_new)
6   foreach G' ∈ M'_MD do
7       G ← min-DNF(G')
8       M_MD ← M_MD + G
```

CO over the complemented tidsets. Note that if $\mathcal{D}$ is sparse and large it is better to mine it directly instead of the complement database $\overline{\mathcal{D}}$, which will be large and dense. In such cases it is easy to modify BLOSOM-MO to perform intersections of tidsets instead of unions.

Minimal DNF and CNF Expressions

Following the structural characterization of minimal DNF expressions outlined earlier, BLOSOM-MD follows a two-phase approach. It first extracts all the minimal AND-clauses and then finds the minimal DNF expressions using those. Figure 6 shows the pseudo-code of BLOSOM-MD. We use BLOSOM-MA to get all minimal AND-clause generators ($\mathcal{M}_{MA}$) using the original dataset (line 1). Then, we regenerate the tidsets for each entry in $\mathcal{M}_{MA}$ (lines 2-3). Using the minimal AND-clause generators along with their tidsets, we create a new dataset $\mathcal{D}^T_{new}$ (line 4). The CreateNewDB method assigns a new label to each generator $G \in \mathcal{M}_{MA}$ and uses the tidsets $\mathbf{t}(G)$ as the new dataset. Note that, as an optimization, only one label is used per distinct tidset, even though there may be multiple corresponding minimal generators. Next, we call BLOSOM-MO to get all OR-clause generators over the new labels using the new dataset $\mathcal{D}^T_{new}$ (line 5), which in fact represent the initial set of minimal DNF generators. Note that in this step we implicitly replace each label by the original set of minimal generators. Finally, in lines 6-8, we delete produce the final set of min-DNF generators $\mathcal{M}_{MD}$.

Another point that needs further elaboration is the effect *minsup* when mining DNF expressions. If we use the *minsup* constraint while mining AND-clauses, then it is possible that we will miss some frequent DNF expressions, which have support more than *minsup*, after considering the tidset unions. For example, with *minsup*=3 in our running example, only the minimal AND-generators $C \times 123$ and $\{A, D\} \times 134$ remain frequent from Table 1. It is clear that in this case, we will fail to generate minimal DNF generators like $(AB \mid E, BD \mid E) \times 345$, $(B \mid AE, B \mid DE) \times 234$, and so on. To guarantee completeness, one has to set *minsup*=1 to mine all possible DNF expressions. Nevertheless, from a practical view-point this may generate too many AND-clauses to mine the DNF expressions. One may adopt a compromise solution, whereby we impose a new lower minimum support constraint *minsup'* < *minsup* for the AND-clauses, and use the *minsup* constraint for the DNF expressions. This is not unreasonable, since it imposes the constraint that any individual clause in the DNF expression is above *minsup*'. Note that *maxsup* constraint does not impact AND-clause mining (in line 1), and really applies only for the DNF expression mining (in line 5).

An Example: Figure 7 shows the steps involved in mining DNF expressions, using the example dataset in Figure 1. The figure shows as starting point the dataset $\mathcal{D}^T_{new}$ obtained after applying BLOSOM-MA on $\mathcal{D}$ and regenerating the tidsets (i.e., after steps 1-4 in Figure 6). Mining the AND-clauses yields the seven unique tidsets, and the corresponding minimal generators shown in Table 1. After sorting the tidsets in decreasing order of cardinality, these AND-clauses form the initial level in Figure 7 (note that for simplicity, original minimal generator groups are shown, as opposed to using new labels to represent them). Applying BLOSOM-MO on the new dataset yields the full DFS tree, as shown in Figure 7. As an example, consider the extensions under node *A,D*. New candidates are created by performing tidset unions will all other siblings. However, if for a node X, its tidset is a subset of $\mathbf{t}(A, D)$, then X is discarded, since it will yield a non-minimal generator. Thus the only valid children under *A,D* are *C* (with *extraset* 123-134=2), *E* (with *extraset* 45-134 = 4), and *B* (with *extraset* 23-134 = 2). Due to the sibling merging technique, *C* and *B* are collapsed into a single node *B,C* (with *extraset* 2). Note that the complete DNF expression at each node can be obtained by adding the OR operator to the sets on the path from the root to that node. Likewise the complete tidset can be obtained by taking the union of the *extraset*s on the path from the root to that node. For example, the child *B,C* under *A,D*, represents the composite DNF expression $\{A, D\} \mid \{B, C\}$, which when expanded, yields the four minimal DNF generators *A|B, A|C, B|D, C|D* characterizing the tidset *1234*. Note that as each new composite expression is produced it is added the hash-table $\mathcal{M}'_{MD}$, and fast-subsumption checking is used to prune non-minimal branches. For example, consider the child node *AE,DE* under *C*, representing the generators $C \mid \{AE, DE\}$ with tidset *1234*. When we try to add this to the hash-table, we find that there is already an entry in the cell with tidset *1234*, inserted earlier (in DFS order) for the generators $\{A, D\} \mid \{B, C\}$. Since $C \mid \{A, D\} \subseteq C \mid \{AE, DE\}$, the new node does not represent a minimal generator and is pruned. Note that even after such pruning we need to finally generate min-DNF expressions, once we expand the composite generators. For example, for the tidset *12345*, we have the following entries in the hash table: $\{A, D\} \mid \{B, C\} \mid E$, and *C|E*. When we expand the first composite entry, we obtain the generators *A|B|E, A|C|E, D|B|E, D|C|E*. However, *A|C|E* and *D|C|E* are not minimal, due to *C|E*. Thus, the final min-DNF generators for *12345* are *A|B|E*, and *B|D|E*, and *C|E*.

MINING CLOSED DNF, AND CNF EXPRESSIONS

For mining the minimal CNF expressions, the roles of BLOSOM-MA and BLOSOM-MO are reversed. BLOSOM-MC starts by mining the minimal OR-clauses and then computes the minimal AND-clauses by treating each of them as a new item. Finally any subsumed generators are purged to obtain the min-CNF forms.

Finally, consider the approach for mining the closed DNF and CNF expressions. For mining closed DNF expressions, we follow the same pseudo-code as for BLOSOM-MD, replacing

Figure 7. DNF Mining Example: DNF expressions are added to $\mathcal{M}'_{MD}$ in DFS manner. Shaded boxes represent expressions that are not minimal, and are thus pruned. The extrasets are shown under each expression.

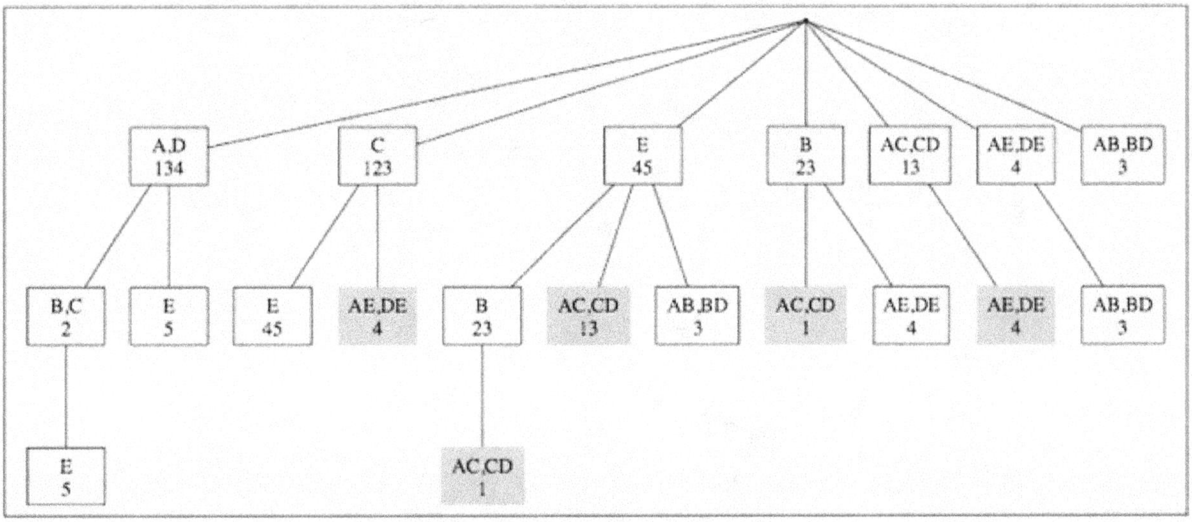

BLOSOM-MA and BLOSOM-MO with BLOSOM-CA and BLOSOM-CO, respectively. That is, we assume we know the closed AND-clauses, and then treating each of these as new items, we mine the closed OR-clauses. Finally, convert each such expression in the maximal min-DNF form. Note that the roles of *minsup* and *maxsup* are also reversed in CNF mining. That is while mining OR-clauses, we use *maxsup*' $\geq$ *maxsup* to extract the base OR-generators, and use the original *maxsup* to mine the CNF generators. For closed CNF expressions, reverse the roles of BLOSOM-CO and BLOSOM-CA, and output the maximal min-CNF expressions.

Experiments

All experiments were done on an Ubuntu virtual machine (over WindowsXP \& VMware) with 448MB memory, and a 1.4GHz Pentium-M processor. We used both synthetic and real datasets to evaluate BLOSOM. The synthetic datasets are generated with three parameters: the number of items $|\mathcal{I}|$, the number of transactions $|\mathcal{T}|$ and dataset density, δ. The size of the dataset is $|\mathcal{I}| \times |\mathcal{T}|$. For each item i, the average size of its transaction set $\mathbf{t}(i)$, is given as $\delta \times |\mathcal{T}|$. The average size is distributed uniformly in the interval $[0, |\mathcal{T}|]$, and the tids are distributed in $\mathbf{t}(i)$ with equal probability.

PERFORMANCE STUDY

Effect of Optimizations: We first study the effect of various proposed optimizations on the performance of BLOSOM-MO, as shown in Figure 8. The x -axis shows the number of items $|\mathcal{I}|$ in the synthetic datasets. The number of transactions were generated as $|\mathcal{T}| = 2|\mathcal{I}|$. Each curve in the figure shows the running time after applying the optimizations specified in succession. Thus the final curve for *no-transaction-set* includes all previous optimizations. In the legends, *original* stands for the unoptimized version, *hash* means using a hash table for

Figure 8. The effects of the speedup optimizations

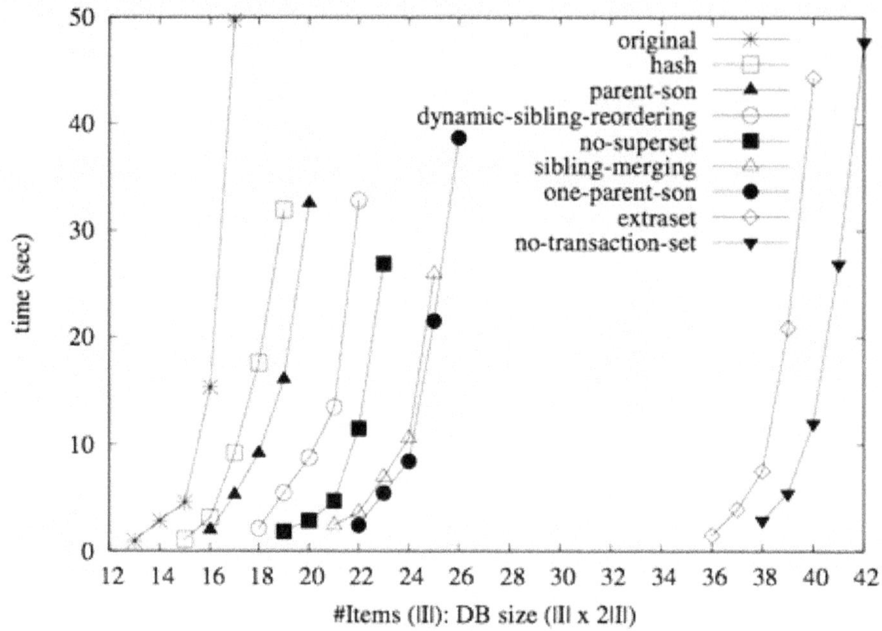

subsumption, *parent-son* means doing parent-child pruning, *dynamic-reordering* is the reordering of nodes, *no-superset* stands for one directional subsumption checking, *sibling-merging* is self explanatory, *one-parent-son* means we check only the left parent in parent-child relationship pruning due to the application of dynamic-sorting, *extrasets* is obvious, and finally *no-transaction-set* means we avoid storing the entire tidset. We can see that the cumulative effect of the optimizations is substantial; BLOSOM-MO can process a dataset around 10 times $((38 \times 76) / (12 \times 24) = 10)$ larger than the base algorithm can in the same running time. Thus all the optimizations together deliver a speedup of over an order of magnitude compared to the base version.

Effect of Parameters: The base-line parameters for the synthetic datasets are shown in Table 7. Note that we set *minsup*=1 and *maxsup*= $|\mathcal{T}|$, which means the entire set of all possible expressions will be mined. In the experiments we vary one parameter at a time, and study the effect. Figure 9 shows how the mining time varies with $|\mathcal{I}|$ (top row), with $|\mathcal{T}|$ (middle row), and with density δ (bottom row). The figure shows the effect for minimal clauses (MO/MA; left col.), minimal CNF/CNF expressions (MC/MD; middle col.), and closed clauses (CO/CA;

Table 7. Common experimental parameters

Parameters	MO/MA	MC/MD
$\|\mathcal{I}\|$	50	15
$\|\mathcal{T}\|$	300	30
δ	0.5	0.5
min_sup	1	1
max_sup	$\|\mathcal{T}\|$	$\|\mathcal{T}\|$
max_item	4	3

Figure 9. Synthetic data: effect of number of items, number of transactions, and density

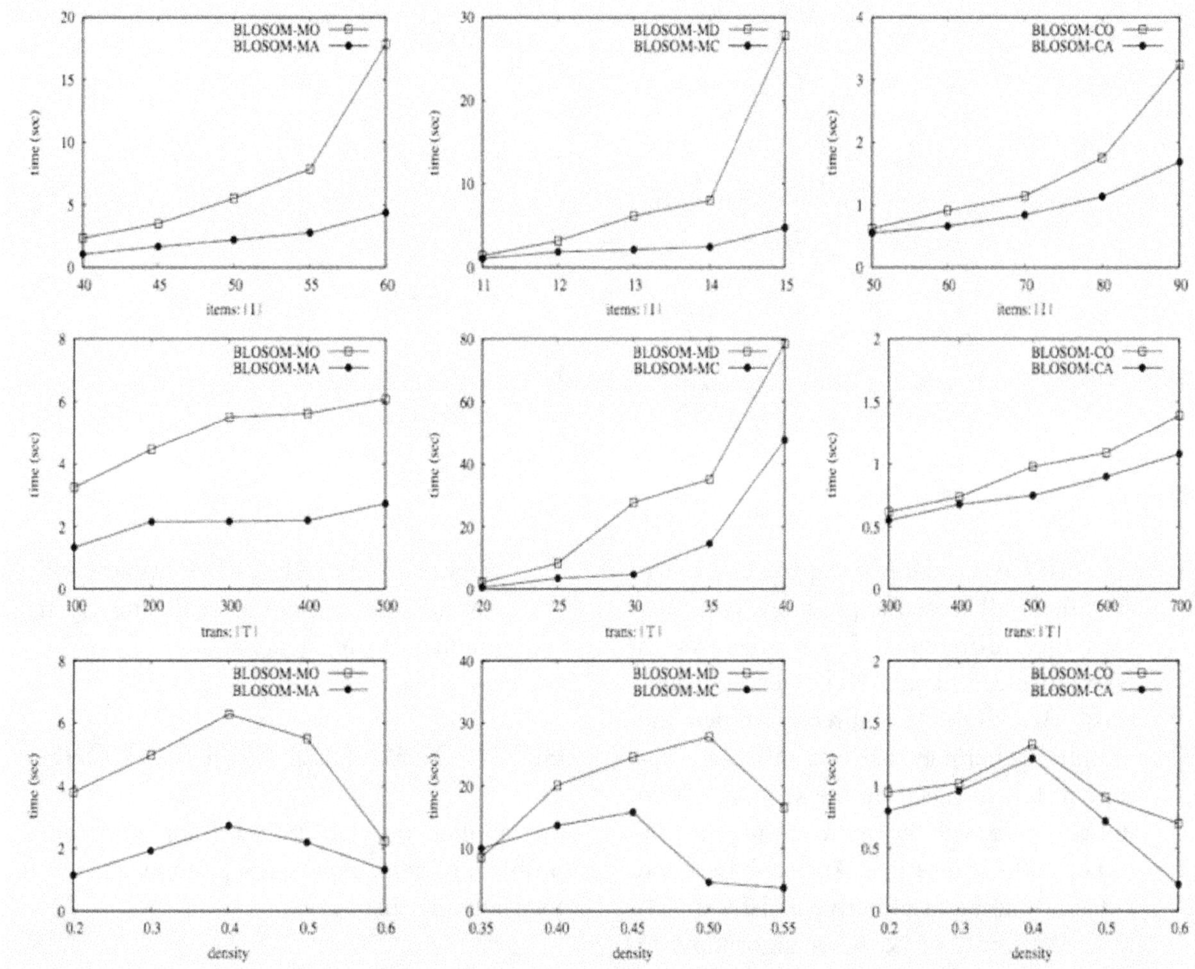

right col.). From these graphs we observe several trends. Notice that as we increase number of items/transactions and density, the disjunction times (for CO, MO, MD) tend to be higher than the conjunction times (CA, MA, MC). This is mainly because "unions" of tidsets tend to produce more distinct tidsets than "intersections", resulting in a larger search space. The trend for increasing | $\mathcal{I}$ | and | $\mathcal{T}$ | are as expected; the more the number of items or transactions, the longer the running time. However, with increasing density, the running time reaches the peak and then comes down. This is mainly because when density is closer to 50% we tend to mine patterns in the middle of the lattice, resulting in larger search spaces.

Comparison with CHARM-L: We compared BLOSOM-MA with CHARM-L (Zaki & Ra-makrishnan, 2005), which can also mine the minimal generators for AND-clauses (i.e., itemsets). We used the chess dataset, from the UCI machine learning repository, which has 3196 rows and 75 items. From Figure 10 we can see that BLOSOM-MA can be about ten times faster than CHARM-L, and the gap is increasing with decreasing

Figure 10. BLOSOM-MA vs. CHARM-L

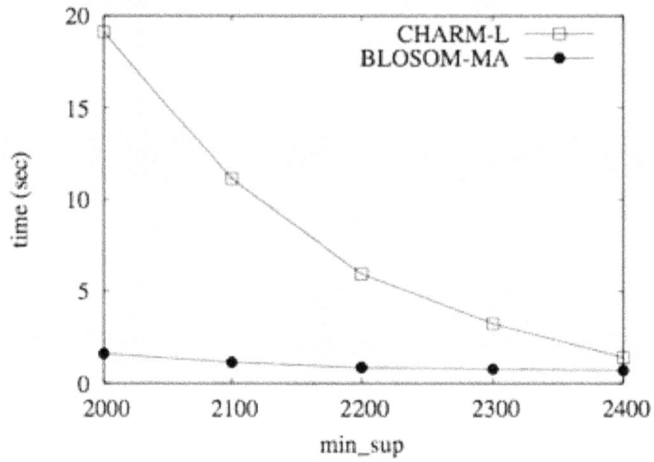

support. This is mainly because CHARM-L first finds all closed expressions and then uses their differential lower shadows to compute the minimal AND-clauses. In contrast, BLOSOM-MA directly mines the minimal generators, and uses effective optimizations to speed up the search. We do not compare with the approach in (Bastide et al., 2000), since it is Apriori-based and not likely to be as effective as BLOSOM-MA. We know of no other algorithms to mine minimal OR-clauses, and closed/minimal CNF and DNF expressions. However, note that is should be possible to modify several of the newly proposed closed itemset mining algorithms in the FIMI (Goethals & Zaki, 2003) repository to give effective algorithms for minimal clauses. Our work is the first step in this direction.

Scalability: Finally, to test the scalability of our framework with respect to the number of transactions, we show used the IBM synthetic dataset generator (Agrawal et al., 1996), to generate datasets with up to 5 million transactions, and with 50 base items, and all other default parameters. Figure 11 shows the scalability of the different instances of the BLOSOM framework. In essence, the methods all scale linearly in the number of transactions.

BIOINFORMATICS APPLICATIONS

We applied our BLOSOM framework on the problem of gene expression patterns and regulatory network discovery.

Gene Expression: We applied BLOSOM to mine frequent boolean expressions as well as redescriptions between descriptors on a gene expression dataset from (Ramakrishnan et al., 2004). This dataset involves 74 genes participating in 824 descriptors, derived from Gene Ontology (GO; www.geneontology.org) categories, gene expression bucketing, and k-means clustering. We specifically focus on finding minimal generators to help redescribe the descriptors corresponding to k-means clusters. There are 70 clusters, and we are interested if any of these can be redescribed by the other terms, at a particular Jaccard level (i.e., the Jaccard coefficient of the corresponding tidsets should be above

Figure 11. Scalability of BLOSOM

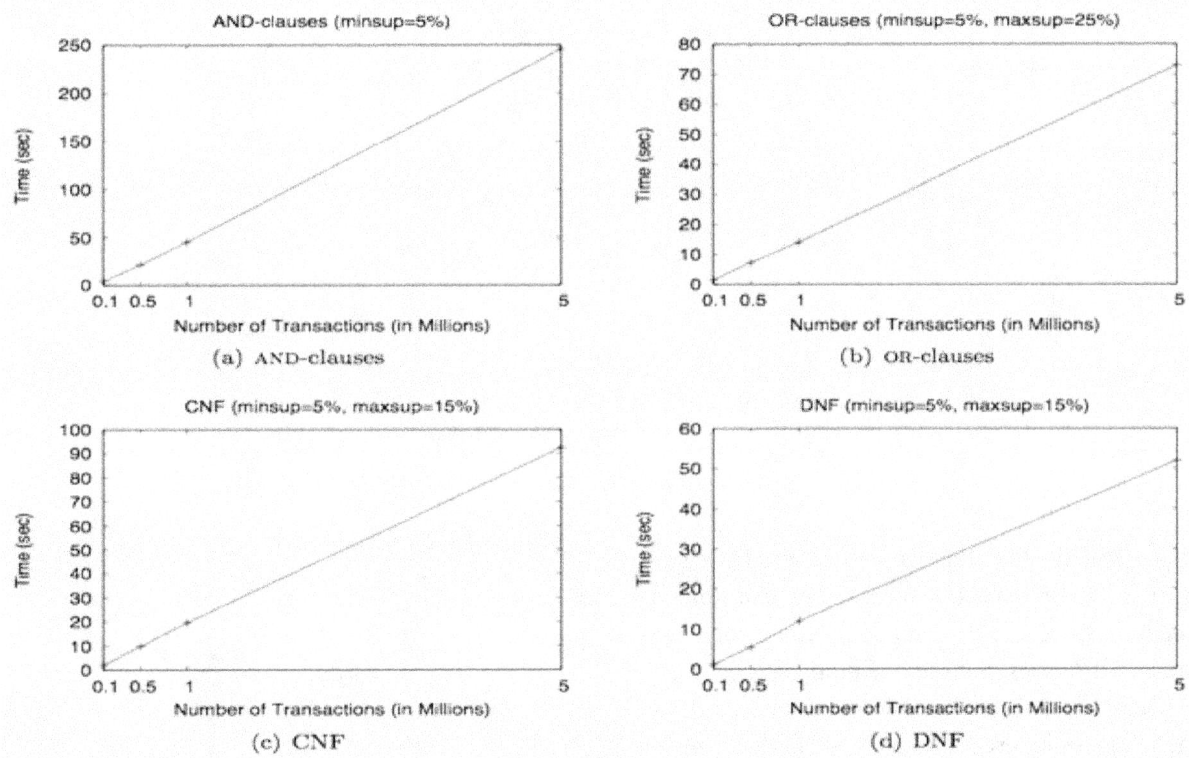

a certain level). One potential application is to obtain more expressive functional enrichments of these clusters in terms of GO categories. Figure 12 shows the running time with respect to *minsup*. With the minimum support set to 1, we were able to extract redescriptions for 9 of the 70 clusters, involving 221 minimal OR-generators at Jaccard level 0.75. For instance, a set of six genes participating in a k-means cluster, represented by descriptor *d512* was approximately redescribed in DNF form as: *d512* ⇔ *d507* OR *d685* OR *d700*, with Jaccard's coefficient 0.83 (i.e., $\frac{\mathbf{t}(d512) \cap \mathbf{t}(d507 \mid d685 \mid d700)}{\mathbf{t}(d512) \cup \mathbf{t}(d507 \mid d685 \mid d700)} = \frac{5}{6} = 0.83$), where *d507* denotes genes in the GO biological process category 'response to heat,' *d685* corresponds to genes in GO cellular location category 'extracellular,' and *d700* corresponds to genes in GO molecular function category 'exopeptidase.' This redescription shows that the concerted activity of a set of genes in a heat shock experiment derives from their role as either heat shock factors, extracellular signaling, or the (downstream) catalytic removal of an amino acid from a polypeptide chain. This showcases the power of BLOSOM to uncover meaningful biological descriptions of gene clusters.

Gene Regulatory Networks: Another application of BLOSOM is in finding complex gene regulatory networks, which can be represented in a simplified form, as boolean networks (Akutsu et al., 1998). Consider the network involving 16 genes, taken from (Akutsu et al., 1998), shown in Figure 13.

Here $\oplus$ and $\ominus$ denote gene *activation* and *deactivation*, respv. For example, genes B, E, H, J, and M are expressed if their parents are not

Figure 12. Gene expression: running time

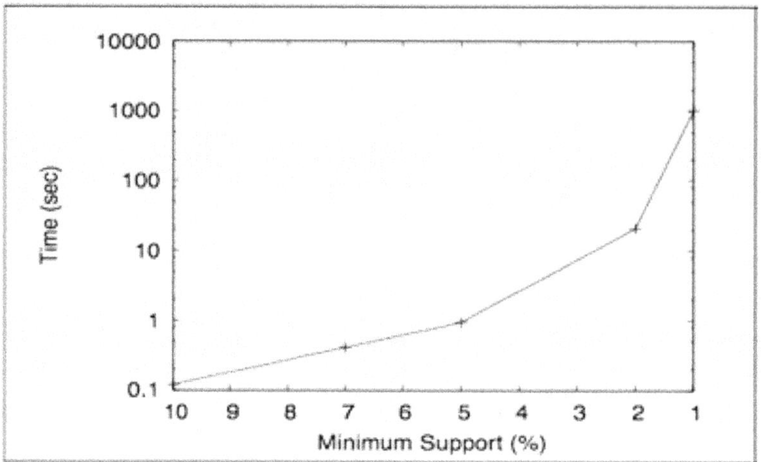

expressed. On the other hand G, L, and D express if all of their parents express. For example, D depends on C, F, X1 and X2. Note that F expresses if A does, but not L. Finally A, C, I, K, N, X1 and X2 do not depend on anyone, and can thus be considered as *input* variables for the boolean network. We generated the truth table corresponding to the 7 input genes but BLOSOM was provided the values for all genes, without explicit instruction about which are inputs and which are outputs. This yields a dataset with 128 rows and 16 items (genes).

We then ran BLOSOM to discover the boolean expression corresponding to this gene network. We mined using *minsup*=1 and extracted only those patterns with support 100%, since we want to find expressions that are true for the entire set of assignments. BLOSOM output 65 expressions in 0.36s, which hold true for the entire dataset. After simplification these can be reduced to the equivalent expression, as shown in Figure 14. We verified that indeed this expression is true for all the rows in the dataset! It also allows us to reconstruct the boolean gene network shown in Figure 13. For example, the first component of the expression in the first row

Figure 13. Gene network

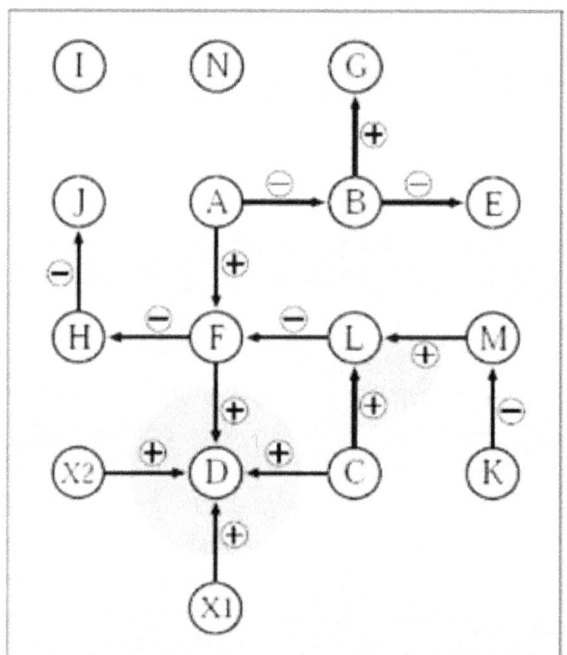

$$\bar{D} \mid (A\bar{B}CEF\bar{G}\bar{H}JK\bar{L}\bar{M}X1X2)$$

can be converted into the implication

$$D \Rightarrow (A\bar{B}CEF\bar{G}\bar{H}JK\bar{L}\bar{M}X1X2),$$

Figure 14. Boolean network expression

$$
\begin{array}{l}
(\overline{D} \mid (A\,\overline{B}\,C\,E\,F\,\overline{G}\,\overline{H}\,J\,K\,\overline{L}\,\overline{M}\,X1\,X2)) \text{ AND} \\
(\overline{L} \mid (C\,\overline{F}\,H\,\overline{J}\,\overline{K}\,M)) \text{ AND} \\
((\overline{A}\,B\,\overline{E}\,G) \mid \overline{C} \mid D \mid L \mid \overline{X1} \mid \overline{X2}) \text{ AND} \\
((\overline{A}\,B\,\overline{E}\,G) \mid (C\,L) \mid (F\,\overline{H}\,J)) \text{ AND} \\
((\overline{F}\,H\,\overline{J}) \mid (A\,\overline{B}\,\overline{C}\,E\,\overline{G}) \mid (A\,\overline{B}\,E\,\overline{G}\,K\,\overline{M}))
\end{array}
$$

which means that D depends on the variables on the right hand side (RHS). If, at this point, we supply any partial knowledge about the input variables or of the maximum fan-out of the network, we could project the RHS only on those variables to obtain (A C K X1 X2), which happens to be precisely the relationship given in Figure 13. The second row tells us that L depends on the activation of C and inactivation of K, i.e., $\overline{K}$, if we restrict ourselves to the input variables. Also C and $\overline{K}$ give the values for the remaining variables in the second row. Note that other dependencies in the boolean network are also included in the mined expression. For example, we find that B and A always have opposite values, and so do B and E, and K and M. G and B always have the same values, and so on. Thus this example shows the power of BLOSOM in mining gene regulatory networks.

CONCLUSION

In this paper we present the first algorithm, BLOSOM, to simultaneously mine closed boolean expressions over attribute sets and their minimal generators. Our four-category division of the space of boolean expressions yields a compositional approach to the mining of arbitrary expressions, along with their minimal generators. The pruning operators employed here have resulted in orders of magnitude speedup, producing highly efficient implementations.

There are still many interesting issues to consider. The first one involves the effective handling of negative literals without being overwhelmed by dataset density. The second issue is to push tautological considerations deeper into the mining algorithm by designing new pruning operators. Finally, given a general propositional reasoning framework, we are interested in mining the simplest boolean expressions necessary for inference in that framework.

ACKNOWLEDGMENT

This work was supported in part by NSF grants EMT-0829835 and EIA-0103708, and NIH grant 1R01EB0080161-01A1.

REFERENCES

Agrawal, R., Mannila, H., Srikant, R., Toivonen, H., & Verkamo, A. I. (1996). Fast discovery of association rules. In *Advances in Knowledge Discovery and Data Mining* (pp. 307–328). Menlo Park, CA: AAAI Press.

Akutsu, T., Kuhara, S., Maruyama, O., & Miyano, S. (1998). Identification of gene regulatory networks by strategic gene disruptions and gene overexpressions. In *Proceedings of the ACM-SIAM Symposium on Discrete Algorithms*.

Antonie, M.-L., & Zaiane, O. (2004). Mining positive and negative association rules: An approach for confined rules. In *Proceedings of the European PKDD Conf.*

Bastide, Y., Taouil, R., Pasquier, N., Stumme, G., & Lakhal, L. (2000). Mining frequent patterns with counting inference. *SIGKDD Explorations*, 2.

Bayardo, R. J., & Agrawal, R. (1999). Mining the most interesting rules. In *Proceedings of the ACM SIGKDD Conf.*

Bshouty, N. (1995). Exact learning boolean functions via the monotone theory. *Information and Computation, 123*, 146–153. doi:10.1006/inco.1995.1164

Calders, T., & Goethals, B. (2003). Minimal k-free representations of frequent sets. In *Proceedings of the European PKDD Conf.*

Calders, T., & Goethals, B. (2005). Quick inclusion-exclusion. In *Proceedings of the ECML-PKDD Workshop on Knowl-edge Discovery in Inductive Databases.*

Davey, B. A., & Priestley, H. A. (1990). *Introduction to lattices and order*. Cambridge, UK: Cambridge University Press.

Dong, G., Jiang, C., Pei, J., Li, J., & Wong, L. (2005). Mining succinct systems of minimal generators of formal concepts. In *Proceedings of the Int'l Conf. Database Systems for Advanced Applications.*

Ganter, B., & Wille, R. (1999). *Formal concept analysis: Mathematical foundations*. New York: Springer-Verlag.

Goethals, B., & Zaki, M. (2003). Advances in frequent itemset mining implementations: report on FIMI'03. *SIGKDD Explorations, 6*, 109–117. doi:10.1145/1007730.1007744

Gunopulos, D., Khardon, R., Mannila, H., Saluja, S., Toivonen, H., & Sharma, R. (2003). Discovering all most specific sentences. *ACM Transactions on Database Systems, 28*, 140–174. doi:10.1145/777943.777945

Jaroszewicz, S., & Simovici, D. A. (2002). Support approximations using bonferroni-type inequalities. In *Proceedings of the 6th European Conference on Principles of Data Mining and Knowledge Discovery.*

Kryszkiewicz, M. (2001). Concise representation of frequent patterns based on disjunction-free generators. In *Proceedings of the Int'l Conf. on Data Mining.*

Kryszkiewicz, M. (2005). Generalized disjunction-free representation of frequent patterns with negation. *Journal of Experimental & Theoretical Artificial Intelligence, 17*, 63–82. doi:10.1080/09528130512331315882

Mannila, H., & Toivonen, H. (1996). Multiple uses of frequent sets and condensed representations. In *Proceedings of the International Conference on Knowledge Discovery and Data Mining.*

Mielikainen, T. (2003). Intersecting data to closed sets with constraints. In *Proceedings of the Workshop on Frequent Itemset Mining Implementations.*

Mitchell, T. (1982). Generalization as search. *Artificial Intelligence, 18*, 203–226. doi:10.1016/0004-3702(82)90040-6

Nanavati, A., Chitrapura, K., Joshi, S., & Krishnapuram, R. (2001). Association rule mining: Mining generalised disjunctive association rules. In *Proceedings of the ACM CIKM Conf.*

Pasquier, N., Bastide, Y., Taouil, R., & Lakhal, L. (1999). Discovering frequent closed itemsets for association rules. In *Proceedings of the 7th Intl. Conf. on Database Theory.*

Pfaltz, J., & Jamison, R. (2001). Closure systems and their structure. *Information Sciences*, *139*, 275–286. doi:10.1016/S0020-0255(01)00169-4

Ramakrishnan, N., Kumar, D., Mishra, B., Potts, M., & Helm, R. (2004). Turning cartwheels: An alternating algorithm for mining redescriptions. In *Proceedings of the ACM SIGKDD Conf.*

Savasere, A., Omiecinski, E., & Navathe, S. (1998). Mining for strong negative associations in a large database of customer transactions. In *Proceedings of the ICDE Conf.*

Shima, Y., Mitsuishi, S., Hirata, K., & Harao, M. (2004). Extracting minimal and closed monotone dnf formulas. In *Proceedings of the Int'l Conf. on Discovery Science*.

Wu, X., Zhang, C., & Zhang, S. (2004). Efficient mining of both positive and negative association rules. *ACM Transactions on Information Systems*, *22*, 381–405. doi:10.1145/1010614.1010616

Yuan, X., Buckles, B. P., Yuan, Z., & Zhang, J. (2002). Mining negative association rules. In *Proceedings of the 7th International Symposium on Computers and Communications*.

Zaki, M., & Ramakrishnan, N. (2005). Reasoning about sets using redescription mining. In *Proceedings of the ACM SIGKDD Intl. Conf. on Knowledge Discovery and Data Mining*.

Zaki, M. J., & Hsiao, C.-J. (2005). Efficient algorithms for mining closed itemsets and their lattice structure. *IEEE Transactions on Knowledge and Data Engineering*, *17*, 462–478. doi:10.1109/TKDE.2005.60

Zhao, L., Zaki, M., & Ramakrishnan, N. (2006). Blosom: A framework for mining arbitrary boolean expressions. In *Proceedings of the ACM SIGKDD Intl. Conf. on Knowledge Discovery and Data Mining*.

This work was previously published in the International Journal of Knowledge Discovery in Bioinformatics, Volume 1, Issue 3, edited by Xiao-Li Li and See-Kiong Ng, pp. 68-96, copyright 2010 by IGI Publishing (an imprint of IGI Global).

Chapter 9
Bioinformatics Methods for Studying MicroRNA and ARE-Mediated Regulation of Post-Transcriptional Gene Expression

Richipal Singh Bindra
New Jersey Institute of Technology, USA

Jason T. L. Wang
New Jersey Institute of Technology, USA

Paramjeet Singh Bagga
Ramapo College of New Jersey, USA

ABSTRACT

MicroRNAs (miRNAs) are short single-stranded RNA molecules with 21-22 nucleotides known to regulate post-transcriptional expression of protein-coding genes involved in most of the cellular processes. Prediction of miRNA targets is a challenging bioinformatics problem. AU-rich elements (AREs) are regulatory RNA motifs found in the 3' untranslated regions (UTRs) of mRNAs, and they play dominant roles in the regulated decay of short-lived human mRNAs via specific interactions with proteins. In this paper, the authors review several miRNA target prediction tools and data sources, as well as computational methods used for the prediction of AREs. The authors discuss the connection between miRNA and ARE-mediated post-transcriptional gene regulation. Finally, a data mining method for identifying the co-occurrences of miRNA target sites in ARE containing genes is presented.

DOI: 10.4018/978-1-4666-1785-8.ch009

1. INTRODUCTION

The expression of protein-coding genes is regulated via a variety of mechanisms. While regulation of gene transcription is central to the cellular metabolism, post-transcriptional control of gene expression has also gained significant attention in the recent years. Post-transcriptional processes like translational regulation and mRNA turnover are fairly complex mechanisms and are being recognized as equally if not more sophisticated than the transcriptional regulation of gene expression.

Several *cis*-acting signals are known to regulate mRNA stability. In the 3' untranslated regions, AU-rich elements (AREs) play dominant roles in the regulated decay of short-lived human mRNAs via specific interactions with proteins. While some of these proteins promote ARE-mediated mRNA degradation (AMD) (Stoecklin et al., 2002), others delay it (Peng, Chen, Xu, & Shyu, 1998). The mRNAs containing AREs generally code for proteins that are involved in transient processes such as cellular proliferation, stress response or development, which require delicate but strict regulation. The expression of these genes requires stringent and prompt controls which can be best achieved at the post-transcriptional level (Barreau, Paillard, & Osborne, 2005).

The post-transcriptional expression of a large number of metazoan genes is also influenced by microRNAs (miRNAs) which have received significant attention recently as important regulators of translation and destabilization of mRNAs. For reviews, see Bushati and Cohen (2007), Filipowicz, Bhattacharyya, and Sonenberg (2008), Bartel (2009), and Zhang and Su (2009). The impact on gene expression is mediated by annealing of the miRNAs to complementary segments of the targeted mRNAs, especially in the 3' untranslated regions (UTRs). Approximately 30-50% of the human protein-coding genes appear to be regulated by miRNAs (John et al., 2004; John, Sander, & Marks, 2006; Filipowicz, Bhattacharyya, & Sonenberg, 2008; Friedman, Farh, Burge, & Bartel, 2009). Since miRNAs are considered to be involved in almost all types of cellular processes, an alteration in their own expression could lead to human diseases (Esquela-Kerscher & Slack, 2006; Kloosterman & Plasterk, 2006; Chang & Mendell, 2007; Rodriguez et al., 2007). Therefore, investigating the mechanism of miRNA-mediated regulation of gene expression is critical to the understanding of cellular function and the diseases associated with altered miRNA function.

Several studies in the recent years have suggested that miRNAs can regulate post-transcriptional gene expression by targeting ARE-mediated mRNA turnover (reviewed in von Roretz & Gallouzi, 2008; Asirvatham, Magner, & Tomasi, 2009). For example, expression of cytokine genes could be regulated via miRNA targeting of proteins involved in AMD (Asirvatham, Gregorie, Hu, Magner, & Tomasi, 2008). Although more than 3,000 human genes are predicted to contain AREs (Bakheet, Williams, & Khabar, 2006), what proportion of these *cis*-acting elements require interactions with miRNAs or miRNPs for mRNA destabilization is not clear.

Owing to a lack of full length complementarity between the ~22 nucleotide miRNAs and their corresponding target sites in the animal mRNAs, accurate prediction of miRNA binding sites has been challenging, especially when attempted at a genomic scale. Prediction of miRNA targeted AREs in the 3'UTR is even trickier because the ARE-complementary region may not be in the 5'-region of the miRNA. It has been widely accepted that successful target recognition in animal mRNAs is dependent on pairing with bases in the 5'-region of the miRNAs (Bartel, 2004; Brennecke, Stark, Russell, & Cohen, 2005; Mendes, Freitas, & Sagot, 2009). This principle has been adopted by several miRNA target prediction tools.

In this article we review well-known methods for the prediction of AREs and miRNA target sites on the mRNAs, and discuss a bioinformatics approach for genome wide investigation of the correlation between miRNA target sites and AREs.

2. AU-RICH ELEMENTS

In this section we first discuss the biology of AREs. We then review some computational methods for ARE prediction.

2.1 Biology of AREs

AU-**R**ich **E**lements (AREs) are regulatory RNA motifs found in the 3' untranslated regions of mRNAs (Bakheet, Frevel, Williams, Greer, & Khabar, 2001). These *cis*-acting elements are commonly found in the mRNAs that are known to undergo rapid degradation through the 3'-5' exosome pathway following deadenylation. However, depending upon their interaction with proteins and cellular conditions, AREs can also help stabilize mRNAs (Ford, Watson, Keene, & Wilusz, 1999; Anderson et al., 2000; Mobarak et al., 2000). Loss of regulated stability in ARE-mRNAs has been associated with disease (Khabar, 2005). In addition to mRNA turnover, AREs are known to regulate other post-transcriptional processes such as mRNA export and translation (Wilusz, Wormington, & Peltz, 2001; Espel, 2005). Roughly 3,700 human genes with diverse functions are known to code for ARE-containing mRNAs (Halees, El-Badrawi, & Khabar, 2008). Majority of these genes, such as growth factors and oncogenes, are involved in transient processes and therefore require strict expression control (Khabar, Bakheet, & Williams, 2005).

ARE sequences can range from 50 to 150 bases in length. A pentamer, 'AUUUA', which can be found in multiple repeats amidst U-rich regions in the 3'UTR, is considered as the identifying motif of the ARE (Chen & Shyu, 1994). However, a nonamer 'UUAUUUAUU' is minimally needed for destabilization of the mRNA and, therefore, is considered the functional ARE motif (Zubiaga, Belasco, & Greenberg, 1995). AREs can be categorized into multiple classes, reviewed in (Barreau, Paillard, & Osborne, 2005). Extensions of the pentamer are at the core of the first two classes which differ on the basis of the repetition pattern of this sequence motif. Class II has been further sub-divided into clusters based on the number of pentamer repeats. The third class of AREs, although U rich, is not known to contain a well-defined sequence motif.

ARE-mediated regulation of post-transcriptional gene expression requires interactions with cellular proteins. Analysis of ARE-binding proteins (ARE-BPs) has gained significant attention since alteration in their expression can lead to cancer (Audic & Hartley, 2004). Most AREs can bind to multiple proteins (Barreau, Paillard, & Osborne, 2005). Overall interactions of an ARE containing mRNA with ARE-BPs could be complex and eventually result in either degradation or destabilization of the mRNA. Some of these proteins, such as Tristetraprolin (TTP), Butyrate Response Factor (BRF1) and KH-type Splice Regulatory Protein (KSRP), are known to decrease the mRNA stability (Stoecklin et al., 2002; Gherzi et al., 2004; Barreau, Paillard, & Osborne, 2005; Cao, Deterding, & Blackshear, 2007). On the other hand, members of the *Hu* family of ARE-binding proteins that are related to ELAV, generally stabilize the mammalian mRNAs (Barreau, Paillard, & Osborne, 2005). For example, HuR, which is present during cellular proliferation, has been found to bind to mRNAs of all three ARE classes (Peng, Chen, Xu, & Shyu, 1998; Barreau, Paillard, & Osborne, 2005). AUF1, which was one of the first identified functional ARE-BP, is known to enhance or inhibit mRNA decay rate depending upon its own alternative expression, cell type and the class of targeted AREs (Xu, Chen, & Shyu, 2001; Sarkar, Xi, He, & Schneider, 2003; Lal et al., 2004; Raineri, Wegmueller, Gross, Certa, & Moroni, 2004).

Interactions between ARE-BPs and the mRNA decay enzymes result in rapid mRNA degradation. Mammalian AMD is dependant on deadenylation, which is followed by 3'-5' degradation by a multimeric complex of exonucleases called the exosome (reviewed in Wilusz & Wilusz, 2004;

Barreau, Paillard, & Osborne, 2005). Degradation of mRNA can be enhanced by direct interaction of exosome components to either AREs (Mukherjee et al., 2002), or ARE-BPs like KSRP and TTP (Chen et al., 2001). Several ARE-BPs are also capable of influencing translation of mRNAs. For example, while TIA-1 is known to inhibit translation of TNF-alpha and Cox-2 mRNAs (Piecyk et al., 2000; Dixon et al., 2003), Hel-N1, a member of the *Hu* family, can increase translation of GLUT1 mRNAs (Jain et al., 1995). On the other hand, HuR itself can increase or down-regulate translation (Galban et al., 2003; Katsanou et al., 2005). The mechanism of ARE-dependent regulation of translation requires further investigation.

2.2 Computational Approaches for ARE Prediction

In order to understand the biological role of AREs in regulating the functioning of transient transcriptome, it would be essential to delineate the genes that contain these *cis*-acting elements. Computational strategies for genome-wide identification of AREs are full of challenges. While the pentamer discussed above forms the core of majority of AREs, the minimal functional ARE is a nonamer with many variations. Attempts have been made to search mammalian genomes for AREs using computational approaches and to classify them (Bakheet, Frevel, Williams, Greer, & Khabar, 2001; Bakheet, Williams, & Khabar, 2003; Bakheet, Williams, & Khabar, 2006; Halees, El-Badrawi, & Khabar, 2008). While very useful databases of ARE containing transcripts have been established as the result of these efforts, these collections are not expected to represent the entire repertoire of AREs in the trancriptome.

Bakheet and coworkers (Bakheet, Frevel, Williams, Greer, & Khabar, 2001), used a list of known 57 ARE containing mRNAs as a training set for computation of the ARE motif. This analysis resulted in a 13-bp pattern WWWUAUUUAUWWW which was then used to search CDS and 3'UTR sequences of 13,057 human mRNA sequences from NCBI-GenBank (Karsch-Mizrachi & Ouellette, 2001). The 897 ARE-mRNAs thus identified were curated in a web accessible database, named ARED 1.0 (http://brp.kfshrc.edu.sa/ARED/). AREs in this database were clustered in 5 groups based on the subsequence patterns and phylogenic relationship. ARED has been updated several times successively. All the versions are accessible at the same URL. The second generation database, named ARED 2.0, was established in 2003 (Bakheet, Williams, & Khabar, 2003). It contained 1,500 non-redundant human full-length ARE-mRNAs. Several inconsistencies of the first version were also removed.

Khabar and co-authors (Khabar, Bakheet, & Williams, 2005) focused on complementing the ARED 1.0 and 2.0 databases (discussed above) with ARE-mRNAs that were initially missed because of inadequate 3'UTRs, partial sequence information, or complexities involving alternatively expressed isoforms. Two similar computational approaches were used to search for AREs in the validated 3' ends of human ESTs. The pentamer ATTTA motif was at the core of search patterns for mining and prediction. The first approach searched 3.4 million human ESTs and resulted in 673 known mRNA records and 26 novel alternatively spliced transcripts containing AREs. A total of 199 ARE-genes and 140 novel ARE-mRNAs were identified. The second approach used an updated version of dbEST with approximately 5.5 million records and involved ARE EST clusters with two or more 3' ends. This experiment resulted in 1,096 ARE-mRNAs including more than 350 novel alternative ARE-mRNA forms. Their combined efforts from both the experiments identified over 1,500 human ARE genes including those which are alternatively spliced, alternatively polyadenylated, or with variant 3'UTRs.

ARED 3.0 (Bakheet, Williams, & Khabar, 2006) contains over 4,000 human ARE-mRNAs. Transcript sequences from NCBI databases: RefSeq (Pruitt, Tatusova, & Maglott, 2007), GenBank, including dbEST (Boguski, Lowe, & Tolstoshev, 1993), were screened with a new refined pattern WWWT(ATTTA)TTTW as the ARE motif for the third version of the database. Eventually, all the versions of the databases were combined, along with ARED 4.0, to generate an ARED integrated database of ARE-containing human mRNAs (Halees, El-Badrawi, & Khabar, 2008). This expanded and updated database contains 3,700 ARE-genes and 6,153 RefSeq transcripts. Another ARED organism database (Halees, El-Badrawi, & Khabar, 2008), which is obtained from the analysis of mRNAs in ENSEMBL (Hubbard et al., 2007), contains human, mouse and rat AREs.

3. MICRORNAs

After describing the biology of AREs and computational approaches for ARE prediction, we now turn to microRNAs. In this section we first discuss the biology of microRNAs. We then review several microRNA target prediction tools and online data sources.

3.1 Biology of MicroRNAs

MicroRNAs (miRNAs) have emerged as a large class of noncoding RNAs which can regulate important biological processes in eukaryotes and other species. Since they have been the subject of many reviews recently (Filipowicz, Bhattacharyya, & Sonenberg, 2008; Asirvatham, Magner, & Tomasi, 2009; Bartel, 2009; Zhang & Su, 2009; Fabian, Sonenberg, & Filipowicz, 2010; Newman & Hammond, 2010; Siomi & Siomi, 2010), we will provide only a brief overview of miRNA biogenesis and function in this article.

MicroRNAs are short singe-stranded RNA molecules with 21-22 nucleotides that are known to regulate post-transcriptional expression of protein-coding genes. More than one thousand miRNAs are thought to be expressed in the humans (Filipowicz, Bhattacharyya, & Sonenberg, 2008). Metazoan miRNAs are known to be differentially expressed in different tissues, especially during development and differentiation (Sempere, Sokol, Dubrovsky, Berger, & Ambros, 2003; Biemar et al., 2005; Wulczyn et al., 2007).

Human miRNA genes, which are located in the introns, intergenic regions or other locations, are mostly transcribed by RNA pol II. The primary transcript, a long pri-miRNA, which could be hundreds of bases long, gets cleaved by a multi-protein complex containing an RNAse III enzyme called Drosha in the nucleus. The resulting ~70 nucleotide double stranded stem-loop pre-miRNA is then transported to the cytoplasm with the help of a protein complex containing Exportin 5. The cytoplasmic RNAse III, Dicer, which recognizes the double stranded stem-loop, then cleaves the pre-miRNA to release the double stranded ~22 nucleotide miRNA intermediate. One of the miRNA strands is then incorporated into the RNA-induced silencing complex (RISC). The miRISC subsequently targets 3'UTRs for miRNA:mRNA interactions. The overall impact of this interaction in animals generally is modulation of translation or mRNA destabilization. These mechanisms don't seem to involve cleavage of the mRNAs in animal systems. Biogenesis of miRNAs in plants is somewhat similar to that in animals. However, plant miRNAs are generally known to cleave the target mRNAs.

Over 50% of the human protein-coding genes are thought to be under an evolutionary pressure to maintain miRNA binding sites in their 3'UTRs (Friedman, Farh, Burge, & Bartel, 2009). Therefore, a wide range of biological processes are likely to be influenced by miRNA targeting. For example, miR-155 influences immune response in mice and has been predicted to regulate a wide spectrum of genes including cytokines and transcription factors (Rodriguez et al., 2007). MicroRNAs are thought to affect growth, morphogenesis, apoptosis and cell to cell signaling. Therefore, alteration in miRNA function can lead to disease (Esquela-Kerscher & Slack, 2006;

Figure 1. Illustration of the binding between a miRNA and an mRNA (a) in plants and (b) in animals

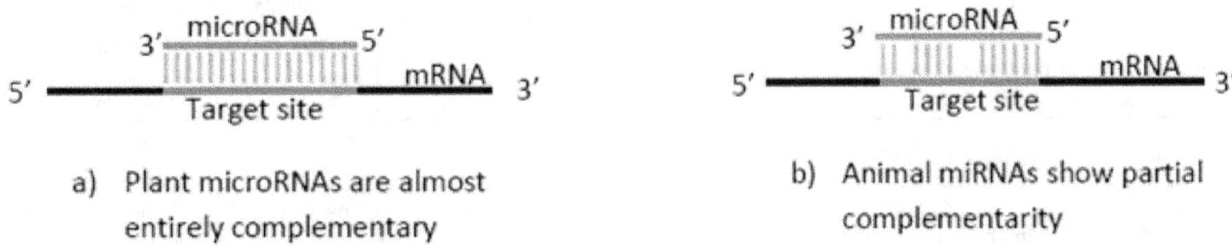

Kloosterman & Plasterk, 2006; Chang & Mendell, 2007; Rodriguez et al., 2007). Understanding the mechanism of their interactions with target mRNAs, regulatory proteins and involvement in molecular pathways is essential.

Several models for regulation of translation and mRNA decay by miRNAs are reviewed in (Filipowicz, Bhattacharyya, & Sonenberg, 2008). However, the more we learn about these complex pathways, more questions are raised. The mechanistic details of many steps are still sketchy and require more research.

3.2 Computational Methods for MicroRNA Target Prediction

In this subsection we first review some miRNA target prediction tools. We then survey four online databases, which store the miRNA target information along with the genes in which miRNA target sites occur, as well as their location information in 3'UTR regions.

3.2.1 MicroRNA Target Prediction Tools

Plant miRNAs are almost entirely complementary to their respective pairing sites in the plant mRNAs (Rhoades et al., 2002) (Figure 1). However, animal miRNAs have only about 7-8 base-pair partial complementarity of their 5' end to the target sites in the animal mRNAs (Brennecke, Stark, Russell, & Cohen, 2005). While it is easier to search for miRNA target sites by looking for complementary sequences in the plant mRNAs (Rhoades et al., 2002), this approach generates many false positives in animals (Bartel, 2009). Therefore, genome-wide miRNA target prediction in metazoans has been challenging. Several methods also use partial complementarity of the 3' end of miRNAs in addition to the 5' end. However, the miRNA target prediction methods that rely solely on pairing of the 5'- and 3'-ends of the miRNAs still result in false positives. The accuracy of predictions can be improved by consideration of mRNA secondary structure, positioning within the 3'UTR, proximity to other miRNA target sites, site conservation and sequence context (Grimson et al., 2007; Bartel, 2009; Mendes, Freitas, & Sagot, 2009).

Table 1 summarizes some widely used miRNA target prediction tools. TargetScan (Lewis, Shih, Jones-Rhoades, Bartel, & Burge, 2003) adopts thermodynamics-based modeling of miRNA:mRNA interactions. Given a miRNA conserved in various organisms, this tool uses comparative sequence analysis to predict miRNA targets conserved across multiple genomes. Specifically, the tool takes a seed-match approach where a seed is defined as a 7-nt segment containing positions 2-8 of a mature miRNA. The TargetScan algorithm consists of the following steps: (i) Search the UTRs of interest for perfect Watson-Crick complementarity (called "seed match") to bases 2-8 of a miRNA (called "miRNA seed"). (ii) Extend each seed match allowing wobble G:U

Table 1. Summary of the miRNA target prediction tools

Name	URL	Availability	Reference
TargetScan	http://targetscan.org/	Online	(Lewis, Shih, Jones-Rhoades, Bartel, & Burge, 2003)
TargetScanS	http://targetscan.org/	Online	(Lewis, Burge, & Bartel, 2005)
RNAHybrid	http://bibiserv.techfak.uni-bielefeld.de/rnahybrid/	Download	(Rehmsmeier, Steffen, Hochsmann, & Giegerich, 2004)
DIANA- microT	http://diana.cslab.ece.ntua.gr/microT/	Online	(Maragkakis et al., 2009)
miRanda	http://www.microrna.org/microrna/home.do	Download	(Betel, Wilson, Gabow, Marks, & Sander, 2008)
PicTar	http://pictar.mdc-berlin.de	Online	(Lall et al., 2006)
MicroTar	http://tiger.dbs.nus.edu.sg/microtar/	Download	(Thadani, & Tammi, 2006)
mirWIP	http://146.189.76.171/query	Online	(Hammell et al., 2008)

pairs. (iii) Optimize base-pairing to 35 bases. (iv) Assign a folding free energy value to each miRNA:target site. (v) Assign a Z score to each miRNA:mRNA binding. (vi) Sort the UTRs and rank them. (vii) Repeat the process for other organisms. TargetScan identified 451 putative miRNA target sites in 400 distinct genes; the false positive rate is estimated to be 31% for human, mouse and rat and 22% for pufferfish.

TargetScanS (Lewis, Burge, & Bartel, 2005) is a simplified version of TargetScan; it includes two more species (dog and chicken). In the new algorithm, the authors improved the target prediction of miRNA by relaxing the score and rank cutoffs. Furthermore, the requirement of the 7-nt match to the seed region was relaxed to a 6-nt (2-7 nucleotides) match followed by an additional 3' match of an adenosine anchor at position 1. High conservation at downstream of a seed match was observed and was usually a conserved A, which can pair to the first nucleotide U of a miRNA. This algorithm does not take into account the thermodynamic stability of pairing, and the presence of multiple complementary sites per UTR. TargetScan and TargetScanS have low false positive rates, but due to the requirement of strict complementarity some targets having 3' compensatory sites are likely to be missed.

RNAHybrid is an extension to the RNA secondary structure (RSS) algorithm (Rehmsmeier, Steffen, Hochsmann, & Giegerich, 2004). RSS uses dynamic programming along with an artificial linker between a miRNA and its potential binding site. However, as per (Stark, Brennecke, Russell, & Cohen, 2003) there are some issues with this approach; the linker may lead to artifacts. RNAHybrid, on the other hand, directly predicts binding sites in target mRNAs; the algorithm finds energetically most favorable hybridizations of a small RNA (miRNA) to a large RNA (mRNA). RNAHybrid does not allow intramolecular hybridizations and multi-loops; the algorithm calculates the minimum free energy of a hybridization of a miRNA and an mRNA. The time complexity of the algorithm is linear with respect to the length of the miRNA.

DIANA-microT (Maragkakis et al., 2009) uses the following features: binding site structural accessibility, nucleotide composition flanking the binding sites or proximity of one binding site to another within the same 3'UTR. The tool groups miRNA driver sequences with presumptive MREs (miRNA recognition elements). The tool first dynamically divides the 3'UTR sequence into 9-nt sections, and then scores these sections with miRNA driver sequences and identifies the one with the highest score. The tool uses Watson-Crick or wobble (G:U) matching, which generates 7-mers, 8-mers and 9-mers, each with a wobble base pair. Second, the tool uses RNA-Hybrid (Rehmsmeier, Steffen, Hochsmann, & Giegerich, 2004) to calculate free binding energy values. Third, the tool finds MREs and checks their sequence conservation, based on sequence alignment. A MRE score is calculated separately for each miRNA, and each binding category. The scores of the MREs are combined through a weighted sum to get the final score. The authors indicated that their tool was found to achieve the highest precision reaching 66%.

The miRanda tool uses dynamic programming to search for maximal local complementarity alignments, corresponding to a double-stranded anti-parallel duplex (John et al., 2004; Betel, Wilson, Gabow, Marks, & Sander, 2008). It uses weighted sum match and mismatch scores for Watson-Crick base pairs. Wobbles (G:U) are allowed but much less scored than perfect pairing; especially favored are pairings between 2-8 nucleotides; gaps are penalized. In the final score the presence of 3' pairing is incorporated along with the free energy of the miRNA:mRNA duplex. Less conserved sites are excluded from prediction. The tool uses the PhastCons conservation score to filter out less conserved anticipated sites. In addition, a conservation of the miRNA:mRNA relationship is considered which depends on (i) conservation of the mature miRNA; (ii) conservation of the relevant parts of the mRNA; and (iii) the presence of a homology between the miRNA and the relevant parts of the mRNA.

PicTar (Krek et al., 2005; Lall et al., 2006) identifies potential targets by using a perfect 7-mer seed match as Watson-Crick or G:U pairing at positions 1-7 or 2-8 from the miRNA 5' end. Imperfect seed pairs are admitted but are scored poorly. A free energy calculation of thc miRNA:mRNA duplex is performed and the results are further filtered; the sites are then checked for conservation. These potential targets are then scored using a hidden Markov model (maximum likelihood score), which assigns probabilities to 3'UTR subsequences as a binding site for a miRNA. The results are scored for each individual species and are combined to get the final score, the outcome of which is the final prediction of miRNA target sites. The false positive rate estimate is 30%, and each vertebrate miRNA has about 200 target transcripts.

MicroTar (Thadani & Tammi, 2006) assumes that (i) miRNA target specificity is caused by 7-mer seed sequences, and (ii) if miRNA:mRNA duplex development is energetically favorable targets are functional (Thadani & Tammi, 2006). MicroTar does not depend on evolutionary conservation. The tool calculates the minimum free energy for each molecule of mRNA using mfold (Zuker, 2003). The energy obtained is denoted as G_1. Next, the tool performs a seed search within mRNA 3'UTRs for Watson-Crick or wobble (G:U) complementary to a seed sequence. This step yields several seed matches, where the seed matches are again folded using cofold (Bernhart et al., 2006). The energy obtained is denoted as G_2 The output of these two steps is a list of seed matches. If miRNA:mRNA heterodimer is energetically more stable than free mRNA (i.e. $G_2 < G_1$) then the seed matches reflect functional targets. To decide the importance of each match of miRNA:mRNA, statistic analysis is done, which results in the final list of miRNA targets.

The mirWIP tool (Hammell et al., 2008) uses RNAHybrid (Rehmsmeier, Steffen, Hochsmann, & Giegerich, 2004) with modifications, to find an initial set of miRNA targets in organisms *C.*

Table 2. Summary of the online data sources

Name	URL	Availability	Reference
miRBase	http://www.mirbase.org/	Online	(Griffiths-Jones, Saini, van Dongen, & Enright, 2008)
TarBase	http://microrna.gr/tarbase	Online	(Sethupathy, Corda, & Hatzigeorgiou, 2006; Papadopoulos, Reczko, Simossis, Sethupathy, & Hatzigeorgiou, 2009)
miRNAMap	http://miRNAMap.mbc.nctu.edu.tw/	Online	(Hsu et al., 2006; Hsu et al., 2008)
miRGator	http://genome.ewha.ac.kr/miRGator/	Online	(Nam, Kim, Shin, & Lee, 2008)

elegans and *C. Briggsae*. The tool adopts several liberal filters. For example, a limit was set for the hybrid minimal free energy (MFE) of the duplex, referred to as the hybrid interaction energy. Some heuristics were implemented, including (i) up to three G:U wobble pairs in a 6-mer beginning at miRNA position 2 were allowed; (ii) up to one bulge on the mRNA side of the duplex in a 7-mer beginning at miRNA position 2 was allowed; and (iii) perfect 6, 7, and 8-mers beginning at miRNA position 3 were allowed. Scoring of the targets was done based on three features: seed topology (S), structural accessibility (A), and total interaction energy (E). A site score was calculated as $S \times A \times E$. This gave an initial set of conserved *C. elegans* miRNA binding sites which were further analyzed using related information and scores.

3.2.2 Online Data Sources

Table 2 summarizes some well-known miRNA target databases. The miRBase database (Griffiths-Jones, Saini, van Dongen, & Enright, 2008) is an online repository with a searchable interface. It stores microRNA nomenclature, sequence data, annotation and target predictions, which allows download of files in various formats including MySQL Dump and FASTA. The miRBase database uses miRanda along with a pipeline for prediction of miRNA targets. The database (release 10.0) contains 5,071 miRNA loci from 58 species, expressing 5,922 unique mature miRNA sequences. The current release 15.0 has over 3,000 new hairpin sequences and more than 4,000 new mature sequences. The web site link is http://www.mirbase.org/.

TarBase 1.0 (Sethupathy, Corda, & Hatzigeorgiou, 2006) stores information of miRNA targets in human, mouse, fruit fly, worm, and zebrafish. The database is functionally linked to Gene Ontology (GO) and UCSC Genome Browser. There are more than 550 target site entries in total, where each target site has information of binding miRNA, the gene in which it occurred, and the location information in the 3' UTR region. TarBase 5.0 (Papadopoulos, Reczko, Simossis, Sethupathy, & Hatzigeorgiou, 2009), which is an updated and extended version of TarBase 1.0, contains data extracted from more than 200 scientific papers and has over 1,300 miRNA target site entries. This version not only contains all the information from the older version, but has additional information such as (i) specific cell lines, (ii) cell-type-specific expression of the gene, (iii) developmental or pathological events that a specific miRNA is involved in, and (iv) HGNC symbols as defined by HUGO. This database is functionally linked to other databases including ENSEMBL, HUGO, UCSC and SwissProt. TarBase5.0 can be accessed at http://microrna.gr/tarbase.

The miRNAMap database (Hsu et al., 2006) is a collection of known miRNA genes, putative miRNA genes, known miRNA targets and putative miRNA targets in human, mouse, rat and other metazoan genomes. It provides cross links to many other biological databases. The miRNAMap database has two versions. Version 1.0 (Hsu et al., 2006) uses known miRNAs from miRBase version 6.0 together with the miRanda target prediction tool and has no filtering criteria. Version 2.0 (Hsu et al., 2008) adopts miRBase version 9.2 for known miRNAs. It uses three target prediction tools, namely miRanda, RNAHybrid and TargetScan. It houses experimentally verified miRNAs and miRNA target genes for human, mouse, rat and other metazoan genomes. In addition, version 2.0 uses three criteria to reduce the false positive rate of prediction of miRNA targets along with Q-PCR miRNA profiling of 224 human miRNAs. It contains more than 2,400 miRNAs in total, of which about 540 are in Homo sapiens (hsa) with 6,750 conserved genes in hsa. The miRNAMap database can be accessed online at http://miRNAMap.mbc.nctu.edu.tw/.

The miRGator database (Nam, Kim, Shin, & Lee, 2008) integrates gene expression data, target prediction, functional analysis and genome annotation in a single repository. For miRNA target prediction, three algorithms are used, which include miRanda, PicTar and TargetScanS. The miRGator database provides a statistical enrichment test of miRNA targets in functional categories like GO function, GenMAPP and KEGG pathways, and various diseases. Search can be performed based on the functional categories or expression profiling. The miRGator database supports the exploration of human and mouse genomes and can be accessed online at http://genome.ewha.ac.kr/miRGator/.

4. DISCUSSION

Is there a connection between miRNA- and ARE-mediated post-transcriptional gene regulation? As discussed above, mRNA turnover and regulation of translation, both biologically important post-transcriptional processes, can be significantly influenced by interactions of regulatory RNA binding proteins with *cis*-regulatory elements in 3'UTRs of the mRNAs. Do ARE- and miRNA-mediated gene regulatory pathways intersect at any point? Is there a relationship between them?

Both of the pathways have been found to share some components. For example, human Dicer, a protein normally involved in miRNA maturation, and an Argonaute (AGO) protein, which is part of the miRNA-induced silencing complex (miRISC), are also required for ARE-mediated mRNA decay in HeLa cells (Jing et al., 2005). Mammalian KSRP, an ARE-BP that promotes AMD, is found in the Dicer and Drosha complexes and is involved in regulating the biogenesis of miRNAs (Trabucchi et al., 2009). ARE-BPs like TTP (tristetraprolin), AUF1 and *Hu* family members could be targeted by miRNAs. In fact, TTP has been shown to interact indirectly with miR16 through the AGO family members for destabilization of TNF-alpha mRNA (Jing et al., 2005).

Over 200 genes of miR16 targets contain AU-rich sequences in their 3'UTRs (Asirvatham, Gregorie, Hu, Magner, & Tomasi, 2008). Grimson and coworkers have found with the help of experimental and bioinformatics studies that efficacy of the miRNA target sites is substantially affected by local AU sequence context (Grimson et al., 2007). Presence of miRNA target sites in the AU-rich regions of the 3'UTRs would presumably keep the sites accessible to regulatory proteins. Therefore, it is not unreasonable to speculate that miRNA target sites may overlap AREs or occur in their vicinity in at least a subset of mRNAs.

Asirvatham and coworkers have found that in three genes (SOCS2, AGO1, and RARB), potential miR16 target sites were associated with 3'UTR AREs (Asirvatham, Gregorie, Hu, Magner, & Tomasi, 2008). In fact, miR16, a human miRNA which contains a sequence complementary to the 3'UTR ARE, has been shown to be required for ARE-mediated turnover of the TNF-alpha mRNA (Jing et al., 2005), raising the possibility that miRNAs can also bind to AREs. However, since the ARE-complementary sequence of miRNAs doesn't seem to be large enough for efficient ARE-targeting, the association may require assistance from other proteins. That is why perhaps miR16 targeting to the ARE sequences requires assistance from tristetraprolin (TTP), an ARE binding protein (Jing et al., 2005).

MicroRNAs can be involved in destabilization or stabilization of ARE-mRNAs. MicroRNA-4661 has been found to upregulate IL-10 expression by binding to its 3'UTR ARE and interfering with TTP-mediated AMD (Ma et al., 2010). The IL-10 ARE sequence was found to be complementary to the seed sequence of miR-4661. Ma and coworkers also discovered miR-4661 targets on many other ARE-mRNAs and found some of them to be down regulated in a gene reporter system (Ma et al., 2010). Several factors, including the type of gene and the cells in which they are expressed, may determine how miRNA targeting will impact ARE-mediated mRNA turnover.

Cooperation between miRNAs and ARE-BPs can also regulate TNF-alpha mRNA translation. The miR369-3p sequence has two complementary binding sites within the 34 nucleotide ARE in the 3'UTR of TNF-alpha mRNA, and has been shown to be responsible for translational up- or down-regulation (Steitz & Vasudevan, 2009). This miRNA is responsible for directing ARE-BPs, fragile X mental retardation related protein 1 (FXR1) and AGO2, to the ARE for translational up-regulation.

Understanding the relationship between miRNA and ARE-mediated regulated gene expression requires detailed investigation. As discussed above and reviewed in von Roretz and Gallouzi (2008), Asirvatham, Magner, and Tomasi (2009), emerging evidence points towards links between the two pathways. The observed cooperation involves sharing of protein factors as well as participation of co-located *cis*-elements. One could speculate that miRNA target sites and AREs may have co-evolved along with the pathways. Studying the genome-wide distribution patterns of ARE and miRNA target site co-occurrences in the 3'UTRs can help develop a better picture of these two post-transcriptional regulation pathways and their relationship with each other.

We have conducted an experimental study to gain a better understanding of the association between AREs and miRNA target sites. We selected the miRNA target prediction tool, DIANA-microT, surveyed in Section 3, and the database of miRNA targets predicted by DIANA-microT, which can be downloaded from the website http://diana.cslab.ece.ntua.gr/microT/. The dataset contains information of miRNA target sites and their known locations in 3'UTR regions of human genes, as well as the ENSEMBL ID and Gene ID of each sequence. Another dataset we selected was ARED downloaded from the website http://brp.kfshrc.edu.sa/ARED/. This website contains flat files of ARED database and cluster groups, where the ARED files are in GenBank flat file view (i.e., nucleotide sequence with annotation).

We used the Apriori algorithm (Agrawal, Imielinski, & Swami, 1993) to find association rules between miRNA target sites in ARE-containing genes. Apriori is a well-known data mining algorithm which generates association rules from a given set of transactions. We modeled the ~3200 ARE containing genes in the ARED database as transaction records and items in the transactions by the miRNA target sites and AREs from the datasets we selected. We found that whenever a hsa-miR-30 target site occurs, a hsa-miR-38 target site and an ARE co-occur with a support of 1.3% and confidence of 37.3% (Figure 2). In Figure 2, *cis*-acting elements, including two miRNA target sites and an ARE, both found in the 3'UTR region of an mRNA sequence, can potentially regulate post-transcriptional gene expression. Overlapping elements may suggest cooperation between the two pathways. Other findings include: whenever a hsa-miR-32 target site occurs, a hsa-miR-36 target site and an ARE co-occur with a support of 1% and confidence of 16.5%; whenever a hsa-miR-41 target site occurs, a hsa-miR-88 target site and an

Figure 2. Illustration of the co-occurrence of two miRNA target sites and an ARE region

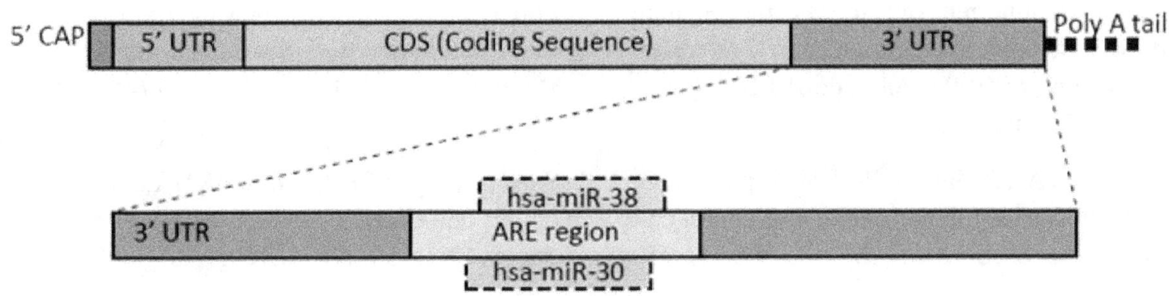

ARE co-occur with a support of 1.3% and confidence of 12.1%; whenever a hsa-miR-51 target site occurs, a hsa-miR-52 target site and an ARE co-occur with a support of 1.2% and confidence of 29.5%; whenever a hsa-miR-54 target site occurs, a hsa-miR-55 target site and an ARE co-occur with a support of 2.5% and confidence of 22.7%.

5. CONCLUSION

AU-rich elements have been mapped to a huge diversity of genes that are involved in important biological processes like development, growth, RNA metabolism, transcriptional regulation, cell to cell signaling, stress response and immune defense. Understanding the mechanisms of ARE and miRNA-mediated mRNA turnover and translational control is vital to unraveling the complexities of post-transcriptional gene expression. Although several computational methods for miRNA target prediction have been developed, it may not be possible to accurately identify AREs as targets. Most miRNA target prediction methods have shortcomings. While some programs can miss legitimate targets, many of them are known to have high false-positive rates. Besides, miRNAs may require assistance from sequence specific RNA binding proteins to bind to their targets. Therefore, studying the role of miRNAs in ARE-mediated mRNA decay will likely require evaluation of miRNA associations with other proteins.

We surveyed here a number of miRNA target prediction tools and online databases of miRNA target sites. We also provided an overview of AREs and available online databases of AREs. We discussed the correlation between AREs and miRNA target sites, and identified co-occurrences between some miRNA target sites in ARE containing genes using a data mining method. Future work includes further analysis of the co-occurrences between AREs and miRNA target sites to understand their biological significance, as well as their role in post-transcriptional gene regulation.

REFERENCES

Agrawal, R., Imielinski, T., & Swami, A. N. (1993). Mining association rules between sets of items in large databases. In *Proceedings of the 1993 ACM SIGMOD International Conference on Management of Data* (pp. 207-216).

Anderson, K. D., Morin, M. A., Beckel-Mitchener, A., Mobarak, C. D., Neve, R. L., & Furneaux, H. M. (2000). Overexpression of HuD, but not of its truncated form HuD I+II, promotes GAP-43 gene expression and neurite outgrowth in PC12 cells in the absence of nerve growth factor. *Journal of Neurochemistry*, *75*, 1103–1114. doi:10.1046/j.1471-4159.2000.0751103.x

Asirvatham, A. J., Gregorie, C. J., Hu, Z., Magner, W. J., & Tomasi, T. B. (2008). MicroRNA targets in immune genes and the Dicer/Argonaute and ARE machinery components. *Molecular Immunology, 45*, 1995–2006. doi:10.1016/j.molimm.2007.10.035

Asirvatham, A. J., Magner, W. J., & Tomasi, T. B. (2009). miRNA regulation of cytokine genes. *Cytokine, 45*, 58–69. doi:10.1016/j.cyto.2008.11.010

Audic, Y., & Hartley, R. S. (2004). Post-transcriptional regulation in cancer. *Biology of the Cell, 96*, 479–498. doi:10.1016/j.biolcel.2004.05.002

Bakheet, T., Frevel, M., Williams, B. R., Greer, W., & Khabar, K. S. (2001). ARED: human AU-rich element-containing mRNA database reveals an unexpectedly diverse functional repertoire of encoded proteins. *Nucleic Acids Research, 29*, 246–254. doi:10.1093/nar/29.1.246

Bakheet, T., Williams, B. R., & Khabar, K. S. (2003). ARED 2.0: an update of AU-rich element mRNA database. *Nucleic Acids Research, 31*, 421–423. doi:10.1093/nar/gkg023

Bakheet, T., Williams, B. R., & Khabar, K. S. (2006). ARED 3.0: the large and diverse AU-rich transcriptome. *Nucleic Acids Research, 34*, D111–D114. doi:10.1093/nar/gkj052

Barreau, C., Paillard, L., & Osborne, H. B. (2005). AU-rich elements and associated factors: are there unifying principles? *Nucleic Acids Research, 33*, 7138–7150. doi:10.1093/nar/gki1012

Bartel, D. P. (2004). MicroRNAs: genomics, biogenesis, mechanism, and function. *Cell, 116*, 281–297. doi:10.1016/S0092-8674(04)00045-5

Bartel, D. P. (2009). MicroRNAs: target recognition and regulatory functions. *Cell, 136*, 215–233. doi:10.1016/j.cell.2009.01.002

Bernhart, S. H., Tafer, H., Muckstein, U., Flamm, C., Stadler, P. F., & Hofacker, I. L. (2006). Partition function and base pairing probabilities of RNA heterodimers. *Algorithms for Molecular Biology; AMB, 1*, 3. doi:10.1186/1748-7188-1-3

Betel, D., Wilson, M., Gabow, A., Marks, D. S., & Sander, C. (2008). The microRNA.org resource: targets and expression. *Nucleic Acids Research, 36*, D149–D153. doi:10.1093/nar/gkm995

Biemar, F., Zinzen, R., Ronshaugen, M., Sementchenko, V., Manak, J. R., & Levine, M. S. (2005). Spatial regulation of microRNA gene expression in the Drosophila embryo. *Proceedings of the National Academy of Sciences of the United States of America, 102*, 15907–15911. doi:10.1073/pnas.0507817102

Boguski, M. S., Lowe, T. M., & Tolstoshev, C. M. (1993). dbEST--database for "expressed sequence tags". *Nature Genetics, 4*, 332–333. doi:10.1038/ng0893-332

Brennecke, J., Stark, A., Russell, R. B., & Cohen, S. M. (2005). Principles of microRNA-target recognition. *PLoS Biology, 3*, e85. doi:10.1371/journal.pbio.0030085

Bushati, N., & Cohen, S. M. (2007). microRNA functions. *Annual Review of Cell and Developmental Biology, 23*, 175–205. doi:10.1146/annurev.cellbio.23.090506.123406

Cao, H., Deterding, L. J., & Blackshear, P. J. (2007). Phosphorylation site analysis of the anti-inflammatory and mRNA-destabilizing protein tristetraprolin. *Expert Review of Proteomics, 4*, 711–726. doi:10.1586/14789450.4.6.711

Chang, T. C., & Mendell, J. T. (2007). microRNAs in vertebrate physiology and human disease. *Annual Review of Genomics and Human Genetics, 8*, 215–239. doi:10.1146/annurev.genom.8.080706.092351

Chen, C. Y., Gherzi, R., Ong, S. E., Chan, E. L., Raijmakers, R., & Pruijn, G. J. (2001). AU binding proteins recruit the exosome to degrade ARE-containing mRNAs. *Cell*, *107*, 451–464. doi:10.1016/S0092-8674(01)00578-5

Chen, C. Y., & Shyu, A. B. (1994). Selective degradation of early-response-gene mRNAs: functional analyses of sequence features of the AU-rich elements. *Molecular and Cellular Biology*, *14*, 8471–8482.

Dixon, D. A., Balch, G. C., Kedersha, N., Anderson, P., Zimmerman, G. A., Beauchamp, R. D., & Prescott, S. M. (2003). Regulation of cyclooxygenase-2 expression by the translational silencer TIA-1. *The Journal of Experimental Medicine*, *198*, 475–481. doi:10.1084/jem.20030616

Espel, E. (2005). The role of the AU-rich elements of mRNAs in controlling translation. *Seminars in Cell & Developmental Biology*, *16*, 59–67. doi:10.1016/j.semcdb.2004.11.008

Esquela-Kerscher, A., & Slack, F. J. (2006). Oncomirs - microRNAs with a role in cancer. *Nature Reviews. Cancer*, *6*, 259–269. doi:10.1038/nrc1840

Fabian, M. R., Sonenberg, N., & Filipowicz, W. (2010). Regulation of mRNA translation and stability by microRNAs. *Annual Review of Biochemistry*, *79*, 351–379. doi:10.1146/annurev-biochem-060308-103103

Filipowicz, W., Bhattacharyya, S. N., & Sonenberg, N. (2008). Mechanisms of post-transcriptional regulation by microRNAs: are the answers in sight? *Nature Reviews. Genetics*, *9*, 102–114. doi:10.1038/nrg2290

Ford, L. P., Watson, J., Keene, J. D., & Wilusz, J. (1999). ELAV proteins stabilize deadenylated intermediates in a novel in vitro mRNA deadenylation/degradation system. *Genes & Development*, *13*, 188–201. doi:10.1101/gad.13.2.188

Friedman, R. C., Farh, K. K., Burge, C. B., & Bartel, D. P. (2009). Most mammalian mRNAs are conserved targets of microRNAs. *Genome Research*, *19*, 92–105. doi:10.1101/gr.082701.108

Galban, S., Martindale, J. L., Mazan-Mamczarz, K., Lopez de Silanes, I., Fan, J., & Wang, W. (2003). Influence of the RNA-binding protein HuR in pVHL-regulated p53 expression in renal carcinoma cells. *Molecular and Cellular Biology*, *23*, 7083–7095. doi:10.1128/MCB.23.20.7083-7095.2003

Gherzi, R., Lee, K. Y., Briata, P., Wegmuller, D., Moroni, C., Karin, M., & Chen, C. Y. (2004). A KH domain RNA binding protein, KSRP, promotes ARE-directed mRNA turnover by recruiting the degradation machinery. *Molecular Cell*, *14*, 571–583. doi:10.1016/j.molcel.2004.05.002

Griffiths-Jones, S., Saini, H. K., van Dongen, S., & Enright, A. J. (2008). miRBase: tools for microRNA genomics. *Nucleic Acids Research*, *36*, D154–D158. doi:10.1093/nar/gkm952

Grimson, A., Farh, K. K., Johnston, W. K., Garrett-Engele, P., Lim, L. P., & Bartel, D. P. (2007). MicroRNA targeting specificity in mammals: determinants beyond seed pairing. *Molecular Cell*, *27*, 91–105. doi:10.1016/j.molcel.2007.06.017

Halees, A. S., El-Badrawi, R., & Khabar, K. S. (2008). ARED Organism: expansion of ARED reveals AU-rich element cluster variations between human and mouse. *Nucleic Acids Research*, *36*, D137–D140. doi:10.1093/nar/gkm959

Hammell, M., Long, D., Zhang, L., Lee, A., Carmack, C. S., & Han, M. (2008). mirWIP: microRNA target prediction based on microRNA-containing ribonucleoprotein-enriched transcripts. *Nature Methods*, *5*, 813–819. doi:10.1038/nmeth.1247

Hsu, P. W., Huang, H. D., Hsu, S. D., Lin, L. Z., Tsou, A. P., & Tseng, C. P. (2006). miRNAMap: genomic maps of microRNA genes and their target genes in mammalian genomes. *Nucleic Acids Research*, *34*, D135–D139. doi:10.1093/nar/gkj135

Hsu, S. D., Chu, C. H., Tsou, A. P., Chen, S. J., Chen, H. C., & Hsu, P. W. (2008). miRNAMap 2.0: genomic maps of microRNAs in metazoan genomes. *Nucleic Acids Research*, *36*, D165–D169. doi:10.1093/nar/gkm1012

Hubbard, T. J., Aken, B. L., Beal, K., Ballester, B., Caccamo, M., & Chen, Y. (2007). Ensembl 2007. *Nucleic Acids Research*, *35*, D610–D617. doi:10.1093/nar/gkl996

Jain, R. G., Andrews, L. G., McGowan, K. M., Gao, F., Keene, J. D., & Pekala, P. P. (1995). Hel-N1, an RNA-binding protein, is a ligand for an A + U rich region of the GLUT1 3' UTR. *Nucleic Acids Symposium Series*, 209–211.

Jing, Q., Huang, S., Guth, S., Zarubin, T., Motoyama, A., & Chen, J. (2005). Involvement of microRNA in AU-rich element-mediated mRNA instability. *Cell*, *120*, 623–634. doi:10.1016/j.cell.2004.12.038

John, B., Enright, A. J., Aravin, A., Tuschl, T., Sander, C., & Marks, D. S. (2004). Human microRNA targets. *PLoS Biology*, *2*, e363. doi:10.1371/journal.pbio.0020363

John, B., Sander, C., & Marks, D. S. (2006). Prediction of human microRNA targets. *Methods in Molecular Biology (Clifton, N.J.)*, *342*, 101–113.

Karsch-Mizrachi, I., & Ouellette, B. F. (2001). The GenBank sequence database. *Methods of Biochemical Analysis*, *43*, 45–63. doi:10.1002/0471223921.ch3

Katsanou, V., Papadaki, O., Milatos, S., Blackshear, P. J., Anderson, P., Kollias, G., & Kontoyiannis, D. L. (2005). HuR as a negative posttranscriptional modulator in inflammation. *Molecular Cell*, *19*, 777–789. doi:10.1016/j.molcel.2005.08.007

Khabar, K. S. (2005). The AU-rich transcriptome: more than interferons and cytokines, and its role in disease. *Journal of Interferon & Cytokine Research*, *25*, 1–10. doi:10.1089/jir.2005.25.1

Khabar, K. S., Bakheet, T., & Williams, B. R. (2005). AU-rich transient response transcripts in the human genome: expressed sequence tag clustering and gene discovery approach. *Genomics*, *85*, 165–175. doi:10.1016/j.ygeno.2004.10.004

Kloosterman, W. P., & Plasterk, R. H. (2006). The diverse functions of microRNAs in animal development and disease. *Developmental Cell*, *11*, 441–450. doi:10.1016/j.devcel.2006.09.009

Krek, A., Grun, D., Poy, M. N., Wolf, R., Rosenberg, L., & Epstein, E. J. (2005). Combinatorial microRNA target predictions. *Nature Genetics*, *37*, 495–500. doi:10.1038/ng1536

Lal, A., Mazan-Mamczarz, K., Kawai, T., Yang, X., Martindale, J. L., & Gorospe, M. (2004). Concurrent versus individual binding of HuR and AUF1 to common labile target mRNAs. *The EMBO Journal*, *23*, 3092–3102. doi:10.1038/sj.emboj.7600305

Lall, S., Grun, D., Krek, A., Chen, K., Wang, Y. L., & Dewey, C. N. (2006). A genome-wide map of conserved microRNA targets in C. elegans. *Current Biology*, *16*, 460–471. doi:10.1016/j.cub.2006.01.050

Lewis, B. P., Burge, C. B., & Bartel, D. P. (2005). Conserved seed pairing, often flanked by adenosines, indicates that thousands of human genes are microRNA targets. *Cell*, *120*, 15–20. doi:10.1016/j.cell.2004.12.035

Lewis, B. P., Shih, I. H., Jones-Rhoades, M. W., Bartel, D. P., & Burge, C. B. (2003). Prediction of mammalian microRNA targets. *Cell*, *115*, 787–798. doi:10.1016/S0092-8674(03)01018-3

Ma, F., Liu, X., Li, D., Wang, P., Li, N., Lu, L., & Cao, X. (2010). MicroRNA-4661 upregulates IL-10 expression in TLR-triggered macrophages by antagonizing RNA-binding protein tristetraprolin-mediated IL-10 mRNA degradation. *Journal of Immunology (Baltimore, MD.: 1950)*, *184*, 6053–6059. doi:10.4049/jimmunol.0902308

Maragkakis, M., Alexiou, P., Papadopoulos, G. L., Reczko, M., Dalamagas, T., & Giannopoulos, G. (2009). Accurate microRNA target prediction correlates with protein repression levels. *BMC Bioinformatics*, *10*, 295. doi:10.1186/1471-2105-10-295

Mendes, N. D., Freitas, A. T., & Sagot, M. F. (2009). Current tools for the identification of miRNA genes and their targets. *Nucleic Acids Research*, *37*, 2419–2433. doi:10.1093/nar/gkp145

Mobarak, C. D., Anderson, K. D., Morin, M., Beckel-Mitchener, A., Rogers, S. L., & Furneaux, H. (2000). The RNA-binding protein HuD is required for GAP-43 mRNA stability, GAP-43 gene expression, and PKC-dependent neurite outgrowth in PC12 cells. *Molecular Biology of the Cell*, *11*, 3191–3203.

Mukherjee, D., Gao, M., O'Connor, J. P., Raijmakers, R., Pruijn, G., Lutz, C. S., & Wilusz, J. (2002). The mammalian exosome mediates the efficient degradation of mRNAs that contain AU-rich elements. *The EMBO Journal*, *21*, 165–174. doi:10.1093/emboj/21.1.165

Nam, S., Kim, B., Shin, S., & Lee, S. (2008). miRGator: an integrated system for functional annotation of microRNAs. *Nucleic Acids Research*, *36*, D159–D164. doi:10.1093/nar/gkm829

Newman, M. A., & Hammond, S. M. (2010). Emerging paradigms of regulated microRNA processing. *Genes & Development*, *24*, 1086–1092. doi:10.1101/gad.1919710

Papadopoulos, G. L., Reczko, M., Simossis, V. A., Sethupathy, P., & Hatzigeorgiou, A. G. (2009). The database of experimentally supported targets: a functional update of TarBase. *Nucleic Acids Research*, *37*, D155–D158. doi:10.1093/nar/gkn809

Peng, S. S., Chen, C. Y., Xu, N., & Shyu, A. B. (1998). RNA stabilization by the AU-rich element binding protein, HuR, an ELAV protein. *The EMBO Journal*, *17*, 3461–3470. doi:10.1093/emboj/17.12.3461

Piecyk, M., Wax, S., Beck, A. R., Kedersha, N., Gupta, M., & Maritim, B. (2000). TIA-1 is a translational silencer that selectively regulates the expression of TNF-alpha. *The EMBO Journal*, *19*, 4154–4163. doi:10.1093/emboj/19.15.4154

Pruitt, K. D., Tatusova, T., & Maglott, D. R. (2007). NCBI reference sequences (RefSeq): a curated non-redundant sequence database of genomes, transcripts and proteins. *Nucleic Acids Research*, *35*, D61–D65. doi:10.1093/nar/gkl842

Raineri, I., Wegmueller, D., Gross, B., Certa, U., & Moroni, C. (2004). Roles of AUF1 isoforms, HuR and BRF1 in ARE-dependent mRNA turnover studied by RNA interference. *Nucleic Acids Research*, *32*, 1279–1288. doi:10.1093/nar/gkh282

Rehmsmeier, M., Steffen, P., Hochsmann, M., & Giegerich, R. (2004). Fast and effective prediction of microRNA/target duplexes. *RNA (New York, N.Y.)*, *10*, 1507–1517. doi:10.1261/rna.5248604

Rhoades, M. W., Reinhart, B. J., Lim, L. P., Burge, C. B., Bartel, B., & Bartel, D. P. (2002). Prediction of plant microRNA targets. *Cell*, *110*, 513–520. doi:10.1016/S0092-8674(02)00863-2

Rodriguez, A., Vigorito, E., Clare, S., Warren, M. V., Couttet, P., & Soond, D. R. (2007). Requirement of bic/microRNA-155 for normal immune function. *Science*, *316*, 608–611. doi:10.1126/science.1139253

Sarkar, B., Xi, Q., He, C., & Schneider, R. J. (2003). Selective degradation of AU-rich mRNAs promoted by the p37 AUF1 protein isoform. *Molecular and Cellular Biology*, *23*, 6685–6693. doi:10.1128/MCB.23.18.6685-6693.2003

Sempere, L. F., Sokol, N. S., Dubrovsky, E. B., Berger, E. M., & Ambros, V. (2003). Temporal regulation of microRNA expression in Drosophila melanogaster mediated by hormonal signals and broad-Complex gene activity. *Developmental Biology*, *259*, 9–18. doi:10.1016/S0012-1606(03)00208-2

Sethupathy, P., Corda, B., & Hatzigeorgiou, A. G. (2006). TarBase: A comprehensive database of experimentally supported animal microRNA targets. *RNA (New York, N.Y.)*, *12*, 192–197. doi:10.1261/rna.2239606

Siomi, H., & Siomi, M. C. (2010). Posttranscriptional regulation of microRNA biogenesis in animals. *Molecular Cell*, *38*, 323–332. doi:10.1016/j.molcel.2010.03.013

Stark, A., Brennecke, J., Russell, R. B., & Cohen, S. M. (2003). Identification of Drosophila microRNA targets. *PLoS Biology*, *1*, E60. doi:10.1371/journal.pbio.0000060

Steitz, J. A., & Vasudevan, S. (2009). miRNPs: versatile regulators of gene expression in vertebrate cells. *Biochemical Society Transactions*, *37*, 931–935. doi:10.1042/BST0370931

Stoecklin, G., Colombi, M., Raineri, I., Leuenberger, S., Mallaun, M., & Schmidlin, M. (2002). Functional cloning of BRF1, a regulator of ARE-dependent mRNA turnover. *The EMBO Journal*, *21*, 4709–4718. doi:10.1093/emboj/cdf444

Thadani, R., & Tammi, M. T. (2006). MicroTar: predicting microRNA targets from RNA duplexes. *BMC Bioinformatics*, *7*(5), S20. doi:10.1186/1471-2105-7-S5-S20

Trabucchi, M., Briata, P., Garcia-Mayoral, M., Haase, A. D., Filipowicz, W., & Ramos, A. (2009). The RNA-binding protein KSRP promotes the biogenesis of a subset of microRNAs. *Nature*, *459*, 1010–1014. doi:10.1038/nature08025

von Roretz, C., & Gallouzi, I. E. (2008). Decoding ARE-mediated decay: is microRNA part of the equation? *The Journal of Cell Biology*, *181*, 189–194. doi:10.1083/jcb.200712054

Wilusz, C. J., & Wilusz, J. (2004). Bringing the role of mRNA decay in the control of gene expression into focus. *Trends in Genetics*, *20*, 491–497. doi:10.1016/j.tig.2004.07.011

Wilusz, C. J., Wormington, M., & Peltz, S. W. (2001). The cap-to-tail guide to mRNA turnover. *Nature Reviews. Molecular Cell Biology*, *2*, 237–246. doi:10.1038/35067025

Wulczyn, F. G., Smirnova, L., Rybak, A., Brandt, C., Kwidzinski, E., & Ninnemann, O. (2007). Post-transcriptional regulation of the let-7 microRNA during neural cell specification. *The FASEB Journal*, *21*, 415–426. doi:10.1096/fj.06-6130com

Xu, N., Chen, C. Y., & Shyu, A. B. (2001). Versatile role for hnRNP D isoforms in the differential regulation of cytoplasmic mRNA turnover. *Molecular and Cellular Biology*, *21*, 6960–6971. doi:10.1128/MCB.21.20.6960-6971.2001

Zhang, R., & Su, B. (2009). Small but influential: the role of microRNAs on gene regulatory network and 3'UTR evolution. *Journal of Genetics and Genomics = Yi Chuan Xue Bao*, *36*, 1–6. doi:10.1016/S1673-8527(09)60001-1

Zubiaga, A. M., Belasco, J. G., & Greenberg, M. E. (1995). The nonamer UUAUUUAUU is the key AU-rich sequence motif that mediates mRNA degradation. *Molecular and Cellular Biology*, *15*, 2219–2230.

Zuker, M. (2003). Mfold web server for nucleic acid folding and hybridization prediction. *Nucleic Acids Research*, *31*, 3406–3415. doi:10.1093/nar/gkg595

This work was previously published in the International Journal of Knowledge Discovery in Bioinformatics, Volume 1, Issue 3, edited by Xiao-Li Li and See-Kiong Ng, pp. 97-112, copyright 2010 by IGI Publishing (an imprint of IGI Global).

Section 3
Protein Interaction Analysis

Chapter 10
Prediction of Protein-Protein Interactions Between Human Host and Two Mycobacterial Organisms

Oruganty Krishnadev
Indian Institute of Science, India

Shveta Bisht
Indian Institute of Science, India

Narayanaswamy Srinivasan
Indian Institute of Science, India

ABSTRACT

The genomes of many human pathogens have been sequenced but the protein-protein interactions across a pathogen and human are still poorly understood. The authors apply a simple homology-based method to predict protein-protein interactions between human host and two mycobacterial organisms viz., M.tuberculosis and M.leprae. They focused on secreted proteins of pathogens and cellular membrane proteins to restrict to uncovering biologically significant and feasible interactions. Predicted interactions include five mycobacterial proteins of yet unknown function, thus suggesting a role for these proteins in pathogenesis. The authors predict interaction partners for secreted mycobacterial antigens such as MPT70, serine proteases and other proteins interacting with human proteins, such as toll-like receptors, ras signalling proteins and immune maintenance proteins, that are implicated in pathogenesis. These results suggest that the list of predicted interactions is suitable for further analysis and forms a useful step in the understanding of pathogenesis of these mycobacterial organisms.

DOI: 10.4018/978-1-4666-1785-8.ch010

INTRODUCTION

Tuberculosis (TB) has adversely affected human affairs since biblical times. WHO estimates that there were 1.7 million cases of death due to tuberculosis in 2007 and around 13 million cases of Tuberculosis globally, most of them from Africa and South East Asian regions (www.who.int/tb/publications/global_report/2009/key_points/en/index.html). The standard medication for TB followed worldwide is isoniazid and rifampicin. However, the emergence of multi drug resistant strains (LoBue, 2009) has prompted a worldwide attempt to develop newer and more effective drugs for the treatment of TB (Kamal, Azeera, Malik, Shaik, & Rao, 2008; Guy & Mallampalli, 2008). Further complications in tackling TB are due to the persistence of *M.tuberculosis* bacteria inside the macrophages, which makes drug delivery inefficient. Due to these reasons, there is an increased need to understand the molecular mechanism of persistence of *M.tuberculosis* inside the human macrophages and to evolve strategies to disrupt the host-pathogen interplay which leads to this persistence of the bacteria inside the host (Houben, Nguyen & Pieters, 2006; Pieters, 2008; Zahrt, 2008). Recent advances in understanding mycobacterium physiology (Young, Stark & Kirschner, 2008; Tailleux et al, 2008) including a computational study of protein-protein interactions in the pathogen (Cui, Zhang, Wang & He, 2008) may lead to a better understanding. However the molecular details of host-pathogen interactions for this organism are still not very well understood.

The host-pathogen interplay in the initial stages of infection in TB, is mainly due to glycolipids and/or lipoproteins (Józefowski, Sobota, & Kwiatkowska, 2008) with crucial roles believed to be mediated by secreted proteins. Thus, there is a need to understand the molecular details of protein-protein interactions occurring between the human host and bacterial pathogen.

Another diseases similar to TB is Leprosy, it affects mainly the neuronal cells and causes degeneration of the nerves. The organism causing leprosy is *Mycobacterium leprae* and is closely related to *M.tuberculosis*.

The detection of protein-protein interactions is not very efficient in such systems and novel procedures have been proposed for experimental detection of protein-protein interactions occurring among the proteins of *M.tuberculosis* (O'hare, Juillerat, Dianiskov_, & Johnsson, 2008). Such approaches and previously proposed methods (Suter, Kittanakom, & Stagljar, 2008; Lalonde et al., 2008) can be extended to detect protein -protein interactions occurring between the host and pathogen also, but to our knowledge has not been attempted so far on a genome scale. We cannot understand the pathophysiology of TB without a comprehensive list of protein-protein interactions (PPI) across the host and pathogen organisms, and in order to bridge the gap we applied a method we developed earlier (Krishnadev & Srinivasan, 2008) for the prediction of PPI across the host and pathogen organisms.

The prediction of PPI is being extensively attempted in the recent years at the level of an organism. Various methods have been proposed for prediction of PPI within an organism (e.g., Wojcik, Boneca & Legrain, 2002; Zhong & Sternberg, 2006; Persico et al., 2005; Rhodes et al., 2005; Huang et al., 2004). In the recent years fewer methods have been proposed to predict PPI across organisms (Dyer, Murali, & Sobral, 2007; Dyer, Murali & Sobral, 2008; Davis, Barkan, Eswar, McKerrow, & Sali, 2008; Krishnadev & Srinivasan, 2008, Lee et al., 2008). Typically, the approach used in such predictions is to employ a set of known PPIs as template and use homology to infer PPI occurring between target proteins. Our approach to the problem is to identify proteins of host and pathogen with high sequence similarity to interacting proteins in the template dataset. The high sequence similarity ensures that PPI properties are conserved, and it is especially necessary since it has been shown earlier that interaction properties can be different for proteins showing sequence similarity range in the twilight zone (Rekha, Machado, Narayanan, Krupa, & Srinivasan, 2005). In this study, we have focused on predicting interactions involving *M.tuberculosis* and *M.leprae* proteins which fall into the following categories:

1. Known to be important for pathogenesis
2. Found on the membrane of the pathogen
3. Known to be secreted into the host cell

We believe that by focusing on the secreted/transmembrane proteins of the pathogen, we can derive a set of PPIs which are biologically meaningful and significant for pathogenesis.

RESULTS AND DISCUSSION

The details of the approach used for prediction is given in the Methods section. In brief, the predictions are made considering that the interaction properties of proteins with very similar sequences (>30% sequence identity) are very similar (Aloy, Ceulemans, Stark, & Russell, 2003). We use two databases which report known protein-protein interactions (mostly physically interacting proteins), namely DIP (Salwinski et al., 2004) and iPfam (Finn, Marshall, & Bateman, 2005). Mapping of protein sequences of host and pathogen to template (DIP, iPfam) is done by sensitive profile based approaches, and additional refinements (as explained in Methods) are made to ensure the accuracy of predicted interactions. The subcellular localization information is then used to further refine the *'in silico'* interactions to give a set of plausible *'in vivo'*' interactions. Literature mining is done manually to determine the significance of each of the predicted interactions.

Proteins of *M.Leprae* and *M.Tuberculosis* Involved in Predicted Interactions

The interactions were predicted using two different template datasets which provide extensive lists of known protein-protein interactions between proteins from the same organism. One of the datasets used is DIP (Salwinski et al., 2005), which is a database of interacting proteins recognised using large-scale yeast-two hybrid and similar methods. The prediction of PPI between host and pathogen proteins using the templates in DIP database was achieved by recognizing homologs of the DIP sequences in the host and pathogen genome using PSI-BLAST (Altschul et al., 1997; Ye, McGinnis & Madden, 2008). The other dataset used is iPfam (Finn, Marshall & Bateman, 2005), a structure-based database of interactions between protein families (or domains). Using RPS-BLAST (Altschul et al., 1997) and the MulPSSM database (Gowri, Krishnadev, Swamy, & Srinivasan, 2007) Pfam domains were mapped on to the protein sequences of both the pathogen and host genome. The details on employing the two datasets are given in the Methods section.

Out of 1605 genes present in the *M.leprae* genome, 709 (44.2%) gene products have been predicted to interact with 6805 human proteins, amounting to a total of 33491 interactions. In the case of *M.tuberculosis* genome, 1458 (36.5%) proteins out of 3991 genes present in the complete pathogen genome were predicted to interact with 7328 human proteins leading to 62962 interactions. Around 100 proteins in *M.tuberculosis* have been identified to be secreted into the host cell, out of these, 25 proteins were predicted to interact with human proteins. For 146 proteins with at least one transmembrane region, we could predict interactions with human proteins. The topology of these proteins is hard to predict due to the unusual nature of the *M.tuberculosis* membrane.

Domains in Secreted and Extracellular Proteins

The domain assignment for each protein in the *M.tuberculosis* and *M.leprae* genome was made against MulPSSM database of Pfam domains using RPS-BLAST. The functional domain assignment can lead to useful information on the functions performed by the pathogen proteins. The subcellular localization of *M.leprae* proteins was taken from the Psortb database (Gardy et al., 2005) since there is very little experimental data available. In case of *M.tuberculosis*, the list of secreted proteins was taken from an experimental data which gives secreted proteins of *M.tuberculosis* based on mass spectrometry methods (Braunstein et al., 2000). The membrane protein prediction was taken from Psortb database (Gardy et al., 2005) for both the pathogens. The accuracy of topology of the membrane proteins is difficult to assess due to the unusual cell wall structure of these pathogens. Supplementary Table S1 lists the Pfam domains assigned to the proteins for which we predict an interaction partner in the host. Out of all the Pfam domains assigned to the proteins predicted to be involved in interactions, it is interesting to note that there are many metabolic enzymes, such as short chain dehydrogenases, Asparagine synthases, FAD utilising oxidases and Redoxins. Although we have predicted interactions for each of the proteins listed in Supplementary Table S1, the presence of these proteins in itself is suggestive of a role in pathogenesis. It is quite likely that the fatty acid utilizing enzymes are involved in either degrading or synthesizing lipids which are crucial for the survival of the pathogen. The redox state of the host is correlated with the generation of nitrogen free radicals and oxygen burst, which are essential for killing the pathogen. Thus, it is probable that the secreted redox regulating enzymes will act to reduce the damage to pathogen by disrupting the normal redox state of the host cell.

Apart from these Pfam domains, in both the pathogens, there are a number of proteases and other enzymes present in the list of secreted proteins. The role of these proteins in disrupting host processes should be investigated further to gain a better understanding of the disease process. In the next section, we discuss some of the examples which according to us are biologically significant and can lead to novel hypotheses on the pathogenesis.

Biologically Significant Examples of Interactions

In this section, a few examples of predicted interactions which are significant in understanding the pathogenesis of the disease are presented. Wherever possible, suitable references are indicated which support the significance of the protein for pathogenesis. In the first section, *M.leprae* protein interactions with human proteins are discussed and in the subsequent section *M.tuberculosis* protein interactions with human proteins are discussed. The examples discussed in these sections are summarized in Table 1 and 2.

Examples of Predicted Protein-Protein Interactions Between *M.leprae* and Human

The first protein in the list given in Table 1 is a possible secreted serine protease from *M.leprae*. The serine protease is predicted to interact with two isoforms of ERBB2 interacting protein from human genome. The involvement of *M.leprae* proteins with ERBB2 is well documented (Tapinos, Ohnishi, & Rambukkana, 2006). The interaction we predict can be seen as an accessory in the pathogenesis of the disease. The serine protease is also predicted to interact with several ras-related signaling proteins (Noon & Llyod, 2007). The implication of this interaction can be understood from recent reports that the leprosy pathogen induces cell damage by disrupting cell signaling events which are mediated by ras-raf-MEK-ERK pathway proteins (Noon & Lloyd, 2007). Thus,

Table 1. Selected examples of predicted protein-protein interactions with appropriate biological context for the pair M.leprae and Human. The details and full list of predicted interactions for each of these proteins is given in Supplementary Table 2. The pathogen proteins in this table are either secreted into the host or are localized to the cellular membrane of the pathogen.

M.leprae gene code	M.leprae protein	Human gene code	Annotation for the human protein
15827528	possible serine protease	55770895	ERBB2 interacting protein isoform 7
		56550077	RAB40A, member RAS oncogene family
		8923909	ERBB2 interacting protein isoform 2
		8922964	synaptojanin 2 binding protein
		88981168	similar to Rap guanine nucleotide exchange factor 2 (Neural RAP guanine nucleotide exchange protein)
15827779	trigger factor	10863901	mitogen-activated protein kinase-activated protein kinase 2 isoform 1
		11321597	kinase insert domain receptor (a type III receptor tyrosine kinase)
		10835065	protein kinase, Y-linked
		113429623	similar to brain protein 239-like
		18765758	dual-specificity tyrosine-(Y)-phosphorylation regulated kinase 1A isoform 1
		24307911	acid sphingomyelinase-like phosphodiesterase 3A
15827592	hypothetical protein ML1173	40254422	nitric oxide synthase 3 (endothelial cell)
15827671	hypothetical protein ML1300	13027612	RAS-like family 11 member B
		15718761	c-K-ras2 protein isoform b
		31559833	small GTPase protein E-Ras
15826884	hypothetical protein ML0021	19924175	axonal transport of synaptic vesicles
		29893564	microspherule protein 1 isoform 1
		113425229	similar to StAR-related lipid transfer protein 9 (StARD9)
		41393559	kinesin family member 1B isoform alpha
15827966	hypothetical protein ML1808	14702169	plasminogen activator
		20302166	toll-like receptor 8 isoform 1
		20302168	toll-like receptor 8 isoform 2
		4504557	intercellular adhesion molecule 2 precursor
		4504879	killer cell lectin-like receptor subfamily B, member 1
		59889560	neurotrophic tyrosine kinase, receptor, type 3 isoform b precursor

the biological context of this interaction makes it an interesting target for further studies. It is interesting to note that this protein is also predicted to interact with various phosphatases thus enabling it to mediate disruption of cell signaling through multiple targets. The template interaction for these predictions and other information about the prediction are given in detail in Supplementary Table S2, which also gives the subcellular localization annotation from Psortb.

The second protein listed in Table 1 is a possible molecular chaperone from *M.leprae*.

Table 2. Selected examples of predicted protein-protein interactions with appropriate biological context for the pair M.tuberculosis and Human. The details and full list of predicted interactions for each of these proteins is given in Supplementary Table 3. The pathogen proteins in this table are either secreted into the host or are localized to the cellular membrane of the pathogen.

M.tuberculosis gene code	M.tuberculosis protein name	Human gene code	Human protein annotation
15607267	Serine protease	10835173	nitric oxide synthase 1
		11321587	complement factor H-related 1
		18104986	protein tyrosine phosphatase, non-receptor type 3
		13129124	caspase recruitment domain protein 14 isoform 1
		14149671	microtubule associated serine/threonine kinase 2
15607631	putative two component sensor histidine kinase SENX3	16418413	solute carrier family 26, member 9 isoform a
		28872812	MORC family CW-type zinc finger 3
		4505911	postmeiotic segregation 1
		5901922	CDC37 homolog
156100012	major secreted immunogenic protein MPT70	4507467	transforming growth factor, beta-induced, 68kDa
		5453834	periostin, osteoblast specific factor
15607161	hypothetical protein Rv0019c	113425593	similar to StAR-related lipid transfer protein 9 (StARD9)
		39725952	angiogenic factor VG5Q
15609444	hypothetical protein Rv2307c	34594664	G-protein signalling modulator 1 (AGS3-like, C. elegans)
15608215	hypothetical protein Rv1075c	76096366	phospholipase B1
15609606	hypothetical protein Rv2469c	62860046	zinc finger, RAN-binding domain containing 3

The molecular chaperones from *M.leprae* and *M.tuberculosis* are known to be important players in the pathogenesis of the organism (Qamra, Mande, Coates, & Henderson, 2005). We have predicted interactions between various protein kinases of human genome with *M.leprae* molecular chaperones. Whether this interaction proves to be clinically significant can be deciphered only after rigorous experimentation and hence is an interesting hypothesis.

In addition to the above mentioned examples, there are a number of conserved hypothetical proteins for which our method was able to detect interactions. Some of these are given in Table 1. The targets for the hypothetical proteins are nitric oxide synthases (which are essential for killing the pathogens), signaling proteins (for e.g. 15827671 and 15827592) and other enzymes which can be instrumental in generating reactive oxygen molecules that are crucial for eliminating this intracellular bacterium. Disruption of the pathways which lead to the death of the pathogen can be one way by which the pathogen is able to successfully survive the harsh environment of the macrophages and other cells of the human tissues.

Examples from Predicted Interactions Involving *M.tuberculosis* Proteins

The examples discussed in this section, from the tuberculosis genome, are given in Table 2. A more detailed list with all the predictions along with their details is given as Supplementary Table S3. In the first example, a serine protease (15607267) has been predicted to interact with various cellular targets in the human. The serine protease is exported into the host cell and thus, the biological context for the interaction is quite appropriate for the interaction to take place. The serine protease has a N-terminal trypsin domain and a C-terminal PDZ domain. Both the domains have been predicted to be involved in interactions with the human host. The trypsin domain of the pathogen proteins could be involved in an enzymatic function cleaving some of the host proteins. The targets for the pathogen protein in the human cell are numerous and selected examples are given in Table 2. These involve phosphatases, receptors involved in immune system function among others. Further exploration of the interactions of this secreted serine protease can lead to a better understanding of the role played by this protein in the pathogenesis of tuberculosis.

The next example of predicted interaction is that of the SENX3 protein which is involved in a two-component system in the pathogen. The two-component systems provide a system to modulate the pathogen physiology according to the host conditions. The SENX3 protein and its partner are known to be important for virulence (Rickman et al., 2004) and according to experimental data available, it is secreted into the host cell (Braunstein et al., 2000). The interaction of this protein with transport proteins and a CDC37 homolog is thus interesting and can provide a novel hypothesis on the mechanism by which the bacteria is able to disrupt host cell division. Another two-component system for which we have predicted interactions with human proteins is the MprB sensor kinase protein which is secreted into the human cell. This protein show slightly different domain architecture than the SENX3 protein (sequence identity 34% in the HATPase domain) and is also known to be important for pathogenesis. The difference in architecture and the divergence of the proteins can explain why two different two component systems are exported to the host.

Another protein predicted to interact with human proteins is the major secreted immunogenic protein MPT70 (15610012) with four proteins in the host including a 68kDa protein induced by TGFβ. The pathogen and the human protein have a single fasciclin domain which has been shown to dimerise in Drosophila (PDB: 1O70, Clout, Tisi & Hohenester, 2003). Since both the human and pathogen protein contain fasciclin domains, it can be hypothesized that the two proteins can interact. When seen with the fact that TGFβ has a major role to play in monocyte differentiation, the interaction assumes a greater significance. Another protein containing the fasciclin domain is the cell surface lipoprotein mpt83. This protein is also predicted to interact with the host proteins (the predictions are the same for both pathogen proteins). Thus, delineating which of the proteins acts at what stage using experiments can lead us to refine our understanding of the attachment of the pathogen to the host and also in understanding if the secreted protein plays a role in the disruption of signal transduction events.

Examples of Predictions Involving Proteins of as Yet Unknown Function

The use of both iPfam and DIP datasets has enabled us to predict interactions for proteins whose function is as yet unknown (annotated as hypothetical proteins). In the case of *M.tuberculosis*, we predicted interactions for 5 proteins which are secreted into the host and 15 proteins which have a membrane spanning region. In case of *M.leprae,* we predicted interactions for 6 proteins which have a membrane spanning region.

The five secreted proteins annotated as hypothetical proteins are described here in more detail. The hypothetical protein Rv1419 (gi code 15608557) is predicted to interact with enzymes which transfer glycosyl moieties. The protein might be involved in the stabilization or maturation of *M.tuberculosis* glycoproteins. The protein Rv0019c (gi code 15607161) is predicted to interact with many lipid transfer proteins in the human host cell. Thus, this protein could be involved in maintaining the pool of lipids in the pathogen which is then used for replication of the pathogen. The proteins Rv1075c (gi code 15608215) and Rv1910c (gi code 15609047) are predicted to interact with phospholipase and other lipid binding proteins and could be involved in lipid metabolism. The protein Rv2469c (gi code 15609606) is predicted to interact with only one protein the zinc finger, RAN binding domain protein of the host. The RAN protein is involved in nuclear transport and hence this interaction assumes significance as it is known that *M.tuberculosis* impairs host maturation using as yet poorly understood mechanisms.

Accuracy of Predicted Interactions

Our objective in this work is to provide a reasonably accurate list of predicted protein-protein interactions between a pathogen and host. When discussing the examples of predictions mentioned in the previous sections, we have assumed that each of the prediction could at least occur in an *in vitro* condition. Our previous work in this area suggests that such an assumption is not inaccurate. Structural analysis based on conservation of interface residues suggests that interaction properties are conserved when the sequence identity between template and target interaction is quite high. Thus, we have used stringent cutoffs to ensure that sequence identity is high over a large proportion of the template and target sequences. As an example, we modeled the structure of sensor histidine kinase SENX3 complex with human TNF receptor associated protein. The modeled complex was generated using swissmodel (Arnold, Bordoli, Kopp, & Schwede, 2006; Schwede, Kopp, Guex, & Peitsch, 2003; Guex & Peitsch, 1997) based on the template 1THN. The interfacial residues are shown in Figure 1 and it shows that the interaction is quite possible since, hydrophobic amino acids are predicted to interact with hydrophobic residues and polar residues are in close proximity to polar residues from the other polypeptide. At this juncture, it is not possible to generate a complete model of the interaction between the two proteins, since the structures of neither of these proteins have been solved experimentally. It is only possible to model the structure of the interacting domains and find out the feasibility of interaction. Thus, it seems very likely that our predicted interactions could occur at least in an *'in vitro'* condition. The rigorous assessment of *'in*

Figure 1. The modeled structure of a complex between SENX3 protein of M.tuberculosis with human protein (gi code: 16418413, solute carrier family 26, isoform A protein), In the figure, only the interface residues are shown, with the human protein residues shown in pink and the pathogen protein residues shown in blue. The residues at the interface are indicated.

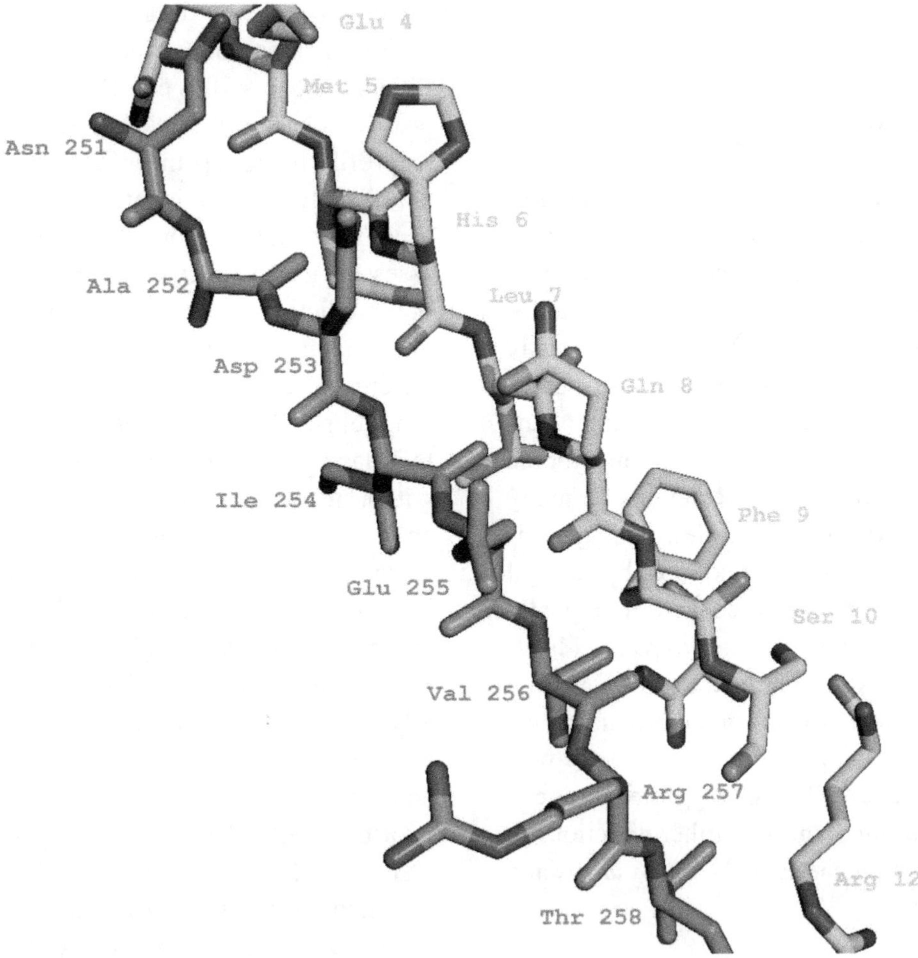

vivo' occurrence of each of the predictions cannot be done reliably as yet since the data on protein -protein interactions is quite minimal and hence statistical methods of pattern recognition cannot be directly applied to this case. Thus, methods such as ours are at present very helpful in bridging this gap in our knowledge and proposing novel hypothesis regarding the pathogenesis. This is most clearly shown by the predictions we have made for hypothetical proteins in the pathogen.

The template databases we chose are both generated by expert curation and literature surveys and hence can be believed to be highly accurate. In case of the DIP database, the interface region is not known in most cases and hence it is difficult to do structural analysis on prediction based on interactions in the DIP database. Most of the interactions present in the DIP database are taken from large scale yeast-two hybrid experiments and thus there is an inherent error associated with these cases, since yeast two hybrid can report indirect interactions. But, by focusing on secreted protein, we have reduced the error rate since, these are present in the cytoplasm of the host cell and hence can participate in indirect interactions. Thus, we believe that our predictions are definitely possible in an *'in vitro'* condition and quite likely to occur under *'in vivo'* conditions.

CONCLUSION

In this article we report prediction of protein-protein interactions of human host with *Mycobacterium tuberculosis* and *Mycobacterium leprae*. We have employed a simple homology-search approach that was developed by us previously (Krishnadev & Srinivasan, 2008). In the present study we focus on secreted proteins of pathogens and pathogenic proteins having a transmembrane region. We have predicted interactions for many secreted proteins which have a role in pathogenesis and these interactions can be explored by detailed experiments. The lists of predicted biologically significant interactions for the two pairs of host-pathogen systems include interactions for some of the hypothetical proteins of hitherto unknown function suggesting possible roles of these proteins in pathogenesis. The list suggested by us can be a good starting point for further directions in research on pathogenesis of *M. tuberculosis* and *M. leprae*.

METHOD

The generation of dataset of putative interactions (*'in silico'* interactions) was done by taking the genome sequences of the host and pathogen (taken from NCBI ftp site) and querying them using PSI-BLAST and RPS-BLAST (Altschul et al., 1997) against the DIP (Salwinski et al., 2004) and Pfam (Finn, Marshall, & Bateman, 2005) databases respectively. Since iPfam lists interacting Pfam domains, the mapping of Pfam domains to each of the pathogen and host proteins is done against the MulPSSM database using RPS-BLAST. In order to decrease the chances of false positives due to short alignments, we additionally use a length coverage cutoff. For similarity to DIP sequences, we use a cutoff that the alignment should have a 80% coverage on the DIP sequence (since the region of interaction is not known). In case of iPfam, we use a 70% cutoff of the Pfam family profile to ensure that we get accurate sequence or profile mapping. In either case, the e-value cutoff of 10^{-5} and an identity cutoff of 25% was used. These cut-off values have been obtained by assessing the accuracy of protein-protein interaction prediction for yeast proteome using PDB as the template and this exercise was performed at different e-values and alignment cut-off values (unpublished). The putative interactions found at this stage can be called as *'in silico'* interactions with a strong likelihood that they will interact in an *'in vitro'* condition since, they have significant sequence similarity to a template interaction.

The predicted subcellular localization was taken from Psortb database which is around 95% accurate for bacterial proteins. Additionally, for the *M.tuberculosis* proteins, we used an experimentally derived set of secreted proteins and this was merged with the set obtained from Psortb database. The putative list generated using PSI-BLAST and RPS-BLAST of pathogen and host genomes against DIP and iPfam databases respectively were pruned using the subcellular localization to generate a set of interactions which according to the method employed will interact under *'in vivo'* conditions. Manual analysis of the interactions was done using literature surveys.

ACKNOWLEDGMENT

This research is supported by Department of Biotechnology, Government of India as well by Microsoft corporation. O. Krishnadev is supported by a fellowship from CSIR. We would like to acknowledge the help from Ms. Chitra and Mr. Dhanush for their contributions to a preliminary study and Mr. Sairam Swamy for his help in structural analysis of protein-protein interactions.

REFERENCES

Aloy, P., Ceulemans, H., Stark, A., & Russell, R. B. (2003). The relationship between sequence and interaction divergence in proteins. *Journal of Molecular Biology*, *332*, 989–998. doi:10.1016/j.jmb.2003.07.006

Altschul, S. F., Madden, T. L., Schaffer, A. A., Zhang, J., Zhang, Z., & Miller, W. (1997). Gapped BLAST and PSI-BLAST: a new generation of protein database search programs. *Nucleic Acids Research*, *25*, 3389–3402. doi:10.1093/nar/25.17.3389

Arnold, K., Bordoli, L., Kopp, J., & Schwede, T. (2006). The SWISS-MODEL Workspace: A web-based environment for protein structure homology modelling. *Bioinformatics (Oxford, England)*, *22*, 195–201. doi:10.1093/bioinformatics/bti770

Braunstein, M., Griffin, T. J. IV, Kriakov, J. I., Friedman, S. T., Grindley, N. D., & Jacobs, W. R. Jr. (2000). Identification of genes encoding exported Mycobacterium tuberculosis proteins using a Tn552'phoA in vitro transposition system. *Journal of Bacteriology*, *182*, 2732–2740. doi:10.1128/JB.182.10.2732-2740.2000

Clout, N. J., Tisi, D., & Hohenester, E. (2003). Novel fold revealed by the structure of a Fas1 domain pair from the insect cell adhesion molecule Fasciclin I. *Structure (London, England)*, *11*, 197–203. doi:10.1016/S0969-2126(03)00002-9

Cui, T., Zhang, L., Wang, X., & He, Z. G. (2009). Uncovering new signaling proteins and potential drug targets through the interactome analysis of Mycobacterium tuberculosis. *BMC Genomics*, *10*, 118. doi:10.1186/1471-2164-10-118

Davis, F. P., Barkan, D. T., Eswar, N., McKerrow, J. H., & Sali, A. (2007). Host pathogen protein interactions predicted by comparative modeling. *Protein Science*, *16*, 2585–2596. doi:10.1110/ps.073228407

Dyer, M. D., Murali, T. M., & Sobral, B. W. (2007). Computational prediction of host-pathogen protein-protein interactions. *Bioinformatics (Oxford, England)*, *23*, i159–i166. doi:10.1093/bioinformatics/btm208

Dyer, M. D., Murali, T. M., & Sobral, B. W. (2008). The landscape of human proteins interacting with viruses and other pathogens. *PLoS Pathogens*, *4*, e32. doi:10.1371/journal.ppat.0040032

Finn, R. D., Marshall, M., & Bateman, A. (2005). iPfam: visualization of protein-protein interactions in PDB at domain and amino acid resolutions. *Bioinformatics (Oxford, England)*, *21*, 410–412. doi:10.1093/bioinformatics/bti011

Gardy, J. L., Laird, M. R., Chen, F., Rey, S., Walsh, C. J., & Ester, M. (2005). PSORTb v.2.0: expanded prediction of bacterial protein subcellular localization and insights gained from comparative proteome analysis. *Bioinformatics (Oxford, England)*, *21*, 617–623. doi:10.1093/bioinformatics/bti057

Gowri, V. S., Krishnadev, O., Swamy, C. S., & Srinivasan, N. (2006). MulPSSM: a database of multiple position-specific scoring matrices of protein domain families. *Nucleic Acids Research*, *34*, D243–D246. doi:10.1093/nar/gkj043

Guex, N., & Peitsch, M. C. (1997). SWISS-MODEL and the Swiss-PdbViewer: An environment for comparative protein modelling. *Electrophoresis*, *18*, 2714–2723. doi:10.1002/elps.1150181505

Guy, E. S., & Mallampalli, A. (2008). Managing TB in the 21st century: existing and novel drug therapies. *Therapeutic Advances in Respiratory Disease*, *2*, 401–408. doi:10.1177/1753465808099522

Houben, E. N., Nguyen, L., & Pieters, J. (2006). Interaction of pathogenic mycobacteria with the host immune system. *Current Opinion in Microbiology*, *9*, 76–85. doi:10.1016/j.mib.2005.12.014

Huang, T. W., Tien, A. C., Huang, W. S., Lee, Y. C., Peng, C. L., & Tseng, H. H. (2004). POINT: a database for the prediction of protein-protein interactions based on the orthologous interactome. *Bioinformatics (Oxford, England)*, *20*, 3273–3276. doi:10.1093/bioinformatics/bth366

Józefowski, S., Sobota, A., & Kwiatkowska, K. (2008). How Mycobacterium tuberculosis subverts host immune responses. *Bioessays, 30*, 943-954.

Kamal, A., Azeeza, S., Malik, M. S., Shaik, A. A., & Rao, M. V. (2008). Efforts towards the development of new antitubercular agents: potential for thiolactomycin based compounds. *Journal of Pharmacy & Pharmaceutical Sciences, 11*, 56s–80s.

Krishnadev, O., & Srinivasan, N. (2008). A data integration approach to predict host-pathogen protein-protein interactions: application to recognize protein interactions between human and a malarial parasite. *In Silico Biology*, *8*, 235–250.

Lalonde, S., & Ehrhardt, D.W., Loqu_, D., Chen, J., Rhee, S.Y., & Frommer, W.B. (2008). Molecular and cellular approaches for the detection of protein-protein interactions: latest techniques and current limitations. *The Plant Journal*, *53*, 610–635. doi:10.1111/j.1365-313X.2007.03332.x

Lee, S. A., Chan, C. H., Tsai, C. H., Lai, J. M., Wang, F. S., & Kao, C. Y. (2008). Ortholog-based protein-protein interaction prediction and its application to inter-species interactions. *BMC Bioinformatics*, *9*(Suppl 12), S11. doi:10.1186/1471-2105-9-S12-S11

LoBue, P. (2009). Extensively drug-resistant tuberculosis. *Current Opinion in Infectious Diseases*, *22*, 167–173. doi:10.1097/QCO.0b013e3283229fab

Noon, L. A., & Lloyd, A. C. (2007). Traeting leprosy:an Erb-al remedy? *Trends in Pharmacological Sciences*, *28*, 103–105. doi:10.1016/j.tips.2007.01.004

O'Hare, H., & Juillerat, A., Dianiskov_, P., & Johnsson, K. (2008). A split-protein sensor for studying protein-protein interaction in mycobacteria. *Journal of Microbiological Methods*, *73*, 79–84. doi:10.1016/j.mimet.2008.02.008

Persico, M., Ceol, A., Gavrila, C., Hoffmann, R., Florio, A., & Cesareni, G. (2005). HomoMINT: an inferred human network based on orthology mapping of protein interactions discovered in model organisms. *BMC Bioinformatics*, *6*(Suppl 4), S21. doi:10.1186/1471-2105-6-S4-S21

Pieters, J. (2008). Mycobacterium tuberculosis and the macrophage: Maintaining a balance. *Cell Host & Microbe*, *3*, 399–407. doi:10.1016/j.chom.2008.05.006

Qamra, R., Mande, S. C., Coates, A. R., & Henderson, B. (2005). The unusual chaperonins of Mycobacterium tuberculosis. *Tuberculosis (Edinburgh, Scotland)*, *85*, 385–394. doi:10.1016/j.tube.2005.08.014

Rekha, N., Machado, S. M., Narayanan, C., Krupa, A., & Srinivasan, N. (2005). Interaction interfaces of protein domains are not topologically equivalent across families within superfamilies: Implications for metabolic and signaling pathways. *Proteins*, *58*, 339–353. doi:10.1002/prot.20319

Rhodes, D. R., Tomlins, S. A., Varambally, S., Mahavisno, V., Barrette, T., & Kalyana-Sundaram, S. (2005). Probabilistic model of the human protein-protein interaction network. *Nature Biotechnology*, *23*, 951–959. doi:10.1038/nbt1103

Rickman, L., Saldanha, J. W., Hunt, D. M., Hoar, D. N., Colston, M. J., & Millar, J. B. (2004). A two-component signal transduction system with a PAS domain-containing sensor is required for virulence of Mycobacterium tuberculosis in mice. *Biochemical and Biophysical Research Communications*, *314*, 259–267. doi:10.1016/j.bbrc.2003.12.082

Salwinski, L., Miller, C. S., Smith, A. J., Pettit, F. K., Bowie, J. U., & Eisenberg, D. (2004). The Database of Interacting Proteins: 2004 update. *Nucleic Acids Research*, *32*, D449–D451. doi:10.1093/nar/gkh086

Schwede, T., Kopp, J., Guex, N., & Peitsch, M. C. (2003). SWISS-MODEL: an automated protein homology-modeling server. *Nucleic Acids Research*, *31*, 3381–3385. doi:10.1093/nar/gkg520

Suter, B., Kittanakom, S., & Stagljar, I. (2008). Two-hybrid technologies in proteomics research. *Current Opinion in Biotechnology*, *19*, 316–323. doi:10.1016/j.copbio.2008.06.005

Tailleux, L., Waddell, S. J., Pelizzola, M., Mortellaro, A., Withers, M., & Tanne, A. (2008). Probing host pathogen cross-talk by transcriptional profiling of both Mycobacterium tuberculosis and infected human dendritic cells and macrophages. *PLoS ONE*, *3*, e1403. doi:10.1371/journal.pone.0001403

Tapinos, N., Ohnishi, M., & Rambukkana, A. (2006). ErbB2 receptor tyrosine kinase signaling mediates early demyelination induced by leprosy bacilli. *Nature Medicine*, *12*, 961–966. doi:10.1038/nm1433

Wojcik, J., Boneca, I. G., & Legrain, P. (2002). Prediction, assessment and validation of protein interaction maps in bacteria. *Journal of Molecular Biology*, *323*, 763–770. doi:10.1016/S0022-2836(02)01009-4

Ye, J., McGinnis, S., & Madden, T. L. (2008). BLAST: improvements for better sequence analysis. *Nucleic Acids Research*, *34*, W6–W9. doi:10.1093/nar/gkl164

Young, D., Stark, J., & Kirschner, D. (2008). Systems biology of persistent infection: tuberculosis as a case study. *Nature Reviews Microbiology*, *6*, 520–528. doi:10.1038/nrmicro1919

Zahrt, T. C. (2003). Molecular mechanisms regulating persistent Mycobacterium tuberculosis infection. *Microbes and Infection*, *5*, 159–167. doi:10.1016/S1286-4579(02)00083-7

Zhong, W., & Sternberg, P. W. (2006). Genome-wide prediction of C. elegans genetic interactions. *Science*, *311*, 1481–1484. doi:10.1126/science.1123287

This work was previously published in the International Journal of Knowledge Discovery in Bioinformatics, Volume 1, Issue 1, edited by Xiao-Li Li and See-Kiong Ng, pp. 1-13, copyright 2010 by IGI Publishing (an imprint of IGI Global).

Chapter 11
A Transfer Learning Approach and Selective Integration of Multiple Types of Assays for Biological Network Inference

Tsuyoshi Kato
AIST Computational Biology Research Center, Japan & Ochanomizu University, Japan

Kinya Okada
KO Institute for Medical Bioinformatics, Japan

Hisashi Kashima
IBM Research, Japan

Masashi Sugiyama
Tokyo Institute of Technology, Japan

ABSTRACT

Inferring the relationship among proteins is a central issue of computational biology and a diversity of biological assays are utilized to predict the relationship. However, as experiments are usually expensive to perform, automatic data selection is employed to reduce the data collection cost. Although data useful for link prediction are different in each local sub-network, existing methods cannot select different data for different processes.

This paper presents a new algorithm for inferring biological networks from multiple types of assays. The proposed algorithm is based on transfer learning and can exploit local information effectively. Each assay is automatically weighted through learning and the weights can be adaptively different in each local part.

The authors' algorithm was favorably examined on two kinds of biological networks: a metabolic network and a protein interaction network. A statistical test confirmed that the weight that our algorithm assigned to each assay was meaningful.

DOI: 10.4018/978-1-4666-1785-8.ch011

BACKGROUND

A thorough understanding of cellular processes is the central goal of molecular biology. For this purpose, it is required to understand not only individual functions but also relationships among their components. Nowadays, a huge amount of data on molecular relationships can be generated through high throughput assays and analyzed as molecular networks. Although each of the assays contains useful information, molecular networks reconstructed from a single assay often have too much noise. To cope with this problem, several groups have tried to integrate data obtained from multiple types of assays for reliable network reconstruction (Pavlidis et al., 2002; Kato et al., 2005; Yamanishi et al., 2005).

How can we integrate data from multiple types of assays? A primitive method is to average data from the multiple types of assays and reconstruct a network from the averaged data (Yamanishi et al., 2005; Vert & Yamanishi, 2005). This method in a sense treats all types of assays equivalently, even if they may not be equivalent; that is, the assays may differ from each other in quality and resolution. To take non-equivalence among multiple types of assays into account in data integration, several groups have proposed methods to optimize the weight assigned to each type of assay (Kato et al., 2005; Lanckriet et al., 2004).

Although the existing methods allow each type of assay to be weighted, the weights are *global* in the entire network. Namely, every edge is predicted using weights common to the whole data. Since the mechanisms underlying cellular processes are complicated and heterogeneous, a type of assay may help shed light on some local cellular processes even if a low weight is assigned to the type of assay. Consequently, due to the heterogeneity of the relationship between assays and cellular mechanisms, the *global weighting* is too coarse, and finer modeling is desired.

Bleakley et al. (2007) have proposed a method that uses *local models* to cope with the heterogeneity issue. Their method builds a local model for each node (they call it a *target node*) and trains it with only its local (i.e., neighboring) information, resulting in scoring functions that are not corrupted by the irrelevant effects of distant parts of its molecular network. One shortcoming of their method is that the amount of local information is often limited, because most of the nodes have a few edges due to the fact that the node connectivity in molecular networks follows a power law distribution (Caldarelli, 2007). Insufficient information for training adversely affects the generalization ability of the method and restricts our ability to automatically acquire the genuine weights of the assays.

We present herein a novel algorithm for edge prediction. Our algorithm is based on Bleakley et al.'s approach, but our algorithm is extended to enjoy two remarkable features. First, in order to address the loss of generalization ability due to the scarcity of local information, we apply transfer learning to building local models. Transfer learning is an approach to machine learning that learns a task together with other related tasks simultaneously. This often leads to a better model for the target task, because it allows the learner to share appropriate information across the tasks. The use of this approach is motivated by the observation that node A is likely to be linked with node B when they share common neighbors in molecular networks (Chua et al., 2006; Samanta & Liang, 2003; Okada et al., 2005). According to this observation, the task of building the local model of a target node is likely to be related to that of its neighboring nodes. Transfer learning builds the local model of a target node with the help of its neighboring nodes. Furthermore, transfer learning optimizes weights assigned to the types of assays when building local models, offering selected assays to each local part of a molecular network. In this study, we reconstruct molecular networks using this method and validate the efficiency of our method.

The rest of this section summarizes abbreviations which are used frequently in this paper. SVM is the abbreviation of the support vector machine (Schölkopf & Smola, 2002). RBF is the radial basis function (Schölkopf & Smola, 2002). BBVK, TPPK, and MLPK are the BBV kernel (Bleakley et al., 2007), the tensor product pairwise kernel (Ben-Hur & Noble, 2005), and the metric learning pairwise kernel (Vert et al., 2007), as detailed in the section of Methods. PIF is the protein interaction profile defined in the section of Discussion. GO is the gene ontology. We will show the full name again at the first occurrence except this paragraph.

RESULTS

A number of large-scale biological networks have become available through high-throughput experiments, even though they are rough and incomplete. In order to refine them, more detailed surveys/studies need to be carried out. Even if some edges are already known for a given node, it is necessary to verify whether the node has other edges or not in the biological network. For such cases, we consider a network reconstruction task that involves building a scoring function to predict the existence of edges of a specified node (which is the one called target node in the previous section) with input nodes, in addition to its known edges. Let us cite an example. The adjacency matrix of the undirected graph in Figure 1(a) is shown in the bottom of Figure 1(b). Basically, a feature vector is given to each node as in the top of (b). We consider the case where a part of the adjacency matrix is missing, such as shown in Figure 1(c). We pose a binary classification problem which is to predict whether an input node is linked with the target node using the given feature vectors. This is same as what BBV algorithm attempts to do (See the section of Methods). For example, we assume that the target node is node 1. The first row of (b) implies that the training set includes only node 3, and the other nodes 2, 4, 5, 6, 7, and 8 are unknown samples in the classification task that we denote by $\mathcal{T}_1$. We also consider another classification task whose target node is node 3, and denote the task by $\mathcal{T}_3$. With the help of the fact that the node 3 is already known to be a neighbor of node 1, we regard $\mathcal{T}_3$ as a related task of $\mathcal{T}_1$, and our algorithm performs transfer learning of $\mathcal{T}_1$ with $\mathcal{T}_3$ to obtain a better solution of $\mathcal{T}_1$.

Our method is tested on two kinds of biological networks in *Saccharomyces cerevisiae*. One is the metabolic network in which enzymes involved in successive metabolic reactions are linked. This network, which contains 769 nodes (mainly enzymes) and 3,702 undirected edges (functional links), is identical to the one used by Yamanishi et al. (2005). The other is the protein interaction network of von Mering et al. (2002). Each edge in the network indicates the physical interaction of two proteins and is rated with a confidence level: high, middle or low. We use only high confidence edges as in previous studies (Kato et al., 2005; Bleakley et al., 2007), because they are supported by multiple experiments. By removing proteins without any edges, we obtain a network of 984 proteins and 2,438 edges. Actually, the two networks are quite different: they share only 188 proteins and 41 edges.

In order to predict the edges, we use a diverse collection of protein data, and convert them into kernel matrices. Three of the kernel matrices are from sequence similarities. Sequence similarities are computed using three biological sequence alignment tools: BLAST (Altschul et al., 1990), ALN (Gotoh, 1982), and Pfam (Finn et al., 2008). The matrix formed with similarity scores is not guaranteed to be positive semi-definite, although the kernel matrices must be positive semi-definite for most of kernel methods (Schölkopf & Smola, 2002). To generate positive semi-definite kernel matrices, feature vectors are made from scores of each protein against all the proteins and the inner

Figure 1. This paper deals with an undirected graph. Figure (a) shows an example of the undirected graph. In this example, each node has a 4-dimensional feature vector, $\mathbf{x}_1,\ldots,\mathbf{x}_8$, *as shown in the top row of (b). The gray levels indicate the value of each element. The bottom of (b) is its adjacency matrix. Task* $\mathcal{T}_1$ *predicts the existence of edges in pairs (1, 2), (1, 4), (1, 5), (1, 6), (1, 7), and (1, 8). BBV algorithm (see the section Methods) cannot work well due to the extremely small training set* $\mathcal{V}_{1,+} = \{3\}$, $\mathcal{V}_{1,-} = \varnothing$. *We regard* $\mathcal{T}_3$ *as the related task to perform transfer learning, where node 3 is the known neighbor of node 1.*

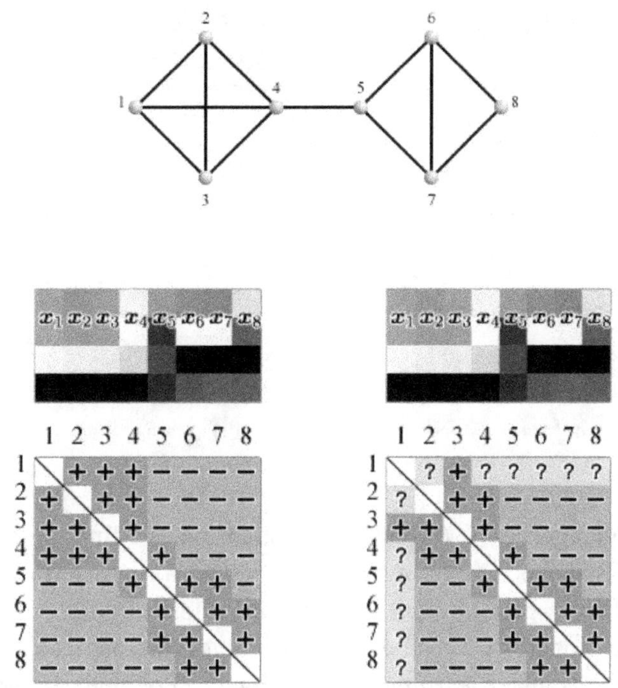

(b) True adjacency matrix (c) Adjacency matrix with unknown elements

product among the feature vectors is taken. The scores for Pfam are the expected values derived from probabilistic models. The next dataset for network inference contains microarray gene expressions. Gene expressions are measured under various experimental conditions. A feature vector consists of the gene expressions. We use an RBF kernel function to generate a kernel matrix from the feature vectors. The resultant kernel matrices are available from the following website: http://noble.gs.washington.edu/proj/sdp-svm/. Following the website, we denote the five kernel matrices by 'blast,' 'expr,' 'fft,' 'pfam_hmm,' and 'sw,' respectively. Each kernel can be used for the protein interaction network and the metabolic network.

We use three pairwise kernels that are computed from node kernels: BBVK (Bleakley et al., 2007), TPPK (Ben-Hur & Noble, 2005), and MLPK (Vert et al., 2007). Then, we combine transfer learning with each of the three pairwise kernels. We determine the hyper-parameters of the algorithms by cross-validation over the training set, and take 300 nodes with the highest degree-of-nodes as the target gene to pose 300 tasks. Then, we randomly select 50% of proteins for the training set of each task.

Figure 2. Averaged AUC of ROC. 'Transfer Learning, Integrated' indicates our algorithm that integrates multiple data sources and predicts the links of networks with transfer learning. 'Conventional Learning, Integrated' is the case without transfer learning, i.e. BBV algorithm. 'Transfer Learning, Individual' means the transfer learning from an individual data source, and the average is plotted. 'Conventional Learning, Individual' is the averaged ROC score of conventional learning from each data source. 'Learning from the Entire Network, Integrated' is the approach to pick training samples from the entire network. Namely, that is the global model.

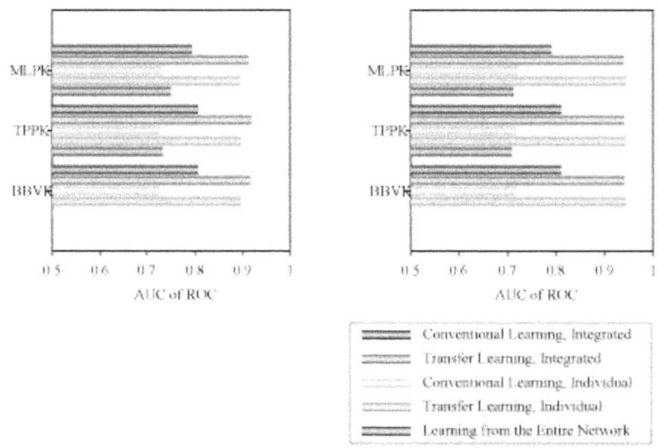

(a) Metabolic network (b) Protein interaction

Figure 2 compares the averaged AUC (Area under the curve) of ROC (Receiver Operating Characteristic) to predict metabolic gene network and protein interactions. The plotted AUC value is the average over 300 tasks. Transfer learning improves the prediction performance substantially for all the cases compared to conventional learning (BBV algorithm), and the differences are more than 0.1. AUCs among the three pairwise kernels do not differ very much.

To demonstrate the effect of data integration, we apply the learning algorithms to each data source and evaluate the ROC scores. The averages are also plotted in Figure 2. We observe improvements due to data integration especially in the results of conventional learning. For the metabolic network, transfer learning performance is improved by data integration. However, almost the same ROC scores are obtained for the protein interaction network. That may be because every data source we used is sufficiently informative to predict protein interactions well. For further investigation, we find the algorithm that achieves the best performance among the six combinations for each task. Figure 3 shows how many times each algorithm achieves the best prediction performance. For both networks, transfer learning yields the best AUC much more frequently than non-transfer learning. The differences among pairwise kernels are not so large.

Our problem setting is different from that of the global models used by Ben-Hur & Noble (2005) and Vert et al. (2007). Global models attempt to infer the existence/absence of unknown edges in the whole network, whereas our algorithm predicts edges that link a specified target node to the other nodes. Despite differences in problem setting, it is still possible to use global models in our setting. In the training stage, global models use all the known edges as positive samples, and they randomly pick an equal number of absent edges as negative samples. From our experiences in problem setting, we find that improvements can be achieved by slightly modifying the way to

Figure 3. The number of times each algorithm achieves the best prediction performance among the six methods. In the problem we tackle, we can assign each node to a prediction task. For each task, we perform the six methods: 'TL, BBVK', 'CL, BBVK', 'TL, TPPK', 'CL, TPPK', 'TL, MLPK', and 'CL, MLPK' where TL and CL are the abbreviations for transfer learning and conventional learning, respectively. BBVK, TPPK, and MLPK are the types of pairwise kernels. For most of the tasks, transfer learning achieves a best prediction performance.

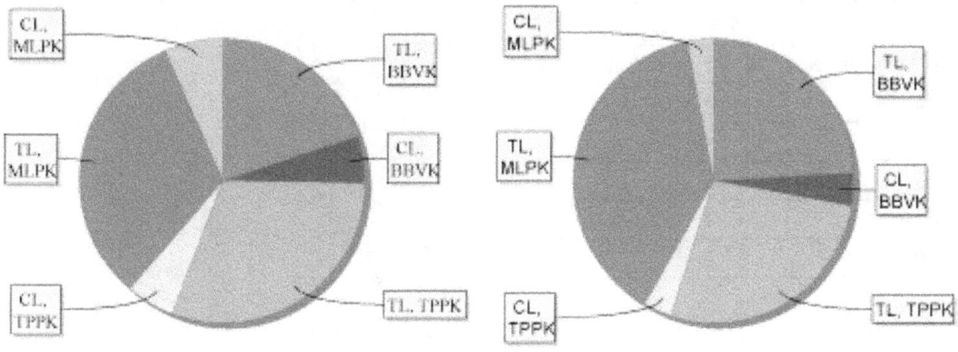

(a) Metabolic network (b) Protein interaction

pick negative samples, namely, to choose all the absent edges of the target node and to pick the other absent edges randomly so that the number of negative training samples is equal to the number of positive ones. One disadvantage is that their approaches need much more computational time for training compared to our method, because the number of training samples is much larger. Due to the heavy computation, we abandon the cross-validation within the training set to determine the value of the regularization parameter of SVM for comparison of global models with ours in terms of prediction performance. Instead, we try several values of the regularization parameter and report the best performance. Although the way of evaluation is advantageous to their approaches, their approaches could not obtain performance superior to our transfer learning (See Figure 2).

DISCUSSION

In building a local model for each node through our algorithm, the weight of each type of assay data is optimized. To find the extent to which each assay data is effective in the entire local model building process, we count the number of resulting local models that have high weight of each data. Because the results do not remarkably vary among any of the kernels (MLPK, TPPK or BBVK) and/or any of cutoff thresholds of weight (0.5, 0.6, 0.7 or 0.8) (Tables 2–7 in Supplemental Information http://www.net-machine.net/~kato/pdf/t-kato-ijkdb2009a-suppl.pdf), hereafter we describe the results of using BBVK and 0.8 as the cutoff threshold.

Regarding the local models for the protein interaction network, expression data and Pfam data are highly weighted compared to other data (Table 1 and Figure 4). Although the multiple selection as shown in this case – some local models prefer the expression data while others prefer the Pfam data – may indicate that the protein interaction network is a superimposition of various interactions induced by distinct mechanisms, it may be simply derived from biologically meaningless artifacts, e.g., experimental properties, including sampling bias, protocol, and experimental prin-

Table 1. Number of local models that have a high weight of each assay data. blast: sequence similarity data calculated by BLAST, expr: expression data, fft: hydropathy profile, pfam hmm: Pfam data. sw: sequence similarity data calculated by ALN. Local models that have high weight (≥ 0.8) in the case of using BBVK were counted.

	Metabolic Network	Protein Interactions
blast	8	5
expr	0	40
fft	0	0
pfam_hmm	149	15
sw	0	0

ciple. Therefore, we next describe the pattern of local models with polarized preference for those assay data.

In the case of weighting highly the expression data in the local model of node *u*, it is expected that not only *u* links with a node or nodes (i.e., neighbors of *u*) whose protein interaction profile (PIF) is similar to that of *u* but also the PIF of the neighbor(s) is highly correlated with its/their expression profile. To verify this, we count the gene ontology (GO) terms of proteins whose local models have high weight of the expression data. As a result, GO terms, 'ribosome biogenesis' (GO: 0042254) and 'RNA metabolic process' (GO: 0016070) of the GO biological process terms, are significantly over-represented (*p*-values are 2.3E-19 and 1.2E-06, respectively), indicating that the proteins are involved in ribosome function. In addition to the fact that ribosome-related proteins are densely connected in the protein interaction network, as described previously (Han et al., 2004), they are highly correlated with the expression profile (Jansen et al., 2002).. Those findings well explain the assignment of a high weight to gene expression data for those local models through the transfer learning framework.

Similarly, we analyze proteins whose local models have high weights of Pfam data, and find that the GO term, 'protein catabolic process' (GO: 0030163), is significantly over-represented (*p*-value: 3.3E-10). As most of them (12/15) indeed comprise proteasome complexes well known as protein catabolizers (i.e., they are subunits of proteasome complexes), they are likely to be densely connected in the protein interaction network according to the basic idea that protein complexes are detected as densely connected subnetworks. In addition, since most of the proteasome subunits share common domains (alpha/beta subunit domains), their Pfam profiles are likely to be similar to each other. Those observations justify the weight generated through data selection in our algorithm.

Figure 4. Weights of multiple types of assays for each target gene as determined by transfer learning with BBVK.

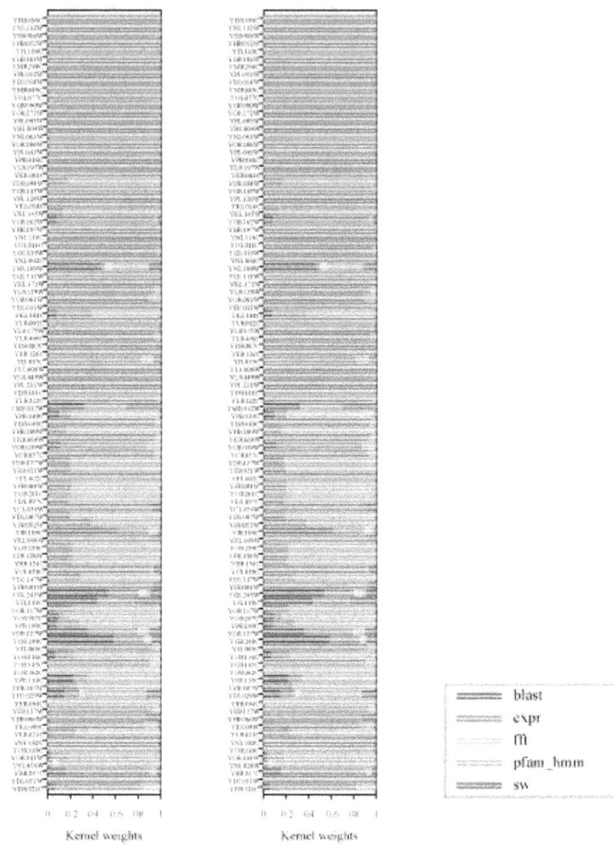

On the other hand, data indicating sequence similarity, particularly Pfam data, are highly weighted in the entire local model building process for the metabolic network (Table 1). Here we conduct pathway category analysis, instead of GO analysis, of proteins whose local models assign high weights to Pfam data. As a result, we find that categories indicating signal transduction, 'cell cycle - yeast,' 'MAPK signaling pathway - yeast,' and 'phosphatidylinositol signaling system,' are significantly over-represented (p-values are 7.0E-03, 9.6E-03, and 1.1E-02, respectively). Following the basic idea that signaling/interaction targets are usually recognized by domains, the preference for Pfam data at least in these pathways seems to be biologically reasonable.

CONCLUSION

In this paper, we discussed the effect of transfer learning in network inference. Our algorithm utilizes the small-world property of related learning tasks and we experimentally showed that the proposed algorithm substantially improves prediction accuracy. We also illustrated that multiple kernel learning works well and yields biologically plausible data selection.

Our algorithm uses only partial (i.e., local) information of assay data when building each local model. As described in the previous section, the optimized weight reflects biologically meaningful properties, at least in part, of protein interaction and metabolic networks. In future studies, we will further investigate biological interpretation of networks. The current data are noisy so they may not be believed absolutely, but we have to resort to using the current data to evaluate the prediction performance. We will investigate the performance again when the more reliable data become available.

METHODS

This section presents our algorithm that infers a biological network from multiple assays. The description shown in this section uses the following notations. Vectors are denoted by boldface lower-case letters and matrices, by boldface upper-case letters. Elements of vectors and matrices are not printed in boldface. The transposition of matrix $\mathbf{A}$ is denoted by $\mathbf{A}^\top$. The $n \times n$ identity matrix is denoted by $\mathbf{I}_n$. The n-dimensional column vector all of whose elements are one is denoted by $\mathbf{1}_n$. We use $\mathbb{R}$ to denote the set of real numbers, $\mathbb{R}^n$ to denote the set of n-dimensional real column vectors, and $\mathbb{R}^{m \times n}$ to denote the set of $m \times n$ real matrices. The set of real nonnegative numbers is denoted by $\mathbb{R}_+$, and the set of n-dimensional real nonnegative vectors is denoted by $\mathbb{R}^n_+$. We use $\mathbb{S}^n$ to denote the set of $n \times n$ symmetric matrices, $\mathbb{S}^n_+$ to denote the set of $n \times n$ symmetric *positive semi-definite* matrices. $\mathbb{N}$ is the set of natural numbers. $\mathbb{N}_n$ is a subset of $\mathbb{N}$, and is defined by $\mathbb{N}_n \equiv \{i \in \mathbb{N} \,|\, i \leq n\}$. Symbols $\leq$ and $\geq$ are used to denote not only the standard inequalities between scalars, but also the componentwise inequalities between vectors.

BBV Algorithm

Our algorithm is based on the approach using local models proposed by Bleakley et al. (2007). We refer to their approach as BBV, taking the first letters of the three authors' names. BBV builds a classifier for a given node with index $u \in \mathbb{N}_n$ where n is the number of nodes. Node u is what we have referred to as the *target node*. The purpose of the classifier is to predict the existence or the absence of links with the given target node u. BBV assumes that a feature vector is given to every node. The feature vectors are used for SVM learning. Additionally, we are given two exclusive subsets of nodes $\mathcal{V}_{u,+}$ and $\mathcal{V}_{u,-}$; that is, $\mathcal{V}_{u,+} \cap \mathcal{V}_{u,-}$

is the null set. Every node in $\mathcal{V}_{u,+}$ is linked with the target node u, and every node in $\mathcal{V}_{u,-}$ is not linked with u. Members in $\mathcal{V}_{u,+}$ are termed neighbors in graph theory. We call the subset $\mathcal{V}_u \equiv \mathcal{V}_{u,+} \cup \mathcal{V}_{u,-}$ the training *set*. The set of the other nodes $\mathbb{N}_n \setminus \mathcal{V}_u$ is called the *test set*. In summary, the BBV algorithm predicts the existence or the absence of the link between the target node and the node in the test set by using feature vectors. Note that we cannot use any class labels of the test set for SVM learning. In the case of Figure 1(c), the training set of $\mathcal{T}_1$ is $\mathcal{V}_1 = \{3\}$. The set of positives $\mathcal{V}_{1,+}$ in the training set consists only of node 3, and the set of negatives $\mathcal{V}_{u,-}$ is empty. Therefore, we can not train SVM because SVM learning requires at least one sample for each class.

The BBV algorithm consists of the learning stage and the prediction stage. In the learning stage, the classifier learns the classification rule of the class labels from the feature vectors in the training set $\mathcal{V}_u$. In the prediction stage, the classifier predicts the labels for the nodes in the test set $\mathbb{N}_n \setminus \mathcal{V}_u$, which corresponds to predicting the existence or absence of links to the target node.

Outline of Our Algorithm

We introduce the idea of transfer learning into the BBV algorithm. Transfer learning is a framework to improve performance of a task using related learning tasks. This idea is useful in the current context since we can find the task relations in the network inference problem, as mentioned in Background. We will formulate the transfer learning approach. Let us denote the local model of a target node *u* by $\mathcal{V}_u$. We assume that the learning task $\mathcal{T}_u$ is related to the tasks associated with the neighbors of *u*, say $\mathcal{T}_v, \forall v \in \mathcal{V}_{u,+}$. Thus, we propose to employ the transfer learning of $\mathcal{T}_u$ with the help of $\mathcal{T}_v, \forall v \in \mathcal{V}_{u,+}$.

Support Vector Machine

BBV employs the support vector machine (SVM) as a classifier. Basically, SVM finds a hyper-plane that distinguishes positive training examples from negative ones in a feature space. For that purpose, SVM formulates a scalar-valued linear score function. The parameters of the score function consist of the normal vector of the hyper-plane and the bias term that shifts the hyper-plane. If the norm of the normal vector is minimized subject to the constraints that the minimum score among positive examples is $+1$ and the maximum score among negative ones is -1, the *margin* between the positive set and the negative set is maximized. However, a hyper-plane satisfying those constraints does not always exist. To cope with this problem, the so-called *soft margin* SVM relaxes the constraints so that some examples can lie in the margin, but are given penalties according to the Hinge loss. Hereafter, we refer to the soft margin SVM as simply SVM.

Transfer Learning with Feature Vectors

To describe our algorithm, we begin with the learning algorithm of BBV using numerical formulas, and introduce the transfer learning framework to the algorithm. Given *d*-dimensional feature vectors $\mathbf{x}_v \in \mathbb{R}^d$ for all nodes $v = 1, \ldots, n$, the BBV learning algorithm for a task $\mathcal{T}_u$ is expressed as

$$\begin{aligned}
&\text{min} && \frac{1}{2}\|\mathbf{w}_u\|^2 + C_\alpha \sum_{v \in \mathcal{V}_u} \xi_{u,v}, \\
&\text{wrt} && \mathbf{w}_u \in \mathbb{R}^d, \quad \forall v \in \mathcal{V}_u : \ \xi_{u,v} \in \mathbb{R}_+, \\
&\text{subj to} && \forall v \in \mathcal{V}_{u,+} : \ \mathbf{w}_u^\top \mathbf{x}_v + b \geq +1 - \xi_{u,v}, \\
& && \forall v \in \mathcal{V}_{u,-} : \ \mathbf{w}_u^\top \mathbf{x}_v + b \leq -1 - \xi_{u,v},
\end{aligned}$$

where C_α is a constant to be determined in advance. In the context of multi-task learning, Evgeniou et al. (2005) suggested to append to the objective function a term that forces $\mathbf{w}_v$ of every related task to be close to $\mathbf{w}_u$. We borrow this idea and formulate our transfer learning algorithm as follows:

$$\begin{array}{lll} \text{min} & \frac{1}{2}\|\mathbf{w}_u\|^2 + C_\alpha \sum_{v\in\mathcal{V}_u} \xi_{u,v} + \frac{1}{2}\sum_{v\in\mathcal{V}_u}\left(\|w_v\|^2 + C_\rho \|w_v - w_u\|^2\right) & \\ \text{wrt} & \forall v_1 \in \mathcal{V}_u \cup \{u\}, \forall v_2 \in \mathcal{V}_{v_1} : & \mathbf{w}_{v_1} \in \mathbb{R}^d,\ \xi_{v_1,v_2} \in \mathbb{R}_+, \\ \text{subj to} & \forall v_1 \in \mathcal{V}_u \cup \{u\}, \forall v_2 \in \mathcal{V}_{v_1,+} : & \mathbf{w}_{v_1}^\top \mathbf{x}_{v_2} + b \geq +1 - \xi_{v_1,v_2}, \\ & \forall v_1 \in \mathcal{V}_u \cup \{u\}, \forall v_2 \in \mathcal{V}_{v_1,-} : & \mathbf{w}_{v_1}^\top \mathbf{x}_{v_2} + b \leq -1 + \xi_{v_1,v_2}, \end{array} \quad (1)$$

where C_ρ is a constant, trading off the strength of transferring for regularization and loss.

Learning with Kernels

We now extend our learning algorithm to a kernel method. What are kernel methods? Kernel methods work on kernel matrices instead of feature vectors. Kernel matrices are alternative representation to feature vectors, and are positive semidefinite similarity matrices. The simplest kernel matrix consists of the inner product between feature vectors. Kernel methods are designed so that when the inner product between feature vectors is used to generate a kernel matrix, the score function learned from the kernel matrix yields exactly the same function as the one obtained directly from the feature vectors.

The BBV algorithm is readily kernelized because it directly uses SVM, which is a typical kernel method. However, since we incorporate transfer learning in our algorithm, kernelization of our algorithm may not be straightforward. In order to feed a kernel matrix to our algorithm, we need to re-formulate the original learning algorithm (1). We will present the kernelized version of our learning algorithm and show that the kernelized learning problem can be reduced to an optimization problem that can be solved efficiently using conventional techniques. Finally, we will further extend the algorithm so that it works on multiple kernel matrices.

Transfer Learning from a Single Kernel Matrix

We now present the kernelized version of our algorithm. Let us denote the original kernel matrix among n nodes by $\mathbf{K}^{\text{node}}$. Our kernelized learning algorithm requires the submatrices of the original kernel matrix; for $\forall a, \forall b \in \mathcal{V}_u \cup \{u\}$, we denote by $\mathbf{K}^{\text{node}}_{(a,b)}$ the submatrix in which the rows and columns correspond to $\mathcal{V}_a$ and $\mathcal{V}_b$, respectively. Each element of the submatrix $\mathbf{K}^{\text{node}}_{(a,b)}$ is given by

$$\forall i \in \mathbb{N}_{\ell_a}, \forall j \in \mathbb{N}_{\ell_b} : \left[\mathbf{K}^{\text{node}}_{(a,b)}\right]_{ij} = K^{\text{node}}_{v_{a,i}, v_{b,j}}$$

where $\mathcal{V}_a \equiv \{v_{a,1}, \ldots, v_{a,\ell_a}\}$ and $\mathcal{V}_b \equiv \{v_{b,1}, \ldots, v_{b,\ell_b}\}$. We assume that the elements are arranged in ascending order, i.e., $v_{a,1} < \cdots < v_{a,\ell_a}$ and $v_{b,1} < \cdots < v_{b,\ell_b}$. For each task $\mathcal{T}_a$ with the training set $\mathcal{V}_a$, we define the label vector $\mathbf{y}_a \in \mathbb{R}^{\ell_a}$ by

$$[y_a]_i \equiv \begin{cases} +1 & \text{if } v_{a,i} \in \mathcal{V}_{a,+}, \\ -1 & \text{if } v_{a,i} \in \mathcal{V}_{a,-}, \end{cases}$$

for $i = 1, \ldots, \ell_a$. Let the positive training set of task $\mathcal{T}_u$ be $\mathcal{V}_{u,+} = \{v_1, \ldots, v_M\}$ where $M = |\mathcal{V}_{u,+}|$. Using the Lagrangian multiplier technique (Boyd & Vandenberghe, 2004), we can derive the kernel version of our algorithm as a minimization problem with the following objective function:

$$\begin{aligned} J(\bar{\alpha}) \equiv\ & \frac{1}{2} F_0 \alpha_u^\top \text{diag}(\mathbf{y}_u) \mathbf{K}^{\text{node}}_{(u,u)} \text{diag}(\mathbf{y}_u) \alpha_u \\ & + \sum_{i=1}^{M} F_1 \alpha_u^\top \text{diag}(\mathbf{y}_u) \mathbf{K}^{\text{node}}_{(u,v_i)} \text{diag}(\mathbf{y}_{v_i}) \alpha_{v_i} \\ & + \frac{1}{2} \sum_{i=1}^{M} F_2 \alpha_{v_i}^\top \text{diag}(\mathbf{y}_{v_i}) \mathbf{K}^{\text{node}}_{(v_i,v_i)} \text{diag}(\mathbf{y}_{v_i}) \alpha_{v_i} \\ & + \sum_{i=1}^{M} \sum_{j=i+1}^{M} F_3 \alpha_{v_i}^\top \text{diag}(\mathbf{y}_{v_i}) \mathbf{K}^{\text{node}}_{(v_i,v_j)} \text{diag}(\mathbf{y}_{v_j}) \alpha_{v_j} \\ & - \mathbf{1}_{\ell_u}^\top \alpha_u - \sum_{i=1}^{M} \mathbf{1}_{\ell_{v_i}}^\top \alpha_{v_i} \end{aligned}$$

where

$$F_0 \equiv \frac{1+2C_\rho}{1+2(1+M)C_\rho}, \quad F_1 \equiv \frac{2C_\rho}{1+2(1+M)C_\rho},$$

$$F_2 \equiv \frac{1}{1+2C_\rho} + \frac{4C_\rho^2}{(1+2(1+M)C_\rho)(1+2C_\rho)},$$

$$F_3 \equiv \frac{1}{1+2C_\rho}.$$

The derivation is in the supplemental document. The variable for this optimization problem is

$$\bar{\alpha} \equiv \left[\alpha_{v_1}^\top, \ldots, \alpha_{v_M}^\top, \alpha_u^\top\right]^\top,$$

and the constraints are

$$\mathbf{0} \le \bar{\alpha} \le C_\alpha \mathbf{1}, \qquad \left\langle \bar{\mathbf{y}}, \bar{\alpha} \right\rangle = 0,$$

where we concatenate the label vectors as

$$\bar{\mathbf{y}} \equiv \left[\mathbf{y}_{v_1}^\top, \ldots, \mathbf{y}_{v_M}^\top, \mathbf{y}_u^\top\right]^\top.$$

Pairwise Kernels

As the BBV algorithm deals with feature vectors of nodes directly, the kernelized BBV algorithm can handle only kernels among nodes (*node kernels*). In the context of predicting the existence of an edge in a node pair, however, it is natural and would be more powerful to use kernels among node pairs (*pairwise kernels*). In BBV, one of the elements in every node pair should always be the target node. Since various node pairs need to be dealt with in the context of transfer learning, the use of the pairwise kernels is not allowed in BBV.

On the other hand, our kernelized algorithm can handle the pairwise kernels, which is an advantage over the original BBV algorithm. To our knowledge, there are two useful pairwise kernels proposed so far: the tensor product pairwise kernel (TPPK) (Ben-Hur & Noble, 2005) and the metric learning pairwise kernel (MLPK) (Vert et al., 2007). For $\mathcal{V}_a = \{v_{a,1}, \ldots, v_{a,\ell_a}\}$ and $\mathcal{V}_b = \{v_{b,1}, \ldots, v_{b,\ell_b}\}$ the elements of TPPK and MLPK are defined by

$$\forall i \in \mathbb{N}_{\ell_a}, \forall j \in \mathbb{N}_{\ell_b}:$$

$$\left[\mathbf{K}_{(a,b)}^{\text{tppk}}\right]_{ij} \equiv K_{a,b}^{\text{node}} K_{v_{a,i},v_{b,j}}^{\text{node}} + K_{a,v_{b,j}}^{\text{node}} K_{b,v_{a,i}}^{\text{node}},$$

$$\left[\mathbf{K}_{(a,b)}^{\text{mlpk}}\right]_{ij} \equiv \left(K_{a,b}^{\text{node}} - K_{a,v_{b,j}}^{\text{node}} - K_{v_{a,i},b}^{\text{node}} + K_{b,v_{b,j}}^{\text{node}}\right)^2,$$

respectively. TPPK is based on a tensorization of the original feature space, whereas MLPK is derived from convex optimization of a distance metric learning problem. In order to distinguish the above pairwise kernels from the kernel used by the original BBV algorithm, we refer to the original one as the BBV kernel (BBVK), which is expressed as

$$\mathbf{K}_{(a,b)}^{\text{bbvk}} \equiv \mathbf{K}_{(a,b)}^{\text{node}}.$$

Hereafter, we use the superscript 'pw' to denote a pairwise kernel.

Connection to SVM

An attractive advantage of our algorithm is that through slight modification of the original kernel matrices, the kernel version of our algorithm is reduced to an optimization problem that shares the same form as the original SVM. Since there are a number of studies for efficient SVM training, we can directly exploit these techniques or the software for our algorithm. Next, we show this fact. We modify the kernel matrices as

$$\begin{aligned}
&\forall i \in \mathbb{N}_M, & \mathbf{G}_{(v_i,v_i)} &\equiv F_2\mathbf{K}^{\mathrm{pw}}_{(v_i,v_i)},\\
&\forall i \in \mathbb{N}_M, \forall j \in \mathbb{N}_M \setminus \{i\}, & \mathbf{G}_{(v_i,v_j)} &\equiv F_3\mathbf{K}^{\mathrm{pw}}_{(v_i,v_j)},\\
&\forall i \in \mathbb{N}_M, & \mathbf{G}_{(v_i,u)} &\equiv F_1\mathbf{K}^{\mathrm{pw}}_{(v_i,u)},\\
& & \mathbf{G}_{(u,v_j)} &\equiv F_1\mathbf{K}^{\mathrm{pw}}_{(u,v_i)},\\
& & \mathbf{G}_{(u,u)} &\equiv F_0\mathbf{K}^{\mathrm{pw}}_{(u,u)},
\end{aligned}$$

for $\mathcal{V}_{u,+} = \{v_1,\ldots,v_M\}$. We arrange these matrices as follows:

$$\mathbf{G}_{\mathrm{fea}} \equiv \begin{bmatrix} \mathbf{G}_{(v_1,v_1)} & \cdots & \mathbf{G}_{(v_1,v_M)} & \mathbf{G}_{(v_1,u)} \\ \vdots & \ddots & \vdots & \vdots \\ \mathbf{G}_{(v_M,v_1)} & \cdots & \mathbf{G}_{(v_M,v_M)} & \mathbf{G}_{(v_M,u)} \\ \mathbf{G}_{(u,v_1)} & \cdots & \mathbf{G}_{(u,v_M)} & \mathbf{G}_{(u,u)} \end{bmatrix}.$$

Note that if we let $\ell \equiv \ell_{v_1} + \ldots + \ell_{v_M} + \ell_u$, the size of $\mathbf{G}_{\mathrm{fea}}$ is $\ell \times \ell$ and the length of $\mathbf{y}$ is ℓ. Then, the objective function can be rearranged as

$$J(\bar{\alpha}) \equiv \frac{1}{2}\bar{\alpha}^\top \mathrm{diag}(\bar{\mathbf{y}})\mathbf{G}_{\mathrm{fea}}\mathrm{diag}(\bar{\mathbf{y}})\bar{\alpha} - \mathbf{1}_\ell^\top\bar{\alpha}.$$

This optimization algorithm turns out to be a special case of the one appearing in Evgeniou et al. (2005). If we regard $\bar{\mathbf{y}}$ and $\mathbf{G}_{\mathrm{fea}}$ as a class label vector and a kernel matrix, respectively, in a standard binary classification problem, the objective function of our algorithm will exactly coincide with the objective function of SVM. We can actually prove that $\mathbf{G}_{\mathrm{fea}}$ is positive semi-definite. The constraints of $\bar{\alpha}$ are also the same as those of SVM. Hence, by a slight modification of kernel matrices as described above, standard SVM software can be used to obtain the solution to our transfer learning problem.

Transfer Learning from Multiple Kernel Matrices

Let us consider the situation where we have multiple information sources. Suppose the number of information sources is n_{kern}, and the data from each information source is represented by a kernel matrix. Let us denote the n_{kern} pairwise kernel matrices by $K^1,\ldots,K^{n_{\mathrm{kern}}}$. The algorithm we develop in this paper allows us to optimize the coefficients in a linear combination of the pairwise kernel matrices. There is a sophisticated algorithm called SDP/SVM (Lanckriet et al., 2004), which chooses the coefficients so that useful data are emphasized in SVM classification. Here we show that some rearrangement allows us to directly use SDP/SVM in the current context. Letting $\mu \in \mathbb{R}^{n_{\mathrm{kern}}}$ be the coefficient vector, we can express the linear combination of the kernel matrices as

$$\mathbf{K}^{\mathrm{int}} = \sum_{k=1}^{n_{\mathrm{kern}}} \mu_k \mathbf{K}^k.$$

We modify the integrated kernel matrix $\mathrm{K}^{\mathrm{int}} \in \mathbb{S}^n_+$ to obtain $\mathbf{G}^{\mathrm{int}} \in \mathbb{S}^n_+$ as

$$\begin{aligned}
&\forall i \in \mathbb{N}_M, & \mathbf{G}^{\mathrm{int}}_{(v_i,v_i)} &\equiv F_2\mathbf{K}^{\mathrm{int}}_{(v_i,v_i)},\\
&\forall i \in \mathbb{N}_M, \forall j \in \mathbb{N}_M \setminus \{i\}, & \mathbf{G}^{\mathrm{int}}_{(v_i,v_j)} &\equiv F_3\mathbf{K}^{\mathrm{int}}_{(v_i,v_j)},\\
&\forall i \in \mathbb{N}_M, & \mathbf{G}^{\mathrm{int}}_{(v_i,u)} &\equiv F_1\mathbf{K}^{\mathrm{int}}_{(v_i,u)},\\
& & \mathbf{G}^{\mathrm{int}}_{(u,v_j)} &\equiv F_1\mathbf{K}^{\mathrm{int}}_{(u,v_i)},\\
& & \mathbf{G}^{\mathrm{int}}_{(u,u)} &\equiv F_0\mathbf{K}^{\mathrm{int}}_{(u,u)},
\end{aligned}$$

for $\mathcal{V}_{u,+} = \{v_1,\ldots,v_M\}$.. Once the optimal values of the coefficients μ are found, the composite kernel matrix is just fed into our learning algorithm. We can also consider the modified kernel matrix $\mathbf{G}^k \in \mathbb{S}^n_+$ obtained from each kernel matrix $\mathbf{K}^k \in \mathbb{S}^n_+$ in a similar way. It can easily be shown that the matrix $\mathbf{G}^{\mathrm{int}}$ associated with the integrated kernel matrix $\mathbf{K}^{\text{int}}$ is expressed as a linear combination of the modified kernel matrices:

$$\mathbf{G}^{\text{int}} = \sum_{k=1}^{n_{\text{kern}}} \mu_k \mathbf{G}^k.$$

Hence, we can straightforwardly employ $\left\{\mathbf{G}^k\right\}_{k=1}^{n_{\text{kern}}}$ as inputs of the SDP/SVM algorithm. The learning problem of SDP/SVM results in a quadratically constrained quadratic program (QCQP) that becomes intractable as the number of training samples increases. Recent studies (Bach et al., 2004; Sonnenburg et al., 2006; Rakotomamonjy et al., 2008) have addressed this issue and these new algorithms allow us to tackle large-scale problems such as our learning problem.

Gene Ontology and Kegg Pathway Category Analysis

The GO terms (Gene Ontology Consortium, 2004) assigned to yeast gene products were downloaded from the website of the Gene Ontology Consortium (http://www.geneontology.org/). After mapping the GO terms into the more general parent GO Slim terms (Generic GO Slim; obtained from the same website), we assigned the GO Slim terms to each yeast gene product. On the other hand, the Kegg pathway category assigned to yeast gene products was downloaded from the website of the Kegg pathway database (http://www.genome.jp/kegg/pathway.html). P-values for over-represented GO Slim terms/Kegg pathway categories were calculated based on hypergeometric distribution by our Perl script.

ACKNOWLEDGMENT

We thank W. Noble for making the data public. TK would like to thank Dr Goro Terai for fruitful discussion.

REFERENCES

Altschul, S. F., Gish, W., Miller, W., Myers, E. W., & Lipman, D. J. (1990). Basic local alignment search tool. *Journal of Molecular Biology*, *215*(3), 403–410.

Bach, F. R., Lanckriet, G. R. G., & Jordan, M. I. (2004). Multiple kernel learning, conic duality, and the SMO algorithm. In *Proceedings of the 21st international conference on machine learning*. New York.

Ben-Hur, A., & Noble, W. S. (2005). Kernel methods for predicting protein-protein interactions. *Bioinformatics (Oxford, England)*, *21*(Suppl. 1), i38–i46. doi:10.1093/bioinformatics/bti1016

Bleakley, K., Biau, G., & Vert, J.-P. (2007). Supervised reconstruction of biological networks with local models. *Bioinformatics (Oxford, England)*, *23*(13), i57–i65. doi:10.1093/bioinformatics/btm204

Boyd, S., & Vandenberghe, L. (2004). *Convex optimization*. Cambridge University Press.

Caldarelli, G. (2007). *Scale-free networks: Complex webs in nature and technology*. Oxford Univ Press.

Chua, H. N., Sung, W. K., & Wong, L. (2006). Exploiting indirect neighbours and topological weight to predict protein function from protein-protein interactions. *Bioinformatics (Oxford, England)*, *22*(13), 1623–1630. doi:10.1093/bioinformatics/btl145

Evgeniou, T., Micchelli, C. A., & Pontil, M. (2005). Learning multiple tasks with kernel methods. *Journal of Machine Learning Research, 6*, 615–637.

Finn, R. D., Tate, J., Mistry, J., Coggill, P. C., Sammut, S. J., & Hotz, H. R. (2008). The Pfam protein families database. *Nucleic Acids Research, 36*, D281–D288. doi:10.1093/nar/gkm960

Gene Ontology Consortium. (2004). The gene ontology (GO) database and informatics resource. *Nucleic Acids Research, 32*(Database issue), D258–D261. doi:10.1093/nar/gkh036

Gotoh, O. (1982). An improved algorithm for matching biological sequences. *Journal of Molecular Biology, 162*(3), 705–708. doi:10.1016/0022-2836(82)90398-9

Han, J.-D. J., Bertin, N., Hao, T., Goldberg, D. S., Berriz, G. F., & Zhang, L. V. (2004). Evidence for dynamically organized modularity in the yeast protein-protein interaction network. *Nature, 430*(6995), 88–93. doi:10.1038/nature02555

Jansen, R., Greenbaum, D., & Gerstein, M. (2002). Relating whole-genome expression data with protein-protein interactions. *Genome Research, 12*, 37–46. doi:10.1101/gr.205602

Kato, T., Tsuda, K., & Asai, K. (2005). Selective integration of multiple biological data for supervised network inference. *Bioinformatics (Oxford, England), 21*(10), 2488–2495. doi:10.1093/bioinformatics/bti339

Lanckriet, G. R. G., Bie, T. D., Cristianini, N., Jordan, M., & Noble, W. (2004). A statistical framework for genomic data fusion. *Bioinformatics (Oxford, England), 20*(16), 2626–2635. doi:10.1093/bioinformatics/bth294

Okada, K., Kanaya, S., & Asai, K. (2005). Accurate extraction of functional associations between proteins based on common interaction partners and common domains. *Bioinformatics (Oxford, England), 21*(9), 2043–2048. doi:10.1093/bioinformatics/bti305

Pavlidis, P., Weston, J., Cai, J., & Noble, W. S. (2002). Learning gene functional classifications from multiple data types. *Journal of Computational Biology, 9*(2), 401–411. doi:10.1089/10665270252935539

Rakotomamonjy, A., Bach, F., Canu, S., & Grandvalet, Y. (2008). SimpleMKL. *Journal of Machine Learning Research, 9*, 2491–2521.

Samanta, M. P., & Liang, S. (2003). Predicting protein functions from redundancies in largescale protein interaction networks. *Proceedings of the National Academy of Sciences of the United States of America, 100*(22), 12579–12583. doi:10.1073/pnas.2132527100

Schölkopf, B., & Smola, A. J. (2002). *Learning with kernels*. MIT Press, Cambridge MA.

Sonnenburg, S., Rätsch, G., Schäfer, C., & Schölkopf, B. (2006). Large scale multiple kernel learning. *Journal of Machine Learning Research, 7*, 1531–1565.

Vert, J. P., Qiu, J., & Noble, W. S. (2007). A new pairwise kernel for biological network inference with support vector machines. *BMC Bioinformatics, 10*(Suppl 10), S8. doi:10.1186/1471-2105-8-S10-S8

Vert, J.-P., & Yamanishi, Y. (2005). Supervised graph inference. In L. K. Saul, Y. Weiss, & L. Bottou (Eds.), *Advances in neural information processing systems 17* (pp. 1433-1440). Cambridge, MA: MIT Press.

von Mering, C., Krause, R., Snel, B., Cornell, M., Oliver, S. G., Fields, S., & Bork, P. (2002). Comparative assessment of large-scale data sets of protein-protein interactions. *Nature*, *417*(6887), 399–403. doi:10.1038/nature750

Yamanishi, Y., Vert, J.-P., & Kanehisa, M. (2005). Supervised enzyme network inference from the integration of genomic data and chemical information. *Bioinformatics (Oxford, England)*, *21*(Suppl 1), i468–i477. doi:10.1093/bioinformatics/bti1012

This work was previously published in the International Journal of Knowledge Discovery in Bioinformatics, Volume 1, Issue 1, edited by Xiao-Li Li and See-Kiong Ng, pp. 66-80, copyright 2010 by IGI Publishing (an imprint of IGI Global).

Chapter 12
Incorporating Network Topology Improves Prediction of Protein Interaction Networks from Transcriptomic Data

Peter E. Larsen
Argonne National Laboratory, USA

Frank Collart
Argonne National Laboratory, USA

Yang Dai
University of Illinois at Chicago, USA

ABSTRACT

The reconstruction of protein-protein interaction (PPI) networks from high-throughput experimental data is one of the most challenging problems in bioinformatics. These biological networks have specific topologies defined by the functional and evolutionary relationships between the proteins and the physical limitations imposed on proteins interacting in the three-dimensional space. In this paper, the authors propose a novel approach for the identification of potential protein-protein interactions based on the integration of known PPI network topology and transcriptomic data. The proposed method, Function Restricted Value Neighborhood (FRV-N), was used to reconstruct PPI networks using an experimental data set consisting of 170 yeast microarray profiles. The results of this analysis demonstrate that incorporating knowledge of interactome topology improves the ability of transcriptome analysis to reconstruct interaction networks with a high degree of biological relevance.

DOI: 10.4018/978-1-4666-1785-8.ch012

INTRODUCTION

As the number of published descriptions of protein-protein interactions (PPIs) continues to grow, it becomes necessary to distill this increasingly vast number of facts into useful and biologically relevant associations. Although high-throughput methods for synthesizing and identifying large numbers of PPIs are beyond the reach of many research laboratories, profiling mRNA expression under various conditions is an effective and attainable approach to uncover underlying cellular machinery. In the context of this study, we define a PPI as one of two kinds of interactions: a direct physical interaction between proteins or two proteins that may or may not come into direct contact with one another but are parts of the same macromolecular complex. These types of interactions are distinct from 'functionally associated' proteins which may be co-expressed and working in the same or related biological process but do not require any form of physical interaction with one another for their function. It is well established that genes whose mRNA expression patterns are correlated across many diverse conditions can often be inferred to be functionally associated and these transcriptional co-expression patterns have been used in inferring physical protein interactions (Deane, Salwinski, Xenarios, & Eisenberg, 2002; Jansen et al., 2003). Although transcriptional data cannot consistently distinguish between direct protein binding and membership in a protein complex, strongly co-expressed mRNAs are more likely to indicate long-lived interactions (Ge, Liu, Church, & Vidal, 2001; Jansen, Greenbaum, & Gerstein, 2002; Simonis, Gonze, Orsi, van Helden, & Wodak, 2006). Recently, a supervised machine learning method (support vector machine) was proposed to predict physical PPIs from microarray gene expression data (Soong, Wrzeszczynski, & Rost, 2008). This method used gene expression profiles across 349 yeast microarray experiments to derive hidden factors that underlie the predicted PPI observations. The method was shown to recover more experimentally observed physical interactions than a conventional correlation-based approach. A recent augmentation of this approach used network topology to refine co-expression networks and uncover potential PPIs (Xulvi-Brunet & Li, 2009). The authors noted, however, that gene co-expression by itself does not necessarily correlate with known PPIs. This is not unexpected as for a given biological condition; the number of known interactions comprises only a small subset of all possible interactions likely to actually occur. In addition, only those interactions observed in the experimental condition and also differentially expressed in the context of a transcriptomic experiment have the potential to be detected. A PPI network then should be specifically associated with a biological condition and matched with differentially expressed relevant genes observed via the transcriptomic experiment. Several methods have been proposed to identify condition-specific biological networks or pathways from microarray data (Dittrich, Klau, Rosenwald, Dandekar, & Muller, 2008; Guo et al., 2007; Ideker, Ozier, Schwikowski, & Siegel, 2002) which map condition-specific interactions differentially expressed genes. However, the network structure of these methods is constrained to a pre-assembled PPI network and therefore they cannot discover novel interactions.

An ideal method for identifying PPI's from transcriptomic data should have the ability to use the large and growing databases of known interactions, and also have the ability to identify novel interactions not previously reported in those databases. This is particularly important for those organisms with a minimal number of experimentally observed PPIs.

The method presented here is an extension of the Asymmetric-N method (Chen, Larsen, Almasri, & Dai, 2008), which constructs a co-expression network of genes through the analysis of microarray gene expression data. This method differs from other correlation-based methods such as, relevance networks (Butte, Tamayo, Slonim,

Golub, & Kohane, 2000), clustering coefficient threshold method (Elo, Jarvenpaa, Oresic, Lahesmaa, & Aittokallio, 2007), nearest neighbor networks (Huttenhower et al., 2007), and ARACNE (Margolin et al., 2006), which are also used for the construction of co-expression networks. Asymmetric-N utilizes a global topology of an existing genome-scale interaction network, e.g., scale-free property, to reduce the false positives. More specifically, the method employs different neighborhood sizes for the "core" and "peripheral" nodes so that the former are more densely connected than the latter. The problem with this method is that the two sizes have to be empirically determined. However, the general framework underlying Asymmetric-N permits the incorporation of different types of topological constraints derived from different biological networks. To overcome this limitation in Asymmetric-N, we propose a novel approach, the Function Restricted Value N (FRV-N) method. In FRV-N, a protein is assigned a neighborhood size according to that protein's molecular function annotation and the interacting partners from a known PPI network. Characteristic neighborhood sizes for a given molecular function are taken from observations of large numbers of known PPIs.

To validate the ability of the FRV-N method for predicting physical interactions from transcriptomic data, we selected subset of the experimental data collected by Knijnenburg et al. (2009). In this experiment, yeast was cultured under a variety of oxidation conditions, carbon sources, limiting nutrients, temperatures, and pH conditions. This experimental set is expected to identify interacting proteins involved in metabolism, nutrient transport, response to stress, and cellular respiration. Protein neighborhood sizes used here were informed by two sources of prior observations: known macromolecular complexes (MCs) and known binary physical interactions (BPIs). For molecular function annotation of proteins, we selected Gene Ontolgy Molecular Function (GO-MF) annotations. The predicted PPI networks are judged by the frequency and type of annotations shared between predicted PPI pairs and the frequency at which predicted PPIs have been previously identified in the publically available databases. The FRV-N method performs far better than the method based on a threshold for the Pearson correlation coefficient for a pair of proteins in terms of the frequency and distribution with which predicted interacting proteins share a specific annotation. Assigning a constant neighborhood size to proteins did not perform as well as neighborhoods informed from prior biological observations. MC-informed neighborhoods preformed better than BPI-informed neighborhoods.

MATERIALS AND METHODS

Publically Available Datasets

Microarray Experiment

A previously reported transcriptomic dataset in which gene expression in *Saccharomyces cerevisiae* was measured while cultured with multiple combinations of nutrients, limiting elements, temperatures, and aeration types (Knijnenburg et al., 2009) was used to validate the method FRV-N proposed here. The data consist of results from 170 Affymertix 'GeneChips'. A subset of this dataset was used, selecting those genes that were differentially expressed under different growth conditions. Genes whose expression measurements varied by more than two standard deviations from the average expression intensities in more than eight (5%) of 'GeneChips' were considered differentially expressed. This procedure identified a total of 1748 genes, which comprise approximately 30% of the *Saccharomyces cerevisiae* genome. The complete set of genes used in this study can be found in Supplemental Table 1.

Table 1. Neighborhood sizes by GO-slim MF annotation

		Complex			Binary		
GO-MF Annotation		**0.75**	**0.90**	**0.95**	**0.75**	**0.90**	**0.95**
GO:0003677	DNA binding	4	2	2	10	4	3
GO:0030234	Enzyme regulator activity	5	3	3	11	5	3
GO:0004386	Helicase activity	10	4	4	6	2	2
GO:0016787	Hydrolase activity	6	3	3	5	3	2
GO:0016853	Isomerase activity	6	5	5	4	3	3
GO:0016874	Ligase activity	12	3	2	7	4	2
GO:0008289	Lipid binding	3	2	2	13	9	5
GO:0016829	Lyase activity	24	7	7	5	2	2
GO:0003774	Motor activity	4	2	2	12	6	5
GO:0016779	Nucleotidyltransferase activity	10	4	3	15	6	3
GO:0016491	Oxidoreductase activity	2	2	2	3	2	2
GO:0008233	Peptidase activity	8	5	5	5	3	2
GO:0004721	Phosphoprotein phosphatase activity	4	4	2	8	5	3
GO:0005515	Protein binding	6	3	2	15	7	4
GO:0004672	Protein kinase activity	3	2	2	19	12	6
GO:0003723	RNA binding	13	9	5	25	8	4
GO:0004871	Signal transducer activity	12	12	12	11	5	3
GO:0005198	Structural molecule activity	40	12	6	14	6	4
GO:0030528	Transcription regulator activity	10	4	3	12	6	3
GO:0016740	Transferase activity	6	3	2	6	3	2
GO:0045182	Translation regulator activity	5	3	2	7	3	3
GO:0005215	Transporter activity	4	3	2	4	3	2
	Other	6	3	2	9	4	2
	Average	8.8	4.3	3.5	9.8	4.8	3.0
	Standard Deviation	8.3	2.9	2.4	5.4	2.5	1.1

Macromolecular Complexes

One of the sources of the known network used for this study is the set of known macromolecular complexes, which was downloaded from the "Macromolecular Complex Terms" from the Saccharomyces Genome Database (SGD) (www.yeastgenome.org). In this set, there are 328 macromolecular complexes consisting of 1,699 proteins. The average cluster size is 12.8. The largest cluster, "ribonucleoprotein complex", consists of 569 members while the smallest clusters have 2 members. The dataset consists of 136,502 total interactions when an interaction is considered between all protein pairs within a complex. Intuitively, this set of complexes has many topological characteristics typically represented in a predicted PPI network. Most complexes are small sets of interactions between proteins working together to perform a single biological function rather than a large knot of interactions comprising most of the complete set of known proteins.

Known Binary Physical Interactions

The other source of prior biological knowledge used here is a dataset of experimentally derived binary physical interactions. The strength of these interactions is that they represent a set of experimentally validated, direct interactions between proteins. The potential drawback is that the observed neighborhood of a specific protein may be influenced by frequency of appearance of the protein in prior scientific publications rather than true biologically-driven phenomenon. The set of binary interactions used in this study was taken from the unique set of known interactions identified as "physical interactions" in the file "interaction_data.tab" made available from the SGD. Out of the set of 142,557 interactions, there are 44,705 unique interactions between 5,249 genes annotated as "physical interactions". Of these known interactions, there are 2,981 unique known interactions for the set of 1,748 differentially expressed genes in the set for PPI prediction.

GO-Slim Annotation

We used Gene Ontology (GO) annotation to assign specific annotations to *S. cerevisiae* proteins. GO has been developed to offer controlled vocabularies for aiding in the annotation of molecular attributes for different model organisms. Three structured ontologies have been proposed, which allow the description of molecular function (MF), biological process (BP) and cellular component (CC). In this paper, we use the GO MF annotations. GO Slim Mapper (http://db.yeastgenome.org/cgi-bin/GO/goTermMapper) terms represent the less specialized terms in yeast in the three ontologies selected by expert curators according to annotation statistics and biological significance to best represent major biological categories found in *S. cerevisiae*. There are 26, 40 and 26 GO Slim Mapper terms for GO-MF, GO-BP and GO-CC, respectively.

For this study, we have selected GO-slim MF annotations as the most relevant annotation to describe a protein's characteristic neighborhood size. Not only does GO-MF describe a protein's dominant molecular characteristics better than GO-BP or GO-CC, but it is also most likely to be deduced for an otherwise uncharacterized protein by sequence information alone. GO-BP is less suitable for describing neighborhood size since proteins with the same GO-BP annotation may be radically different in their molecular characteristics. For example, a transcription factor and a pheromone receptor protein may both be correctly annotated with the BP annotation of "Cell Cycle", but that annotation does not imply any molecular similarity between those two proteins. Likewise, two proteins may both be correctly annotated with the CC annotation of "Cytoplasm", though that information does not define the variety of proteins and functions that may be found in that cellular location.

Identify Network Topology from Prior Biological Observations

Identifying Neighborhood Sizes by GO-slim MF Annotations

The distribution of observed neighborhood sizes of proteins by their GO-slim MF annotation was calculated from the set of MCs and the set of unique BPIs in SGD. For a protein in a MC, the neighborhood is considered to be equal to the number of members of the complex to which it is annotated, minus one. The neighborhood size of a protein in the set of BPIs is equal to the number of unique interactions that include that protein. The annotation "Other", though cannot usefully

Figure 1. Distribution of neighborhood sizes by GO-slim MF annotation

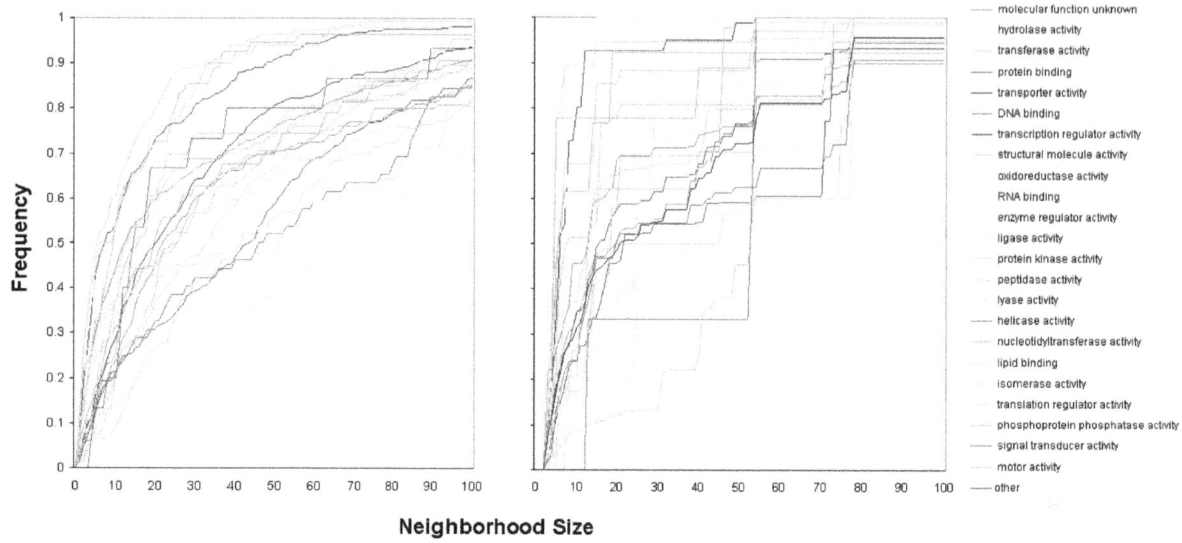

describe anything about the molecular nature of a protein, is included to assign a neighborhood to otherwise un-annotated proteins and permit them to participate in predicted PPIs. The frequency distribution of observed neighborhood sizes by GO-MF annotation is summarized in Figure 1. From this distribution of neighborhood sizes, the threshold for neighborhood of a protein for a specific annotation can be expressed as a percentile. Percentiles of 0.75, 0.90 and 0.95 were selected for neighborhood thresholds for this study (Table 1). The average neighborhood sizes at each threshold for the BPI-informed neighborhoods are 9.8 (standard deviation 5.4), 4.8 (standard deviation 2.5), and 3.0 (standard deviation 1.1) for percentiles 0.75, 0.90 and 0.95 respectively. The average neighborhood sizes at each threshold for the MC-informed neighborhoods are 8.8 (standard deviation 8.3), 4.3 (standard deviation 2.9), and 3.5 (standard deviation 2.4) for percentiles 0.75, 0.90 and 0.95 respectively. Although the average neighborhood sizes for both methods are within one neighbor or less of one another at each threshold, the annotation-specific neighborhoods can be quite different. The largest MC-informed neighborhood sizes belong to proteins with the annotations "Structural molecule activity" followed by "Lyase activity", "Signal transducer activity", and "RNA binding". The largest neighborhoods are found for RNA binding (25) followed by "Protein kinase activity" and "Lipid binding activity" for the BPI-informed neighborhoods. The annotation "other" has neighborhood sizes slightly smaller than the average neighborhood derived from either set of known interactions and at any given threshold.

Constant N Neighborhoods

To examine the requirement of annotation-specific neighborhoods on improvement in predicted PPI networks or if restriction on neighborhood size would improve PPI network predictions over the method based on PCC method, the FRV-N method was also calculated using a constant neighborhood size. Physical limitations on the number of neighbors of a protein were modeled by selecting a value of 12 as the largest possible neighborhood. This neighborhood size is derived from the assumption that proteins can be modeled as identically-sized spheres, and 12 is the maximum number of neighbors in a hexagonal close packed

Figure 2. Distribution of PCC values in networks

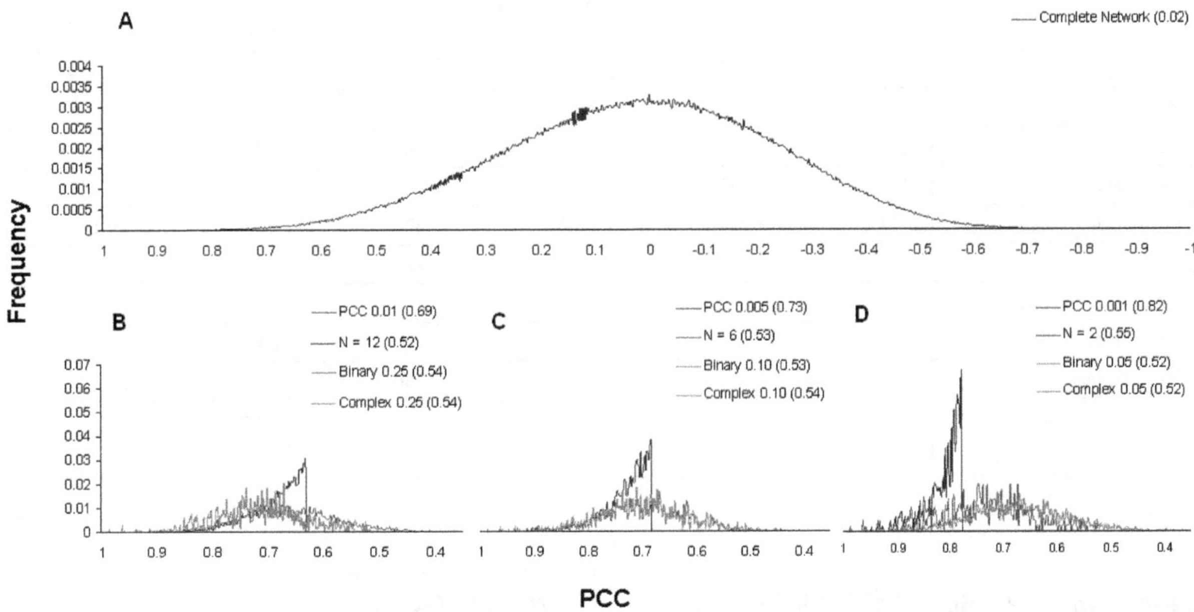

lattice. Restricted neighborhood sizes of 6 and 2 were also selected for this study.

Predicting PPI Networks from Co-expression Data

Pearson Correlation Coefficients

Pearson correlation coefficient (PCC) was calculated between all pairs of expression profiles in the set of 1748 differentially expressed genes across 170 data points in the transcriptomic data set. As their distribution approximates the normal distribution (Figure 2A), one minus the percentile of PCC values can be considered approximately equivalent to a p-value. Only positive PCC values are considered for possible PPIs. Following the definition of PPI used in this study, only those proteins that are co-expressed in response to a change in biological condition are expected to partake in direct, physical interactions.

FRV-N Algorithm

Given a vector N in which each entry N(i) is the neighborhood size determined for gene i, our new algorithm FRV-N can be described as follows.

Algorithm FRV-N

Step 1:

for each gene i
organize SortedNeighbor[i] to be a list of genes whose
PCCs with gene i are in descending order.

Step 2:

for a pair of genes i and j, connect genes i and j if gene
i's rank in SortedNeighbor[j] is smaller than N[i]; and
gene j's rank in SortedNeighbor[i] is smaller than N[j].

Table 2. Predicted PPI network sizes

	# Interactions	# Proteins	Interactions / Protein	# Clusters	Largest Cluster	Ave Cluster
PCC(0.01)	8177	1262	13.0	13	1236	97.1
PCC(0.005)	3658	979	7.5	30	910	32.6
PCC(0.001)	531	299	3.6	48	73	6.2
FRV-N(12)	4188	1532	5.5	14	1502	109.4
FRV-N(6)	1862	1348	2.8	67	1205	20.1
FRV-N(2)	501	770	1.3	275	13	2.8
FRV-NBinary (0.75)	3073	1475	4.2	20	1423	73.8
FRV-NBinary (0.90)	1251	1188	2.1	136	640	8.7
FRV-NBinary (0.95)	638	882	1.4	266	20	3.3
FRV-NComplex (7.25)	2324	1403	3.3	42	1296	33.4
FRV-NComplex (0.90)	1019	1091	1.9	180	173	6.1
FRV-NComplex (0.95)	682	900	1.5	258	33	3.5

Generating Predicted PPI Networks

Four methods with variable thresholds were used to predict PPI networks from the subset of 1728 genes obtained from the 170 yeast metabolism microarrays. PCC method was used to generate three networks, PCC(0.99), PCC(0.995) and PCC(0.999) using thresholds at percentiles 0.99, 0.995, and 0.999, respectively. Three networks, PPI networks FRV-N(12), FRV-N(6), and FRV-N(2), were generated using FRV-N method with constant neighborhood sizes of 12, 6, and 2, respectively. FRV-N using neighborhood sizes by GO-slim MF annotation with the MC-informed and BPI-informed neighborhoods to generated PPI networks FRV-$N_{Complex}$(threshold) and FRV-N_{Binary} (threshold) respectively. The neighborhood sizes were determined from thresholds 0.75, 0.90 and 0.95 of percentiles. The number of predicted PPIs, number of proteins in each predicted PPI network, and number of unique clusters of at least 2 proteins were determined (see Table 2).

RESULTS

Certainly what constitutes a "good" predicted PPI network is difficult to define or quantify by assignment of a numerical value. Commonly used metric determines how many predicted interactions occur between proteins that share a common annotation or how many predicted interactions appear in databases of experimentally validated interactions. Both of these useful metrics are considered in this study. If, however, only these metrics are considered then there is little value in predicting the PPI network. Protein annotations are already known and many databases of known PPIs are readily available. A better measure of a predicted PPI network is its utility to further the understanding of some biological phenomenon under specific experimental conditions and the application of the PPI network prediction for the design of new and novel biological experiments. Thus, a method for predicting PPI networks should be informed by prior observations but not limited to them. A method should be able to identify key characteristics of a "good" PPI network from prior observations and apply those characteristics to predicted networks. In this study, we will also consider the frequency at which specific annotations are shared in sets of known interactions and measure how well predicted PPI networks reflect the distribution of those annotations.

Neighborhood Sizes and Structure

The size and connectivity of the predicted networks varied widely by method and specific threshold (Table 2). PCC method generated the largest networks at lower stringencies with FRV-$N_{Complex}$ generated the largest networks at the highest stringencies. For PCC(0.99) and PCC(0.995), FRV-N(12) and FRV-N(6), FRV-N_{Binary}(0.75) and FRV-$N_{Complex}$(0.75), the largest cluster comprises about 90% or more of all proteins, indicating that these methods and stringencies did not do well in identifying discrete molecular complexes. For FRV-N(2) and FRV-N_{Binary}(0.95), the average cluster sizes are 13 and 20 respectively, demonstrating that these stringencies are more likely to produce very small clusters.

Distribution of PCC Values in Predicted PPI Networks

The predicted PPI networks can also be differentiated by the average and the distribution of PCC scores for its included interactions (see Figure 2 B-D). A simple PCC threshold leads to an abrupt termination of PCC score distribution and always leads to a higher average PCC score for its predicted interactions. Restriction of neighborhood sizes by any FRV-N method leads to a wider distribution and lower average of PCC values for its included, predicted interactions. From these distributions it can also be inferred that the specific predicted interactions are quite different between a simple PCC threshold and a restricted neighborhood. For example, for PCC(0.995), no interaction has a PCC score less than 0.684. For the restricted neighborhood methods, 54%, 58%, and 60% of included interactions have PCC scores less than 0.684 for FRV-N(6), FRV-N_{Binary}(0.90), and FRV-$N_{Complex}$(0.90), respectively. This indicates that restriction of neighborhood size does not improve predicted PPI networks by taking a smaller subset of the larger PCC network, but rather by including biologically relevant interactions with lower PCC scores.

Shared Annotations

If one of the metrics for measuring the quality of predicted PPI networks is the proportion of interactions that share an annotation term or are present in the set of known interactions, then the set of expectations may be established for the range of observed shared annotations that are considered good. We examined the two sets of known interactions: the set of macromolecular complexes and the set of known binary physical interactions of SGD. Recall that for macromolecular complexes, all members of a complex are considered to interact with every other member of the complex. For these two known sets of interactions and for the complete network of all possible interactions in the 1,748 protein dataset, we identified the frequency of interacting proteins that shared an annotation term. GO annotations of "other", "not yet annotated", and "unknown" were not included as a shared annotation (Table 3). Over 70% of interactions in the known sets and about 30% in the complete network share a GO-slim CC annotation. About half of the interactions in the known sets share a GO-slim BP annotation while only less than 10% do in the complete graph. About 25% of known interactions share a GO-slim MF annotation, up from less than 5% in the complete network. 22% of known physical interactions are found in the same complex while about 9% of proteins in the same macromolecular complex are in the set of known physical interactions. About 33% of all interactions in the complete network share at least one form of annotation. These observations suggest parameters for evaluation of predicted PPI networks. A "good" predicted PPI network should have a significant increase in interactions

Table 3. Frequency with which interacting proteins share annotations in set of known PPIs

	Complete	Complex	Published
Shared GO-CC:	**30.5%**	76.5%	71.1%
Shared GO-BP:	**9.9%**	55.1%	47.8%
Shared GO-MF:	**4.5%**	27.5%	22.8%
Same Complex:	**0.5%**	100.0%	22.2%
Previously Published:	**0.2%**	9.3%	100.0%
Any Shared:	**33.9%**	100.0%	100.0%

that share an annotation over the complete graph, but are not necessarily expected to contain interactions with shared annotations more frequent than sets of known interactions.

The predicted PPI networks can be evaluated in the context of the observations from the complete network and the two known sets of interactions (see Table 4). All methods and thresholds considered increase the frequencies of shared interactions and known interactions relative to the complete network. Similar to the observations from known interactions, GO-slim CC annotations tends to have the largest proportion of shared interactions, and GO-slim MF the lowest. For PCC method, increasing stringency generally elevates the frequencies of shared annotation and known interactions as the predicted network size decreases. For methods with constant neighborhood sizes and MC-informed neighborhood sizes, the best frequencies of shared annotation and known interactions occur at the intermediate levels of stringency. The percent of interactions sharing any annotations in the predicted networks obtained from the BPI-informed neighborhood continues to improve with increased stringency. The method with constant neighborhood size shows an improvement over PCC method, and FRV-N performs better than the method with constant neighborhood sizes in percent of interactions sharing any annotation. FRV-$N_{Complex}$ performs better than FRV-N_{Binary}. Only PCC(0.999) shows the increased frequency of shared GO-slim MF annotations in predicted PPIs over other methods, although overall PCC(0.999) does not perform as well as the worst neighbor-restricted network. By the metric of percent interactions that share any annotation, the best performing method is FRV-$N_{Complex}$(0.90).

Correlation with Specific Shared Annotations in Known Physical Interactions and Macromolecular Complexes

The improvements in frequency of shared GO-slim annotations in predicted PPIs obtained from FRV-N over PCC method and constant neighborhood size method, though significant, are sometimes slight. More interesting than the overall frequency of shared annotations in predicted PPI is the observation of specific GO-slim annotations that are shared in predicted PPIs. How well those specific annotations match the specific shared annotations from the sets of known interactions is an indication of the biological relevance of a predicted network.

To examine the patterns of specific GO-slim annotations shared in the predicted PPIs, we considered the frequency at which specific annotations occur in the sets of interactions that share an annotation. Each GO-slim annotation branch, molecular function, biological process, and cellular compartment, was considered separately. An example of this form of comparison using PCC method and method with the MC-informed neighborhoods is shown in Figure 3. A PCC was

Table 4. Frequency with which interacting proteins share annotations in predicted PPIs

	Shared GO-CC	Shared GO-BP	Shared GO-MF	Same Complex	Previously Published	Any Shared
PCC(0.01)	39.2%	18.9%	8.4%	2.2%	0.9%	44.4%
PCC(0.005)	38.8%	20.2%	9.8%	2.8%	1.1%	44.0%
PCC(0.001)	40.7%	29.8%	20.0%	5.6%	0.9%	46.3%
FRV-N(12)	47.9%	23.0%	10.4%	3.0%	0.9%	53.2%
FRV-N(6)	48.2%	24.3%	10.8%	3.6%	1.3%	53.5%
FRV-N(2)	49.4%	24.9%	10.0%	3.2%	1.4%	55.1%
FRV-NBinary (0.75)	48.9%	24.8%	10.7%	3.8%	1.2%	54.6%
FRV-NBinary (0.90)	51.1%	27.4%	12.6%	4.3%	1.8%	57.0%
FRV-NBinary (0.95)	51.1%	28.6%	13.7%	4.1%	2.0%	57.5%
FRV-NComplex (7.25)	52.1%	28.6%	14.5%	6.7%	1.5%	58.1%
FRV-NComplex (0.90)	53.1%	30.1%	14.9%	7.2%	2.3%	58.7%
FRV-NComplex (0.95)	51.9%	28.4%	14.4%	5.9%	2.1%	58.0%

calculated for the pattern of shared annotations between all predicted PPI networks and the sets of known interactions. This measurement indicates how well the shared annotations in the predicted PPI networks resemble the shared interaction in the sets of known interactions. If the distribution of specific shared annotations in a predicted PPI closely resembles the distribution in a set of known interactions, then the PCC should be closer to 1. A 'good' PCC score should also be higher than the observed PCC for the complete network.

Evaluation of the correlations with the set of known interactions (see Table 5) demonstrates poor performance of PCC, frequently with PCC values less than those for the complete network. PCC(0.999), though the 'best' of the predicted networks by PCC method considering the frequency of shared annotations, shows a precipitous drop in PCC values relative to the less stringent PCC(0.99) and PCC(0.995). The best correlations with known interactions are observed in PPI networks predicted with FRV-N_{Binary}, followed by FRV-$N_{Complex}$. For all predicted PPI networks, PCC with shared GO-slim CC annotations in the set of known interactions is close to that of the complete network, reflecting the fact that about 75% of PPIs in the predicted, known, and complete networks have the shared but not very specific annotation of "cytoplasm".

Correlations between specific shared annotations show far more dramatic differences for the set of macromolecular complexes. Again, PCC method does the worst of all predicted networks with PCC(0.999) having the lowest correlations with shared annotation in the set of macromolecular complexes. FRV-$N_{Complex}$ performs the best, followed by FRV-N_{Binary}, although this time the increase in PCC scores are far larger than for the similar measurement using the set of known interactions. The correlation for the PPI networks predicted by PCC method is frequently low, close to zero for PCC(0.999) with GO-slim BP and GO-slim MF, indicating that the networks bear little resemblance in their shared annotations to known macromolecular clusters in spite of higher frequencies of shared GO-BP and GO-MF annotations.

Correlations between specific shared annotations show far more dramatic differences for the set of macromolecular complexes. Again, PCC method does the worst of all predicted networks with PCC(0.999) having the lowest correlations

Figure 3. Specific annotations shared between interacting proteins in PPI networks

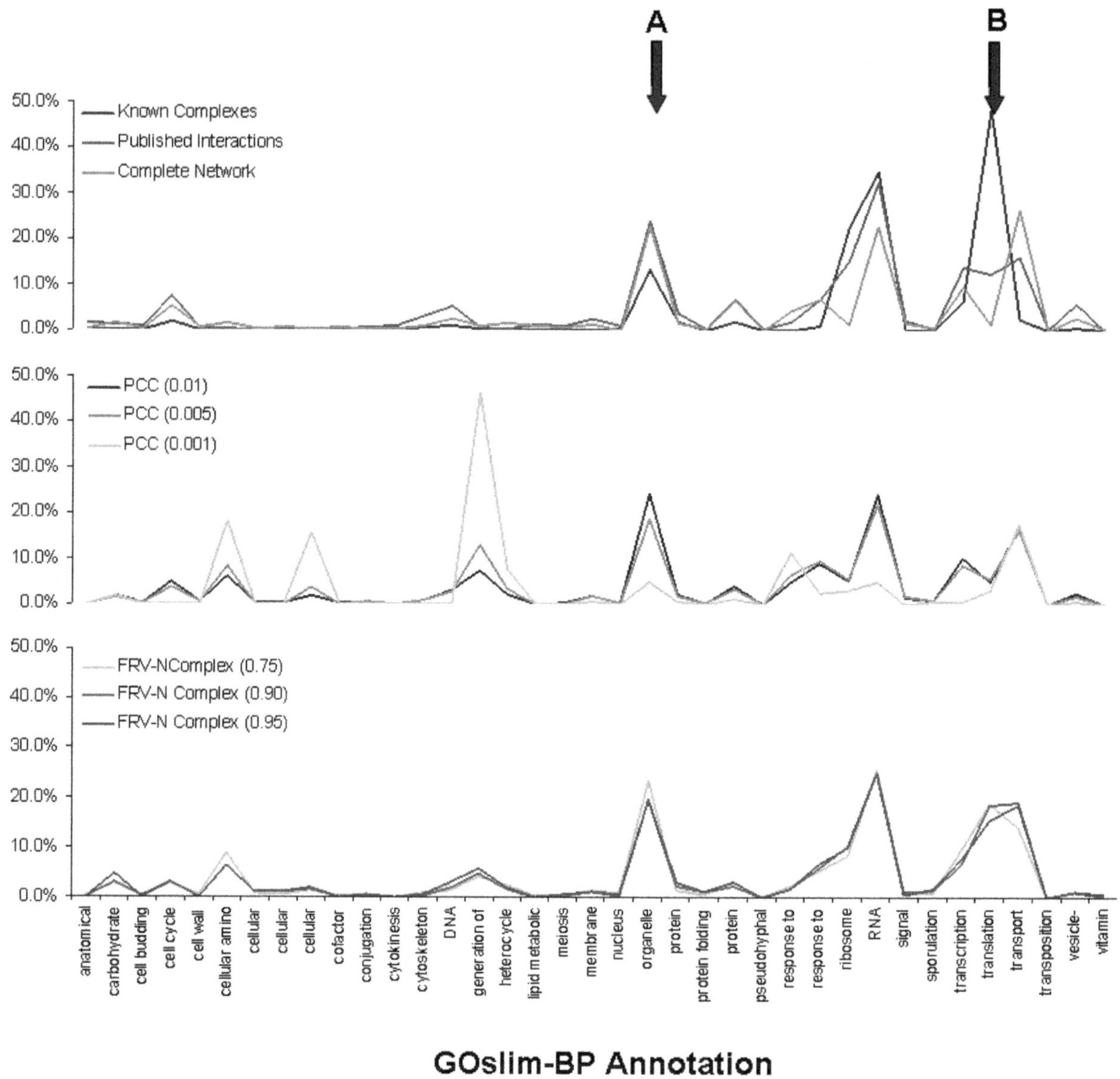

with shared annotation in the set of macromolecular complexes. FRV-$N_{Complex}$ performs the best, followed by FRV-N_{Binary}, although this time the increases in PCC scores are far larger than for the similar measurement using the set of known interactions. The correlation for the PCC networks is frequently low, close to zero for PCC(0.999) with GO-BP and GO-MF, indicating that the networks show little resemblance in shared annotations to known macromolecular clusters in spite of higher frequencies of shared GO-BP and GO-MF annotations.

To interpret these results, for example, PCC(0.999) has the highest observed frequency of shared GO-slim MF annotations of all predicted PPIs considered, yet it has the lowest GO-slim BP correlations observed. Looking at the specific shared annotation terms (see Figure 3) we

Table 5. Correlation with known physical interactions

	BINARY			COMPLEX		
	Shared GOBP	**Shared GOMF**	**Shared GOCC**	**Shared GOBP**	**Shared GOMF**	**Shared GOCC**
PCC(0.01)	0.90	0.67	0.95	0.50	0.17	0.78
PCC(0.005)	0.83	0.56	0.93	0.48	0.14	0.76
PCC(0.001)	0.05	0.14	0.68	0.00	-0.02	0.57
FRV-N(12)	0.90	0.70	0.95	0.58	0.24	0.79
FRV-N(6)	0.91	0.68	0.95	0.59	0.27	0.79
FRV-N(2)	0.86	0.57	0.94	0.63	0.23	0.78
FRV-NBinary (0.75)	0.95	0.93	0.98	0.64	0.48	0.80
FRV-NBinary (0.90)	0.96	0.86	0.98	0.65	0.42	0.80
FRV-NBinary (0.95)	0.94	0.85	0.96	0.66	0.47	0.80
FRV-NComplex (7.25)	0.91	0.89	0.96	0.78	0.83	0.84
FRV-NComplex (0.90)	0.91	0.84	0.95	0.78	0.78	0.83
FRV-NComplex (0.95)	0.92	0.74	0.96	0.74	0.57	0.82

see that of the shared GO-slim BP annotations in PCC(0.999), 46.2% of those shared annotations are "generation of precursor metabolites and energy" and 18.4% are "cellular amino acid and derivative metabolic process". For the set of known interactions, only 0.75% and 0.3% of interactions shares those interactions respectively. For the set of complex interactions, only 0.17% and 0.04% share those annotations respectively. This suggests that although the PCC method did in fact successfully identify a large number of co-expressed genes with same and seemingly relevant annotations, proteins expressed with those annotations are infrequently observed to participate in PPIs. The frequency in the PPIs predicted by FRV-$N_{Complex}$ is robust to the level of the percentile, indicating that networks predicted by this method have a high degree of biological relevance at any percentile threshold.

DISCUSSION

The ultimate goal of this method and methods like this is to identify specific PPIs present in specific experimental conditions and to discover novel, previously unknown interactions from new experimental data. The distinction between a false positive and novel discovery can only be made by further biological experimentation. Given that FRV-N method has been demonstrated to produce interaction networks enriched for biologically relevant interactions, then predictions made with this method could be expected to be an excellent source for generating hypothesis-driven biological experiments. Such experiments are beyond the scope of this study. However, we can contemplate strategies that use these results to generate testable hypotheses and propose experiments. For example, of the PPI networks predicted here, FRV-$N_{Complex}$ (0.90) preformed the best using the metrics considered in this study. Taking all predicted PPIs for complexes of at least 5 members (see Figure 4), the predicted PPI network is found to consist of many, discrete predicted protein complexes. As seen in an example (Figure 5), protein nodes can be highlighted by their relative expression at four different biological conditions. In this instance, aerobic or anaerobic growth and limiting carbon or limiting nitrogen were used. By considering

Figure 4. FRV-$N_{complex}$(0.90) predicted PPI network

the relative expression of each protein cluster under different conditions, even in this simplified figure it is possible to observe that some predicted protein clusters are differentially expressed at specific conditions. These observations can be readily translated into functional predictions for specific clusters, and then can be validated through additional biological experiments.

As expected, PPI networks predicted by FRV-N methods did not recover all possible known interactions. While it is possible for previously observed interactions between analyzed proteins to be true, many of those interactions might not detectable in the experimental data. Interactions between proteins expressed at specific stages of the cell cycle, for example, are likely present under most experimental conditions, but those

Figure 5. Expression of selected, predicted complexes under different biological conditions

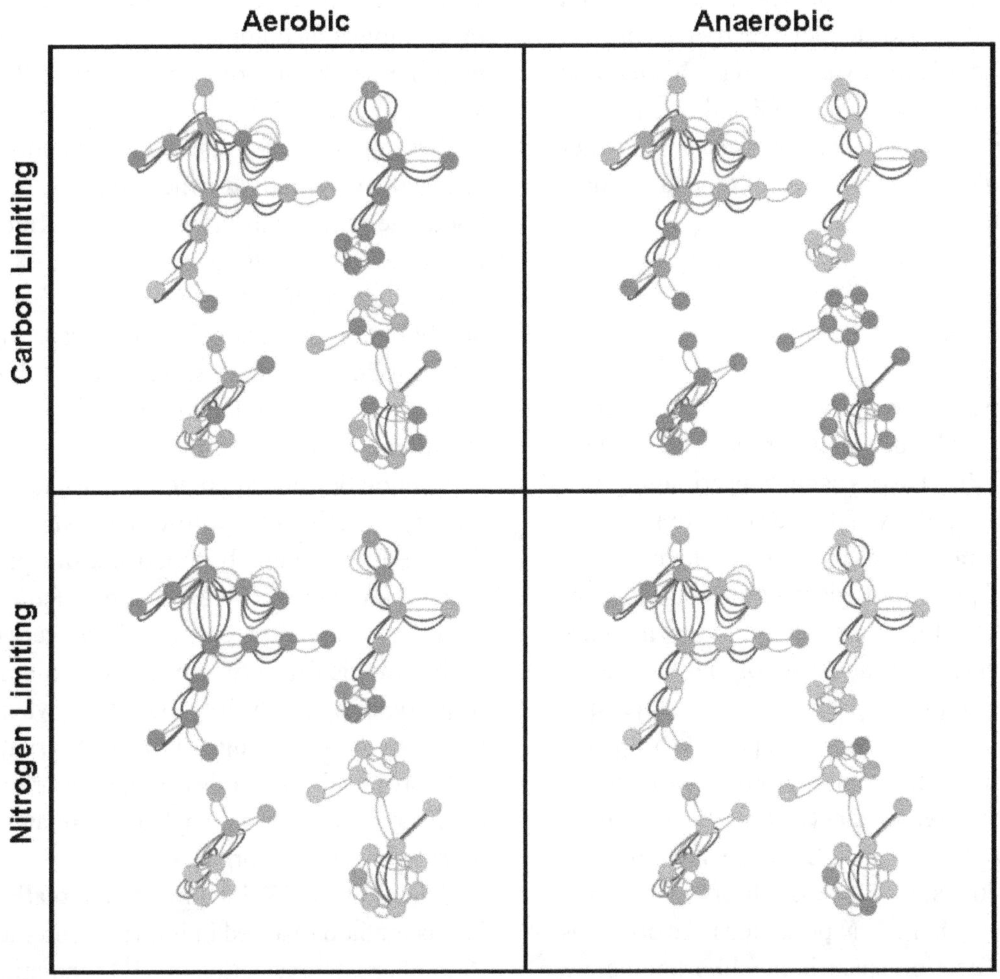

PPIs could not be observed in a culture of non-synchronized cells. Also, predicted interactions are limited in this study to proteins whose genes are differentially expressed. It is possible for a differentially expressed protein to interact with a stably expressed protein which would not be detectable by methods that use correlated expression patterns. Experimental limitation may also generate known interactions which are not observed in the biological experiment. For example, proteins interaction clusters for metal ion pumps that respond to heavy metal toxicity would not likely appear in experimental data that did not include exposure to heavy metals.

More than half of PPIs predicted by FRV-N method had some supporting evidence for the interaction either by shared annotation or agreement with prior observations. However, many predicted interactions were not so supported by the metrics considered here. Some proteins are currently without annotation, so no interaction with these proteins will share annotations. This might be expected to improve over time as the annotation of the yeast genome and proteome evolves. Proteins might be predicted to interact when they in fact interact with a common, non-differentially expressed partner. For example, protein A and B might both in fact interact with protein C. If C is not among the set of differentially expressed proteins in the experimental data however, protein A might be predicted to interact directly with protein C. Of course it is the expectation of this method that previously unknown, predicted interactions are candidates for novel discoveries, but only additional biological experimentation can validate these predictions.

CONCLUSION

In this study, we have proposed a method, Function Restricted Value Neighborhood (FRV-N) for predicting PPI networks using transcriptomic data and network topology of PPI networks drawn from prior experimental observations. We considered two sets of prior biological knowledge to construct expected topologies: a set of known binary physical interactions and a set of known macromolecular complexes. Relative to co-expression data alone, incorporation of topology from sets of prior biological observations improved the quality of the predicted networks. By comparing results from FRV-N with the one using constant neighborhood sizes, it was observed that the improvements in FRV-N predicted networks was due to the function-specific neighborhood sizes derived from the sets of known interactions. The frequency at which specific annotations shared in networks predicted with FRV-N more closely resemble the frequency of shared annotations from the set of known interactions, in comparison to that obtained from a method without using network topology. Not only does inclusion of topology increase the frequency of predicted interactions between proteins that share an annotation, but the specific annotations shared are of increased biological relevance. Using neighborhoods derived from the set of binary physical interactions, the predicted PPI networks show greater similarity to the known PPI networks, while the predicted PPI networks using neighborhoods informed by the known macromolecular complexes show a greater resemblance to known complexes. In this study, we conclude that the neighborhoods derived from macromolecular complexes are more informative than the neighborhoods derived from the set of known binary PPIs.

It is important to note that the only information used in FRV-N method is the maximum neighborhood size of a protein based on its broad molecular characteristics and transcriptomic data. No other information about the nature of a specific protein's neighbors or collections of prior transcriptomic data is used. Nonetheless, inclusion of this information produces predicted PPIs that are enriched for interactions between proteins of the same GO-slim annotation and interactions that had been previously experimentally described. This suggests that the derived neighborhood size of a protein in this work is a powerful and simple characteristic for improving the prediction of PPI networks of high biological utility and that this method can be particularly applicable to systems where there is little prior knowledge or prior observations of PPIs provided that a relevant neighborhood size distribution can be taken from a related, more thoroughly studied organism.

Although the 23 GO Slim MF annotations used in this method resulted in improved prediction of PPIs, alternative selections of GO annotations may prove even more applicable to identifying proteins whose molecular functions possess characteristic neighborhood sizes. It is not necessary to limit investigations to GO annotations at all as there are a number of potential annotations for molecular function, such as KEGG pathway annotations could be used in FRV-N. The method proposed here can be generalized, and is not restricted to protein interactions. There are many other types of interactions such as expression regulation, protein modification, and signaling cascades. Each type of interaction network is likely to have a unique network topology that can be exploited for analysis.

The maximum number of neighbors for a protein with a specific GO-slim MF annotation at significance thresholds is identified at percentiles 0.75, 0.90, and 0.95 respectively. Neighborhood sizes are informed by the set of known macromolecular complexes or the set of known binary PPIs.

Predicted PPI network sizes for different methods and different thresholds: Pearson Correlation Coefficient (PCC), Constant Neighborhood Size (N), and Macromolecular Complex-informed Neighborhood Sizes (FRN). "# Total Interactions": is the total number of PPIs in network. "# Total Proteins" is the total number of proteins with at least one neighbor in network. "Interactions/Proteins" is the number of interactions per number of proteins in network. "# Clusters" is the total number of discontinuous clusters of at least two members in the PPI network. "Largest Cluster" is the number of proteins in largest cluster in predicted PPI network. "Ave Cluster" is the average number of proteins in predicted clusters

In order to establish parameters to judge success of predicted PPI networks, the characteristics of sets of known interactions are considered. The PPI networks used here are the complete network formed using all 1748 differentially expressed genes in transcriptomic data set ("Complete"), the network derived from the set of known macromolecular complexes ("Complex"), and known physical interactions from SGD ("Binary"). "Shared GO-CC, BP, and MF" is the frequency of a PPI sharing a GO-slim Cellular Compartment, Biological Process, or Molecular Function annotation respectively. "Same Complexes": Frequency of a PPI between proteins annotated to be in the same macromolecular complex. "Known BPI" is frequency of PPIs that are present in the set of known physical interactions. "Any Shared" is frequency of a PPI sharing at least one annotation term from the above.

Frequency that PPIs in predicted networks share an annotation for three methods at multiple thresholds: Pearson Correlation Coefficient (PCC), Constant Neighborhood Size (N), and Macromolecular Complex informed Neighborhood Sizes (FRV-N). "Shared GO-CC, BP, and MF": Frequency of a PPI sharing a GO-slim Cellular Compartment, Biological Process, or Molecular Function annotation respectively.

"Same Complex": Frequency of a PPI between proteins annotated to the same macromolecular complex.

"Known BPI": Frequency of PPIs that are present in the set of known physical interactions.

"Any Shared": Frequency of proteins in a predicted PPI sharing at least one annotation term from the above.

Correlations between frequency of specific shared GO-slim annotations in PPIs between set of known physical interactions and those for the complete network ("Complete") and three PPI network prediction methods at multiple thresholds: Pearson Correlation Coefficient (PCC), Constant Neighborhood Size (N), and Macromolecular Complex informed Neighborhood Sizes (FRV-$N_{Complex}$). Correlations are calculated for specific GO-slim annotation branches, Biological Process ("GO-BP"), Molecular Function ("GO-MF"), and Cellular Compartment ("GO-CC").

Two sets of known PPI networks were used in this study for generation of a network topology. "Binary" is the set of known experimentally validated binary PPIs. "Complex" is from the set of known macromolecular complexes. The frequency at which a protein of a particular GO-slim MF annotation has a neighborhood less than a given integer value is used to assign neighborhood size to a protein. If a protein has more than one annotation, then its neighborhood is considered for all of its annotations.

The distribution of PCC scores in various networks provides insights into their biological significance. Each network is identified by is name and average PCC score in parentheses.

(A) The distribution of all PCC scores in the complete graph between each pair of 1701 differentially expressed genes follows an approximated normal distribution curve.

(B–D) The distribution of PCC scores in predicted PPI networks at (B) lowest, (C) middle, and (D) highest stringencies. Average PCC values for all interactions included in a PPI network are indicated in parenthesis.

The frequency at which specific annotations are shared between proteins in a PPI pair were calculated for all networks and each branch of Gene Ontology, but only a selection are pictured here. The arrow 'A' indicates an annotation which is highly prevalent in PCC(0.999), but not in the sets of known interactions, indicating that although PCC(0.999) is in fact enriched for interactions that share a common annotation, that specific annotation is unlikely to occur in known PPIs. Arrow 'B' identifies an annotation that is conversely highly enriched in the sets of known Complexes and detected in FRV-$N_{Complex}$ PPI networks, but was not identified in PCC(0.999).

The predicted PPI network from method FRV-$N_{Complex}$(0.90). Predicted interactions between proteins are in grey. Additional colored lines between proteins indicate that the interactions proteins share some form of annotation: Purple = known physical interaction, Gold = same macromolecular complex, Blue = Same GO-slim CC annotation, Red = Same GO-slim BP annotation, and Green = Same GO-slim MF annotation. Protein nodes are colored according to the relative expression levels at a specific condition, relative to the average expression of that gene across all experimental measurements. Green indicates an increase in relative expression, red indicates a decrease in relative expression. Clusters indicated by an "*" are also highlighted in Figure 5.

Proteins selected from predicted FRV-$N_{Complex}$(0.90) complexes are colored by their relative gene expression under different biological conditions: Aerobic or anaerobic, and carbon limiting or nitrogen limiting. Green indicates an increase in relative expression, red indicates a decrease in relative expression. Patterns of regulation under different biological conditions are one way to predict the function of predicted complexes.

REFRENCES

Butte, A. J., Tamayo, P., Slonim, D., Golub, T. R., & Kohane, I. S. (2000). Discovering functional relationships between RNA expression and chemotherapeutic susceptibility using relevance networks. *Proceedings of the National Academy of Sciences of the United States of America*, *97*(22), 12182–12186. doi:10.1073/pnas.220392197

Chen, G., Larsen, P., Almasri, E., & Dai, Y. (2008). Rank-based edge reconstruction for scale-free genetic regulatory networks. *BMC Bioinformatics*, *9*(1), 75. doi:10.1186/1471-2105-9-75

Deane, C. M., Salwinski, L., Xenarios, I., & Eisenberg, D. (2002). Protein Interactions: Two Methods for Assessment of the Reliability of High Throughput Observations. *Molecular & Cellular Proteomics*, *1*(5), 349–356. doi:10.1074/mcp.M100037-MCP200

Dittrich, M. T., Klau, G. W., Rosenwald, A., Dandekar, T., & Muller, T. (2008). Identifying functional modules in protein-protein interaction networks: an integrated exact approach. *Bioinformatics (Oxford, England)*, *24*(13), i223–i231. doi:10.1093/bioinformatics/btn161

Elo, L. L., Jarvenpaa, H., Oresic, M., Lahesmaa, R., & Aittokallio, T. (2007). Systematic construction of gene coexpression networks with applications to human T helper cell differentiation process. *Bioinformatics (Oxford, England)*, *23*(16), 2096–2103. doi:10.1093/bioinformatics/btm309

Ge, H., Liu, Z., Church, G. M., & Vidal, M. (2001). Correlation between transcriptome and interactome mapping data from Saccharomyces cerevisiae. *Nature Genetics*, *29*, 482–486. doi:10.1038/ng776

Guo, Z., Li, Y., Gong, X., Yao, C., Ma, W., & Wang, D. (2007). Edge-based scoring and searching method for identifying condition-responsive protein protein interaction sub-network. *Bioinformatics (Oxford, England)*, *23*(16), 2121–2128. doi:10.1093/bioinformatics/btm294

Huttenhower, C., Flamholz, A., Landis, J., Sahi, S., Myers, C., & Olszewski, K. (2007). Nearest Neighbor Networks: clustering expression data based on gene neighborhoods. *BMC Bioinformatics*, *8*(1), 250. doi:10.1186/1471-2105-8-250

Ideker, T., Ozier, O., Schwikowski, B., & Siegel, A. F. (2002). Discovering regulatory and signalling circuits in molecular interaction networks. *Bioinformatics (Oxford, England)*, *18*(1), S233–S240.

Jansen, R., Greenbaum, D., & Gerstein, M. (2002). Relating whole-genome expression data with protein-protein interactions. *Genome Research*, *12*, 37–46. doi:10.1101/gr.205602

Jansen, R., Yu, H., Greenbaum, D., Kluger, Y., Krogan, N. J., & Chung, S. (2003). A Bayesian networks approach for predicting protein-protein interactions from genomic data. *Science*, *302*, 449–453. doi:10.1126/science.1087361

Knijnenburg, T. A., Daran, J.-M. G., Broek, M. A. d., Daran-Lapujade, P. A., Winde, J. H. d., & Pronk, J. T. (2009). Combinatorial effects of environmental parameters ontranscriptional regulation in Saccharomyces cerevisiae: Aquantitative analysis of a compendium of chemostat-basedtranscriptome data. *BMC Genomics*, *10*(53).

Margolin, A., Nemenman, I., Basso, K., Wiggins, C., Stolovitzky, G., & Favera, R. (2006). ARACNE: An Algorithm for the Reconstruction of Gene Regulatory Networks in a Mammalian Cellular Context. *BMC Bioinformatics*, *7*(1), S7. doi:10.1186/1471-2105-7-S1-S7

Simonis, N., Gonze, D., Orsi, C., van Helden, J., & Wodak, S. J. (2006). Modularity of the Transcriptional Response of Protein Complexes in Yeast. *Journal of Molecular Biology*, *363*(2), 589. doi:10.1016/j.jmb.2006.06.024

Soong, T.-t., Wrzeszczynski, K. O., & Rost, B. (2008). Physical protein-protein interactions predicted from microarrays. *Bioinformatics (Oxford, England)*, *24*(22), 2608–2614. doi:10.1093/bioinformatics/btn498

Xulvi-Brunet, R., & Li, H. (2009). Co-expression networks: graph properties and topological comparisons. *Bioinformatics (Oxford, England)*, *26*(2), 205–214. doi:10.1093/bioinformatics/btp632

This work was previously published in the International Journal of Knowledge Discovery in Bioinformatics, Volume 1, Issue 3, edited by Xiao-Li Li and See-Kiong Ng, pp. 1-19, copyright 2010 by IGI Publishing (an imprint of IGI Global).

Chapter 13
Mining Protein Interactome Networks to Measure Interaction Reliability and Select Hub Proteins

Young-Rae Cho
Baylor University, USA

Aidong Zhang
State University of New York, Buffalo, USA

ABSTRACT

High-throughput techniques involve large-scale detection of protein-protein interactions. This interaction data set from the genome-scale perspective is structured into an interactome network. Since the interaction evidence represents functional linkage, various graph-theoretic computational approaches have been applied to the interactome networks for functional characterization. However, this data is generally unreliable, and the typical genome-wide interactome networks have a complex connectivity. In this paper, the authors explore systematic analysis of protein interactome networks, and propose a k-round signal flow simulation algorithm to measure interaction reliability from connection patterns of the interactome networks. This algorithm quantitatively characterizes functional links between proteins by simulating the propagation of information signals through complex connections. In this regard, the algorithm efficiently estimates the strength of alternative paths for each interaction. The authors also present an algorithm for mining the complex interactome network structure. The algorithm restructures the network by hierarchical ordering of nodes, and this structure re-formatting process reveals hub proteins in the interactome networks. This paper demonstrates that two rounds of simulation accurately scores interaction reliability in terms of ontological correlation and functional consistency. Finally, the authors validate that the selected structural hubs represent functional core proteins.

DOI: 10.4018/978-1-4666-1785-8.ch013

INTRODUCTION

Understanding functional behaviors of molecular components is an underlying base for biomedical applications. A wide range of computational approaches have been applied to characterize molecular functions from various types of data sources. In the past, sequence or structure analysis of proteins has contributed to characterize their functions. However, they are incapable of systematically analyzing complex functional mechanisms through biochemical reactions or interactions. Proteins typically execute their functions through interactions with other biomolecular units. Comprehensive knowledge of protein-protein interactions is thus essential to understanding the intrinsic mechanisms of biological processes.

Earlier data of protein-protein interactions were obtained via intensive small-scale investigations of restricted sets of proteins of interest, each yielding data sets regarding a limited number of protein-protein interactions. However, recent high-throughput techniques, such as yeast two-hybrid systems and mass spectrometry, involve genome-wide detection of protein-protein interactions (Uetz et al., 2000; Ito, Chiba, Ozawa, Yoshida, Hattori, & Sakaki, 2001; Gavin et al., 2002; Ho et al., 2002; Giot et al., 2003; Li et al., 2004). The yeast two-hybrid system (Parrish, Gulyas, & Finley, 2006) seeks feasible binary interactions between any two proteins encoded in the genome of interest. The interaction of two proteins transcriptionally activates a reporter gene. This reaction tracks the interaction, revealing "prey" proteins that interact with a known "bait" protein. The mass spectrometry (Aebersold & Mann, 2003) analyzes the composition of a partially purified protein complex. It uses an affinity tag attached to target "bait" proteins for purifying complexes. Comprehensive protein-protein interaction data sets in model organisms, generated by the high-throughput experiments, are publicly available in a number of open databases such as BioGRID (Breitkreutz, Stark, Reguly, Boucher, Breitkreutz, Livstone, Oughtred, Lackner, Bahler, Wood, Dolinski, & Tyers, 2008), MIPS (Mewes, Dietmann, Frishman, Gregory, Mannhaupt, Mayer, Munsterkotter, Ruepp, Spannagl, Stumptflen, & Rattei, 2008), DIP (Salwinski, Miller, Smith, Pettit, Bowie, & Eisenberg, 2004), MINT (Chatr-aryamontri, Ceol, Montecchi-Palazzi, Nardelli, Schneider, Castagnoli, & Cesareni, 2007), IntAct (Aranda et al., 2010), and HPRD (Prasad et al., 2009). However, accurate analysis of protein-protein interactions has been limited due to unreliable interaction data. The large-scale experimental data sets are susceptible to false positives, i.e., some fraction of the putative interactions detected should be considered spurious because they cannot be confirmed to occur in vivo (von Mering, Krause, Snel, Cornell, Oliver, Fields, & Bork, 2002; Sprinzak, Sattath, & Margalit, 2003).

The protein-protein interaction data sets from a genome-scale perspective are structured into an interactome network, which is represented as an undirected, un-weighted graph having proteins as a set of nodes and interactions between them as a set of edges. Specific topological features have been observed in the networks, such as scale-free distribution (Barabasi & Albert, 1999), small-world property (Watts & Strogatz, 1998), and hierarchical modularity (Ravasz, Somera, Mongru, Oltvai, & Barabasi, 2002). Various graph-theoretic computational approaches (Sharan, Ulitsky, & Shamir, 2007) have been applied to the interactome networks for functional characterization. Connection pattern is a major attribute to predict functions of unknown proteins in the networks (Oliver, 2000). However, accurate analysis of interactome networks has been challenged because of complex connectivity. In general, this complexity is caused by multiple functions of each protein and numerous cross-links between functional modules (Hartwell, Hopfield, Leibler, & Murray, 1999). A functional module is defined as a group of proteins that participate in the same functional activities (Spirin & Mirny, 2003). When handling such complex interactome networks, classical graph-theoretic algorithms are inefficient. A novel algorithm to extract functional connection patterns effectively from the complex systems is ungently needed.

Hubs play a central role in structural behaviors of scale-free networks which have power-law degree distributions (Barabasi & Oltvai, 2004). It has been observed that a protein interaction network is composed of a collection of functional modules hierarchically distributed, and hubs exist in each functional module. Based on this observation, we can achieve the hierarchical distribution of hubs, from local hubs to global hubs. Local hubs mostly include intra-module hubs which have high connectivity to the members in each functional module. In contrast, global hubs typically bridge different functional modules as intermodule hubs (Han, Bertin, Hao, Goldberg, Berriz, Zhang, Dupuy, Walhout, Cusick, Roth, & Vidal, 2004). A protein is able to perform multiple functions according to its environmental condition. The nodes connecting to multiple functional modules become the global hubs which are located in upper levels of the hierarchy. This hub-oriented tree structure thus provides better understanding of functional organizations in a complex interaction network.

In this article, we explore systematic analysis of protein interactome networks. We first propose a novel algorithm to measure interaction reliability from connectivity of the interactome networks. This algorithm quantitatively characterizes functional linkage between proteins by simulating the propagation of information signals through complex connections. Next, we present an algorithm for mining the complex interactome network structure. The algorithm restructures the network by hierarchical ordering of nodes. That is, it converts complex relationships between nodes into well-structured relationships between them. This structure re-formatting process reveals hub proteins in the interactome networks and finally uncovers hierarchically organized functional modules. We demonstrate that the structural hubs selected by our algorithm are functional core proteins.

METHOD

Interaction Reliability Measurement

Curating a large amount of erroneous interaction data is a critical preprocess in analysis of protein interactome networks. We solve this problem by quantifying the reliability of interactions. We propose a novel unsupervised scheme based on the global connection patterns across the network. In our scheme, the reliability of an interaction is described as the strength of alternative paths between the two interacting proteins. Then, how can we compute the path strength? How can we enumerate all possible alternative paths? To compute the strength of a path, we designed a path strength model. To approximate the strength of all alternative paths, we developed a dynamic flow simulation algorithm. The interaction reliability is then estimated by k-rounds of the flow simulation.

Path Strength Model

In a path $S(p) = \lambda \prod_{i=0}^{n-1} P^{wt}(v_i, v_i + 1)$ with v_0 as the start node and v_n as the end node, each node v_i selects its succeeding node v_{i+1} by a weighted probability $P^{wt}(v_i, v_i + 1)$ which is the ratio of the weight between v_i and v_{i+1} to the sum of the weights between v_i and its neighboring nodes.

$$P^{wt}(v_i, v_i + 1) = \frac{w_i(i+1)}{\sum v_j \in N(v_i) w_{ij}} \quad (1)$$

where $w_{i(i+1)}$ is the weight of the edge between two nodes v_i and $v_{i(i+1)}$, and $N(v_i)$ denotes a set of neighboring nodes of v_i, i.e., a set of nodes directly connected to v_i. The term $\sum v_j \in N(v_i) w_{ij}$ is called weighted degree of v_i. We then define the strength of the path p as

$$-\log P \quad (2)$$

λ is a scale parameter which depends on the network properties. This formula signifies that the strength of p is estimated by the weighted probability of random walk through the path from v_0 to v_n.

To make the path strength in Formula (2) symmetric, we delete the weighted degree of the starting node v_0 from Formula (1). By setting $p^{wt}(v_0, v_i) = w_{0,1}$, the path strength of a path p then becomes

$$S(p) = \lambda \cdot w_{0,1} \prod_{i=1}^{n-1} p^{wt}(v_i, v_i + 1) \qquad (3)$$

where $w_{0,1}$ is the weight of the edge between v_0 and v_1. Since $0 \leq w_{i(i+1)} \leq 1$ in Formula (1), the path strength $S(p)$ has inverse relationships with the length of p and the weighted degrees of the nodes on p. It also has a positive relationship with the average weights on p. This model indicates if a path from v_0 to v_n has a higher chance to be connected in weighted probability, it has higher strength.

Signal Flow Simulation Algorithm

We developed a dynamic flow simulation algorithm to estimate the strength of all possible paths between a source node s and a target node t. This algorithm was designed to simulate efficiently information signals flowing from s to t. The primary assumption is that each node contains the self-information which can be propagated through all connections. As notations for this algorithm, $f_s(x \to y)$ denotes the quantity of the signal that travels from x to y where x and y are directly connected. $\inf_s(x)$ denotes the quantity of the signal on x. The initial signal on the source node s, $\inf_s(s)$, traverses all connecting edges as reducing its quantity by weighted probability. In the same way, the signal on each node x, $\inf_s(x)$, is passed into its neighbors y based on weighted probability as Formula (4).

$$f_s(x \to y) = \frac{w_{x,y}}{\sum_{v \in N(x)} w_{x,v}} \cdot \inf_s(x) \qquad (4)$$

where $w_{x,y}$ is the weight of the edge between x and y, and $N(x)$ is the set of neighbors of x. The quantity of the signal on y is then calculated by summing all incoming signals as formula (5).

$$\inf_s(y) = \sum_{z \in N(y)} f_s(z \to y) \qquad (5)$$

During the simulation, the signal quantity on each node is repeatedly updated by Formula (5), and traverses the connected edge by Formula (4). The signal flow on an edge stops if the signal quantity reaches a user-dependent minimum threshold $\theta \min$. The simulation starting from s terminates when there is no more flow of signals in the network. The algorithm finally outputs the cumulative quantity P_s of signals on the target node t. The high-level description of this flow simulation from s to t is shown below.

1. Initialize $\inf_s(s)$
2. Compute the initial flow $f_{init}(s \to y)$ by formula (4) for each y where $e(s, y) \in E$
3. Compute $\inf_s(y)$ by formula (5) for each y
4. Compute the flow $f_s(y \to z)$ by formula (4) for each z where $e(y, z) \in E$
5. Remove the flow $f_s(y \to z)$ if it is less than a threshold θ_{flow}
6. Replace z with y, and repeat 3,4,5 and 6 until there is no more flow in the network.
7. Output cumulative $\inf_s(t)$

k-Rounds of Flow Simulation

We present a novel approach to estimate interaction reliability by k-rounds of flow simulation. Suppose we measure the reliability of the edge e between two nodes, x and y. This method first eliminates e, and then iterates k times the computation of path strength of all possible paths between x and y by the signal flow simulation algorithm described above. The source node and target node for flow simulation should be x and y, respectively. The signal flow starts from x and the cumulative quantity of the signal on y is produced as an output. The output value indicates the approximation of the sum of strength of all possible paths between x and y. Since the edge e was eliminated before the simulation process, this method approximates the strength of alternative paths of e.

At the initial round, the signal flow is simulated on the unweighted input graph. It is conducted under the assumption that the signal at a node travels to a neighboring node chosen by the same chance. In other words, we use 1 as all edge weights for the computation of Formula (4) in line (2) and (4) of the algorithm description above. The algorithm calculates the weight of the edge between x and y. By measuring weights of all edges at this initial round, the unweighted network is converted to a preliminary weighted graph.

At the next round, the preliminary weights are refined by simulating signal flow on the weighted graph. In this step, the signal at a node x chooses a succeeding node y in a path by the weighted chance, i.e., the ratio of the weight between x and y to the sum of the weights between x and all its neighbors. All edge weights produced at the initial round are used for the computation of Formula (4) in line (2) and (4) of the algorithm description above, and updated at this round.

During the additional rounds of signal flow simulation, all edge weights continue to be revised until they converge. How many rounds of flow simulation generate the best edge weights? We evaluate the appropriate value of k for k-rounds of flow simulation in protein interactome networks.

Efficiency Analysis

The efficiency is one of the major strengths of our approach. In general, random walk on a graph is manipulated by the matrix computation. The product of adjacency matrices for each round of random walks runs in $O(n^3)$. For the n rounds, the time complexity grows to $O(n^4)$. The matrix computation for random walks between two ending nodes of an edge is thus very inefficient in the large-scale interactome networks. However, our flow simulation algorithm traces the nodes through only existing links, and prunes the trivial flow which is less than a threshold as early as possible to approximate the weight.

Unlike other graph-theoretic methods, run time of our flow simulation algorithm is obviously unrelated to the network diameter because flow traverses through all possible walks including cycles. Since our algorithm uses a threshold to halt the flow as a user-specified criterion, the theoretical upper-bound of its run time is unknown. However, we investigated the potential factors that affect time complexity of the signal flow simulation algorithm.

In synthetic networks with different features, we monitored the average run time of flow simulation starting from randomly selected 200 source nodes. First, we created the networks by increasing the numbers of nodes, from 500 to 6000, but retained the density as 0.002. The density of a network represents the proportion of the number of actual edges to the number of all possible edges. Next, we used the networks produced by the same change of the number of nodes but a constant average degree of 5. The results are shown in Figure 1 (a). When the density is constant, the rum time increases as adding nodes, because of the squared increase of the number of edges for the constant density. However, when the average degree is constant, the run time is uniform regardless of the network size. We additionally tested the run time with the networks created by the change of

Figure 1. Run time of the signal flow simulation algorithm in synthetic networks. The networks are structured by (a) the change of the number of nodes in constant density or a constant average degree, and by (b) the change of density in the constant number of nodes or a constant average degree. The results indicate that, when the average degree is constant, run time of the flow simulation algorithm is uniform regardless of the network size or density.

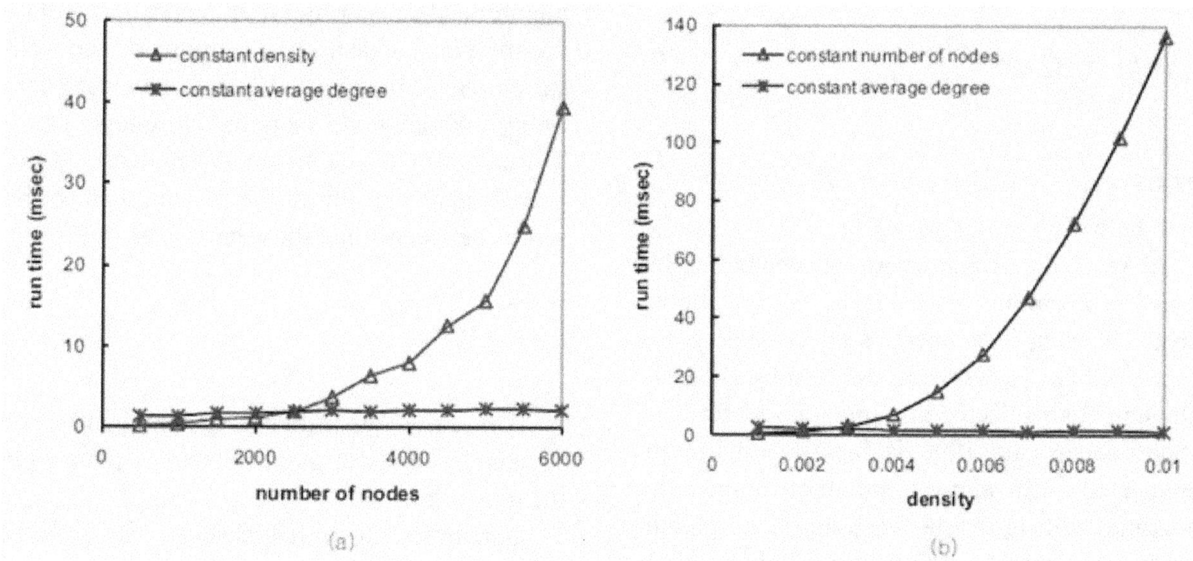

density in the fixed number of nodes as 2000 and in a constant average degree of 5. As shown in Figure 1 (b), when the network size is fixed, the run time increases as the density becomes higher. However, when the average degree is constant, the run time is also uniform regardless of the network density. These results indicate that the average degree of networks is a more critical factor to reduce time complexity of our flow simulation algorithm than their size or density. Since the average degree of complex systems is typically low in a power-law degree distribution, the signal flow simulation algorithm measures reliability of all interactions very efficiently.

Hub Protein Detection

In this section, we present a network restructuring algorithm introduced in our previous work (Cho & Zhang, 2010). It takes a weighted network as an input, and yields the tree structure of the network, a list of structural hubs, and a list of clusters as outputs. The reliability score of each interaction, computed by *k*-rounds of flow simulation, is assigned to the edge as a weight in the protein interactome network. The network restructuring algorithm converts the complex network into the hierarchical ordering of nodes. The generated tree-structured network then provides useful knowledge hidden in the original complex network, such as identifying hubs and functional modules. This approach is composed of four stages: structural closeness measurement, network restructuring, hub confidence measurement, and clustering.

Structural Closeness Measurement

At the first stage, we formulate the structural closeness of a pair of nodes based on the path strength model in the previous section. Two nodes are close in the network structure if there are strong paths between them. All possible paths between them are informative when we calculate the path strength. However, due to computational inefficiency, we use the single strongest path. The structural closeness between two nodes is then described as:

$$F(x,y) = \max_{v_i = x, v_j = y} S(\langle v_i, \ldots v_j \rangle) \quad (6)$$

where $S(\langle v_i, \ldots v_j \rangle)$ represents the strength of a path between v_i and v_j by Formula (3).

To select the strongest path between two nodes x and y, Formula (6) still has a computational problem when it enumerates all possible paths between them. However, it has been observed that any node pair selected in a small world network (Watts & Strogatz, 1998) is directly or indirectly connected with a relatively small path length. The maximal path length between any node pair in biological networks is thus typically limited. To resolve the computational complexity, we restrict the maximum path length up to k. We define the k-length path strength S_k as the maximum strength of all distinct paths with length k between x and y.

$$S_k(x,y) = \max_{v_0 = x, v_k = y} S(\langle v_0, v_1 \ldots v_k \rangle) \quad (7)$$

Using a user-specified threshold θ to set the boundary of k, the structural closeness F between x and y is calculated by the maximum k-length path strength.

$$F(x,y) = \max_{k} S_k(x,y) \quad (8)$$

where $l \leq k \leq l + \theta$ and l is the shortest path length between x and y. We then compute structural closeness for all possible pairs of nodes by Formula (8) to produce a closeness matrix.

Network Restructuring Algorithm

At the next stage, we retrieve a structural hierarchy from the complex interactome network. This network restructuring process does not indicate hierarchical clustering. It indicates converting the complex connectivity of a network into well-structured connections. We first calculate centrality for each node using the closeness matrix. The centrality C of a node x in a network $G(V,E)$ is defined as the sum of the structural closeness scores between x and the other nodes in V.

$$C(x) = \sum_{y \in V} F(x,y) \quad (9)$$

Formula (9) quantifies the potential influence of x on the others through connection patterns in the network.

Our strategy for restructuring a network is to place the nodes with high centrality on the upper level in the hierarchical tree structure. We define the set of ancestor nodes T of x as the nodes that have higher centrality than x.

$$T(x) = \{y \mid C(y) > C(x)\} \quad (10)$$

Among the nodes in $T(x)$, the node that is the most similar with x becomes the parent node $p(x)$ of x.

$$p(a) = \begin{cases} null & \text{if } T(x) = \theta \\ \arg\max_{y \in T(x)} F(x,y) & \text{otherwise.} \end{cases} \quad (11)$$

We then efficiently construct a hierarchical tree structure by selecting a parent node for each node by Formula (11). The node with the highest centrality has no parent and becomes the root. The nodes, which have lower centrality than the root, but have high closeness to the root, are located on the next level as the child nodes of the root. This tree structure including the root and hierarchical levels dynamically evolves on network growth.

Hub Confidence Measurement and Clustering

The hierarchical structure of nodes, representing a protein interactome network, facilitates identifying hubs and functional modules. For each node x, we obtain the set of child nodes $D(x)$ of x.

$$D(x) = \{y \,|\, p(y) = x\} \tag{12}$$

We then recursively trace down the tree structure starting from x and combine all child nodes to produce a set of descendant nodes L_x of x.

$$L_x = (\bigcup_{y \in D(x)} L_y) \cup \{x\} \tag{13}$$

The total list of descendant node sets is finally generated by applying Formula (13) to all the nodes except leaf nodes.

We provide a new definition of a structural hub as a core node in the hierarchical structure, not just high-degree nodes. The structural hubs are selected by estimating hub confidence. The hub confidence H of a node x is calculated by the sum of the structural closeness scores between x and all the members of L_x, divided by the structural closeness score between x and its parent node. If x is the root, we use only the sum of the structural closeness scores between x and all the other nodes as its hub confidence.

$$H(x) = \begin{cases} \sum_{y \in L_x} F(x,y) & \text{if } p(x) = null \\ \sum_{y \in L_x} F(x,y) \;/\; F(x,p(x)) & \text{otherwise.} \end{cases} \tag{14}$$

Hub confidence in Formula (14) assesses how likely the node is to play a role as a hub in the hierarchical structure.

The set of clusters is finally generated from the hierarchy of nodes and their hub confidence scores. We iteratively select a hub x having the highest hub confidence, and produce L_x as a cluster until the hub confidence score of x reaches a user-dependent minimum threshold. The clusters are also hierarchically arranged based on the positions of their hubs in the tree structure. When this approach is applied to a protein interactome network, the output clusters are considered as functional modules, and the structural hubs become the functional core proteins within functional modules or bridging multiple functional modules.

EXPERIMENTAL RESULTS

Data Source

We tested our approach in the protein interactome network of *Saccharomyces cerevisiae*. The protein-protein interaction data are publicly available in several databases. They have been mostly collected from high-throughput experimental results published. In our test, we downloaded the interaction data from DIP, Database of Interacting Proteins (Salwinski, Miller, Smith, Pettit, Bowie, & Eisenberg, 2004). The full version of the data set consists of 4823 distinct proteins and 17471 interactions. However, it has been curated based on other biological information such as protein sequence and RNA expression profiles to select core interactions (Deane, Salwinski, Xenarios, & Eisenberg, 2002). Since the core version, including 2314 proteins and 5663 interactions between them, is a more reliable subset of the full version, we used it for accurate analysis.

Assessment of Interaction Reliability

We computed reliability of each interaction in the network by up to 3 rounds of flow simulation. Figure 2 shows the distribution of interactions at each round with respect to their reliability scores,

Figure 2. Distribution of interactions with respect to their reliability scores. The interaction reliability was computed by k-rounds of signal flow simulation in the protein interactome network of S. cerevisiae when k =1, 2, and 3.

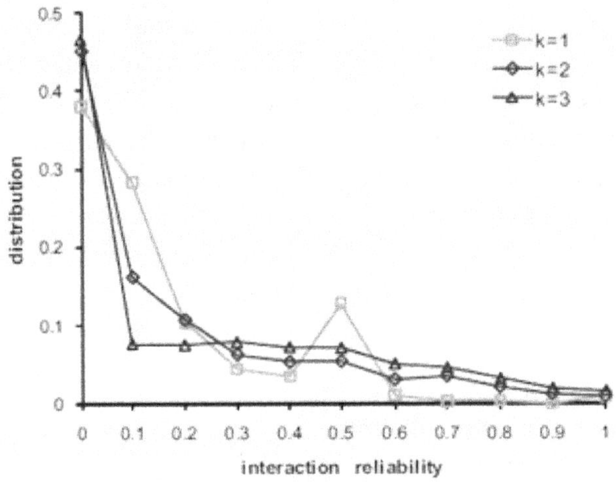

i.e., edge weights. At the first round of simulation on the original un-weighted network, most of the interactions have the weights in the range between 0 and 0.5, and a small peak on 0.5 occurs because a significant number of degree-2 nodes were assigned the weight of 0.5. The second and third round of simulation on the weighted network made an effect on refining the edge weights. This iterative simulation assigned low weights below 0.1 into more than 45% of edges. They might be considered as potential false positive interactions in the protein interactome network. All the other edges have a uniform distribution between 0.1 and 1.0 of their weights. To assess the interaction reliability scores in terms of functional association, we made use of the information in annotations on Gene Ontology (The Gene Ontology Consortium, 2008) and MIPS functional catalogues (Mewes, Dietmann, Frishman, Gregory, Mannhaupt, Mayer, Munsterkotter, Ruepp, Spannagl, Stumptflen, & Rattei, 2008).

Comparison with Semantic Similarity

For each interacting protein pair, we first compared the edge weight with semantic similarity derived from the annotations in Gene Ontology (The Gene Ontology Consortium, 2008), which is currently the most comprehensive biological ontology database. It provides ontology terms, which are the shared concepts across different organisms, in three categories: biological processes, molecular functions, and cellular components. The ontology terms are structured as a directed acyclic graph (DAG) according to the "is-a" and "part-of" relationships among them. Genes and proteins are annotated on the terms based on the published evidence. Semantic similarity of two terms in the taxonomy can be measured by their commonality, i.e., more common information two terms share, more similar they are.

Previous studies (Cho, Hwang, Ramanathan, & Zhang, 2007; Wang, Du, Payattakool, Yu, & Chen, 2007) have applied semantic similarity measures to quantification of functional links between annotated genes or proteins in Gene Ontology.

Suppose we measure semantic similarity between two proteins x and y. The annotations follow the transitivity property, i.e., if a gene is annotated on a term, then it is also included in the annotations of all parent terms on the paths towards the root in the taxonomy. In Information Theory, the information content of an event is calculated by the log negation of the probability of the event. In the same way, we compute the common information of x and y by the negative log of the proportion of the number of proteins annotated on the most specific term shared by them. That is, less proteins are annotated on the most specific shared term, more similar they are. Semantic similarity between x and y is thus formulated as:

$$S(x,y) = -\log(\min_i P_i(x,y)) \tag{15}$$

where

$$P_i(x,y) = \frac{|G_i(x,y)|}{G_{root}} \tag{16}$$

$G_i(x,y)$ is a set of genes or proteins annotated on the term, whose annotation includes both x and y, and G_{root} is the set of genes or proteins annotated on the root term.

The distribution of interacting protein pairs with respect to their semantic similarity is displayed in Figure 3 (a). It has a similar pattern to the distribution of interactions with respect to their reliability when k=2 in Figure 2. This result supports that around 45% of interactions are potentially false positives, and two rounds of flow simulation captured the similar amount of erroneous data. Figure 3 (b) shows the average semantic similarity scores of interacting proteins with respect to their edge weights after the first-round (k=1), second-round (k=2), and third-round (k=3) of flow simulation. The first-round weights do not have a monotonically decreasing pattern with semantic similarity scores. In particular, the interactions with the weight of 0.5 have unusually low semantic similarity scores because a significant amount of degree-2 proteins were connected to the proteins in different ontological terms. However, re-weighting by repeated simulation effectively smoothed the noisy weights at the first-round to make an explicit monotonic relationship with the semantic similarity scores. Two plots from the second-round and third-round weights do not have any critical difference. However, additional re-weighting lowered the semantic similarity scores across all the edges. Figure 3 explicitly demonstrates that multiple rounds of flow simulation improve the correctness of scoring interaction reliability in terms of ontological correlation. Moreover, two rounds of simulation are good enough to measure the interaction reliability.

Comparison with Functional Consistency

Next, for each interacting protein pair, we compared the edge weight with functional consistency from MIPS functional catalogues. MIPS database (Mewes, Dietmann, Frishman, Gregory, Mannhaupt, Mayer, Munsterkotter, Ruepp, Spannagl, Stumptflen, & Rattei, 2008) provides the functional descriptions and their annotations manually created. These descriptions are hierarchically organized. We extracted the annotation data on the functional catalogues in three different levels of the taxonomy. The first-level functions represent the length-1 nodes from the root in the tree structure. They thus indicate the most general description of molecular functions. Each first-level function is divided into several functions more specific on the second level. In the same way, the third level includes the functions more specifically defined than the second level.

In general, a single protein has multiple functions according to its environmental conditions. The functional consistency between two proteins has been calculated by the ratio of common

Figure 3. (a) Distribution of interactions with respect to their semantic similarity. It has a similar pattern to the distribution of interactions with respect to their reliability when k=2 in Figure 2. (b) The relationship between average semantic similarity and reliability scores of interactions. The reliability was computed by the first-round (k=1), second-round (k=2), and third-round (k=3) of signal flow simulation.

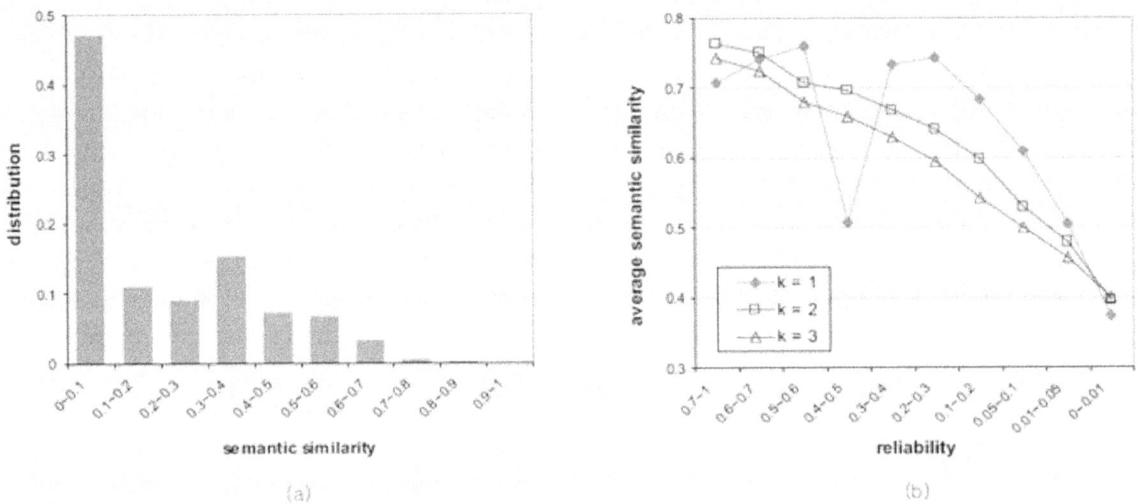

functions. The distribution of interacting protein pairs with respect to their functional consistency is displayed in Figure 4 (a). Its general pattern is also closest to the distribution of interactions with respect to their reliability when *k*=2 in Figure 2. This result verifies again that the edge weights computed by two rounds of flow simulation (*k*=2) correctly match functional consistency. Figure 4 (b) shows the average functional consistency scores of interacting proteins with respect to their edge weights after two rounds of flow simulation (*k*=2). The edge weights have a monotonic positive relationship with functional consistency from all three levels in the functional hierarchy, although the interacting pairs in more specific functions have lower consistency. These monotonic plots indicate that the signal flow simulation algorithm properly quantified functional linkage between interacting proteins, which can be interpreted as interaction reliability.

Assessment of Hub Proteins

We assessed topological significance of structural hubs in the protein interactome network. We built a weighted network by two rounds of signal flow simulation, transformed the network into a hierarchical tree structure, and measured the hub confidence for each node by Formula (14).

Network Vulnerability Test

For the topological assessment of structural hubs, we tested network vulnerability on random attacks and hub attacks. It has been investigated that typical scale-free networks are robust on random attacks, but vulnerable on targeted attacks to the hubs. To evaluate the network vulnerability on random attacks, we monitored the fractions of the largest component when we repeatedly disrupted a randomly selected node. In the same way, for the network vulnerability on hub attacks, we observed the fractions of the largest component when we

Figure 4. (a) Distribution of interactions with respect to their functional consistency. It has a similar pattern to the distribution of interactions with respect to their reliability when k=2 in Figure 2. (b) The relationship between average functional consistency and reliability scores of interactions. The reliability was computed by the second-round (k=2) of signal flow simulation. The annotating proteins on functional categories from MIPS were used to compute functional consistency. In the functional hierarchy, the most general functions are located on the level-1, whereas the level-3 has more specific functions than the level-2 and level-1.

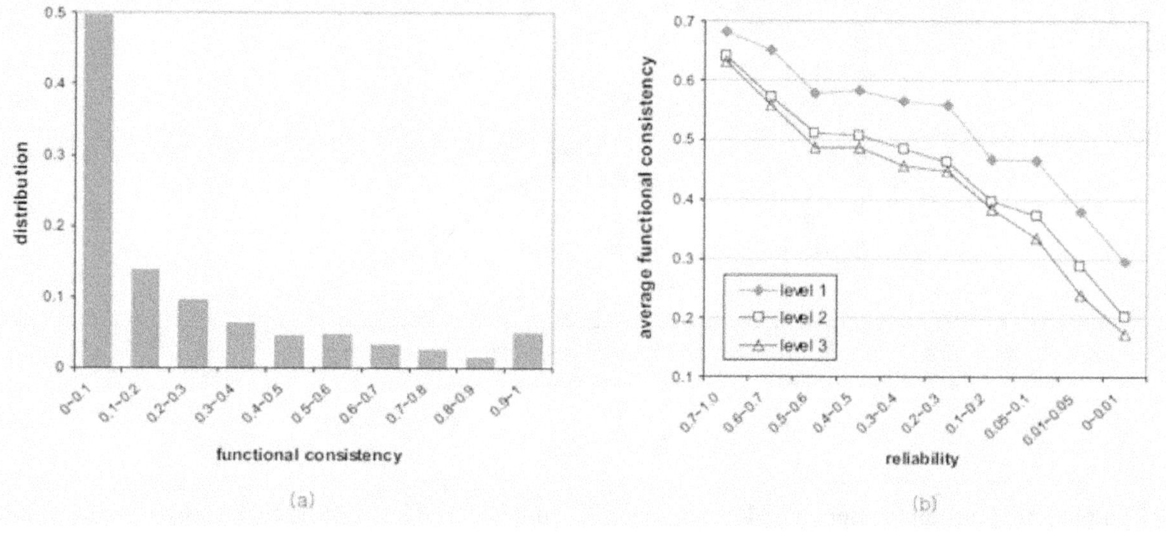

repeatedly disrupted the node with the highest hub confidence score in Formula (14).

Figure 5 shows the results of the network vulnerability test. Because all nodes in the network are directly or indirectly connected with each other, the fraction of the largest component is 1 in the original network, and decreases by removing nodes. In Figure 5, we observe that the repeated structural hub attacks more rapidly decrease the fraction of the largest component than random attacks. This result indicates that the topologically significant nodes in the complex network were scored high by the hub confidence in Formula (14). A previous study (Albert, Jeong, & Barabasi, 2000) investigated that the repeated highest-degree node attacks had the same effect to the structural hub attacks. It has also been observed that the degree of nodes in protein interactome networks is likely to represent the essentiality of the protein in biological functions. Overall, it implies that the structural hubs selected by our method correspond to functionally essential proteins.

As further validation, we compared the hub confidence by Formula (14) to node degree. Since all degree-1 nodes are the leaf nodes in the tree structure, their hub confidence score is 0. For the nodes whose degree is greater than 1, the hub confidence has a monotonic increase by the increment of their degree. It indicates that high-degree nodes are generally located on the upper levels (the levels closer to the root) in the tree structure.

Assessment by Clusters

We implemented clustering of proteins from the tree-structured network. We first enumerated 300 structural hubs from the highest hub confidence score in Formula (14), and generated 300 clusters by combining all descendent nodes of each hub in the tree structure. To evaluate clustering accuracy,

Figure 5. Network vulnerability test results by random attacks and structural hub attacks. We repeatedly disrupted a randomly selected node and the hub with the highest hub confidence, respectively, from the protein interactome network. The fraction of the largest component decreases more rapidly by the structural hub attacks than the random attacks. It indicates that the hub confidence measure assesses topological significance of nodes in the network.

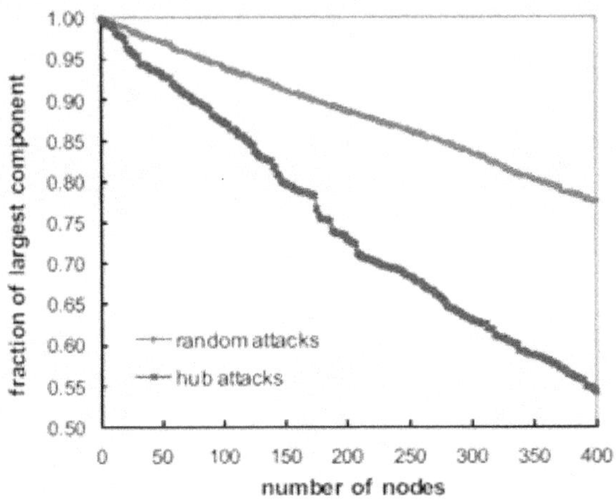

we compared the output clusters with known real functional modules which were obtained from the functional categories on the 3rd-level in the hierarchy in MIPS (Mewes, Dietmann, Frishman, Gregory, Mannhaupt, Mayer, Munsterkotter, Ruepp, Spannagl, Stumptflen, & Rattei, 2008). For statistical evaluation between an output cluster and a real functional module, we used p-value from the hypergeometric distribution.

$$P = 1 - \sum_{i=0}^{k-1} \frac{\binom{|X|}{i}\binom{|V|-|X|}{n-i}}{\binom{|V|}{n}} \qquad (17)$$

For each output cluster, we found the best match to a functional module by searching the maximum value of $-\log P$.

Figure 6 shows the clustering results of our network restructuring approach and their accuracy. For every 50 structural hubs in the descending order of hub confidence scores, we examined the average cluster size and calculated the average of $-\log P$ from Formula (17). The number of output clusters represents the number of selected structural hubs. As shown in Figure 6, more clusters were added, their average size and their average accuracy monotonically decreased. In typical, the structural hubs with higher hub confidence scores are located closer to the root in the tree structure, and produce larger-sized clusters. As a result, the hub with a higher confidence score generates a cluster with higher accuracy in p-value. It demonstrates that our approach is likely to select the cores in real functional modules, i.e., functionally essential proteins, as structural hubs.

CONCLUSION

Protein interactome networks are significant resources for functional characterization of molecular components and their functional pathway identification. In this study, we have introduced

Figure 6. Clustering results and their accuracy in p-value. We implemented the network restructuring and clustering approach with a protein interactome network. Up to 300 structural hubs were selected in a descending order of hub confidence scores. Clustering accuracy was monitored with the set of clusters generated from every 50 structural hubs. Real functional modules from MIPS were used to compare with output clusters in p-value computation.

number of clusters	average cluster size	accuracy ($-\log P$)
50	138	11.78
100	123	10.33
150	86	9.73
200	65	8.30
250	53	7.45
300	46	6.94

a systematic framework of retrieving functional knowledge hidden in complex connectivity. We have proposed the algorithm to measure interaction reliability by *k*-rounds of signal flow simulation. The motivation of this approach is that the interaction evidence between proteins is interpreted as functional links between them. Moreover, a whole protein interactome network represents a functional system within a cell. In other words, a connecting path in the network indicates a functional pathway. We thus analyze the protein interactome network in a system-level. Our path strength model describes the likelihood that proteins on a path are involved in a specific functional pathway. Functional linkage between two proteins is quantitatively approximated by the signal flow simulation algorithm to estimate the strength of alternative paths for their interaction.

In the computational view, efficiency and scalability are the major strengths of this approach. The classical random walk computation and any other probabilistic methods are not able to be applied to the genome-wide interactome networks of higher-level organisms such as *H. sapiens* because they are inefficient in runtime and memory usage. However, because time complexity of our flow simulation algorithm depends on only the average degree of the networks, it performs very efficiently in any typical scale-free networks which have a power-law degree distribution. The outstanding performance of our algorithm is thus validated in the fact that the interactome network of *H. sapiens* follows the scale-free property. Overall, the proposed approach will build a strong foundation for characterizing protein functions in *H. sapiens*.

We also presented the algorithm to find hub proteins by restructuring protein interactome networks. The network restructuring algorithm accurately converts the complex biological network into a well-structured hierarchical ordering of nodes. It is based on our path strength model and centrality measurement. This process of network structure conversion culminates in identifying hubs and modules hierarchically organized. The global picture of the functional cores and structured functional organizations, which are extracted from complex protein interactome networks, will provide insights into the underlying mechanisms of biological processes within a cell.

Currently more than 35,000 protein-protein interactions in *H. sapiens* have been confirmed by high-throughput and small-scale experiments (Prasad et al., 2009). The proposed methods in this paper will make significant contributions in many potential biomedical applications. First, our approach can be used to identify disease-related pathways. Systematic functional analysis of the interactome network will establish a mapping between genotype and phenotype in a system level. The structured information of functional roles of genes or proteins reveals not only functional linkage between them but also pleiotropy which occurs when a single protein influences multiple phenotypes. The results will be thus extended to the discovery of biochemical pathways that mediate a specific phenotype. For example, the potential cell cycle pathways causing cancer cells will be detected by our systematic analysis of interactome networks. Next, our approach can be used to identify the best target molecules for a drug on the specific disease-related pathways. To be an effective drug target, pharmacological intervention at the target molecule should have strong effects on the key processes that the molecule participates in. This problem can be solved by analyzing the functional linkage maps integrated with pharmacological knowledge.

ACKNOWLEDGMENT

This work was partially supported by the grant from Young Investigator Development Program by the Vice Provost for Research at Baylor University.

REFERENCES

Aebersold, R., & Mann, M. (2003). Mass spectrometry-based proteomics. *Nature*, *422*, 198–207. doi:10.1038/nature01511

Albert, R., Jeong, H., & Barabasi, A.-L. (2000). Error and attack tolerance of complex networks. *Nature*, *406*, 378–482. doi:10.1038/35019019

Aranda, B. (2010). The IntAct molecular interaction database in 2010. *Nucleic Acids Research*, *38*, D525–D531. doi:10.1093/nar/gkp878

Barabasi, A.-L., & Albert, R. (1999). Emergence of scaling in random networks. *Science*, *286*, 509–511. doi:10.1126/science.286.5439.509

Barabasi, A.-L., & Oltvai, Z. N. (2004). Network biology: understanding the cell's functional organization. *Nature Reviews. Genetics*, *5*, 101–113. doi:10.1038/nrg1272

Breitkreutz, B.-J., Stark, C., Reguly, T., Boucher, L., Breitkreutz, A., & Livstone, M. (2008). The BioGRID interaction database: 2008 update. *Nucleic Acids Research*, *36*, D637–D640. doi:10.1093/nar/gkm1001

Chatr-aryamontri, A., Ceol, A., Montecchi-Palazzi, L., Nardelli, G., Schneider, M. V., Castagnoli, L., & Cesareni, G. (2007). MINT: the Molecular INTeraction database. *Nucleic Acids Research*, *35*, D572–D574. doi:10.1093/nar/gkl950

Cho, Y.-R., Hwang, W., Ramanathan, M., & Zhang, A. (2007). Semantic integration to identify overlapping functional modules in protein interaction networks. *BMC Bioinformatics*, *8*, 265. doi:10.1186/1471-2105-8-265

Cho, Y.-R., & Zhang, A. (2010). Identification of functional hubs and modules by converting interactome networks into hierarchical ordering of proteins. *BMC Bioinformatics*, *11*(3), S3. doi:10.1186/1471-2105-11-S3-S3

Deane, C. M., Salwinski, L., Xenarios, I., & Eisenberg, D. (2002). Protein interactions: two methods for assessment of the reliability of high throughput observations. *Molecular & Cellular Proteomics*, *1*(5), 349–356. doi:10.1074/mcp.M100037-MCP200

Gavin, A.-C. (2002). Functional organization of the yeast proteome by systematic analysis of protein complexes. *Nature*, *415*, 141–147. doi:10.1038/415141a

Giot, L. (2003). A protein interaction map of Drosophila melanogaster. *Science*, *302*, 1727–1736. doi:10.1126/science.1090289

Han, J.-D. J., Bertin, N., Hao, T., Goldberg, D. S., Berriz, G. F., & Zhang, L. V. (2004). Evidence for dynamically organized modularity in the yeast protein-protein interaction network. *Nature*, *430*, 88–93. doi:10.1038/nature02555

Hartwell, L. H., Hopfield, J. J., Leibler, S., & Murray, A. W. (1999). From molecular to modular cell biology. *Nature*, *402*, c47–c52. doi:10.1038/35011540

Ho, Y. (2002). Systematic identification of protein complexes in Saccharomyces cerevisiae by mass spectrometry. *Nature*, *415*, 180–183. doi:10.1038/415180a

Ito, T., Chiba, T., Ozawa, R., Yoshida, M., Hattori, M., & Sakaki, Y. (2001). A comprehensive two-hybrid analysis to explore the yeast protein interactome. *Proceedings of the National Academy of Sciences of the United States of America*, *98*(8), 4569–4574. doi:10.1073/pnas.061034498

Li, S. (2004). A map of the interactome network of the metazoan C. elegans. *Science*, *303*, 540–543. doi:10.1126/science.1091403

Mewes, H. W., Dietmann, S., Frishman, D., Gregory, R., Mannhaupt, G., & Mayer, K. F. X. (2008). MIPS: analysis and annotation of genome information in 2007. *Nucleic Acids Research*, *36*, D196–D201. doi:10.1093/nar/gkm980

Oliver, S. (2000). Guilt-by-association goes global. *Nature*, *403*, 601–603. doi:10.1038/35001165

Parrish, J. R., Gulyas, K. D., & Finley, R. L. (2006). Yeast two-hybrid contributions to interactome mapping. *Current Opinion in Biotechnology*, *17*, 387–393. doi:10.1016/j.copbio.2006.06.006

Prasad, T. S. K. (2009). Human protein reference database - 2009 update. *Nucleic Acids Research*, *37*, D767–D772. doi:10.1093/nar/gkn892

Ravasz, E., Somera, A. L., Mongru, D. A., Oltvai, Z. N., & Barabasi, A.-L. (2002). Hierarchical organization of modularity in metabolic networks. *Science*, *297*, 1551–1555. doi:10.1126/science.1073374

Salwinski, L., Miller, C. S., Smith, A. J., Pettit, F. K., Bowie, J. U., & Eisenberg, D. (2004). The database of interacting proteins: 2004 update. *Nucleic Acids Research*, *32*, D449–D451. doi:10.1093/nar/gkh086

Sharan, R., Ulitsky, I., & Shamir, R. (2007). Network-based prediction of protein function. *Molecular Systems Biology*, *3*, 88. doi:10.1038/msb4100129

Spirin, V., & Mirny, L. A. (2003). Protein complexes and functional modules in molecular networks. *Proceedings of the National Academy of Sciences of the United States of America*, *100*(21), 12123–12128. doi:10.1073/pnas.2032324100

Sprinzak, E., Sattath, S., & Margalit, H. (2003). How reliable are experimental protein-protein interaction data? *Journal of Molecular Biology*, *327*, 919–923. doi:10.1016/S0022-2836(03)00239-0

The Gene Ontology Consortium. (2008). The Gene Ontology project in 2008. *Nucleic Acids Research*, *36*, D440–D444. doi:10.1093/nar/gkm883

Uetz, P. (2000). A comprehensive analysis of protein-protein interactions in Saccharomyces cerevisiae. *Nature*, *403*, 623–627. doi:10.1038/35001009 von Mering, C., Krause, R., Snel, B., Cornell, M., Oliver, S. G., Fields, S., & Bork, P. (2002). Comparative assessment of large-scale data sets of protein-protein interactions. *Nature*, *417*, 399–403. doi:10.1038/nature750

Wang, J. Z., Du, Z., Payattakool, R., Yu, P. S., & Chen, C.-F. (2007). A new method to measure the semantic similarity of GO terms. *Bioinformatics (Oxford, England)*, *23*(10). doi:10.1093/bioinformatics/btm087

Watts, D. J., & Strogatz, S. H. (1998). Collective dynamics of 'small-world' networks. *Nature*, *393*, 440–442. doi:10.1038/30918

This work was previously published in the International Journal of Knowledge Discovery in Bioinformatics, Volume 1, Issue 3, edited by Xiao-Li Li and See-Kiong Ng, pp. 20-35, copyright 2010 by IGI Publishing (an imprint of IGI Global).

Section 4
Network Mining and Systems Biology

Chapter 14
Finding Minimum Reaction Cuts of Metabolic Networks Under a Boolean Model Using Integer Programming and Feedback Vertex Sets

Takeyuki Tamura
Kyoto University, Japan

Kazuhiro Takemoto
Graduate School of Frontier Sciences, Japan

Tatsuya Akutsu
Kyoto University, Japan

ABSTRACT

In this paper, the authors consider the problem of, given a metabolic network, a set of source compounds and a set of target compounds, finding a minimum size reaction cut, where a Boolean model is used as a model of metabolic networks. The problem has potential applications to measurement of structural robustness of metabolic networks and detection of drug targets. They develop an integer programming-based method for this optimization problem. In order to cope with cycles and reversible reactions, they further develop a novel integer programming (IP) formalization method using a feedback vertex set (FVS). When applied to an E. coli metabolic network consisting of Glycolysis/Glyconeogenesis, Citrate cycle and Pentose phosphate pathway obtained from KEGG database, the FVS-based method can find an optimal set of reactions to be inactivated much faster than a naive IP-based method and several times faster than a flux balance-based method. The authors also confirm that our proposed method works even for large networks and discuss the biological meaning of our results.

DOI: 10.4018/978-1-4666-1785-8.ch014

INTRODUCTION

One of the important features of living organisms is robustness. For example, many organisms have strong adaptability to environmental changes. As another example, many organisms can live even if some of their genes are mutated. However, the origin of this robustness is not yet fully understood. Therefore, in order to understand the principles of living organisms, it is quite important to reveal the origin of the robustness. Studying robustness is also important from a medical point of view. It is considered that cancer cells are very robust (Kitano, 2004) and thus cancers are difficult to treat. If we can identify where the robustness of cancer cells comes from, we may develop novel and effective treatment methods by efficiently breaking the robustness of cancer cells.

Of course, the origin of the robustness will not be unique or simple. Many aspects must be taken into account. For example, we should consider dynamics and structure of various kinds of biological networks, which include genetic networks, protein-protein interaction networks, signal transduction networks and metabolic networks. However, it is quite difficult to consider these networks simultaneously. Although the amount of data for structures of genetic networks and protein-protein interaction networks is increasing (Kafri et al., 2005; Levy & Pereira-Leal, 2008), it is not enough yet. On the other hand, we have a lot of knowledge on metabolic networks. Indeed, many network data of various kinds of organisms are available from such databases as KEGG (Kanehisa et al., 2008) and EcoCyc (Karp et al., 2007. Thus, we focus on metabolic networks in this paper. In particular, we focus on structural aspects of metabolic networks because kinetic parameters and/or equations of many reactions are not available.

Various kinds of studies have been done on analysis of structures of metabolic networks. Many studies have been done on analysis of graph-theoretic properties such as degree distribution, network diameter, network motifs, and modularity (Guimerà et al., 2007; Jeong et al., 2000; Milo et al., 2002; Wagner & Fell, 2001. However, in these studies, the effects of substrates and products are not well reflected. Many studies have also been done based on *elementary modes* and *extreme pathways* (Behre et al., 2008; Haus et al., 2008; Klamt & Gilles, 2004; Papin et al., 2003; Stelling et al. 2002. In particular, robustness against mutations, insertions and deletions of enzymes was analyzed in (Schuster et al., 1999; Stelling et al., 2002). The methods used there are based on enumeration of elementary modes or extreme pathways and exhaustive mutations of single or multiple enzymes. Therefore, these methods are not very efficient.

Many studies have been done in order to find a minimum set of reactions such that inactivation of these reactions leads to prevention of production of specific compounds. This kind of problems are called the *minimum knockout* problem or the *reaction cut* problem. Detection of such a reaction cut has potential application to drug design because it might be necessary to inactivate multiple reactions (via knockout of enzymes or genes) so that the target reactions or compounds become inactive. However, Klamt, and Stelling (2006) reported that it is quite difficult to apply these convex analysis-based methods to the minimum knockout problem because of their combinatorial complexities. Though some elegant techniques were applied by Haus et al. (2008) to avoid enumeration of elementary modes, the worst case time complexity still depends on the number of elementary modes. Furthermore, Acuña et al. (2009) proved that the reaction cut problem based on elementary modes is NP-hard. This means that finding the minimum reaction cut is difficult regardless of the number of elementary modes. In order to cope with large-scale networks, Bacardi et al. developed *bilevel programming* methods to find strategies for maximizing or minimizing the production of target chemical compound by knockout of a limited number of genes or reactions (Burgard et al., 2003; Dashika et al., 2006). In their methods, the problems are first formalized as bilevel integer programs and then these are transformed into conventional integer programs by making use of the duality in linear programming.

In order to effectively apply the above mentioned flux balance-based methods, we should know exact or near exact network topology and stoichiometry parameters. However, details of such information may not be available in some cases. Furthermore, in flux balance based methods, reaction rates balance metabolites; that is, for each metabolite, it holds that the sum of the rates of all reactions consuming the metabolites equals the sum of the rates of the reactions producing it (Haus et al., 2008; Klamt & Gilles, 2004. However, if some pathways are not reflected in a given network obtained from a database, the lack of information of pathways (especially, the lack of information of degradation pathways) may affect knock-out strategies too much and misleading solutions may be obtained. Thus, it is worthy to develop another approach which is effective even when some information is lacking.

In order to study structural and combinatorial aspects of metabolic networks, Handorf et al. (2005) introduced the concept of *scope*, which can be regarded as a concept under a Boolean model of metabolic networks. The scope is a set of all possible metabolites obtained from a given set of *seed* compounds and a given structure of a metabolic network. Handorf et al. (2008) also defined the inverse of the scope problem: given a set of metabolites, compute or enumerate minimal sets of seed compounds. Furthermore, they studied robustness of scopes against modifications of enzymes (Handorf et al., 2005). However, the methods for analyzing robustness are based on exhaustive or random disruptions of enzymes and cannot be directly used for measuring the structural robustness of a metabolic network.

Recently, Sridhar et al. (2008) defined a Boolean model of metabolic networks and developed a branch-and-bound algorithm for finding a reaction cut that incurs minimum damage to non-target compounds. Though their model and method are interesting and useful, the concept of maximality of fluxes is not taken into account in their model. Since the importance of considering maximality of fluxes is pointed out in many papers (Acuña et al., 2009; Burgard et al., 2003; Dashika et al., 2006; Haus et al., 2008; Klamt & Gilles, 2004), we should also take into account similar concept(s) in Boolean models of metabolic networks.

Based on the above mentioned works, we study in this paper computational aspects of the reaction cut problem under a Boolean model of metabolic networks and define this problem as follows: given a metabolic network, a set of source compounds and a set of target compounds, determine the minimum number of reactions whose inactivation prevents generation of some of the target compounds (i.e., the set of target compounds is no more a subset of the scope) under a maximal assignment of Boolean value. Since this problem is NP-hard as suggested by Acuña et al. (2009), we develop an *integer programming* (IP) based method in this paper. However, in a naive method, the number of variables appearing in an integer program is $O\left((m+n)^2\right)$ where *m* and *n* are the numbers of compounds and reactions respectively and thus the method cannot be applied to large-scale networks. In order to reduce the number of variables in the IP-based method, we adopt a novel use of a *feedback vertex set* (FVS), which is a concept in graph theory and is a set of vertices whose removal makes a given graph cycle-free. It is well-known that computation of the minimum cardinality FVS is NP-hard for both directed and undirected graphs, whereas several approximation algorithms have been developed (Bar-Yehuda et al., 1998; Even et al., 1998). Since it is not necessary to use the optimal FVS in our application, we adopt a simple greedy algorithm to compute an FVS. By using an FVS, we can reduce the number of variables in IP from $O\left((m+n)^2\right)$ to

$O(f(f+m+n))$ where f is the size of an FVS when compared to a naive method. Note that f is usually very small while the number of cycles is relatively large since many cycles may share one edge. Since IP's computational time usually grows exponentially in the number of variables and f is small (exact numbers are shown on pages 15 and 16), it leads to a significant improvement of practical efficiency. When applied to an *E. coli* metabolic network consisting of Glycolysis/Glyconeogenesis, Citrate cycle and Pentose phosphate pathway extracted from KEGG database, the improved IP method is much faster than the naive IP-based method and several times faster than a flux balance-based method. We also confirmed the efficiency and biological meanings of our method by applying it to larger networks including central metabolism for *E. coli* (eco), *Saccharomyces cerevisiae* (sce), and *Homo sapiens* (hsa).

METHODS

Problem Definition

Here we define the main problems to compute the reaction cut of metabolic networks on a Boolean model. Let $V_c = \{V_{c_1}, \ldots, V_{c_m}\}$ and $V_r = \{V_{r_1}, \ldots, V_{r_n}\}$ be a set of *compound nodes* and a set of *reaction nodes* respectively, where $V_c \cap V_r = \{\}$. It is to be noted that most reactions are catalyzed by enzymes and thus each reaction can be inactivated in most cases by disruption of a gene corresponding to the enzyme catalyzing the reaction. Let $V = V_c \cup V_r$. Let $V_s \subseteq V_c$ and $V_t \subseteq V_c$ are sets of *source nodes* and *target nodes* respectively, where $V_s \cap V_t = \{\}$, and source nodes correspond to seed compounds of Handorf et al. (2005).

A *metabolic network* is defined as a directed graph $G(V,E)$ satisfying the following conditions:

- For each edge $(u,v) \in E$, either $(u \in V_c) \cap (v \in V_r)$ or $(u \in V_r) \cap (v \in V_c)$ holds.
- Each source node does not have an incoming edge.
- Each node $v \notin V_c$ has at least one incoming edge.

Let $V_a \subseteq V_r$ be a set of reaction nodes corresponding to a set of inactivated reactions. We assign either 0 or 1 to each node V. "1" means that the corresponding reaction is active or the corresponding compound is producible. On the other hand, "0" means that the corresponding reaction is inactive or the corresponding compound is nonproducible. Let A be such an assignment (i.e., A is a function from V to $\{0,1\}$). For each node $v \in V$, we write $v=0$ (resp. $v=1$) if 0 (resp. 1) is assigned to v. A is called a *valid assignment* if the following conditions are satisfied: (i) For each $v \in V_s$, $v=1$. (ii) For each $v \in V_c - V_s$, $v=1$ if and only if there is u such that $(u,v) \in E$ and $u=1$. (iii) For each $v \in V_r$, $v=1$ if and only if $v \notin V_a$ holds and $u=1$ holds for all u such that $(u,v) \in E$.

The second condition means that compound nodes correspond to OR nodes. The third condition means that reaction nodes correspond to AND nodes, where the output is forced to be 0 if a node is inactivated (i.e., a node belongs to V_a). Clearly, the following proposition holds.

Proposition 1: *If there is no directed cycle in $G(V,E)$, a valid assignment is uniquely determined for a given V_a.*

If there exist directed cycles, we may have multiple valid assignments. In particular, there exist valid assignments in which the states of compounds in directed cycles are 0 even if there is no inactivated reaction. In order to avoid such cases, we introduce the concept of a maximal assignment, where similar properties are taken into account in many of flux balance-based meth-

Figure 1. is the solution of BN-ReactionCut when $V_c = \{v_{c_1}, \ldots, v_{c_8}\}, V_r = \{v_{r_1}, v_{r_2}, v_{r_3}\}, V_s = \{v_{c_1}, v_{c_3}, v_{c_6}\}$ *and* $V_t = \{v_{c_8}\}$ *are given*

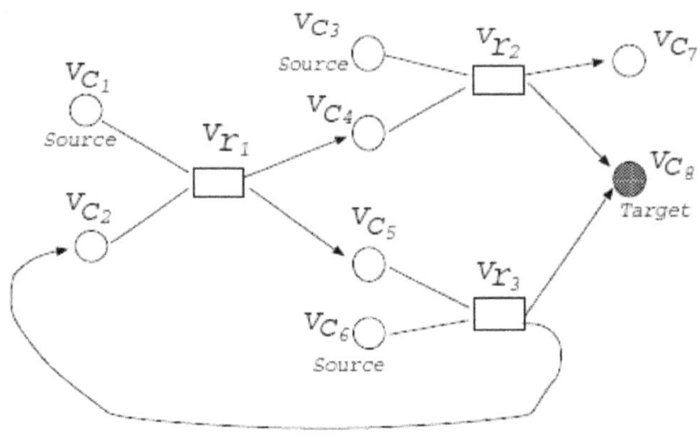

ods (Acuña et al., 2009; Burgard et al., 2003; Dashika et al., 2006; Haus et al., 2008; Klamt & Gilles, 2004). A valid assignment *A* is called *maximal* if *A* is valid and $\{v|v=1, v \in V\}$ is maximal. As to be shown later, a maximal valid assignment is determined uniquely for a given metabolic network *G*(*V*,*E*) and a given set of inactivated reactions V_a.

For example in Figure 1, assume that $V_c = \{v_{c_1}, \ldots, v_{c_8}\}, V_r = \{v_{r_1}, v_{r_2}, v_{r_3}\}$, $V_s = \{v_{c_1}, v_{c_3}, v_{c_6}\}$ and $V_t = \{v_{c_8}\}$ are given. The maximal valid assignment for $V_a = \{v_{r_2}\}$ is $(v_{c_1}, \ldots, v_{c_8}, v_{r_1}, v_{r_2}, v_{r_3})$ =(1,1,1,1,1,1,0,1,1,0,1). Although $(v_{c_1}, \ldots, v_{c_8}, v_{r_1}, v_{r_2}, v_{r_3})$=(1,0,1,0,0,1,0,0,0,0,0) is valid, it is not maximal for $V_a = \{v_{r_2}\}$. Intuitively, a maximal valid assignment is the steady state to which a metabolic network converges assuming all nodes were initially assigned 1 and nodes in V_a were inactivated.

Here we formulate the problem of determining the minimum reaction cut of a metabolic network (**BN-ReactionCut**) as follows:

- **Input:** A metabolic network *G*(*V*,*E*).
- **Output:** A minimum cardinality set V_a for which *v*=0 is satisfied for some $v \in V_t$ in the maximal valid assignment.

Recall that the maximal valid assignment is uniquely determined when V_a is given. However, a minimum cardinality set V_a is not uniquely determined when V_t is given.

In Theorem 12 of Acuña et al. (2009), finding minimum reaction cut on flux balance model was proved as NP-hard by polynomial time reduction from Hitting Set Problem. Almost the same reduction can be used to prove that BN-ReactionCut is NP-hard. It should also be noted that we have only considered irreversible reactions so far for simplicity. Reversible reactions can be taken into account if we replace one reversible reaction with two irreversible reactions. In such a case, special treatment is needed as discussed later since inactivation of single reversible reactions corresponds to inactivation of two irreversible reactions.

In BN-ReactionCut, a set of target compounds is given. However, it is equivalent to the case where only one target compound is given because the original problem can be reduced to $|V_t|$ cases of the single target problem. In highly robust cells such as cancer cells, it may not be enough to prevent generation of a single compound in V_t because there may exist hidden or unknown reactions. In such a case, it may be useful to prevent generation of all compounds in V_t. We also consider this variant, which is referred as **BN-ReactionCutAll**. Note that target nodes may have outgoing edges in BN-ReactionCutAll. Modifications of all IP-based methods in this paper to cope with BN-ReactionCutAll are straight-forward.

IP Formalization for BN-ReactionCut

Since BN-ReactionCut is NP-hard, we employ IP since IP is widely used for solving NP-hard problems. Furthermore, we introduce the notion of time into the IP formulation. If we do not use the notion of time, it is very difficult to avoid that a valid assignment which is not maximal is obtained as a candidate of the solution. Here, we explain our IP formulation using an artificial small example of a metabolic network for simplicity of the presentation. The IP formulation for the network in Figure 1 is as follows:

Minimize

$$FE1(0)+FE2(0)+FE3(0) \quad (1)$$

subject to

$$FC8(m+n)=1 \quad (2)$$

for all $t=0,\ldots,m+n$

$$TR1(t+1)+FC1(t)+FC2(t)+FE1(t)\geq 1,\ FR1(t+1)+TC1(t)\geq 1,\ FR1(t+1)+TC2(t)\geq 1,\ FR1(t+1)+TE1(t)\geq 1 \quad (3)$$

$$TR2(t+1)+FC3(t)+FC4(t)+FE2(t)\geq 1,\ FR2(t+1)+TC3(t)\geq 1,\ FR2(t+1)+TC4(t)\geq 1,\ FR2(t+1)+TE2(t)\geq 1 \quad (4)$$

$$TR3(t+1)+FC5(t)+FC6(t)+FE3(t)\geq 1,\ FR3(t+1)+TC5(t)\geq 1,\ FR3(t+1)+TC6(t)\geq 1,\ FR3(t+1)+TE3(t)\geq 1 \quad (5)$$

$$FC2(t+1)+TR3(t)\geq 1,\ TC2(t+1)+FR3(t)\geq 1 \quad (6)$$

$$FC4(t+1)+TR1(t)\geq 1,\ TC4(t+1)+FR1(t)\geq 1 \quad (7)$$

$$FC5(t+1)+TR1(t)\geq 1,\ TC5(t+1)+FR1(t)\geq 1 \quad (8)$$

$$FC7(t+1)+TR2(t)\geq 1,\ TC7(t+1)+FR2(t)\geq 1 \quad (9)$$

$$FC8(t+1)+TR2(t)+TR3(t)\geq 1,\ TC8(t+1)+FR2(t)\geq 1, TC8(t+1)+FR3(t)\geq 1 \quad (10)$$

$$TE1(t+1)=TE1(t),\ TE2(t+1)=TE2(t),\ TE3(t+1)=TE3(t) \quad (11)$$

$$TC1(t)=TC3(t)=TC6(t)=1 \quad (12)$$

$$TC2(0)=TC4(0)=TC5(0)=TC7(0)=TC8(0)=TR1(0)=TR2(0)=TR3(0)=1 \quad (13)$$

$$TX+FX=1 \textit{ for any } X \quad (14)$$

where every variable takes either 0 or 1, $m+n$ is the number of nodes, and constraints including terms with $t=m+n+1$ should be ignored.

In the above example, $v_{r_i} = 1$ (resp. $v_{r_i} = 0$) is represented by TRi=1 (resp. FRi=1). Therefore, $TRi+FRi$=1 holds. Similarly, values of compound nodes are also represented by TCi and FCi. To represent whether v_{r_i} is inactivated, we utilize an additional node v_{e_i} as one of predecessors of v_{r_i}. Since v_{r_i} is represented by an AND node, $v_{e_i} = 0$

inactivates v_{r_i}. Thus, in the above example, v_{r_1} has three predecessors v_{c_1}, v_{c_2} and v_{e_1} in our IP formalization although v_{r_1} has only two predecessors in Figure 1.

(1) means that the number of inactivated reactions should be minimized. Note that *FEi*=1 causes that a reaction v_{r_i} is inactivated. (2) means that v_{c_8} should be 0 after the 0-1 assignment converges. (3) represents the constraint by v_{r_1}. Note that the constraint of $v_{r_1} = v_{c_1} \wedge v_{c_2} \wedge v_{e_1}$ can be represented by

$$(v_{r_1} \vee \overline{v_{c_1}} \vee \overline{v_{c_2}} \vee \overline{v_{e_1}}) \wedge (\overline{v_{r_1}} \vee v_{c_1}) \wedge (\overline{v_{r_1}} \vee v_{c_2}) \wedge (\overline{v_{r_1}} \vee v_{e_1}).$$

It is also to be noted that $x_1 = x_2 \wedge x_3 \wedge ... \wedge x_k$ is represented by

$$(x_1 \vee \overline{x_2} \vee \overline{x_3} \vee ... \vee \overline{x_k}) \wedge (\overline{x_1} \vee x_2) \wedge (\overline{x_1} \vee x_3) \wedge ... \wedge (\overline{x_1} \vee x_k)$$

and $x_1 = x_2 \vee x_3 \vee ... \vee x_k$ is represented by $(\overline{x_1} \vee x_2 \vee x_3 \vee ... \vee x_k) \wedge (x_1 \vee \overline{x_2}) \wedge (x_1 \vee \overline{x_3}) \wedge ... \wedge (x_1 \vee \overline{x_k})$. (4)-(10) represent the constraints of $v_{r_2}, v_{r_3}, v_{c_2}, v_{c_4}, v_{c_5}, v_{c_7}, v_{c_8}$, respectively. (11) means that the set of inactivated reactions does not change between consecutive time steps. (12) means that v_{c_1}, v_{c_3} and v_{c_6} are 1 for any time step since their indegrees are "0". (13) means that "1" is assigned for every node at the initial time step. Moreover, (14) means that *TX*+*FX*=1 holds for any X. Note that each variable *FX* can be represented by $1-TX$ and then the number of variables in IP can be reduced, but both *TX* and *FX* are used in the above formulation for simplicity. We denote this IP-based method by IP-ReactionCut.

Theorem 1: *IP-ReactionCut correctly solves BN-ReactionCut.*

Proof. From constraint (11), we can assume that V_a is fixed for all *t*. Let *A*(*t*) be a 0-1 assignment on *V* at time step *t*. When *G*(*V*,*E*) and V_a are given, the transition of the states of nodes in *G*(*V*,*E*) is uniquely determined and thus can be written as *A*(*t*+1)=**f**(*A*(*t*)) by using a list **f** of *m*+*n* Boolean functions. Let $A_1 = (\hat{v}_1^1, ..., \hat{v}_{m+n}^1)$ and $A_2 = (\hat{v}_1^2, ..., \hat{v}_{m+n}^2)$ be assignments for *V*. We denote $A_1 \succ A_2$ if $\hat{v}_i^1 \geq \hat{v}_i^2$ holds for all *i* (1≤*i*≤*m*+*n*). Since the network consists of only AND nodes and OR nodes, $\mathbf{f}(A_1) \succ \mathbf{f}(A_2)$ holds if $A_1 \succ A_2$ holds.

The 0-1 assignment at time step *t* in the above IP formalization can be represented by

$$A_{IP}(t) = \mathbf{f}\left(\mathbf{f}\left(\mathbf{f}\left(\cdots \mathbf{f}\left(A_{IP}(0)\right)\cdots\right)\right)\right) = \mathbf{f}^t\left(A_{IP}(0)\right) = \mathbf{f}^t\left(A(0)\right)$$

where **f** is applied to $A_{IP}(0)$ for *t* times and $A_{IP}(0) = A(0) = (1, 1, ..., 1)$. Since *A*(0)≻*A* holds for any assignment *A*, *A*(0)≻**f**(*A*(0)) holds, from which we have $A_{IP}(t) \succ A_{IP}(t+1)$. From this property and the fact that the number of nodes in the network is *m*+*n*,

$$A_{IP}(t) = A_{IP}(t+1) = A_{IP}(t+2) = \cdots$$

holds for some *t*≤*m*+*n*. This implies that $A_{IP}(m+n)$ is a valid assignment.

Let *A* be an arbitrary valid assignment. Since *A*(0)=(1,1,…,1)≻*A* holds,

$$A_{IP}(m+n) = \mathbf{f}^{m+n}\left(A(0)\right) \succ \mathbf{f}^{m+n}\left(A\right) = A$$

holds. This means that $A_{IP}(m+n)$ is the maximal valid assignment.

Since $|V_a|$ is minimized in IP-ReactionCut, the theorem holds. (QED)

This theorem also implies that a maximal valid assignment is uniquely determined.

Corollary 1: *A maximal valid assignment is uniquely determined for G(V,E) and* V_a.

Figure 2. Example of decomposition of cycle: v_{r_1} is decomposed into v_{r_1} and v_{s_1}

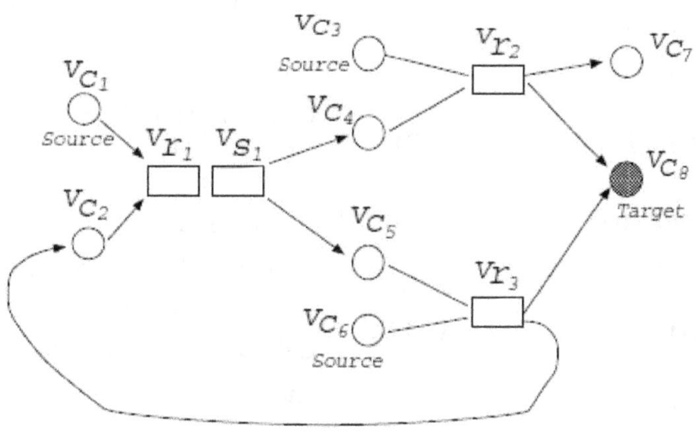

IP Formalization Using Feedback Vertex Set

In the above formalization, the number of variables of IP is $O\left((m+n)^2\right)$. So, this method cannot be applied to large scale networks. To deal with large scale networks, we develop an improved method utilizing the notion of FVS. Recall that an FVS is a set of nodes whose removal makes the network acyclic. Recall also that we need not introduce the notion of time if the network is acyclic. Therefore, we combine these two ideas. In the improved method, we first find an FVS F that consists of reaction nodes, and then apply the IP formulation similar to that of IP-ReactionCut to the obtained acyclic network. We use constraints similar to those in IP-ReactionCut though we use the same time step t for variables except those in F and those corresponding to (in)activation of reactions. In order to take the effect of cycles into account, we divide each reaction node v_{r_i} in an FVS F into two nodes v_{s_i} and v_{r_i} so that v_{r_i} has only in-edges and v_{s_i} has only out-edges (see Figure 2). Furthermore, we put an additional constraint that $v_{s_i}(t+1) = v_{r_i}(t)$. We denote this IP-based method by IP-FVS-ReactionCut. By means of this modification, as shown below, the number of required time steps is reduced from $m+n+1$ to $f+1$, where $f=|F|$. Therefore, the required number of variables in IP-formulation is reduced from $O\left((m+n)^2\right)$ to $O(f(m+n+f))$. Since it is expected that f is much smaller than $m+n$ and the computation time of IP exponentially increases with the number of variables, it is a significant improvement. Though finding an FVS of minimum cardinality is NP-hard, we need not use an FVS of minimum cardinality. Thus, we employ a simple greedy method to select an FVS: detection of a directed cycle and division of a reaction node in the cycle are iterated until there is no directed cycle.

We explain more details using an example. Consider the network shown in Figure 1. Suppose that $F = \{v_{r_1}\}$ is obtained as an FVS. Then, $G(V,E)$ is transformed into the network shown in Figure 2 and we obtain the following IP.

Minimize

$$FE1(0)+FE2(0)+FE3(0) \tag{15}$$

subject to

$$FC8(1)=1 \tag{16}$$

for t=0,1

$TR1(t)+FC1(t)+FC2(t)+FE1(t)\geq 1,$
$FR1(t)+TC1(t)\geq 1,$

$$FR1(t)+TC2(t)\geq 1,\ FR1(t)+TE1(t)\geq 1 \quad (17)$$

$TR2(t)+FC3(t)+FC4(t)+FE2(t)\geq 1,$
$FR2(t)+TC3(t)\geq 1,$

$$FR2(t)+TC4(t)\geq 1,\ FR2(t)+TE2(t)\geq 1 \quad (18)$$

$TR3(t)+FC5(t)+FC6(t)+FE3(t)\geq 1,$
$FR3(t)+TC5(t)\geq 1,$

$$FR3(t)+TC6(t)\geq 1,\ FR3(t)+TE3(t)\geq 1 \quad (19)$$

$$FC2(t)+TR3(t)\geq 1,\ TC2(t)+FR3(t)\geq 1 \quad (20)$$

$$FC4(t)+TS1(t)\geq 1,\ TC4(t)+FS1(t)\geq 1 \quad (21)$$

$$FC5(t)+TS1(t)\geq 1,\ TC5(t)+FS1(t)\geq 1 \quad (22)$$

$$FC7(t)+TR2(t)\geq 1,\ TC7(t)+FR2(t)\geq 1 \quad (23)$$

$FC8(t)+TR2(t)+TR3(t)\geq 1,\ TC8(t)+FR2(t)\geq 1,$
$$TC8(t)+FR3(t)\geq 1 \quad (24)$$

$TE1(t+1)=TE1(t),\ TE2(t+1)=TE2(t),$
$$TE3(t+1)=TE3(t) \quad (25)$$

$$TS1(t+1)=TR1(t) \quad (26)$$

$$TC1(t)=TC3(t)=TC6(t)=1 \quad (27)$$

$$TS1(0)=1 \quad (28)$$

$$TX+FX=1 \text{ for any } X \quad (29)$$

The meaning of each constraint is almost obvious. It is to be noted that 1 is assigned to only v_{s_i} s in FVS at time t=0 (see (28)). It should also be noted that the number of required time steps is only 2 (=$|F|$+1) in IP-FVS-ReactionCut whereas IP-ReactionCut needs 12 (the number of nodes+1) steps. We can show that $|F|$+1 steps is enough to guarantee the convergence of assignments.

Theorem 2: *IP-FVS-ReactionCut correctly solves BN-ReactionCut.*

Proof. As in the proof of Theorem 1, we can assume that V_a is fixed for all t. Let F be an FVS. Let $G(V,E)$ and $G'(V',E')$ be the networks before and after division of each reaction node v_{r_i} in F into two nodes v_{r_i} and v_{s_i} (each v_{r_i} belongs to both V and V', but each v_{s_i} belongs only to V'). Let $A(t)$ and $A^F(t)$ be assignments on V and on v_{s_i} s at time step t in IP-FVS-ReactionCut, respectively. Since $G'(V',E')$ is acyclic, $A(t)$ is uniquely determined from $A^F(t)$. Furthermore, since the state of v_{s_i} at time t+1 is equal to the state of v_{r_i} at time t, $A^F(t+1)$ is uniquely determined from $A^F(t)$. Thus, we can define two functions **g** and **h**, where **g** maps $A^F(t)$ to $A(t)$ and **h** maps $A^F(t)$ to $A^F(t+1)$.

As in the proof of Theorem 1, for assignments A_1^F and A_2^F on F, we can see that $\mathbf{g}(A_1^F) \succ \mathbf{g}(A_1^F)$ and $\mathbf{h}(A_1^F) \succ \mathbf{h}(A_2^F)$ hold if $A_1^F \succ A_2^F$ holds. Since we start with

$$A^F(0) = (1,1,.,,,1), A^F(t) \succ A^F(t+1)$$

holds for all t. Thus, we have $A^F(t) = A^F(t+1) = A^F(t+2) = \cdots$ for some $t \leq |F|$. Hence, we have $A^F(|F|) = A^F(|F|+1)$, from which $A(|F|)=A(|F|+1)$ follows. It means that $A(|F|)$ is a valid assignment.

Next, we show that $A(|F|)$ is maximal. Let $A_{\max}$ be the maximal valid assignment on V. Let $A_{\max}^F$ denote the restriction of $A_{\max}$ to F (i.e., assignments on FVS nodes in $A_{\max}$). Since $A_{\max}^F$ is valid, $\mathbf{h}(A_{\max}^F) = A_{\max}^F$ holds. From $A^F(0) \succ A_{\max}^F$,, we have

Figure 3. Example for IP-FVS-ReactinCut2

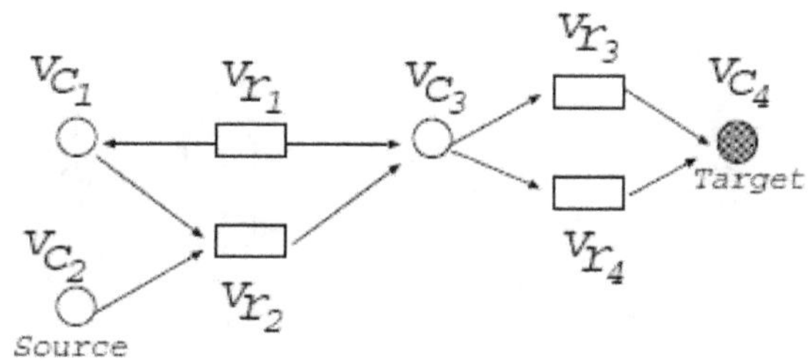

$$\mathbf{h}^{|F|}\left(A^{F}(0)\right) \succ \mathbf{h}^{|F|}\left(A_{\max}^{F}\right).$$

Therefore,

$$A(|F|)=\mathbf{g}\left(\mathbf{h}^{|F|}\left(A^{F}(0)\right)\right) \succ \mathbf{g}\left(\mathbf{h}^{|F|}\left(A_{\max}^{F}\right)\right)=A_{\max}$$

holds. This means that $A(|F|)$ is the maximal valid assignment.

Since $|V_a|$ is minimized in IP-FVS-ReactionCut, the theorem holds. (QED)

IP Formalization for BN-ReactionCut2

We have not paid much attention to reversible reactions in BN-ReactionCut so far. That is, although IP-ReactionCut and IP-FVS-ReactionCut replace each reversible reaction v_{r_i} with two irreversible reactions v_{r_i}, v_{r_i} and the constraint that "$v_{r_i} \in V_a$ iff $v_{r_{i'}} \in V_a$" is added, there is no further special treatment for original reversible reactions.

However, reversible reactions cause more effects in real networks. Therefore, we need to improve the definition of BN-ReactionCut in order to take further effect of reversible reactions into account. Assume that a reversible reaction v_{r_1} of Figure 3 is given. If the maximal valid assignment is defined as the solution, both v_{c_1} and v_{c_3} are initially assigned 1 and they never become 0 unless v_{r_1} is inactivated. If $v_{r_2}=1$ holds, this constraint is reasonable. However, if neither v_{c_1} nor v_{c_3} is provided by another reaction, that is, if $v_{r_2}=0$ in this case, it is more reasonable to assume that v_{c_1} and v_{c_3} eventually attenuate and become $v_{c_1}=v_{c_3}=0$. Therefore we extend the definition of BN-ReactionCut as follows. Let $v_{s_1},\ldots,v_{s_{k_1}}$ and $v_{p_1},\ldots,v_{p_{k_2}}$ be substrates and products of a reversible reaction where products can be exchanged with substrates and vice versa. An assignment A is called an *extended maximal valid assignment* if A is valid, $\{v|v=1,v\in V\}$ is maximal, and both v_{s_1} and v_{p_j} are assigned 1 only if either all v_{s_1} or all v_{p_j} are provided by the other reactions. Now, we formulate the extended version of BN-ReactionCut.

Problem: BN-ReactionCut2:

- **Input:** A metabolic network $G(V,E)$.
- **Output:** A minimum cardinality set V_a for which v=0 is satisfied for some $v \in V_t$ in the extended maximal valid assignment.

To partially take into account the effect of this extension of the problem, we developed IP-FVS-ReactionCut2 which is an extended version of IP-FVS-ReactionCut. IP-FVS-ReactionCut2 decomposes only cycles including at least one irreversible reaction. If such a cycle is found, one of irreversible reactions v_{r_i} is decomposed into two nodes v_{r_i} and v_{s_i} where v_{r_i} has only in-edges and v_{s_i} has only out-edges. For example, assume that a metabolic network of Figure 3 is

Figure 4. After replacing each reversible reaction with two irreversible reactions

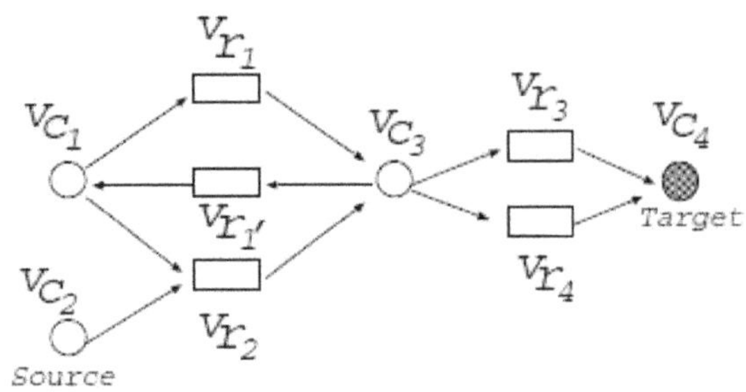

given. Then, IP-FVS-ReactionCut2 replaces each reversible reaction with two irreversible reactions, that is, v_{r_1} of Figure 3 is replaced with v_{r_1} and v_{r_1} of Figure 4. Thus, G=(V,E) is represented as

$$V_c = \{v_{c_1},\ldots,v_{c_4}\}, V_r = \{v_{r_1}, v_{r_1}, v_{r_2}, v_{r_3}, v_{r_4}\}, V_s = \{v_{c_2}\}$$

and $V_t = \{v_{c_4}\}$ as shown in Figure 4.

Although there are two cycles $\{v_{c_1}, v_{r_1}, v_{c_3}, v_{r_1}\}$ and $\{v_{c_1}, v_{r_2}, v_{c_3}, v_{r_1}\}$ in Figure 4, every reaction node in $\{v_{c_1}, v_{r_1}, v_{c_3}, v_{r_1}\}$ is originally a reversible reaction. Therefore, IP-FVS-ReactionCut2 decomposes only $\{v_{c_1}, v_{r_2}, v_{c_3}, v_{r_1}\}$ and then v_{r_2} is decomposes into v_{r_2} and v_{s_2} as shown in Figure 5.

After decomposing cycles including irreversible reactions, the same procedure as IP-FVS-ReactionCut is applied. The IP-formalization for the network of Figure 5 is as follows.

Minimize

$FE1(0)+FE2(0)+FE3(0)+FE4(0)$ (30)

Subject to

$FC4(1)=1$ (31)

for t=0,1

$TR1(t)+FC1(t)+FE1(t)\geq 1$, $FR1(t)+TC1(t)\geq 1$, $FR1(t)+TE1(t)\geq 1$ (32)

$TR1'(t)+FC3(t)+FE1(t)\geq 1$, $FR1'(t)+TC3(t)\geq 1$, $FR1'(t)+TE1(t)\geq 1$ (33)

$TR2(t)+FC1(t)+FC2(t)+FE2(t)\geq 1$, $FR2(t)+TC1(t)\geq 1$,

$FR2(t)+TC2(t)\geq 1$, $FR2(t)+TE2(t)\geq 1$ (34)

Figure 5. After decomposing cycle including original irreversible reaction

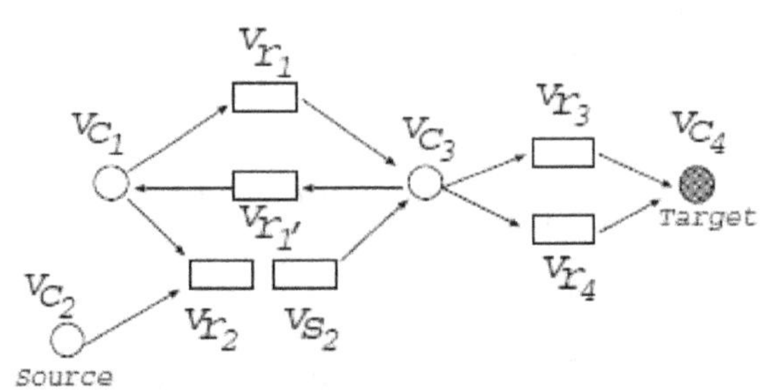

$TR3(t)+FC3(t)+FE3(t)\geq 1,\ FR3(t)+TC3(t)\geq 1,$
$FR3(t)+TE3(t)\geq 1$ (35)

$TR4(t)+FC3(t)+FE4(t)\geq 1,\ FR4(t)+TC3(t)\geq 1,$
$FR4(t)+TE3(t)\geq 1$ (36)

$FC1(t)+TR1'(t)\geq 1,\ TC1(t)+FR1'(t)\geq 1$ (37)

$FC3(t)+TR1(t)+TS2(t)\geq 1,\ TC3(t)+FR1(t)\geq 1,$
$TC3(t)+FS2(t)\geq 1$ (38)

$FC4(t)+TR3(t)+TR4(t)\geq 1,\ TC4(t)+FR3(t)\geq 1,$
$TC4(t)+FR4(t)\geq 1$ (39)

$TE1(t)=TE1(t+1),\ TE2(t)=TE2(t+1)$ (40)

$TS2(t)=TR2(t+1)$ (41)

$TC2(t)=1$ (42)

$TS2(0)=1$ (43)

$TX+FX=1$ *for any X* (44)

Note that FE1(0)=1 causes both $v_{r_1} = 0$ and $v_{r_1'} = 0$ since v_{r_1} and $v_{r_1'}$ were originally one reversible reaction.

Difference between IP-FVS-ReactionCut and IP-FVS-ReactionCut2 is intuitively explained by using Figure 3. Although only $\{v_{c_1}, v_{r_2}, v_{c_3}, v_{r_1}\}$ is decomposed in IP-FVS-ReactionCut2, IP-FVS-ReactionCut decomposes both $\{v_{c_1}, v_{r_2}, v_{c_3}, v_{r_1}\}$ and $\{v_{c_1}, v_{r_1}, v_{c_3}, v_{r_1}\}$. Therefore, $V_a = \{v_{r_1}, v_{r_2}\}$ (or $\{v_{r_3}, v_{r_4}\}$) is necessary to satisfy $v_{c_4}(1) = 0$. However, since 1s are initially assigned only for v_{c_2} and v_{s_2} in IP-FVS-ReactionCut2, nodes included in $\{v_{c_1}, v_{r_1}, v_{c_3}, v_{r_1}\}$ can become 0 if $V_a = \{v_{r_2}\}$. This solution is more reasonable than that of IP-FVS-ReactionCut since neither v_{c_1} nor v_{c_3} is supplied by the other nodes if v_{r_2} is inactivated. Essential factor of IP-FVS-ReactinCut2 is that initial values of nodes other than v_{c_2} and v_{s_2} arenotspecifiedandthennodes $\{v_{c_1}, v_{r_2}, v_{c_3}, v_{r_1}\}$ can be assigned 0 without inactivating v_{r_1}.

RESULTS ON COMPUTATIONAL EXPERIMENTS

Comparison of Methods

We developed IP-ReactionCut and IP-FVS-ReactionCut for BN-ReactionCut and IP-FVS-ReactionCut2 for BN-ReactionCut2. Since IP-FVS-ReactionCut outputs the same or similar solutions as IP-ReactionCut and it is expected that CPU time of IP-FVS-ReactionCut is similar to that of IP-FVS-ReactionCut2, we only examined IP-ReactionCut and IP-FVS-ReactionCut2. In order to compare the efficiencies of IP-ReactionCut and IP-FVS-ReactionCut2, we used an *E. coli* metabolic network consisting of Glycolysis/Glyconeogenesis (eco00010.xml), Citrate cycle (eco00020.xml) and Pentose phosphate pathway (eco00030.xml), which were obtained from KEGG database (Kanehisa et al., 2008). It contains 60 compound nodes and 111 reaction nodes (44 reversible reactions and 23 irreversible reactions). Reversible reactions are represented by two nodes so that each edge has only one direction. Note that 44×2+23=111. The experiment was done on a PC with a Xeon 3GHz CPU and 8GB RAM under the Linux (version 2.6.24) operating system, where CPLEX (Version 10.1.0) was used as the solver of integer programming.

Elapsed times for IP-ReactionCut and IP-FVS-ReactionCut2 for each target compound are shown in Table 1. Pyruvate (C00022), acetyl-CoA (C00024), acetate (C00033), oxaloacetate (C00036), and phosphoenolpyruvate (C00074) are selected as target compounds from the viewpoint of the importance of amino acids. In Table 1, it is seen that IP-FVS-ReactionCut2 is much faster than IP-ReactionCut.

Table 1. Elapsed time of IP-ReactionCut and IP-FVS-ReactionCut2 for each target compound

Target	Computation time	Computation time	Ratio
compound	of IP-ReactionCut	of IP-FVS-ReactionCut2	
C00022	10.15s	0.23s	44.13
C00024	46.88s	4.39s	10.68
C00033	49.93s	4.95s	10.09
C00036	42.62s	4.91s	8.68
C00074	65.62s	0.45s	145.82
BN-ReactionCutAll	39.28s	5.15s	7.63
Number of variables in IP	40698	3263	12.22

The reason why IP-FVS-ReactionCut2 is faster than IP-ReactionCut is that the number of variables in IP-FVS-ReactionCut2 is much less than that for IP-ReactionCut. 40698 variables were used in IP-ReactionCut whereas only 3263 variables were used in IP-FVS-ReactionCut2 and thus their ratio was 12.47. Theoretically, they are $O\left((m+n)^2\right)$ and $O(f(f+m+n))$ respectively, where f was 13 in our computational experiment.

The minimum cardinality sets of inactivated reactions for IP-ReactionCut and IP-FVS-ReactionCut2 for each target compound are shown in Table 2. Since the numbers of inactivated reactions for IP-FVS-ReactionCut2 are less than those for IP-ReactionCut in many cases, it is seen that IP-FVS-ReactionCut2 can output more efficient solutions than IP-ReactionCut.

In order to compare the computation time for calculating the reaction cut in a Boolean model by our proposed methods with existing methods, we calculated elementary modes by using Metatool (on the same CPU) (Kamp & Schuster, 2006). Computation time for calculating elementary modes for each target compound by Metatool is shown in Table 3. Note that elementary modes for each target compound will be different from each other since target compounds are treated as external nodes in Metatool. Though the minimum reaction cut based on the elementary mode model (flux balance model) can be obtained by solving Hitting Set Problem for a set of all elementary modes (Acuña et al., 2009), it was not implemented in Metatool and thus we could not examine. Since Hitting Set Problem is NP-hard, it is expected to take long time. At least, it takes more

Table 2. Solution by IP-ReactionCut and IP-FVS-ReactionCut2 for each target compound

Target	Indegree	Inactivated reactions	Inactivated reactions
compound		by IP-ReactionCut	by IP-FVS-ReactionCut2
C00022	2	R00200,R05605	R00200, R05605
C00024	4	R00235,R00354,R01323,R02569	R00351, R07618
C00033	2	R00235,R00362	R00351, R007618
C00036	4	R00341,R00342,R00354,R00362	R00351, R01518, R02570
C00074	2	R00341,R00658	R00341, R01518
BN-ReactionCutAll		R00235,R00341,R00342,R00354,R00362,	R00351,R01518,R02570,R05605
		R00658,R01323,R02569,R05605	

Table 3. Computation time for calculating elementary modes by Metatool

Target compound	Computation time	Number of elementary mode
C00022	12.40s	628
C00024	30.31s	796
C00033	15.41s	796
C00036	17.55s	2232
C00074	17.89s	1984
BN-ReactionCutAll	15.17s	1088

computation time than shown in Table 3. By comparing Table 1 with Table 3, it is seen that IP-FVS-ReactionCut2 is much faster than using Metatool.

Meaning of Reactions Detected by IP-FVS-ReactionCut2

Here, we explain the meaning of each reaction detected by IP-FVS-ReactionCut2.

- **R01518 (and R00200):** These reactions are included in the Embden-Meyerhof (EM) pathway which is one pathway of glycolysis eventually generating pyruvate (C00022) and phosphoenol-pyruvate (C00074) from glucose, and they relate to the synthesis of acetyl-CoA, which is an important compound in the citrate cycle, and the synthesis of phosphoenol-pyruvate, which are important to synthesize amino acids such as phenylalanine, tyrosine, and tryptophan.
- **R05605:** This reaction is involved in the Entner-Doudoroff (ED) pathway which is the other pathway of glycolysis (independent of the EM pathway), and it is also important to synthesize pyruvate (and phosphoenol-pyruvate) as R01518 (and R00200) on the EM pathway.
- **R00351 (and R00341):** Citrate cycle brings with acetyl-CoA and oxaloacetate, and then it produces biogenic energy. These reactions are necessary to start the citrate cycle.
- **R02570 (and R07618):** These reactions relate to gluconeogenesis eventually resynthesizing glucose from pyruvate, lactate, glycerol, and glucogenic amino acids in the case of depletion of glucose. In particular, these reactions are necessary to resynthesize acetyl-CoA. As a result, biogenetic energy keeps to be produced in such a disadvantageous case for survival.

As discussed above, IP-FVS-ReactionCut2 could find important reactions for the synthesis of amino acids and the production of biogenetic energy and thus the result of our computational experiment is reasonable also from a biological viewpoint.

Comparison of Inactivated Reactions Between IP-ReactionCut and IP-FVS-ReactionCut2

Most inactivated reactions for IP-ReactionCut essentially have few biological meanings because they relate to the synthesis of target compounds by reversible reactions. On the other hand, IP-FVS-ReactionCut2 provides more biologically interesting reactions.

In the case of acetyl-CoA (C00024) and acetate (C00033), for example, IP-ReactionCut outputs the reactions directly connecting to these target compounds. Of course, acetyl-CoA and acetate cannot be synthesized when these reactions are inactive. However, these reactions are not biologically reasonable because these reactions are trivial and we need to inactivate relatively many such reactions. In contrast, IP-FVS-ReactionCut2 provides nontrivial reactions: R00351 and R07618, which do not directly generate the target compound. In the case of *E. coli*, in general, acetyl-CoA and acetate are synthesized in the gluconeogenesis and the citrate cycle. R00351 and R07618 are key reactions that activate these metabolisms. Therefore, this set of reactions is more critical to synthesize acetyl-CoA and acetate.

In the case of the other target compounds, similarly, IP-ReactionCut provides trivial reactions directly linking to these target compounds. On the other hand, IP-FVS-ReactionCut2 gives key reactions in the metabolisms that synthesize the target compounds. As discussed above, IP-FVS-ReactionCut2 can search more biologically important sets of reactions with equivalent or smaller size than IP-ReactionCut.

Efficiency of IP-FVS-ReactionCut2 for Large-Scale Networks

In order to validate the efficiency of IP-FVS-ReactionCut2, we applied IP-FVS-ReactionCut2 to larger metabolic networks of *E. coli* (eco), *Saccharomyces cerevisiae* (sce), and *Homo sapiens* (hsa).

For eco, eco00010.xml, eco00020.xml, eco00030.xml, eco00040.xml, eco00051.xml, eco00052.xml, eco00053.xml, eco00061.xml, eco00062.xml, eco00071.xml, eco00100.xml, eco00120.xml, eco00130.xml of KEGG database were combined. There are 261 compound nodes and 404 reaction nodes (reversible: 150, irreversible: 104). $|FVS|=28$.

For sce, sce00010.xml, sce00020.xml, sce00030.xml, sce00040.xml, sce00051.xml, sce00052.xml, sce00061.xml, sce00062.xml, sce00071.xml, sce00072.xml, sce00100.xml, sce00120.xml, sce00130.xml, of KEGG database were combined. There are 198 compound nodes and 286 reaction nodes (reversible: 97, irreversible: 92). $|FVS|=21$.

For hsa, hsa00010.xml, hsa00020.xml, hsa00030.xml, hsa00040.xml, hsa00051.xml, hsa00052.xml, hsa00053.xml, hsa00061.xml, hsa00062.xml, hsa00071.xml, hsa00072.xml, hsa00100.xml, hsa00120.xml, hsa00130.xml of KEGG database were combined. There are 271 compound nodes and 410 reaction nodes (reversible: 131, irreversible: 148). $|FVS|=34$.

The minimum cardinality set of inactivated reactions for IP-FVS-ReactionCut2 for each target compound and each species is shown in Tables 4, 5 and 6. It is seen that each problem was solved in tolerable time.

The meaning of each reaction is described below. The reactions R00200, R00235, R00658, R01512 and R01518 are included in the EM pathway and they are related to the synthesis of acetate (C00033) and acetyl-CoA (C00024), and phosphoenolpyruvate. In particular, acetyl-CoA and phosphoenolpyruvate are important metabolites to bring the citrate cycle and to synthesize amino acids such as phenylalanine, tyrosine, and tryptophan, respectively.

The reactions R05605 and R08056 are involved in the ED pathway and the "ascorbate and aldarate" metabolism, respectively. They are also important to synthesize pyruvate (and phosphoenolpyruvate) as the reaction R00200 on the EM pathway. In the case of yeast and human (eukaryotes), however, these reactions are excluded from the set of inactivated reactions because such pathways are only observed in prokaryotes such as *E. coli*.

The reactions R00341, R00431, and R00726 in the pyruvate metabolism correspond to the reaction between phosphoenolpyruvate and oxaloacetate (C00036). These reactions are essentially similar, however, their availabilities differ according to species. In addition to these reactions, the reaction R00342 also belongs to the

Table 4. Solution by IP-FVS-ReactionCut2 for E. coli (eco) metabolic network; There are 261 compound nodes and 404 reaction nodes (reversible: 150, irreversible: 104) |FVS|=28

Target	Indegree	Inactivated reactions	Computation
Compound		by IP-FVS-ReactionCut2	Time
C00022	4	R00200, R05605, R08056	2.89s
C00024	12	R00238, R00351, R03270	124.83s
C00033	2	R00235, R00362	41.31s
C00036	4	R00351, R00658, R03316	55.15s
C00074	2	R00341, R01518	8.08s
BN-ReactionCutAll		R00342, R00351, R01175,	18631.16s
		R01518, R05605, R08056	

Table 5. Solution by IP-FVS-ReactionCut2 for Saccharomyces cerevisiae (sce) metabolic network; There are 198 compound nodes and 286 reaction nodes (reversible: 97, irreversible: 92). |FVS|=21

Target	Indegree	Inactivated reactions	Computation
Compound		by IP-FVS-ReactionCut2	Time
C00022	1	R00200	1.71s
C00024	10	R01278, R07618	6.95s
C00033	3	R01279, R07618	6.43s
C00036	2	R00344, R01082	3.18s
C00074	2	R00341, R01512	1.94s
BN-Reaction CutAll		R00238, R00341, R01512	2.09s

Table 6. Solution by IP-FVS-ReactionCut2 for Homo sapiens (hsa) metabolic network; There are 271 compound nodes and 410 reaction nodes (reversible: 131, irreversible: 148). |FVS|=34

Target	Indegree	Inactivated reactions	Computation
compound		by IP-FVS-ReactionCut2	Time
C00022	3	R00200, R00344	271.56s
C00024	4	R00235, R00238, R03270	1260.81s
C00033	4	R00235, R00362	2058.59s
C00036	5	R00344, R00362, R00658, R02570	8782.41s
C00074	3	R00431, R00726, R01518	4579.86s
BN-ReactionCutAll		R00014, R00200, R00235,	14858.87s
		R00238, R01518	

pyruvate metabolism, and it is also related to the synthesis of oxaloacetate.

The reactions R00344 and R00351 are necessary to start the citrate cycle. In addition to these reactions, the reactions R00014, R00342, R00362, R01082, R02570, R03270, R03316, and R07618 play an important role to maintain the citrate cycle. As a result, these reactions are related to the synthesis of metabolites in the citrate cycle such as acetyl-CoA and oxaloacetate.

Acetyl-CoA is also acquired through the beta oxidation of fatty acids in addition to glycolysis. The reactions R00238, R01175, and R01278 belonging to the metabolism related to fatty acids, they accordingly lead to synthesize acetyl-CoA.

The number of inactivated reactions of *E. coli* (prokaryotes) is larger than that of yeast and human (eukaryotes). This is because eukaryotes do not have any metabolic pathways such as the ED pathway and the ascorbate metabolism. Such pathways of higher organisms might be lost in evolutionary history due to inefficient production of biogenetic energy and alternative ways via ingestion intake.

As above, the proposed algorithm can find important reactions for the synthesis of amino acids and the production of biogenetic energy, moreover, it provides appropriate results between different species. Thus, the result of our computational experiment is reasonable also from a biological viewpoint.

CONCLUSION

In this paper, we defined two types of problems BN-ReactionCut and BN-ReactionCut2 for computing minimum reaction cuts of metabolic networks under a Boolean model. Definitions of them are slightly different from each other in terms of treatment of reversible reactions. We developed integer programming based methods IP-ReactionCut and IP-FVS-ReactionCut for BN-ReactionCut and IP-FVS-ReactionCut2 for BN-ReactionCut2, where novel use of feedback vertex sets was introduced in IP-FVS-ReactionCut and IP-FVS-ReactionCut2. We compared IP-ReactionCut with IP-FVS-ReactionCut2 using data from KEGG database. The results suggest that IP-FVS-ReactionCut2 can output more appropriate solutions than IP-ReactionCut. Furthermore, IP-FVS-ReactionCut2 is shown to be much faster than IP-ReactionCut and several times faster than a flux balance-based method.

Though we did not consider damages (i.e., side effects) induced by inactivation of reactions, it is also important to minimize damages. Indeed, Sridhar et al. (2008) defined a problem of finding a reaction cut with minimum damages using a similar but a bit more complex Boolean model of metabolic networks and developed a branch-and-bound type algorithm in order to compute an optimal solution although maximality of assignments was not taken into account in their model. It seems possible to extend our models and methods so that damages are taken into account with keeping the maximality of assignments. Such an extension would also be useful to reduce the ambiguity of minimum reaction cuts and thus is left as an important future work.

In this paper, the size of the minimum reaction cut was regarded as a measure of the structural robustness of metabolic networks. However, this measure is meaningful only when source compounds and target compounds are given. Furthermore, this measure is the worst case measure. Thus, it is also important to study the average case robustness measure when source or target compounds are not specified. Though there are some existing studies (Behre et al., 2008; Handorf, 2005; Jiang et al., 2009), few studies have been done from a computational viewpoint. Therefore, study of such models and methods is also left as an important future work.

REFERENCES

Acuña, V., Chierichetti, F., Lacroix, V., Marchetti-Spaccamela, A., Sagot, M. F., & Stougie, L. (2009). Modes and cuts in metabolic networks: Complexity and algorithms. *Bio Systems*, *95*(1), 51–60. doi:10.1016/j.biosystems.2008.06.015

Bar-Yehuda, R., Geiger, D., Naor, J., & Roth, R. M. (1998). Approximation algorithms for the feedback vertex set problem with applications to constraint satisfaction and Bayesian inference. *SIAM Journal on Computing*, *27*, 942–959. doi:10.1137/S0097539796305109

Behre, J., Wilhelm, T., Kamp, A., Ruppin, E., & Schuster, S. (2008). Structural robustness of metabolic networks with respect to multiple knockouts. *Journal of Theoretical Biology*, *252*, 433–441. doi:10.1016/j.jtbi.2007.09.043

Burgard, A. P., Pharkya, P., & Maranas, D. (2003). OptKnock: a bilevel programming framework for identifying gene knockout strategies for microbial strain optimization. *Biotechnology and Bioengineering*, *84*, 647–657. doi:10.1002/bit.10803

Dasika, M. S., Burgard, A., & Maranas, D. (2006). A computational framework for the topological analysis and targeted disruption of signal transduction networks. *Biophysical Journal*, *91*, 382–398. doi:10.1529/biophysj.105.069724

Even, G., Naor, J., Schieber, B., & Sudan, M. (1998). Approximating minimum feedback sets and multicuts in directed graphs. *Algorithmica*, *20*, 151–174. doi:10.1007/PL00009191

Guimerà, R., Sales-Pardo, M., & Amaral, L. A. N. (2007). A network-based method for target selection in metabolic networks. *Bioinformatics (Oxford, England)*, *23*(13), 1616–1622. doi:10.1093/bioinformatics/btm150

Handorf, T., Christian, N., Ebenhöh, O., & Kahn, D. (2008). An environmental perspective on metabolism. *Journal of Theoretical Biology*, *252*(3), 530–537. doi:10.1016/j.jtbi.2007.10.036

Handorf, T., Ebenhöh, O., & Heinrich, R. (2005). Expanding metabolic networks: scopes of compounds, robustness, and evolution. *Journal of Molecular Evolution*, *61*, 498–512. doi:10.1007/s00239-005-0027-1

Haus, U. U., Klamt, S., & Stephen, T. (2008). Computing knock-out strategies in metabolic networks. *Journal of Computational Biology*, *15*(3), 259–268. doi:10.1089/cmb.2007.0229

Jeong, H., Tombor, B., Albert, R., Oltval, Z. N., & Barabási, A. L. (2000). The large-scale organization of metabolic networks. *Nature*, *407*, 651–654. doi:10.1038/35036627

Jiang, D., Zhou, S., & Chen, Y.-P. P. (2009). Compensatory ability to null mutation in metabolic networks. *Biotechnology and Bioengineering*, *103*, 361–369. doi:10.1002/bit.22237

Kafri, R., Bar-Even, A., & Pilpel, Y. (2005). Transcription control reprogramming in genetic backup circuits. *Nature Genetics*, *37*(3), 295–299. doi:10.1038/ng1523

Kanehisa, M., Araki, M., Goto, S., Hattori, M., Hirakawa, M., & Itoh, M. (2008). KEGG for linking genomes to life and the environment. *Nucleic Acids Research*, *36*, D480–D484. doi:10.1093/nar/gkm882

Karp, P. D., Keseler, I. M., Shearer, A., Latendresse, M., Krummenacker, M., & Paley, S. M. (2007). Multidimensional annotation of the Escherichia coli K-12 genome. *Nucleic Acids Research*, *35*, 7577–7590. doi:10.1093/nar/gkm740

Kitano, H. (2004). Cancer as a robust system: implications for anticancer therapy. *Nature Reviews. Cancer*, *4*, 227–235. doi:10.1038/nrc1300

Klamt, S., & Gilles, E. D. (2004). Minimal cut sets in biochemical reaction networks. *Bioinformatics (Oxford, England)*, *20*(2), 226–234. doi:10.1093/bioinformatics/btg395

Klamt, S., & Stelling, J. (2002). Combinatorial complexity of pathway analysis in metabolic networks. *Molecular Biology Reports*, *29*, 233–236. doi:10.1023/A:1020390132244

Levy, E. D., & Pereira-Leal, J. B. (2008). Evolution and dynamics of protein interactions and networks. *Current Opinion in Structural Biology*, *18*(3), 349–357. doi:10.1016/j.sbi.2008.03.003

Milo, R., Shen-Orr, S., Itzkovitz, S., Kashtan, N., Chklovskii, D., & Alon, U. (2002). Network motifs: simple building blocks of complex networks. *Science*, *298*, 824–827. doi:10.1126/science.298.5594.824

Papin, J. A., Price, N. D., Wiback, S. J., Fell, D. A., & Palsson, B. O. (2003). Metabolic pathways in the post-genome era. *Trends in Biochemical Sciences*, *28*, 250–258. doi:10.1016/S0968-0004(03)00064-1

Schuster, S., Dandekar, T., & Fell, D. A. (1999). Detection of elementary flux modes in biochemical networks: a promising tool for pathway analysis and metabolic engineering. *Trends in Biotechnology*, *17*, 53–60. doi:10.1016/S0167-7799(98)01290-6

Sridhar, P., Song, B., Kahveci, T., & Ranka, S. (2008). Mining metabolic networks for optimal drug targets. *Proceedings of Pacific Symposium on Biocomputing*, *2008*, 291–302. doi:10.1142/9789812776136_0029

Stelling, J., Klamt, S., Bettenbrock, K., Schuster, S., & Gilles, E. D. (2002). Metabolic network structure determines key aspects of functionality and regulation. *Nature*, *420*, 190–193. doi:10.1038/nature01166

Tamura, T., Takemoto, K., & Akutsu, T. (2009). Measuring structural robustness of metabolic networks under a Boolean model using integer programming and feedback vertex sets. *2nd International Workshop on Intelligent Informatics in Biology and Medicine (IIBM2009), A Part of Proceedings of 3rd International Conference on Complex, Intelligent and Software Intensive Systems* (pp. 819–824).

von Kamp, A., & Schuster, S. (2006). Metatool 5.0: Fast and flexible elementary modes analysis. *Bioinformatics (Oxford, England)*, *22*(15), 1930–1931. doi:10.1093/bioinformatics/btl267

Wagner, A., & Fell, D. (2001). The small world inside large metabolic networks. *Proceedings. Biological Sciences*, *268*, 1803–1810. doi:10.1098/rspb.2001.1711

This work was previously published in the International Journal of Knowledge Discovery in Bioinformatics, Volume 1, Issue 1, edited by Xiao-Li Li and See-Kiong Ng, pp. 14-31, copyright 2010 by IGI Publishing (an imprint of IGI Global).

Chapter 15
New Trends in Graph Mining:
Structural and Node-Colored Network Motifs

Francesco Bruno
DEIS, Università della Calabria, Italy

Luigi Palopoli
DEIS, Università della Calabria, Italy

Simona E. Rombo
DEIS, Università della Calabria, Italy

ABSTRACT

Searching for repeated features characterizing biological data is fundamental in computational biology. When biological networks are under analysis, the presence of repeated modules across the same network (or several distinct ones) is shown to be very relevant. Indeed, several studies prove that biological networks can be often understood in terms of coalitions of basic repeated building blocks, often referred to as network motifs. This work provides a review of the main techniques proposed for motif extraction from biological networks. In particular, main intrinsic difficulties related to the problem are pointed out, along with solutions proposed in the literature to overcome them.

Open challenges and directions for future research are finally discussed.

INTRODUCTION

Recent years have witnessed the harvesting of enormous amounts of "rough" biological information which need to be interpreted. Such information include both string-shaped (e.g., nucleic and protein sequences) and complex-structured data, such as protein interaction networks and protein/RNA structures.

One of the main features characterizing biological data is the presence of *regularities* and *repetitions*, often associated to common biological functions and/or properties shared by the cellular components of different organisms. That is, biological data is characterized by the presence

DOI: 10.4018/978-1-4666-1785-8.ch015

of "syntactic" features modeling "semantic" properties common to the associated biological components. Such features become *interesting* when some particular conditions, related to the specific problem under analysis, are satisfied. For example, subsequences repeated more frequently than a fixed threshold in a set of protein sequences may be considered interesting, since this probably denotes that those subsequences embody some specific biological function common to the proteins in the given set.

When the repeated syntactic structures are worth being considered interesting, they are generally called *motifs* for the biological data in which they repeatedly occur, and it is usually important to look for them and single them out.

The problem of motif discovery has been thoroughly addressed during the last few years, specifically in the context of analyzing string-shaped data, and many important results have been obtained (see, e.g., (Brazma et al. 1998; Rombo et al., 2007) for surveys on this topic). Motif extraction from complex biological structures is relevant from the application viewpoint as well (Apostolico et al., 2008b; Apostolico et al. 2008; Cheng et al. 2008), as further discussed below. Indeed, following the huge efforts paid towards completing the genome coding of several organisms, a large deal of attention is now turning towards the analysis of how cellular components interact in order for biological functions to be realized (von Mering et al., 2002; Miller et al., 2005). Interactions among cellular components are most often modeled by *biological networks* (Alm et al., 2002), where the nodes represent components (e.g., proteins, genes, enzymes) and edges among nodes denote interactions amongst corresponding components. Different types of biological networks are currently available for analysis purposes, such as protein-protein interaction networks, gene regulatory networks and metabolic networks. Also, several studies (Cheng et al., 2008; Mangan et al., 2005; Milo et al., 2002; Shen-Orr et al., 2002) lead to the important observation that, similarly as with sequences, biological networks can be often understood in terms of basic repeated building blocks that are *network motifs*. And, for this reason, several techniques have been recently devised to single out motifs of biological networks. The aim of this paper is precisely that of providing a review on the main approaches to network motif extraction presented in the literature.

Preliminarily, it is worth noting that, technically speaking, the problem at hand presents several intrinsic difficulties such as, for instance, the need to deal with (variants of) the sub-graph isomorphism problem (a well known NP-hard problem (Garey et al., 1979)) when searching for identical or quasi-identical sub-graphs within several input graphs. In this respect, it will be interesting to point out how such difficulties have been dealt with in the context of the various proposed techniques. Also, different kinds of network motifs may be defined, referring to both network topology and node similarities, which is conducive to different approaches to motif extraction problems. This issue will also be highlighted in the following.

This is not the first review written on biological network motifs, rather, a few of them can be found in the literature (Alm et al., 2002; Ciriello et al., 2008; Willy, 2009). In more detail, (Alm et al., 2002) deals with transcriptional regulatory networks, discussing two types of network thereof, that are, *sensory networks*, modeling responses to signals such as stresses and nutrients, and *developmental networks*, that guide differentiation events. The author of (Alm et al., 2002) discusses four motif families that seem to cover most of the known interactions in the transcription networks of *E. coli* and *S. cerevisiae*, appearing to be the main building blocks of these sensory networks. In (Ciriello et al., 2008), a comprehensive survey on the main models and algorithms presented in the literature for motif discovery in protein-protein interaction networks is illustrated. In (Ciriello et al., 2008), the authors focus on both algorithmic issues and statistical aspects concerning the significance of repeated sub-graphs and the comparison between biological and random networks. Similar to (Ciriello et al., 2008), (Willy, 2009) also surveys on motif extraction from protein-protein interaction data. In particular, the authors divide the analyzed methods into two categories: statistically-based methods and frequency-based methods.

Differently from the reviews (Alm et al., 2002; Ciriello et al., 2008; Willy, 2009), rather than focusing upon just one network type, we survey all the three principal kinds of biological networks which are the subject of significant studies as reported in the literature, that are, *gene regulatory networks*, *protein-protein interaction networks* and *metabolic networks*. Furthermore, we carry out a comparison amongst analyzed techniques on the basis of the kind of motifs they search for. In fact, differently than in strings, where the structure of the objects we are searching for is standard and well understood, in networks a motif may be considered both as a recurrent structure, in terms of repeated topologies in the network, and a frequent module, where also the biological properties of the nodes associated to cellular components have to be taken into proper account. In this respect, this work should be considered orthogonal to the cited reviews (Alm et al., 2002; Ciriello et al., 2008).

Before closing the introductory section, we report on the structure of the following paper. In Section 2 we give some basic notions useful in the rest of the paper. Section 3 first provides a brief chronology of the main techniques of motif extraction from biological networks presented in the literature, and then categorizes them according to the kind of motifs they search for. In Section 4 a comparative summary of the techniques under analysis is provided, along with a discussion about open issues. Finally, in Section 5, we draw some conclusions.

2 BASICS

This section is devoted to provide some basic knowledge useful to the remaining part of this review.

2.1 Biological Networks

We distinguish here the three main types of biological networks that have been studied in the literature, that are, *gene regulatory networks*, *protein-protein interaction networks* and *metabolic networks* (Chor et al., 2006; Sharan et al., 2006). All of these networks can be represented as graphs $G = <N,E>$, where N is a set of nodes and E a set of (either directed or indirect) edges linking the nodes. In general, links among nodes mean that the correspondent cellular components interact with each other.

Public databases (e.g., (Chatraryamontri et al., 2006; Xenarios et al., 2000; Mewes et al., 1997; Schleif, 2000; Karp et al., 2005) store information on interactions among cellular components, from which it is possible to build the corresponding biological networks. Notice that both high-throughput experimental techniques (Ito et al., 2001; Krogan et al., 2006) and computational methods (Miller et al., 2005; von Mering et al., 2002) contributed to collect interaction data, which cannot be therefore considered completely reliable (Sprinzak et al., 2003), as also testified by several specific studies (Bader et al., 2004; Deng et al., 2003). Clear enough, the unsure reliability of such data may potentially affect any attempt to extract useful information from them.

Gene regulatory networks. Gene regulatory networks control the genes expressed in a cell as a response to biological signals. Gene regulatory networks are also called *Transcription regulation networks*, since they describe the interactions between transcription factor proteins and the genes that they regulate (Cheadle et al., 2005; Levine et al., 2005; Thieffry et al., 1998). Transcription factors respond to biological signals by accordingly changing the transcription rate of genes, thus allowing cells to produce the proteins they need at the appropriate time, pace and amount (Matys et al., 2006).

A gene regulatory network can be represented as a directed graph, in which nodes represent proteins, their corresponding mRNAs, and protein-protein complexes. In particular, the nodes of a gene regulatory network are also associated to genes, by referring to the DNA sequences which are transcribed into the mRNAs that translate into proteins. Arcs between nodes represent individual molecular reactions, that are, protein-protein and protein-mRNA interactions through which the products of one gene affect those of another. When interactions are *inductive*, i.e., an increase in the concentration of one component leads to an increase in another, they are represented by arrowheads; when interactions are inhibitory, i.e., the increase in one component leads to a decrease in another, then they are represented by filled circles. A series of edges indicates a chain of such dependencies, with cycles corresponding to feedback loops. The network structure would serve as an abstraction of the systems chemical dynamics, by describing the manifold ways in which one substance affects the expression of all the other ones to which it is connected.

Alternatively, a gene regulatory network may be also represented as a *connectivity matrix* M, such that Mij =1 if the component associated to node j encodes a transcription factor that transcriptionally regulates the component associated to node i, and Mij =0 otherwise.

Protein-protein interaction networks. Proteins are the basic constituents of living beings. They form the basis for structural components of cells as well as for metabolic processes characterizing living organisms. Understanding protein functions has, therefore, a central role in the analysis of the biological mechanisms underlying life processes. It has been shown that the mere analysis of proteins as single-objects is not sufficient to properly understand their behavior; rather, analyzing how proteins interact inside the cell is necessary to understand the biological processes which they are involved in (von Mering et al., 2002).

The set of all the protein-protein interactions of a given organism is its *interactome*, which is usually modeled as an indirect graph, called *protein-protein interaction network*, whose nodes represent involved proteins and edges encode their interactions.

Many protein-protein interactions have been discovered in the last years (Ito et al., 2001; Krogan et al., 2006, Miller et al., 2005) and several techniques have been proposed to analyze them, both referring to a single organism (Arnau et al., 2004; Adamcsek et al., 2006; Pizzuti et al., 2009) and comparing protein-protein interaction networks of different species (Bandyopadhyay et al., 2006; Singh et al., 2007; Fionda et al., 2009).

Often, a protein-protein interaction network is represented by its *adjacency matrix*, that is, a binary matrix M where both rows and columns are associated to the proteins in the network, such that $M[i,j]$ =1 if protein i interacts with protein j, $M[i,j]$ =0 otherwise.

Metabolic networks. Metabolic networks model the complete set of metabolic and physical processes that determine the physiological and biochemical properties of a cell. They comprise the chemical reactions of metabolism, as well as the regulatory interactions that guide these reactions. Examples of metabolic pathways include glycolysis, Krebs cycle, pentose phosphate pathway, and others (Sharan et al., 2006). Metabolic network reconstruction and simulation allows for an in-depth analysis into molecular mechanisms underlying cell metabolic processes in a given organism, especially correlating the genomic information with molecular physiology (Francke et al., 2005; Stelling et al., 2002).

A metabolic network can be modeled as a bipartite graph, where the two sets of nodes represent, respectively, reactions and chemical compounds, while edges link compounds and corresponding reactions. Alternatively, a metabolic network can be represented as a compound graph, that is, a compact version of the bipartite graph, where only compound vertices are kept and information on the reactions are stored as edge labels. The reaction graph is the symmetric representation of a compound graph (i.e., reaction vertices are kept and information on the compounds is stored as edge labels). When directed versions of these graphs are considered, the direction upon edges expresses the reversibility/irreversibility of some reactions.

2.2 Network Motifs

The concept of *motif* has been exploited in different applications of computational biology (Apostolico et al., 2008; Parida, 2008). Depending on the context, *what is a motif* may assume significantly different meanings. In general, motifs are always associated to *interesting* repetitions in a given data set. Interestingness is the key concept for the definition of motif; for example, a repetition can be considered interesting when its frequency is greater than a fixed threshold, or instead when it is much different than expected (Apostolico et al., 2003).

In the context of biological networks as well, a motif can be defined according to its *frequency* or its *statistical significance* (Ciriello et al., 2008). In the first case, a motif is a sub-graph that appears more than a given number of times in an input network; in the second case, a motif is a sub-graph that appears more often than expected (e.g., w.r.t. a random extraction). In particular, to measure the statistical significance of a motif, several authors compare the number of appearances of the motif in the biological network at hand with the number of it appearing in some randomized networks (Erdos et al., 1959; Erdos et al., 1960), by exploiting suitable statistical indices such as the *p-value* and the *z-score* (Milo et al, 2002).

Despite some relevant similarities, though, the computational complexity figures characterizing

Figure 1. Two sub-graphs which can be considered the same structural motif

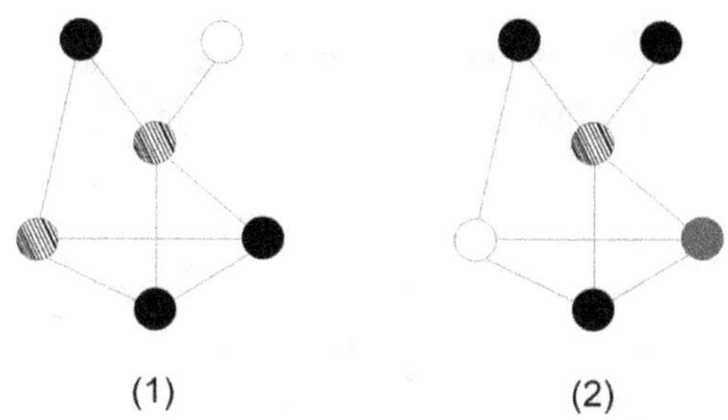

the two problems of extracting motifs from strings and singling them out from biological networks are quite different. Indeed, the problem of motif extraction is polynomial for strings and exponential (in the size of the input) for networks.

In the remaining part of this section, we describe two classes of network motifs, that we found to be the most vastly analyzed in the literature. We call them *structural* and *node-colored* motifs, respectively.

Structural motifs. The first generalization of the notion of motifs in sequences to the level of networks, has been provided in (Shen-Orr et al., 2002), where only the topology of the recurring sub-graphs identified as motifs are taken into account, without making any reference whatsoever to the biological properties and similarities characterizing the nodes of selected sub-graphs. When dealing with such a kind of network motifs, we refer to them as *structural motifs*, meaning that only the *topological structure* of graphs and sub-graphs under consideration is taken into account. Figure 1 shows two indirect sub-graphs that denote two occurrences of the same motif according to this definition. The concept is clearly and immediately extended to the case of direct graphs.

While still taking into consideration only the structure of the motifs to search for, it is possible to distinguish multiple types of interactions within the same biological network (Yeger-Lotem et al., 2004). In such a case, both the network and motifs may be modeled by exploiting two or more

Figure 2. A structural motif with multiple types of interactions

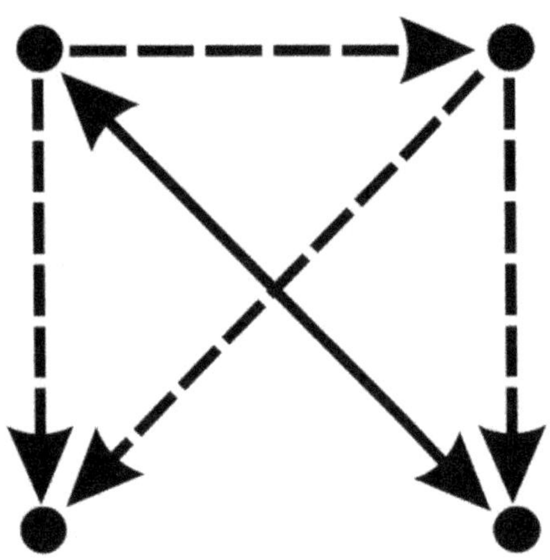

Figure 3. (a) Feedback brick motif. (b) Feedback bridge motif

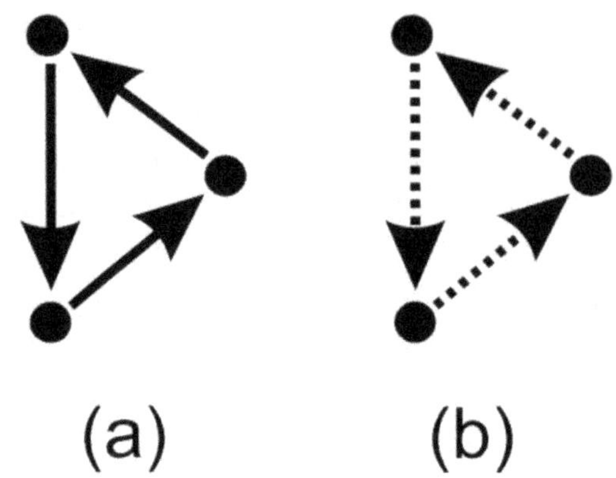

types of edges, to which different colors (or textures) are associated. As an example, Figure 2 shows a motif of a direct network, with edges of two different types, indicated by continuous and dashed arrows, respectively. In particular, the motif the figure displays represents the so called "feed-forward loop", studied in several works (Shen-Orr et al., 2002; Mangan et al., 2003; Mangan et al., 2005).

Moreover, in a network, it is possible to distinguish edges of different *weight* (Cheng et al., 2008). In this case, the modeled interactions are all of the same type, but some are stronger than others, as denoted, for instance, using different kinds of arcs. As an example, Figure 3 shows a *brick* motif, consisting only of strong links that are highlighted by full arrows, and a *bridge* motif, consisting only of weak links represented by dashed arrows.

Node-colored motifs. It has been demonstrated that biological networks are characterized by a highly modular nature (Hartwell et al., 1999; Pereira et al. 2004), such that different cellular components interact together, each providing specific contributes, in order to perform some biological function within the cell.

Structural motifs defined above treat each component of the biological network as a unique and anonymous entity, ignoring any other useful biological information possibly associated with them. In fact, being the components "unlabeled", the motif semantics is solely captured by the topology of the associated sub-graphs.

It is interesting to define an alternative concept of motif focusing on the functional nature of the components forming a given motif. In such a case, the nodes of the input network, as well as the motifs to be searched for, are labeled. Each node label is representative of specific biological properties that can be shared by different nodes in the network. Thus, nodes can be *colored* in such a way that each color is associated to a set of nodes sharing common properties. *Node-colored* motifs can be therefore defined in analogy to the other cases illustrated above, but with the difference that colors associated to occurrence matching nodes should be (approximately) identical. Figure 4 provides an intuitive illustration of the matter.

Recent approaches (Chen et al., 2007; 37, 48, Parida, 2007) exploited the node-color motifs notion, sometimes in a way independent from the sub-graph structure.

3 MOTIF EXTRACTION APPROACHES

When graphs are involved, alignment, search and comparison of sub-graphs is generally difficult, and most often intractable (no exact polynomial-time algorithms are available). This becomes immediately apparent when considering that the most common graph matching problems, such as the sub-graph isomorphism (deciding whether a graph contains a given sub-graph) (Messmer et al., 1998; Ullmann, 1976) or finding the largest common sub-graph of two graphs (Garey et al., 1979), have been proved to be in general NP-hard. This leads to the need of devising heuristic techniques for carrying out motif extraction from biological networks. Some exact techniques have been proposed, but their inherent complexity makes them adoptable only to solve motif extraction problems of small size.

3.1 A Brief Chronology

Research on singling out significant motifs in biological networks probably starts with the pioneering work by Shen-Orr et al. (Shen-Orr et al., 2002), where *network motifs* have been defined as "patterns of interconnections that recur in many different parts of a network at frequencies much higher than those found in randomized networks". In (Shen-Orr et al., 2002) the authors studied the transcriptional regulation network of *Escherichia coli*, searching for small motifs composed by three or four nodes. This way, three highly significant motifs characterizing such a network were discovered, the most famous of which is the "feed-forward loop", whose biological relevance has been pointed out also in subsequent studies (Mangan et al., 2003; Mangan et al., 2005).

The technique presented in (Shen-Orr et al., 2002) lays the foundations for several extensions thereof, the main of which are described in (Berg et al., 2004; Cheng et al., 2008; Yeger-Lotem et al., 2004).

Figure 4. Two sub-graphs which can be considered the same Node-colored motif

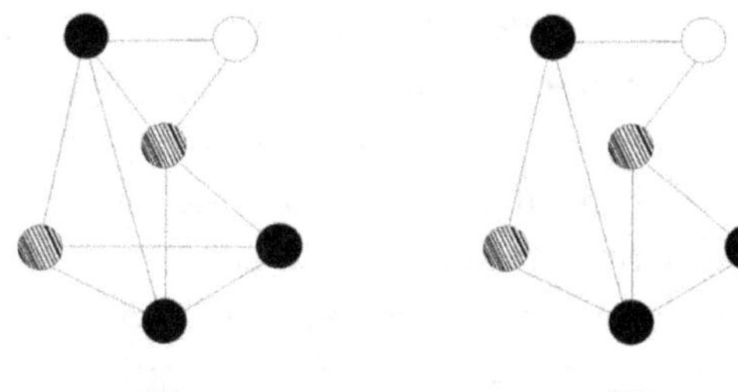

In detail, in (Yeger-Lotem et al., 2004), composite motifs consisting of two kinds of interactions (that are, transcription-regulation and protein-protein interactions) are dealt with by exploiting edges of different colors in the network model.

In (Berg et al., 2004), topological motifs derived from families of mutually similar, but not necessarily identical, patterns are defined and extracted from the gene regulatory network of *Escherichia coli*.

In (Cheng et al., 2008), n-node "bridge" and "brick" motifs are searched for in complex networks, and the presence of such motifs are associated with the network topology, but not with the network size.

All the approaches mentioned above (Berg et al., 2004; Cheng et al., 2008; Shen-Orr et al., 2002; Yeger-Lotem et al., 2004), as well as the other ones described in detail in Section 3.2, deal with structural motifs, related to the network topology only. Thus, for these techniques, nodes denoting sharply different biological components (e.g., two proteins with a very low or null sequence homology) can be identified as far as the occurrence of a motif is concerned.

As observed in (Lacroix et al., 2006), there are biological networks (e.g., metabolic networks) where a purely topological definition of a motif seems to be rather inappropriate, since similar topologies can correspond to very different functional modules. Thus, in (Lacroix et al.,

2006), a new definition of motifs in the context of metabolic networks is introduced, such that the components of the network play a central role in the motif extraction process, while the topology can be used just as an additional constraint to be satisfied by resultant graph substructures.

Similarly to (Lacroix et al., 2006), in (Parida, 2007) the concept of motif is related to both graph topologies and node similarity. In particular, the author presents a three-step exact algorithm based on the application of a notion of maximality, used extensively in strings, to graphs.

The two cited works (Lacroix et al., 2006; Parida, 2007), actually opened the way to the definition of new approaches, of both the exact and approximate kind, where both motif topologies as well as biological properties of motif components are accounted for in motif individuation and extraction.

The following two subsections will describe in greater detail the two typologies of approaches discussed above.

3.2 Approaches Based on Network Topology

For each of the approaches we discuss, we first explain how the methods work and then summarize the main results obtained by using them.

3.2.1 Milo et al., and its Extensions

In (Shen-Orr et al., 2002), the gene regulatory network of *Escherichia coli* has been chosen for analysis purposes, and an algorithm has been proposed to systematically detect network motifs defined as recurring patterns of interconnections, featuring frequencies much higher than those found in randomized networks. The authors represented the network by its connectivity matrix, and define a (non-metric) distance between nodes, based on the number of transcription factors regulating associated components. They exploit their distance to cluster nodes, and then analyze the properties of the discovered clusters to identify suitable repeated patterns in the network. In particular, they scan all the sub-matrices of the connectivity matrix corresponding to connected sub-graphs, and evaluate the p-value of such sub-matrices by comparing the number of times they occur in the real network to the number of times they appear in a randomized ensemble.

Milo et al. (Milo et al, 2002) generalized the approach presented in (Shen-Orr et al., 2002) in order to detect any type of connectivity graph in networks representing a broad range of natural phenomena. In particular, they considered gene regulatory networks, ecosystem food webs (Cohen et al., 1990), where nodes represent groups of species and edges are directed from a node representing a predator to a node representing its prey; neuronal connectivity networks (Kashtan et al., 2004), where nodes represent neurons (or neuron classes), and edges represent synaptic connections between neurons; technological networks, such as sets of sequential logic electronic circuits (Cancho et al., 2001), where the nodes represent logic gates and flip-flops.

Also Berg et al. in (Berg et al., 2004) analyzed the gene regulation network of *E. coli*, following the line of (Shen-Orr et al., 2002). In particular, they developed a search algorithm to extract topological motifs called *graph alignment*, in analogy to sequence alignment, that is based on a scoring function. The authors observed that, in biological networks, functionally related motifs do not need to be topologically identical; thus, they discuss motifs de-rived from families of mutually similar but not necessarily identical patterns. Then, they compare the maximum-likelihood alignment in the *E. coli* data set with suitable random graph ensembles. Two steps are performed within the technique in order to disentangle the significance of the number of internal links, and that of the mutual similarity of patterns found in the data.

In (Yeger-Lotem et al., 2004), composite network motifs have been looked for. Such motifs consist of both transcription-regulation and protein-protein interactions that recur significantly more often than in random networks. The authors developed algorithms for detecting motifs in networks with two or more types of interactions. In more detail, they modeled an integrated cellular interaction network using two edge types (colors), representing protein-protein and transcription-regulation interactions, respectively, and developed algorithms for detecting network motifs in networks with multiple types of edges. Such a study may be considered as a basic reference framework for detecting the building blocks of networks with multiple interaction types.

The most evolved extension of (Shen-Orr et al., 2002) has been presented in (Cheng et al., 2008), where two types of motifs have been defined, that are, *bridge motifs*, consisting of weak links only, and *brick motifs*, consisting of strong links only. In particular, links are considered either *weak* or *strong* according to the strength of the corresponding interaction (Girvan et al., 2002; Newman, 2003). The authors propose a method for performing simultaneously both the detection of global statistical features and local connection structures, and the location of functionally and statistically significant network motifs. They distinguish bridge motifs from brick ones by observing that brick motifs play a central role in defining global topological organization (Dobrin et al., 2004); bridge motifs include isolated motifs that neither interact nor overlap with other motifs.

Cheng et al. expanded the methodology presented in (Milo et al., 2002) to include the following steps: *(i)* calculate the weighted value of each link in a network of interest and an ensemble of random networks to calculate the significance of n-node sub-graphs; *(ii)* label all weighted links in the network of interest and random network ensemble as strong or weak according to a benchmark of two standard deviations from the mean weighted value of all links in the ensemble; links with weighted values below the benchmark are labeled as weak; *(iii)* identify all n-node bridge/brick sub-graph types in the network of interest and random network ensemble; *(iv)* mark all n-node bridge/brick sub-graph types by calculating their numbers in the network of interest and random network ensemble. An n-node bridge/brick sub-graph type is selected as a representative motif only if its frequency in the network of interest far exceeds its frequency in the ensemble.

Cheng et al. examined functional and topological differences between bridge and brick motifs for predicting biological network behaviour and functions.

Results. Shen-orr et al. (Shen-Orr et al., 2002) found that much of the gene regulatory network of *E. coli* consists of repeated occurrences of three highly significant motifs, the most famous of which is the "feed-forward loop". Each network motif has a specific function in determining gene expression, such as generating temporal expression programs and governing the responses to fluctuating external signals. They evaluated the statistical significance of the network motifs by comparison with randomized networks having the same characteristics as the actual *E. coli* network. In particular, for each of the individuated motifs, the probability that a randomized network had an equal or larger number of that motif than the *E. coli* network was determined by enumerating motifs found in 1,000 randomized networks.

The "feed-forward loop" has been further studied by Milo et al. (Milo et al, 2002), showing its importance in the different kinds of networks they analyzed. In particular, the authors found that none of the network motifs shared by the food webs matched the motifs extracted from the gene regulation networks or the World Wide Web. Only one of the food web motif also appeared in the neuronal network. Different motif sets were found in electronic circuits with different functions. Thus, Milo et al. argued in (Milo et al, 2002) that motifs can define broad classes of networks, each with specific types of elementary structures, reflecting the underlying processes that generated that type of network. Furthermore, they characterized the statistical significance of the motifs as a function of the network size, by analyzing pieces (sub-networks) of various sizes within the full network. They found that the concentration of motifs in the sub-networks is about the same as that in the full network, but the concentration of the corresponding sub-graphs in the randomized versions of the sub-networks sharply decreases with size.

In (Berg et al., 2004), a network alignment algorithm has been proposed, which is shown to produce well-defined alignments of maximal relative likelihood, and the probabilistic consensus for motifs occurring in the data for different numbers of nodes is discussed. The authors found that, counting identical motifs rather than following a probabilistic approach as the one they presented, a relevant fraction of relevant sub-graphs occurring in the data would be missed, with this fraction actually increasing with the data size.

The approach proposed in (Yeger-Lotem et al., 2004) has been applied on an integrated data set of protein-protein interactions and transcription regulation of *Saccharomyces cerevisiae*. In particular, by applying the technique presented therein, a two-protein mixed-feedback loop motif, five types of three-protein motifs exhibiting co-regulation and complex formation, and several motifs involving four proteins have been discovered.

Finally, Cheng et al. (Cheng et al., 2008) analyzed the gene regulation networks of *E. coli*, *S. cerevisiae*, *C. elegans* and *D. melanogaster*. They observed brick motif similarities for *E. coli* and *S. cerevisiae* networks, while noting that bridge motifs differentiate *C. elegans* from *Drosophila*. They argued that similarities and (resp., differences) in bridge and brick motifs imply similar (resp., different) key circuit elements of the three organisms. Furthermore, they found that the presence of bridge and brick motifs in a biological network is closely associated with network topological structures (particularly local connections) but not with network size (number of nodes). In particular, bridge motifs can assist in the identification of isolated motifs, and brick motifs can be used to locate motifs whose functions indeed overlap.

3.2.2 Motifs Are "Not-Isolated"

In (Dobrin et al., 2004) Dobrin et al. studied the transcriptional regulatory network of the bacterium *Escherichia coli*. The authors distinguished *coherent* motifs where all the directed links are activating, from *incoherent* links, where one of the links inhibits the activity of its target node. They observed that, in the analyzed network, the vast majority of motifs overlap generating distinct topological units which are referred to as *homologous motif clusters*. Then, by merging all the homologous motif clusters, they found that these form a single large connected component (that is, a *motif supercluster*) in which the previously identified homologous motif clusters are no longer clearly separable.

In (Mazurie et al., 2005) the integrated network of *Saccharomyces cerevisiae*, comprising transcriptional and protein-protein interaction data, has been investigated. A comparative analysis has been carried out with respect to *Candida glabrata*, *Kluyveromyces lactis*, *Debaryomyces hansenii* and *Yarrowia lipolytica*, which belong to the same class of *hemi-ascomycetes* the *S. cerevisiae* belongs to, but span a broader evolutionary range. The fact that the four analyzed organisms share several functional similarities with *S. cerevisiae* and yet span a broader range of evolutionary distances, comparable to the entire phylum of chordates, makes them ideal for protein comparisons. Then, phylogenetic profiles of genes within different forms of motifs have been analyzed, and the functional role, as resulting in vivo, of the motifs was examined for those instances where enough biological information was available.

Results. Dobrin et al. (Dobrin et al., 2004) demonstrated that the two previously identified motif types in the network (i.e., feed-forward loops and bi-fan motifs) do not occur in isolation but, rather, they aggregate into homologous motif clusters that largely overlap with known biological functions. Moreover, these clusters further coalesce into a super-cluster, thus establishing distinct topological hierarchies that show global statistical properties similar to the whole network. Targeted removal of motif links disintegrates the network into small, isolated clusters, while random disruptions of equal number of links do not cause such an effect. The authors also found that coherent motif types are significantly more common than incoherent ones both for feed-forward and bi-fan motifs.

The experiments driven by Mazurie et al. (Mazurie et al., 2005) showed that the statistical abundance of network motifs has not been associated to any evident counterpart at the evolutionary and in vivo functional level. In fact, when different organisms have been compared, the genes composing the motif showed similar profiles as genes in interaction patterns with a normal abundance. Some occurrences have been shown to arise spuriously from the representation of the interaction data in the form of a network, but the great majority of motif occurrences are embedded in larger structures and have important edge and/or node sharing with the rest of the network. Only a small minority is isolated and likely to perform a specific functional task that does not depend on the context. Thus, the authors concluded that the overabundance of the network motifs does not have any immediate functional or evolutionary counterpart.

3.2.3 Other Approaches

In (Prill et al., 2005), Prill et al. performed an exhaustive computational analysis showing that a dynamical property, related to the stability or robustness to small perturbations, is highly correlated with the relative abundance of small sub-networks (network motifs) in several previously determined biological networks. They argued that robust dynamical stability can be considered an influential property that can determine the non-random structure of biological networks.

In (Wernicke, 2006), Wernicke presented *MFinder*, an algorithm overcoming the drawbacks of the sampling technique presented in (Kashtan et al., 2004) to detect network motifs, suffering from a sampling bias and scaling poorly with increasing sub-graph size. The new approach described in (Wernicke, 2006) is based on randomized enumeration, and comprises a new way for estimating the frequency of sub-graphs in random networks that, in contrast to previous approaches, does not require the explicit generation of random networks.

Chen et al. presented *NeMoFinder* (Chen et al., 2006), a network motif discovery algorithm to single out repeated and unique meso-scale network motifs in a large protein-protein interaction network. This specific problem is indeed important, since many of the relevant processes in biological networks have been shown to correspond to such a meso-scale (5-25 genes or proteins) (Spirin et al., 2003). The procedure is based on the search of repeated trees, to be exploited for partitioning the input network into a set of graphs, represented by their adjacency matrices. The concept of *graph cousins* is introduced and exploited in order to facilitate the candidate generation and the frequency counting processes.

In (Grochow et al., 2007), an algorithm for discovering large network motifs has been presented, based on sub-graph querying and symmetry-breaking. The size of the considered motifs would exceed 15 elements, since the exploited symmetry-breaking technique eliminates repeated isomorphism testing. Such a technique reverses the traditional network-based search at the heart of the algorithms for motif-based search, which also eliminates the need to store all motifs of a given size and enables parallelization and scaling.

Finally, a tool for the exploration of network motifs, namely, *MAVisto*, has been described in (Schreiber et al., 2005). Such a tool is based on a flexible motif search algorithm. Different views for the analysis and visualization of network motifs have been provided, such that the frequency of motif occurrences can be compared with randomized networks. Also, a list of motifs along with information about structure and number of occurrences depending on the reuse of network elements showed potentially interesting motifs, and a motif fingerprint revealed the overall distribution of motifs of a given size and the distribution of a particular motif in the network, which can be also visualized.

Results. The authors of (Prill et al., 2005) analyzed the transcriptional regulatory networks of *E. coli*, *S. cerevisiae*, and other kinds of biological networks involving transduction data of *D. melanogaster* and *C. elegans*. Their analysis showed a strong correlation between network motif abundance and structural stability.

In (Wernicke, 2006) the transcriptional network of *E. coli* and *S. cerevisiae*, the neuronal network of *C. elegans* and the food web of the *Y. estuary* have been analyzed, showing that a direct calculation scheme may serve as a fast and accurate alternative to the explicit generation of random networks.

The definition of motif adopted in (Chen et al., 2006) allows for a reduction of the number of graphs to be extracted and, thus, it is possible to handle graphs of larger sizes without paying for an explosion of computation time. On the other hand, some motifs might go overlooked.

The technique described in (Grochow et al., 2007) has been applied to the protein-protein interaction network and transcription regulatory network of *S. cerevisiae*, discovering several large network motifs, including a 29-node cluster of 15-node motifs corresponding to the key transcription machinery of *S. cerevisiae*.

3.3 Approaches Based on Node Colors

The first paper in which the idea of introducing colors on nodes has been described is (Moon et al, 2005), where a colored vertex graph model has been exploited, and different nodes and edges classes have been considered. In particular, the class of a node is determined by its color, while the class of an edge is determined by the colors of the two nodes it links. The authors focused on protein interaction data, and modeled both domains and proteins belonging to such domains as nodes in the graph. Reflexive edges on domains or proteins indicate that they are self-interacting. Relationships denoting connections of a protein with a related domain are indicated by dotted lines, while interactions between domains or proteins are indicated by solid lines. The proposed algorithm searches for sub-graphs of a given graph whose frequency is substantially higher than that found in randomized networks. The algorithm first enumerates all of the possible sub-graphs, then it counts the frequency of each sub-graph and, finally, it compares their frequencies with those in randomized graphs. In order to count the frequency of each sub-graph efficiently, the authors exploited a canonical labeling of a graph (McKay et al., 1978), so that two graphs have the same canonically label graph if and only if they are isomorphic to each other.

In (Lacroix et al., 2006), the authors introduced a definition of motif such that the properties characterizing the components (nodes) of the network play the central role, while the sub-graph topology can be used as an additional constraint that the resultant motifs have to satisfy. The proposed approach has been specialized to metabolic networks, calling the motifs that are to be searched for *reaction motifs*. Here, the hierarchical classification of enzymes developed by the International Union of Biochemistry and Molecular Biology (IUBMB) (Webb, 1992) has been exploited in order to label network nodes. The technique then works on sets of nodes, instead of sub-graphs, this way avoiding the need to deal with sub-graph isomorphism.

In (Chen et al., 2007), a method called *LaMoFinder* has been presented: it labels network motifs with Gene Ontology terms (Asbumer et al., 2000) in a protein-protein interaction network. It has also shown how the resulting labeled network motifs can be used to predict unknown protein functions. The authors followed the line lied down in (Kashtan et al., 2004) and (Chen et al., 2006). The technique is based on first searching for the classes of isomorphic sub-graphs that occur frequently in the input network, and then verifying which of these sub-graph classes are also displayed at a much higher frequency than in random graphs. Unlike the previous approaches, though, Chen et al. in (Chen et al., 2007) exploited a further step, which consists in assigning biological labels to the vertices of the network motifs, so that the resulting labeled sub-graphs are verified also to occur frequently in the underlying labeled input network.

The last approach we discuss here is that presented by Parida in (Parida, 2007), where an exact three-step approach has been presented, that exploits the concept of *maximality*, suitably defined by the author in the context of network motifs. In particular, a compact notation has been introduced to handle the combinatorial explosion arising from the need of dealing with isomorphisms. Such a notation is based on grouping together sets of nodes that can be considered "indistinguishable". Nodes within the same set, in fact, have the same color, and each of them can be considered equivalent w.r.t. nodes in another sets, and also play the same role in the correspondent sub-graph topology.

Results. Moon et al. (Moon et al, 2005) applied their algorithm to interaction data on domains and proteins of several organisms. They obtained that, compared with a random distribution, a substantial part of the protein interactions could be explained by domain-level structural interaction information. Further, they found that there are distinct kinds of protein interaction patterns classified by specific and distinguishable numbers of domains. The intermolecular domain interaction was shown to be the most dominant protein interaction pattern. Finally, despite the coverage of the protein interaction information differing among species, the similarity of their networks indicated shared architectures of protein interaction in living organisms.

Lacoix et al. (Lacroix et al., 2006) applied the proposed approach to the comparative analysis of different amino-acid biosynthesis pathways. They found that, on the average, a motif of size 3 (respectively 4) has 74% (respectively, 92%) of its occurrences that are inter-pathway occurrences. More in general, the distribution of the number of occurrences of motifs of size *k* has been found to be well modeled by a power law, that had already been shown to represent a good model for the distribution of the degrees of the vertices in metabolic networks, indicating the existence of a few hubs (highly connected vertices) and a vast majority of poorly connected vertices.

Chen et al. (Chen et al., 2007) applied LaMoFinder on an experimentally-derived (yeast-two-hybrid) interaction data for *S. cerevisiae*, showing that the extracted labeled network motifs were biologically meaningful. Moreover, they show that their method was able to score better performances than existing protein-protein topology based methods for predicting unknown protein functions.

The approach presented in (Parida, 2007) has been applied to the transcriptional regulation network of *E. coli*, "discovering" motifs with known biological meanings. The main result of (Parida, 2007) is probably that the use of the compact notation decreases the output size substantially with no loss of information. Furthermore, it provides an alternative way of presenting the results as compact location lists, and opens the possibility of evaluating significance of the motifs in a given specific domain, directly on such compact lists.

4 DISCUSSION

Table 1 summarize the main properties of the approaches discussed in this review. In particular, in the first column from left to right, the method under consideration is reported; in the second column the kind of interaction data on which the method has been applied is specified; the third column underlines if the method is exact or approximate, while the last column specifies the type of motif the method searches for

It is possible to observe from the table that most of the discussed approaches are approximate. This is due to the intrinsic difficulty of network motif extraction, that generally has the sub-graph isomorphism as one of its sub-problems. When exact approaches are used, a limit is implicitly imposed on the size of the motifs that can be searched for, or otherwise different constraints limit the search to specific (forms of) motifs. Obviously, in both cases, despite the accuracy of the employed techniques, some information might be missed. For example, motifs of size greater than a dozen of nodes are not to be extracted, or sub-graphs actually characterized by different topologies may be recognized as motifs. The latter is not always considered a drawback, since emerging studies (Koyuturk et al., 2006) showed that gaps and mismatches should also be taken into account when network sub-graphs are compared, since this mirrors changes potentially introduced in the analyzed structure through evolution. Thus, searching for approximate (rather than exact) occurrences of network motifs seems to be sensible in this context. As an example, Koyuturk et al. (Koyuturk et al., 2006) defined an alignment score for comparing PPI networks based on considering duplications, matches and mismatches. In particular, a duplication is represented by the elimination or the emergence of an interaction between two proteins in the network of an organism w.r.t. that of another one, and biologically corresponds to the duplication of a gene in the course of evolution, reflecting divergence of function between the corresponding pairs of proteins. A match holds when an interaction between two proteins is conserved across two species, and it represents conservation of function between orthologous proteins during evolution. Mismatches can be associated to the lack of interactions in the PPI network of one organism between a pair of proteins whose orthologs interact in the other organism. A mismatch may correspond to the emergence of a new interaction, the elimination of a previously existing interaction in one of the species after processes such as cellular split, or experimental errors as well.

Approximate approaches present better performances in terms of computation time when compared with exact ones, and allow for increased motifs sizes, but, on the other hand, they might overlook repeated occurrences of some sub-graphs.

Table 1 also tells that techniques looking for structural motifs are more usual that those dealing with node-colored motifs. This is probably due to the fact that the possibility to significantly associate nodes with colors should be considered as a recent achievement (determined by the availability, in public, of information suitable to this purpose). The consideration of node-similarity has recently been successfully exploited also to deal with other kinds of analysis involving biological networks, such as network alignment (Singht et al., 2007; Singht et al., 2008; Fionda et al., 2009) and network querying (Fionda et al., 2008). Such techniques consider as similar pairs of sub-graphs having not only similar topologies, but also corresponding nodes with sufficiently high sequence similarity; these studies show that this helps in correctly detecting conservations across species.

Figure 5 provides a graphical view of the information contained in Table 1. As it is clear from both the table and the graphics, most of the techniques devoted to searching for structural motifs have been applied on gene regulation data. This is probably due to the fact that recurrent topologies may have significant biological meanings in gene regulation networks since, otherwise, to take into account node biological properties is necessary, especially when protein-protein interaction networks and metabolic networks are under analysis.

Table 1. Summary of the methods features

Method	Data	Exact	Motiftype
Berg et al., 2004	Gene regulation Protein	No	Structural
Chen et al., 2006	Protein interactions	No	Structural
Chen et al., 2007	Protein interactions	No	Node-colored
Cheng et al., 2008	Gene regulation	No	Structural
Dobrin et al., 2004	Gene regulation	No	Structural
Grochow et al., 2007	Gene regulation Protein interactions	No	Structural
Lacoix et al., 2006	Metabolic pathways	Yes	Node-colored
Mazurie et al., 2005	Gene regulation Protein interactions	No	Structural
Milo et al., 2002	Gene regulation Ecosystem food webs Neuronal connectivity net Technological networks	No	Structural
Moon et al., 2005	Protein interactions Domain interactions	No	Node-colored
Parida, 2007	Gene regulation	Yes	Node-colored
Prill et al., 2005	Gene regulation	No	Structural
Shen-orr et al., 2002	Gene regulation	Yes	Structural
Wernike, 2006	Gene regulation Workfood web	Yes	Structural
Yeger et al., 2004	Gene regulation Protein interactions	Yes	Structural

Figure 5. Type of motif at the varying of the network kind

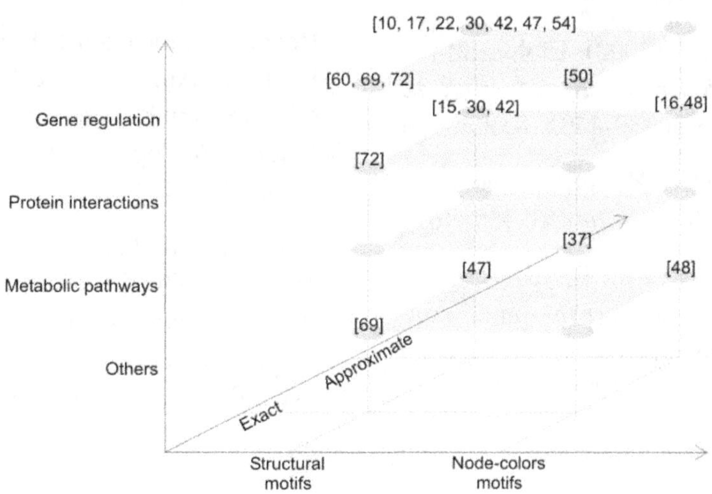

5 CONCLUSION

In this work we provided an overview of the main techniques for extracting motifs from biological networks presented in the literature. To this end, we distinguished two different kinds of motifs in biological networks, the first one taking into account only network topologies, while the latter one exploiting also the biological features of specific network components. Then, we classified the approaches at hand on the basis of the different kind of network motifs they can work on.

Several challenges are still open, notably, among them, the quest for more efficient techniques to extract motifs of consistent sizes, without missing important instances thereof. Furthermore, the problem of motif extraction from sets of biological networks has yet received only too a very limited attention, despite the importance it has when investigating evolution across species.

REFERENCES

Adamcsek, B., Palla, G., Farkas, I. J., Dernyi, I., & Vicsek, T. (2006). Cfinder: locating cliques and overlapping modules in biological networks. *Bioinformatics (Oxford, England)*, *22*(8), 1021–1023. doi:10.1093/bioinformatics/btl039

Alm, E., & Arkin, A. P. (2002). Biological networks. *Current Opinion in Structural Biology*, *13*, 193–202. doi:10.1016/S0959-440X(03)00031-9

Alon, U. (2007). Network motifs: theory and experimental approaches. *Nature*, *8*, 450–461.

Apostolico, A., Bock, M. E., & Lonardi, S. (2003). Monotony of surprise and large-scale quest for unusual words. *Journal of Computational Biology*, *10*(2/3), 283–311. doi:10.1089/10665270360688020

Apostolico, A., Ciriello, G., Guerra, C., Heitsch, C., Hsiao, C., & Williams, L. (2008). Finding 3d motifs in ribosomal rna structures. *Nucleic Acids Reserch.*

Apostolico, A., Parida, L., & Rombo, S. E. (2008). Motif patterns in 2d. *Theoretical Computer Science*, *390*(1), 40–55. doi:10.1016/j.tcs.2007.10.019

Asburner, S. (2000). Gene ontology: tool for the unification of biology. the gene ontology consortium. *Nature Genetics*, *25*, 25–29. doi:10.1038/75556

Bader, J. S., Chaudhuri, A., Rothberg, J. M., & Chant, J. (2004). Gaining confidence in high-throughput protein interaction networks. *Nature Biotechnology*, *22*, 78–85. doi:10.1038/nbt924

Bandyopadhyay, S., Sharan, R., & Ideker, T. (2006). Systematic identification of functional orthologs based on protein network comparison. *Genome Research*, *16*(3), 428–435. doi:10.1101/gr.4526006

Berg, J., & Lassig, M. (2004). Local graph alignment and motif search in biological networks. *Proceedings of the National Academy of Sciences of the United States of America*, *101*(41), 14689–14694. doi:10.1073/pnas.0305199101

Brazma, A., Jonassen, I., Eidhammer, I., & Gilbert, D. (1998). Approaches to the automatic discovery of patterns in biosequences. *Journal of Computational Biology*, *5*(2), 277–304. doi:10.1089/cmb.1998.5.279

Cancho, R. F., Janssen, C., & Sole, R. V. (2008). Topology of technology graphs: Small world patterns in electronic circuits. *Physical Review E: Statistical, Nonlinear, and Soft Matter Physics*, *64*(4), 046119. doi:10.1103/PhysRevE.64.046119

Chatraryamontri, A., Ceol, A., & Montecchi, L., Palazzi, Nardelli, G., Schneider, M. V., Castagnoli, L., & Cesareni, G. (2006). Mint: the molecular interaction database. *Nucleic Acids Research, 35*(Database issue), D572–D574. doi:10.1093/nar/gkl950

Cheadle, C., Fan, J., Cho-Chung, Y. S., Werner, T., Ray, J., & Do, L. (2005). Control of gene expression during t cell activation: alternate regulation of mrna transcription and mrna stability. *BMC Genomics, 6*(1), 75. doi:10.1186/1471-2164-6-75

Chen, J., Hsu, W., & Lee, M. L. (2007). Labeling network motifs in protein interactomes for protein function prediction. In *ICDE*.

Chen, J., Hsu, W., Lee, M. L., et al. (2006). NeMoFinder: dissecting genome-wide protein-protein interac-tions with meso-scale network motifs. In *KDD* (pp. 106–115).

Cheng, C.-Y., Huang, C.-Y., & Sun, C.-T. (2008). Mining bridge and brick motifs from complex biological networks for functionally and statistically significant discovery. *IEEE Transactions on Systems, Man, and Cybernetics–Part B, 38*(1), 17–24. doi:10.1109/TSMCB.2007.908842

Chor, B., & Tuller, T. (2006). Biological networks: Comparison, conservation, and evolutionary trees. In *RECOMB* (pp. 30–44).

Ciriello, G., & Guerra, C. (2008). A review on models and algorithms for motif discovery in protein-protein interaction network. *Briefings in Functional Genomics and Proteomics*.

Cohen, J., Briand, F., & Newman, C. (1990). *Community Food Webs: Data and Theory*. Springer.

Deng, M., Sun, F., & Chen, T. (2003). Assessment of the reliability of protein protein interactions and protein function prediction. In *Pac. Symp. Biocomput. (PSB2003)* (pp. 140–151).

Dobrin, R., Beg, Q. K., Barabasi, A. L., & Oltvai, Z. N. (2004). Aggregation of topological motifs in the escherichia coli transcriptional regulatory network. *BMC Bioinformatics, 5*(10).

Erdos, P., & Renyi, A. (1959). On random graphs. *Publicationes Matematicae, 6*, 290–297.

Erdos, P., & Renyi, A. (1960). On the evolution of random graphs. *Publ. Mat. Inst. Hungarian Acad. Sc., 5*, 17–61.

Fionda, V., Panni, S., Palopoli, L., & Rombo, S. E. (2008). Protein-protein interaction network querying by a "focus and zoom" approach. In *Proceedings of BioInformatics Research and Development (BIRD'08)* (pp. 331-346).

Fionda, V., Panni, S., Palopoli, L., & Rombo, S. E. (n.d.). A technique to search functional similarities in ppi networks. *International Journal of Data Mining and Bioinformatics*.

Francke, C., Siezen, R. J., & Teusink, B. (2005). Reconstructing the metabolic network of a bacterium from its genome. *Trends in Microbiology, 13*(11), 550–559. doi:10.1016/j.tim.2005.09.001

Garey, M., & Johnson, D. (1979). *Computers and intractability: A guide to the theory of NP-completeness*. New York: Freeman.

Girvan, M., & Newman, M. E. J. (2002). Community structure in social and biological networks. *Proceedings of the National Academy of Sciences of the United States of America, 99*(12), 7821–7826. doi:10.1073/pnas.122653799

Grochow, J., & Kellis, M. (2007). Network motif discovery using sub-graph enumeration and symmetry-breaking. *Research in Computational Molecular Biology* (pp. 92-106).

Hartwell, L., Hopfield, J., Leibler, A., & Murray, A. (1999). From molecular to modular cell biology. *Nature, 402*, c47–c52. doi:10.1038/35011540

Ito, T., Chiba, T., Ozawa, R., Yoshida, M., Hattori, M., & Sakaki, Y. (2001). A comprehensive two-hybrid analysis to explore the yeast protein interactome. *Proceedings of the National Academy of Sciences of the United States of America, 98*(8), 4569–4574. doi:10.1073/pnas.061034498

Karp, P. D., Ouzounis, C. A., & Moore-Kochlacs, C. (2005). Expansion of the BioCyc collection of pathway/genome databases to 160 genomes. *Nucleic Acids Research, 19*, 6083–6089. doi:10.1093/nar/gki892

Kashtan, N., Itzkovitz, S., Milo, R., & Alon, U. (2004). Topological generalizations of network motifs. *Physical Review E: Statistical, Nonlinear, and Soft Matter Physics, 70*(3), 031909. doi:10.1103/PhysRevE.70.031909

Koyuturk, M., Kim, Y., Topkara, U., Subramaniam, S., Szpankowski, W., & Grama, A. (2006). Pairwise align-ment of protein interaction networks. *Journal of Computational Biology, 13*(2), 182–199. doi:10.1089/cmb.2006.13.182

Krogan, N. J., & Cagney, G. (2006). Global landscape of protein complexes in the yeast saccharomyces cerevisiae. *Nature, 440*(7084), 637–643. doi:10.1038/nature04670

Lacroix, V., Fernandes, C. G., & Sagot, M.-F. (2006). Motif search in graphs: Application to metabolic networks. *IEEE ACM Transactions on Computational Biology and Bioinformatics, 3*(4), 360–368. doi:10.1109/TCBB.2006.55

Levine, M., & Davidson, E. H. (2005). Gene regulatory networks for development. *Proceedings of the National Academy of Sciences of the United States of America, 102*, 4936–4942. doi:10.1073/pnas.0408031102

Mangan, S., & Alon, U. (2003). Structure and function of the feed-forward loop network motif. *Proceedings of the National Academy of Sciences of the United States of America, 100*(21), 11980–11985. doi:10.1073/pnas.2133841100

Mangan, S., Itzkovitz, S., Zaslaver, A., & Alon, U. (2005). The incoherent feed-forward loop accelerates the response-time of the *gal* system of *escherichia coli*. *Journal of Molecular Biology, 356*(5), 1073–1081. doi:10.1016/j.jmb.2005.12.003

Matys, V., Kel-Margoulis, O. V., Fricke, E., Liebich, I., Land, S., & Barre-Dirrie, A. (2006). Transfac and its module transcompel: transcriptional gene regulation in eukaryotes. *Nucleic Acids Research, 34*(Database issue), D108–D110. doi:10.1093/nar/gkj143

Mazurie, A., Bottani, S., & Vergassola, M. (2005). An evolutionary and functional assessment of regulatory network motifs. *Genome Biology, 6*, R35. doi:10.1186/gb-2005-6-4-r35

McKay, B. (1978). Computing automorphisms and canonical labelling of graphs. *LNM, 686*, 223–232.

Messmer, B. T., & Bunke, H. (1998). A new algorithm for error-tolerant sub-graph isomorphism detection. *IEEE Transactions on Pattern Analysis and Machine Intelligence, 20*(5), 493–504. doi:10.1109/34.682179

Mewes, H. W., Albermann, K., Heumann, K., Liebl, S., & Pfeier, F. (1997). MIPS -a database for protein sequences, homology data and yeast genome information. *Nucleic Acids Research, 25*, 28–30. doi:10.1093/nar/25.1.28

Miller, J.P., Lo, R.S., Ben-Hur, A., Desmarais, C., Stagljar, I., & Stafford, W., Noble, & Fields, S. (2005). Large-scale identification of yeast integral membrane protein interactions. *Proceedings of the National Academy of Sciences of the United States of America, 102*(34), 12123–121. doi:10.1073/pnas.0505482102

Milo, R., Shen-Orr, S., Itzkovitz, S., Kashtan, N., & Chklovskii, D. (2002). Network motifs: Simple building blocks of complex networks. *Science, 298*(5594), 824–828. doi:10.1126/science.298.5594.824

Moon, H. S., Bhak, J., Lee, H. K., & Lee, D. (2005). Architecture of basic building blocks in protein and domain structural interaction networks. *Bioinformatics (Oxford, England)*, *21*(8), 1479–1486. doi:10.1093/bioinformatics/bti240

Newman, M. E. J. (2003). The structure and function of complex networks. *SIAM Review*, *45*(2), 167–2. doi:10.1137/S003614450342480

Parida, L. (2007). Discovering topological motifs using a compact notation. *Journal of Computational Biology*, *14*(3), 46. doi:10.1089/cmb.2006.0142

Parida, L. (2008). *Pattern Discovery in Bioinformatics, Theory and Algorithms*. Chapman and HAll/CRC.

Pereira, J. B., Enright, A.J., & Ouzounis, C.A. (2004). Detection of functional modules from protein interac-tion networks. *Proteins: Structure, Fuctions, and Bioinformatics*, (20), 49.

Pizzuti, C., & Rombo, S. E. (2009). Multi-functional protein clustering in PPI networks. *Biological Data Mining in Protein Interaction Networks*. Hershey, PA, USA: IGI Global.

Prill, R. J., Iglesias, P. A., & Levchenko, A. (2005). Dynamic properties of network motifs contribute to biological network organization. *PLoS Biology*, *3*(11), e343. doi:10.1371/journal.pbio.0030343

Rombo, S. E., & Palopoli, L. (2007). Pattern discovery in biosequences: From simple to complex patterns. *Data Mining Patterns: New Methods and Applications* (pp. 85-105). Hershey, PA, USA: IGI Global.

Schleif, R. (2000). Regulation of the l-arabinose operon of Escherichia coli. *Trends in Genetics*, *16*, 559. doi:10.1016/S0168-9525(00)02153-3

Schreiber, F., & Schwbbermeyer, H. (2005). MAVisto: a tool for the exploration of network motifs. *Bioinformatics (Oxford, England)*, *21*(17), 3572–3574. doi:10.1093/bioinformatics/bti556

Schuster, S., Fell, D. A., & Dandekar, T. (2000). A general definition of metabolic pathways useful for sys-tematic organization and analysis of complex metabolic networks. *Nature Biotechnology*, *18*, 326–332. doi:10.1038/73786

Sharan, R., & Ideker, T. (2006). Modeling cellular machinery through biological network comparison. *Nature Biotechnology*, *24*(4), 427–433. doi:10.1038/nbt1196

Shen-Orr, S. S., Milo, R., Mangan, S., & Alon, U. (2002). Network motifs in the trascriptional regulation network of *escherichia coli*. *Nature*, *31*, 64.

Singh, R., Xu, J., & Berger, B. (2007). Pairwise global alignment of protein interaction networks by matching neighborhood topology. In *RECOMB* (pp. 16–31).

Singh, R., Xu, J., & Berger, B. (2008). IsoRank: Global Alignment of Multiple Protein Interaction Networks with Applications to Functional Orthology Detection. *Proceedings of the National Academy of Sciences of the United States of America*, *105*(35), 12763–12768. doi:10.1073/pnas.0806627105

Spirin, V., & Mirny, L. A. (2003). Protein complexes and functional modules in molecular networks. *Proceedings of the National Academy of Sciences of the United States of America*, *100*(21), 12123–12128. doi:10.1073/pnas.2032324100

Sprinzak, E., Sattath, S., & Margalit, H. (2003). How reliable are experimental protein-protein interaction data? *Journal of Molecular Biology*, *327*(5), 919. doi:10.1016/S0022-2836(03)00239-0

Stelling, J., Klamt, S., Bettenbrock, K., Schuster, S., & Gilles, E. D. (2002). Metabolic network structure determines key aspects of functionality and regulation. *Nature*, *420*, 190–193. doi:10.1038/nature01166

Thieffry, D., Huerta, A. M., Prez-Rueda, E., & Collado-Vides, J. (1998). From specific gene regulation to genomic networks: a global analysis of transcriptional regulation in escherichia coli. *BioEssays*, *20*(5), 433–440. doi:10.1002/(SICI)1521-1878(199805)20:5<433::AID-BIES10>3.0.CO;2-2

Ullmann, J. R. (1976). An algorithm for sub-graph isomorphism. *Journal of the Association for Computing Machinery*, *23*(1), 31–42.

von Mering, D., & Krause, C. (2002). Comparative assessment of a large-scale data sets of protein-protein interactions. *Nature*, *417*(6887), 399–403. doi:10.1038/nature750

Webb, E. C. (1992). *Recommendations of the Nomenclature Committee of the International Union of Biochem-istry and Molecular Biology on the Nomenclature and Classification of Enzymes*. Oxford University Press.

Wernicke, S. (2006). Efficient detection of network motifs. *IEEE, ACM Transactions On Computational Biology And Bioinformatics*, *3*(4), 347–359. doi:10.1109/TCBB.2006.51

Willy, H. (2009). *Discovering interaction motifs from protein interaction networks*. (pp. 99–116). Hershey, PA, USA: IGI Global.

Xenarios, I., Rice, D. W., Salwinski, L., Baron, M. K., Marcotte, E. M., & Eisenberg, D. (2000). DIP: The database of interacting proteins. *Nucleic Acids Research*, *28*, 289–291. doi:10.1093/nar/28.1.289

Yeger-Lotem, E., Sattath, S., Kashtan, N., Itzkovitz, S., Milo, R., & Pinter, R. Y. (2004). Network motifs in integrated cellular networks of transcriptionregulation and proteinprotein interaction. *Proceedings of the National Academy of Sciences of the United States of America*, *101*(16), 5934–5939. doi:10.1073/pnas.0306752101

This work was previously published in the International Journal of Knowledge Discovery in Bioinformatics, Volume 1, Issue 1, edited by Xiao-Li Li and See-Kiong Ng, pp. 81-99, copyright 2010 by IGI Publishing (an imprint of IGI Global).

Chapter 16
Discriminative Subgraph Mining for Protein Classification

Ning Jin
University of North Carolina, Chapel Hill, USA

Calvin Young
University of North Carolina, Chapel Hill, USA

Wei Wang
University of North Carolina, Chapel Hill, USA

ABSTRACT

Protein classification can be performed by representing 3-D protein structures by graphs and then classifying the corresponding graphs. One effective way to classify such graphs is to use frequent subgraph patterns as features; however, the effectiveness of using subgraph patterns in graph classification is often hampered by the large search space of subgraph patterns. In this paper, the authors present two efficient discriminative subgraph mining algorithms: COM and GAIA. These algorithms directly search for discriminative subgraph patterns rather than frequent subgraph patterns which can be used to generate classification rules. Experimental results show that COM and GAIA can achieve high classification accuracy and runtime efficiency. Additionally, they find substructures that are very close to the proteins' actual active sites.

INTRODUCTION

3-D protein structures can be represented as graphs by replacing each amino acid with a node labeled with the amino acid type and connecting two nodes with an edge labeled with the distance between the two corresponding amino acids if the distance is not too long. Such graph representations preserve important structural information of the proteins and thus a classification of these graphs is in effect a protein classification based on their structures. Performing graph classification by hand is computationally intractable because of the complexity of graphs and the rapidly increasing amount of structural data, so much attention has been devoted to developing graph classification methods.

DOI: 10.4018/978-1-4666-1785-8.ch016

One solution to graph classification is to use frequent subgraph patterns as graph features and represents each graph as a vector of features. Thus, the problem of graph classification converts to classification of high dimensional data points and many existing generic classification algorithms can be applied. However, one of the major drawbacks of this approach is that when the frequency threshold is low, the number of features may be so large that no existing frequent subgraph mining algorithm can enumerate them with a reasonable amount of computational resource, which is due to the fact that the number of subgraphs is exponential to the number of nodes and edges in graphs. Using high frequency threshold can lead to significant reduction in the number of frequent subgraph patterns, but the discriminative subgraph patterns with lower frequencies than the threshold will be omitted.

To overcome this problem, many approaches have been proposed to mine only discriminative subgraph patterns as features instead of frequent subgraph patterns because only the discriminative subgraph patterns are of use in generating graph classifiers and the number of discriminative subgraph patterns is considerably less than the number of frequent subgraph patterns. SubdueCL (Gonzalez et al., 2002) uses beam search to look for the most discriminative subgraph pattern iteratively until each positive graph can be covered by at least one subgraph pattern. However, its efficiency is low for it calculates subgraph frequency by computing subgraph-isomorphism, which is an NP-complete problem. In (Kudo et al., 2004), the authors integrate the AdaBoost algorithm with gSpan (Yan & Han, 2002), an efficient frequent subgraph mining algorithm. They adapt gSpan to implement a branch-and-bound search to find the subgraph feature with the highest gain. Leap (Yan et al., 2008) is another algorithm developed to mine the most discriminative subgraph pattern. Leap uses two novel mining techniques, structural proximity pruning and frequency-descending mining, to excel the aforementioned branch-and-bound search algorithm. The structural proximity pruning takes into account the fact that subgraph patterns that share a large subgraph have similar discrimination power and uses it to calculate a tight upper-bound of their discrimination scores. The frequency-descending mining takes advantage of the observation that subgraphs with higher frequency are more likely to be discriminative and thus may reach the optimal solution faster. Another algorithm, gPLS (Saigo et al., 2008), uses subgraph pattern mining to find features for partial least squares regression so that the mining process only searches for patterns that can improve the accuracy of the resulting classifier. The algorithms mentioned above are not efficient because they find one discriminative subgraph pattern at a time and must be invoked repeatedly to find enough discriminative subgraph patterns to generate graph classifiers, thereby limiting their ability to process large datasets.

CORK (Thoma et al., 2009) uses the number of correspondences to measure the discrimination power of a subgraph pattern and thereby achieves a theoretically near-optimal solution. Given a set of subgraph patterns, the number of correspondences for a set of subgraph patterns is the total number of pairs of graphs that these subgraphs cannot discriminate. Although the solution is near-optimal in terms of minimizing the number of correspondences, using the number of correspondences as the measurement of discrimination power may be problematic because subgraphs of significantly different discrimination power can have the same number of correspondences. GraphSig (Ranu & Singh, 2009) tackles the problem of explosive number of subgraph patterns from a different perspective. Instead of looking for discriminative subgraph patterns in the whole dataset, it clusters similar graphs into small groups and searches for frequent subgraph patterns within each group. First, it uses random walk to generate feature vectors for all graphs and groups together graphs with similar feature vectors. Discriminative subgraph patterns with low frequencies among all the graphs can therefore have high frequencies within these groups. Then, graphSig uses relatively high frequency thresholds to mine frequent subgraph patterns from each group. Its efficiency completely depends on the quality of clustering in the first step. If it fails to group similar graphs together because the feature vectors generated by random walk cannot capture the characteristics of the graphs, it will still miss discriminative subgraph patterns of low frequency. GraphSig significantly outperforms Leap (Yan et al., 2008) and the optimal assignment kernel method (Fröhlich et al., 2005) in terms of runtime efficiency and accuracy as is reported in (Ranu & Singh, 2009).

In addition to subgraph pattern mining, another direction for graph classification is using graph kernel methods. In (Kashima & Koyanagi, 2002), feature vectors are composed of the counts of subgraph patterns. In (Gärtner et al., 2003) direct product kernel is defined as the count of the identical walks that can be taken in both graphs. In (Kashima et al., 2003), marginalized kernel is defined as the inner product of the count vectors averaged over all possible paths produced by random walks on graphs. An efficient computation can be performed by solving simultaneous linear equations. (Fröhlich et al., 2005) presents a graph kernel named optimal assignment kernel for chemical compound graphs. The idea is to compute an optimal assignment from nodes in one graph to those in another graph such that the overall matching score is maximized. (Smalter et al., 2008) defines a diffusion kernel for graph classification. They map frequent subgraphs to the graphs and then use "pattern diffusion" to assign values to nodes in the graphs. After "pattern diffusion," a graph alignment algorithm is used to compute the inner product of two graphs.

Generally speaking, when comparing similarity of two graphs, most graph kernel methods concentrate on global structural similarity while subgraph-based methods focus more on local similarity. As a result, subgraph-based methods are more suitable for protein structure classification than most graph kernel methods because proteins with the same function usually have certain highly conserved substructures (Huan et al., 2004; Bandyopadhyay et al., 2006) containing the active sites that most proteins without the function do not have. The conserved substructures are typically very small compared with the whole protein structures and thereby have little impact on the result of graph kernel computation, so they can only be identified by subgraph-based methods.

In this paper, we present two efficient discriminative subgraph pattern mining algorithms for protein classification: COM and GAIA. COM (Jin et al., 2009) uses a heuristic subgraph exploration order to find discriminative patterns faster. It also takes into account co-occurrences of subgraph patterns to boost the discrimination power of features. COM significantly outperforms Leap and graphSig in speed and produces higher accuracy in classifying protein structures. GAIA (Jin et al., 2010) is the most recent development in efficient graph classification. It utilizes evolutionary computation to search for discriminative subgraph patterns and achieves superior performance in terms of classification accuracy and runtime efficiency.

The remainder of this paper is organized as follows. We first introduce the preliminaries about subgraph pattern mining. Then we analyze the problem of graph classification and describe how COM and GAIA search for discriminative subgraph patterns and use them to generate graph classifiers, respectively. Experimental results are shown and analyzed. In the end we conclude the paper.

Preliminaries

Definition 1 (Graph). A graph is denoted by = *(V, E)*, where *V* is a set of nodes (vertices) and *E* is a set of edges connecting the nodes. Both nodes and edges may have labels.

Figure 1. an example of graphs

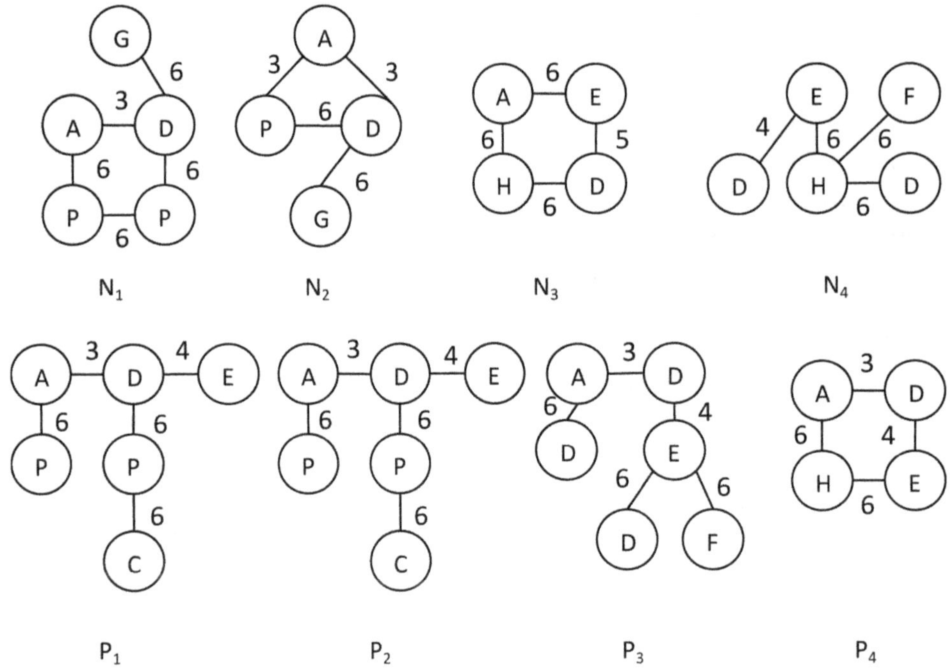

Figure 1 shows an example of two sets of graphs: $\{N_1, N_2, N_3, N_4\}$ and $\{P_1, P_2, P_3, P_4\}$. The node set of a graph is denoted as *V(g)* and the edge set of is denoted as *E()*. For example, in graph N_3, the node set $V(N_3)$ is *{A, D, E, H}* and the edge set $E(N_3)$ is *{A6H, A6E, D5E, D6H}*.

Definition 2 (Subgraph). The label of a node *u* is denoted as *l(u)* and the label of an edge *(u, v)* is denoted as *l((u, v))*. For two graphs g and g', if there is an injection *f: V(g) → V(g')*, such that for any node *v* in *V(g)*, *l(u) = l(f(u))* and for any edge *(u, v)* in *E(g)*, *l((u, v)) = l((f(u), f(v)))*, then g is a subgraph of g' and g' is a supergraph of g, or g' supports g.

For example, in Figure 1, *A3D4E* is a subgraph of graph P_1.

Definition 3 (Frequency). Given a graph set *G*, for a subgraph *p*, let *S* = {g | *g is in G and g supports p*}, then the *frequency* of g is |*S*| / |*G*|.

For example, in the graph set $\{N_1, N_2, N_3, N_4\}$ in Figure 1, the supporting set of *A3D* is $\{N_1, N_2\}$ and its frequency is 2/4 = 0.5.

Definition 4 (Positive frequency and negative frequency). Given two sets of graphs, namely the positive set G_p and the negative set G_n, the positive (negative) frequency of a subgraph pattern *p* is its frequency in the positive (negative) graph set, denoted as *pfreq*(*p*) (*nfreq*(*p*)).

In protein classification, the positive set is usually composed of the graphs representing proteins that exhibit a certain function/attribute and the negative set consists of graphs of proteins that don't exhibit the function/attribute. In most cases, the negative set is much larger than the positive set for only functions/attributes that most proteins do not have are interesting to users.

In Figure 1, let $\{N_1, N_2, N_3, N_4\}$ be the negative set and $\{P_1, P_2, P_3, P_4\}$ be the positive set. The positive frequency of *A3D* is 1.0 and its negative frequency is 0.5.

Definition 5 (Connectivity). For two nodes v_0 and v_n in graph g, if there exists a sequence of nodes $v_0, v_1, v_2, \ldots, v_n$ such that there is an edge in g connecting v_i and v_{i+1}, for any i, $0 \leq i < n$, then v_0 and v_n are connected. For a graph g, if every two nodes in g are connected, then g is connected.

All graphs in Figure 1 are connected graphs. Most subgraph mining algorithms look for connected subgraph patterns only, but COM is able to find disconnected subgraph patterns by assembling connected subgraph patterns into co-occurrences, which is defined as follows.

Definition 6 (Co-occurrence). A co-occurrence of subgraphs, *C*, is defined as a set of subgraph(s) that occur concurrently in a set of graphs. A co-occurrence can consist of only one subgraph pattern.

For example, in the graph set $\{P_1, P_2, P_3, P_4\}$ in Figure 1, $\{A3D, D4E\}$ is a co-occurrence and its frequency is 4/4 = 1.0. $\{A3D\}$ is also a co-occurrence and its frequency is 1.0.

PROBLEM ANALYSIS

The input of a graph classification problem is composed of two sets of graphs, denoted as the positive set G_p and the negative set G_n. Each graph in the input has a label *L*(g) to indicate which graph set it belongs to. If g is in the positive set, *L*(g) is 1; otherwise, *L*(g) is -1.

The goal of graph classification is to learn a classifier function *R*: $G_p \bigcup G_n \rightarrow \{1, -1\}$. The classifier *R* is then used to predict labels of unlabeled graphs. In protein classification, the labels indicate whether the proteins have a certain function/attribute. Graph classification algorithms learn classifiers based on labeled proteins and the classifiers are applied to protein function/attribute prediction. To evaluate the accuracy of a graph classification algorithm, the graph set $G = G_p \bigcup G_n$ is usually partitioned into $G_{training}$ and G_{test}, where $G = G_{training} \bigcup G_{test}$ and $G_{training} \bigcap G_{test} = \varnothing$. The classification algorithm learns a classifier *R* solely based on $G_{training}$ and the accuracy of *R* can be measured by its normalized accuracy in G_{test}.

Definition 7 (Sensitivity, Specificity and Normalized Accuracy).

$$Sensitivity(R) = \frac{|\{g_i | L(g_i) = 1 \, and \, R(g_i) = 1\}|}{|\{g_i | L(g_i) = 1\}|}$$

$$Specificity(R) = \frac{|\{g_i | L(g_i) = -1 \, and \, R(g_i) = -1\}|}{|\{g_i | L(g_i) = -1\}|}$$

$$Normalized\,accuracy(R) = \frac{Sensitivity + Specificity}{2}$$

We use normalized accuracy rather than raw accuracy to prevent the measurement from being heavily biased by imbalance between the sizes of positive set and negative set. For example, if there are 100 negative graphs and only 10 positive graphs, a classifier *R* that always predicts a graph as negative has a high accuracy equal to 100/110=0.91 while *R* is obviously a poor classifier. On the contrary, using normalized accuracy can show the poor performance of *R* as its normalized accuracy is (1.0+0.0)/2 = 0.50.

Subgraph patterns are of great use in graph classification when there exists a strong correlation between the existence of certain subgraph patterns and the labels of graphs. Proteins are typically a good example of this because whether a protein exhibits a certain function/attribute depends heavily on whether it has the active site for this function/attribute. However, there may be a huge number of subgraph patterns, of which only few may be useful in generating classifiers. Subgraph patterns are of interest only if they are discriminative, which means they occur in many graphs in one set but rarely appear in graphs in the other set. In protein classification, discriminative subgraph patterns usually occur frequently in the positive set and infrequently in the negative set because the negative set is so diverse that few frequent subgraph patterns can be found in it. Therefore, both COM and GAIA focus on searching for discriminative patterns in the positive set. However, it is straightforward to extend them to mine discriminative patterns in the negative set if necessary.

The discrimination power a pattern p is can be quantitatively measured by an objective function of its positive and negative frequencies. In this paper, we define the objective function as follows.

Definition 8 (Discrimination score).

$$score(p) = \log\left(\frac{pfreq(p)}{nfreq(p)}\right)$$

When the negative frequency of p is zero, we replace its negative frequency with a very small value ε to avoid being divided by zero.

For example, in Figure 1, the score of pattern *A3D* is log(2) and the score of pattern *A3D4E* is log(1/ε).

The goal of COM and GAIA is to find a set of subgraph patterns with high discrimination scores and use them to generate graph classifiers.

COM: DISCRIMINATIVE SUBGRAPH MINING BASED ON PATTERN CO-OCCURRENCES

Given a graph set $G = G_p \bigcup G_n$, COM searches for co-occurrences whose positive frequencies are no less than t_p and whose negative frequencies are no greater than t_n, where t_p and t_n are user specified parameters. The intuition is to find co-occurrences that are frequent enough in the positive set and infrequent enough in the negative set. The reasons why COM looks for co-occurrences instead of individual subgraph patterns are as follows. First of all, sometimes the discriminative substructures are so flexible that their conformations vary in each graph and thereby can hardly be detected by subgraph mining algorithms. If these flexible substructures can be decomposed into several smaller parts and each part has the same conformation in graphs, then the flexible substructures can be detected as co-occurrences by COM. Secondly, extending small subgraph patterns into large subgraph patterns can be a very expensive operation. Using co-occurrences of small patterns to approximate large patterns can save the expensive pattern extension operation. For example, in Figure 1, subgraph pattern *A3D4E* can be approximated by co-occurrence {*A3D*, *D4E*}. Although the co-occurrence approximation does not include as much structural information as the large pattern, *A3D4E* and {*A3D*, *D4E*} have the same supporting graphs and thus the same discrimination score. Therefore, this approximation provides the same amount of information for graph classification.

To find discriminative co-occurrences, COM explores candidate subgraph pattern space in a heuristic order and assembles individual subgraph patterns into co-occurrences if the assembly can improve discrimination score.

COM terminates when it finds a set of co-occurrences such that these co-occurrences satisfy the frequency requirements and each positive graph supports at least one co-occurrence in the set. The rationale for this termination condition is that the number of co-occurrences that satisfy the frequency requirements is still very large and these co-occurrences are redundant in that many of them are supported by the same or similar set of graphs. For example, in Figure 1, {*A3D4E*} and {*A3D*, *D4E*} are supported by the same set of graphs and thus they are essentially the same feature when only the supporting set is concerned.

Pattern Exploration Order

With a scoring function, we can rank all subgraph patterns by their discrimination scores. We want to organize the order in which subgraph patterns are visited so as to increase the probability that we visit patterns with higher score ranks earlier than those with lower score ranks because co-occurrences composed of subgraph patterns with higher ranks are more likely to satisfy the frequency requirements.

We take advantage of the following observation: let p be a pattern p' be the parent pattern from which p is extended, the score rank of p is correlated with the value of $\Delta(p) = score(p) - score(p')$. For patterns with two nodes, we set their Δ values equal to their scores $score(p)$.

Therefore, when we explore the pattern space, we first enumerate all patterns with 2 nodes as candidates and insert them into a heap structure with the candidate having the highest Δ value at the top. Ties are broken by favoring higher positive frequency. Then we always take the pattern at the top of the heap and generate all of its super-patterns with one more edge by performing the CAM extension operation (Huan et al., 2003). We insert new patterns into the heap structure. In this way, we are able to visit patterns with high score ranks early.

Co-Occurrence Generation

COM uses the resulting co-occurrences to generate classification rules. Each resulting co-occurrence C leads to a classification rule: if a graph supports all the subgraph patterns in $C \rightarrow L(g) = 1$. The final graph classifier produced by COM is composed of a set of such classification rules. Given a graph g, if any rule in the classifier indicates L(g) is 1 (positive), the classifier labels g as 1 (positive); otherwise, g is labeled as -1 (negative).

The goal of COM is to find a set of co-occurrences to maximize the number of graphs that can be classified correctly, where each co-occurrence has positive frequency no less than t_p and negative frequency no greater than t_n, where t_p and t_n are user-specified parameters to describe how discriminative users require the co-occurrences to be.

This problem can be proved to be equivalent to the set cover problem and is therefore NP complete. It is intractable to find an optimal solution in the enormous pattern space. Therefore, we adopt a greedy approach for co-occurrence generation.

Let the candidate co-occurrence set be T_t and the resulting set be T. The algorithm explores the pattern space with the aforementioned heuristic order and whenever it comes to a new pattern p that has not been processed before, if there exists one positive supporting graph of p that does not support, or in other words is not covered by, any resulting co-occurrence generated so far, the algorithm generates a new candidate co-occurrence containing only p and examines the possibility of merging this new co-occurrence into one of the existing candidate co-occurrences. Given a new pattern p and a candidate co-occurrence C_t, $\Delta(p, C_t) = score(C_t \cup \{p\}) - score(p)$. Pattern p is to be inserted into candidate co-occurrence C', $C' = \text{argmax}_{C_t \text{ is in } T_t}(\Delta(p, C_t))$, $\Delta(p, C_t) \geq 0$. If there are patterns in C' whose supporting sets are supersets of the supporting set of p, then inclusion of p into C' will make these patterns redundant. These patterns will be removed from C' when p is inserted. Then, for either the newly generated co-occurrence $\{p\}$ or the updated C', if it has $pfreq \geq t_p$ and $nfreq \leq t_n$ and it can cover at least one positive graph that is not covered by any co-occurrence in T, it will be removed from T_t and inserted into T. The algorithm terminates either when all patterns are explored or when all positive graphs are covered by some resulting co-occurrences. Although in the worst case the algorithm is still exhaustive, experiments show that it is time efficient in practice.

Gaia: Discriminative Subgraph Mining Using Evolutionary Computation

GAIA applies evolutionary computation, which is a randomized searching strategy for optimal solution and simulates biological evolution, to look for discriminative subgraph patterns. GAIA does not consider co-occurrences of subgraphs because the individual subgraph patterns found by GAIA are so discriminative that their discrimination scores can hardly be improved by assembling them into co-occurrences. In addition, although GAIA also uses patterns to generate classification rules as graph classifiers, unlike COM, GAIA adopts a feature selection process after pattern mining to maximize the normalized accuracy of the classifiers in $G_{training}$. These differences enable GAIA to outperform COM in terms of classification accuracy and runtime efficiency.

Evolution of Subgraph Patterns

Evolutionary computation can be viewed as a generic search process for solutions of high quality or fitness, which begins with a set of sample points in the search space and gradually biases to regions of high fitness. In the problem of discriminative subgraph pattern mining, we use discrimination score as the fitness function to measure the potential discrimination power of a pattern and larger patterns that can be generated by subgraph extension. As a result, our evolutionary search process here is directed toward subgraph patterns with high discrimination power.

For each graph g in the positive graph set G_p, we store a representative subgraph pattern and a list of up to s candidate subgraph patterns, where s is a user-specified parameter. The representative pattern is the subgraph pattern with the highest discrimination score among all patterns that are subgraphs of g found during pattern evolution. The candidate lists are initialized with one-edge patterns for they are the most primitive candidates.

The motivation of the design of this framework is to create selection pressure which can significantly speed up the convergence of evolutionary search. When the candidate lists are not large enough to hold all patterns that can be found in positive graphs, one resource that candidate patterns need to compete for is a slot in candidate lists and candidate patterns that fail to find a slot in candidate lists are eliminated from further consideration. Generally speaking, the larger the candidate lists are, the less selection pressure there is and thereby the more patterns are considered in the search. When the candidate lists are infinitely large, the search process becomes an exhaustive search. How the competition is carried out will be discussed later in this subsection.

Another resource that candidate patterns compete for is the opportunity to extend or, analogous to biological evolution, to produce offspring. All subgraph pattern mining algorithms start with small subgraph patterns and then extend them into larger patterns. However, pattern extension is a costly operation and not every pattern extension leads to a discriminative pattern. In an evolutionary search process, candidate patterns compete for the opportunity of pattern extension according to their fitness, which enables the search process to focus on candidate patterns that are more likely to lead to discriminative patterns. This competition is implemented as follows.

Each candidate pattern currently in the candidate lists has a non-zero probability of being selected for pattern extension. To perform pattern evolution, GAIA runs for n iterations, where n is a parameter set by the user. In each iteration, GAIA selects one pattern from each candidate list to extend. The probability for pattern p in the candidate list of graph g to be selected for extension is proportional to the score of p and is calculated as follows:

$$Probability\left(pattern\, p\, gets\, extended\right) = \frac{score(p)}{\sum_{p' \text{is in the candidate list of } g} score(p')}$$

The probability is always between 0 and 1 if only patterns with positive scores are allowed in candidate lists. Note that when $s = 1$, each candidate list only holds 1 pattern. The probability of this pattern being selected for extension is 1. When $s > 1$, multiple patterns may be held in a candidate list. A random number generator is used to determine which pattern is selected for extension according to their probabilities.

For an extension operation of pattern p, GAIA generates a pattern set $X(p)$ and each pattern p' in $X(p)$ has one new edge attached to p. Unlike many previous subgraph pattern mining algorithms that only extend patterns with certain types of edges in order to efficiently maintain their canonical codes, GAIA considers all one-edge extensions of pattern p that occur in the positive graphs. This difference enables GAIA to explore the candidate pattern space in any direction that appears promising.

In most cases, an extension operation on one pattern generates many new patterns and as a result the number of patterns found by the algorithm grows. Sooner or later the number of patterns will exceed the number of available positions in the candidate lists. It is also possible that the number of one-edge patterns already exceeds the number of available positions in the candidate lists at the very beginning if s is small. Therefore some rules are needed to determine which patterns should survive in the candidate lists and which candidate list they should dwell in.

First, a pattern that has already been extended should not stay in the candidate lists any longer because it has served its role in producing new patterns.

Second, some pattern in the candidate list may migrate to the candidate list of another graph if such migration will increase its chance of survival. Let p be the candidate pattern for migration and $S(p)$ be the set of positive graphs supporting p. Let g be the graph in $S(p)$ which has the lowest value of $\sum_{p'}$ is in the candidate list of $g^{score(p')}$

Pattern p will migrate to the candidate list of g. The rationale for this pattern migration is that if a pattern wants to survive then it should go to a candidate list with the least fierce competition. In GAIA, the fierceness of competition of a candidate list is measured by the sum of scores of patterns in the list.

If the candidate list of g still has vacant positions, then p can move into one vacant position directly. However, if the candidate list is already full, then p has to compete with the patterns that already exist in the list. The score of p is compared against the score of a pattern p', which is randomly selected with probability $1/s$ from the candidate list. If the score of p is higher, then p' is discarded from the list and p takes over the position of p'; otherwise, p is eliminated from further consideration.

Again, the exhaustive extension operation is of great importance to allow pattern competition and elimination. When we eliminate a pattern p, we lose not only this pattern but also the patterns generated by extending p. In previous subgraph pattern mining algorithms, such as gSpan (Yan & Han, 2002) and FFSM (Huan et al., 2003), a pattern p can only be extended from one of its subpatterns, p'. If p' is lost, then the algorithms will never find p. These algorithms surely lose many patterns as a result of pattern elimination, some of which are discriminative. But in GAIA, eliminating p' does not necessarily lead to the loss of p because the exhaustive extension operation allows p to be extended from many different patterns. As a result, the risk of missing discriminative patterns is much lower than other subgraph mining algorithms.

Generating Classification Rules

GAIA uses the representative subgraph patterns to generate classification rules. Each representative pattern p corresponds to one classification rule: if a graph g supports pattern $p \rightarrow L(g) = 1$, which is very similar to the rules generated by COM. However, unlike COM's graph classifiers consisting of all resulting co-occurrences, GAIA has a feature selection process to choose a subset of all representative patterns to compose its graph classifiers in order to maximize the normalized accuracy of classifiers in $G_{training}$.

Given the representative subgraph patterns from pattern evolution, GAIA's rule generation process selects patterns to compose the graph classifier by sequential coverage. First GAIA sorts the representative patterns by their scores in decreasing order. Then it traverses all representative patterns in the sorted order (that is GAIA visits patterns with higher scores first). For each representative pattern p, GAIA evaluates whether inclusion of a new classification rule "if a graph g supports pattern $p \rightarrow L(g) = 1$" in the classifier can increase the normalized accuracy of classification of the training set. The new rule "if a graph g supports pattern $p \rightarrow L(g) = 1$" is to be included in the classifier if and only if the normalized accuracy increases. The rule generation process terminates when all representative patterns have been evaluated.

Improving Performance by Generating Consensus Model with Parallel Computing

Because GAIA is a randomized algorithm (when $s > 1$), each single run of pattern evolution may generate different representative patterns and consume varying amount of CPU time. Some runs of pattern evolution may find better representative patterns than others and thus lead to classifiers with higher normalized accuracy. Therefore, if we run many instances of pattern evolution in parallel and generate a classifier based on all representative patterns found by these instances of pattern evolution, it is very likely that we can build a better classifier than using representative patterns from one instance of pattern evolution alone. Therefore, by generating a consensus model based on many parallel instances of pattern evolution and only using the fastest instances of pattern evolution, we can improve the classification accuracy and expected response time by taking advantage of parallel computing. We choose to start classification rule generation before all instances complete because, if we wait for all the pattern evolution instances, the runtime of GAIA will be determined by the runtime of the slowest instance (The time for sequential coverage is trivial). Let c be the number of parallel instances of pattern evolution, which is a user specified parameter. We start generating classification rules as soon as $c / 2$ instances of pattern evolution have terminated and only use the representative patterns found by these $c / 2$ instances. The classification rule generation process is the same as described in the previous subsection except that now we consider representative patterns found by $c / 2$ instances rather than only one instance.

Experiments

The algorithms were implemented in C++ and compiled with g++. The experiments were per-

Table 1. List of selected SCOP families

SCOP ID	*Family name*	*# of selected proteins*
46463	***Globins***	***51***
47617	***Glutathione S-transferase (GST)***	***36***
48623	***Vertebrate phospholipase A2***	***29***
48942	***C1 set domains***	***38***
50514	***Eukaryotic proteases***	***44***
51012	***alpha-Amylases, C-terminal beta-sheet domain***	***26***
51487	***beta-glycanases***	***32***
51751	***Tyrosine-dependent oxidoreductases***	***65***
51800	***Glyceraldehyde-3-phosphate dehydrogenase-like***	***34***
52541	***Nucleotide and nucleoside kinases***	***27***
52592	***G proteins***	***33***
53851	***Phosphate binding protein-like***	***32***
56251	***Proteasome subunits***	***35***
56437	***C-type lectin domains***	***38***
88634	***Picornaviridae-like VP***	***39***
88854	***Protein kinases, catalytic subunit***	***41***

formed on a 2.20 GHz dual core and 3.7 GB memory PC running Ubuntu Linux 9.10. We analyze the performance from two perspectives: runtime efficiency and normalized accuracy in protein classification. All experiments are performed with 5-fold cross-validation: each dataset is partitioned into 5 subsets with the ratio of the number of positives to the number of negatives being maintained. For each dataset, we run the algorithms 5 times and each time we use 1 subset as the test set and the rest as the training set. The classifiers are learned solely based on the test set and the average performance is reported.

The protein datasets consist of protein structures from RCSB Protein Data Bank (http://www.rcsb.org/pdb/) classified by SCOP (Structural Classification of Proteins: http://scop.mrc-lmb.cam.ac.uk/scop/). We use 16 protein datasets generated from all the large SCOP families with more than 25 members (listed in Table 1). In each dataset, protein structures in a selected family are taken as the positive set. Unless otherwise specified, we randomly select 256 other proteins (i.e., not members of the 16 families) as a common negative set used by all protein datasets. In order to remove redundancy and possible bias in graph sets, we only use proteins with pairwise sequence identity less than 90% from the culled PDB list created by Dunbrack Lab (http://dunbrack.fccc.edu/PISCES.php). To generate a protein graph, each graph node denotes an amino acid, whose location is represented by the location of its alpha carbon. We perform 3-D Almost Delaunay Tessellation (Bandyopadhyay & Snoeyink, 2004) on locations of all alpha carbons in the protein to generate the edges. Nodes are labeled with their amino acid types and edges are labeled with the distances between the alpha carbons (we discretized distances between alpha carbons into 7 bins and thereby there are 7 edge labels). We only consider edges shorter than 11.5 angstroms because amino acids have little long-distance interaction. On average, each protein graph has 250 nodes and 2700 edges.

Performance Analysis

We first compare the performance of COM (Jin et al., 2009), GAIA (Jin et al., 2010) and graphSig (Ranu & Singh, 2009). The implementation of graphSig can only process balanced datasets that have equal number of positives and negatives, so we generate balanced protein datasets, whose numbers of positives are listed in Table 1 and whose numbers of negatives are the same as the numbers of positives, for this comparison. The parameters used by the three algorithms are summarized in Table 2.

Figure 2 shows the normalized accuracy comparison between GAIA, COM and graphSig for all 16 protein datasets. GAIA has higher accuracy than COM in 14 out of 16 datasets. On average, GAIA has a normalized accuracy 5.7% higher than that of COM. We performed a paired t-test to evaluate the statistical significance of the difference between the accuracies of GAIA and COM. The p-value for GAIA's improvement over COM in normalized accuracy is 0.008 and thereby the difference in accuracy is significant at the 0.01 level. GraphSig is not comparable in processing protein datasets. Its average normalized accuracy is 52.5%.

Table 2. Summary of parameters for experiments

Algorithm	Parameters for protein datasets
GAIA	s=100, n=10, c=32
COM	t_p=30%, t_n=0%
graphSig	maxPvalue=0.1, minFreq=10%

Figure 3 compares the runtime performance of GAIA, COM and graphSig. On average, GAIA is ~1.2 times faster than COM and ~90 times faster than graphSig.

The poor performance of graphSig can be due to its failure to group similar graphs together in its first step. Once dissimilar graphs are grouped together, frequent subgraph patterns cannot be found to characterize the graphs in a group since they are so diverse. Even if graphSig manages to cluster similar graphs into small groups and finds subgraph patterns with relatively high frequency, the number of subgraph patterns may still be too large for LIBSVM (Chang & Lin, 2001), which is recommended by the authors, to generate classifiers effectively. GraphSig has high performance in chemical compound datasets, but protein graphs are much more complex than chemical compound graphs. First of all, protein graphs have more nodes and edges than chemical compound graphs. The protein graphs used in this paper have 250 nodes and 2700 edges on average while the chemical compound graphs used in (Ranu & Singh, 2009) have ~55 nodes and ~57 edges. Secondly, protein graphs have more labels than chemical compound graphs. There are 20 node

Figure 2. Normalized accuracy comparison for balanced protein datasets between GAIA, COM and graphSig

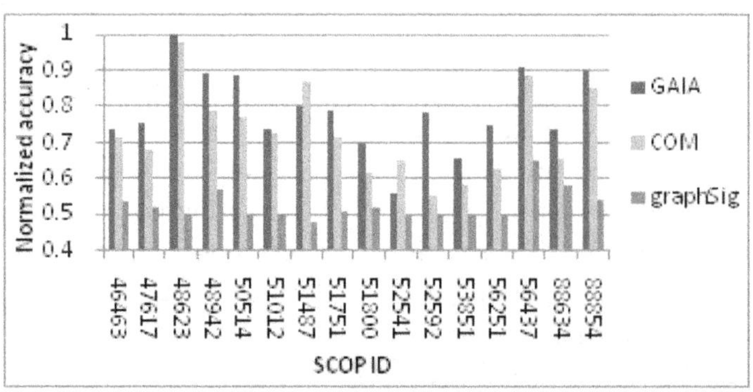

Figure 3. Runtime comparison for balanced protein datasets between GAIA, COM and graphSig

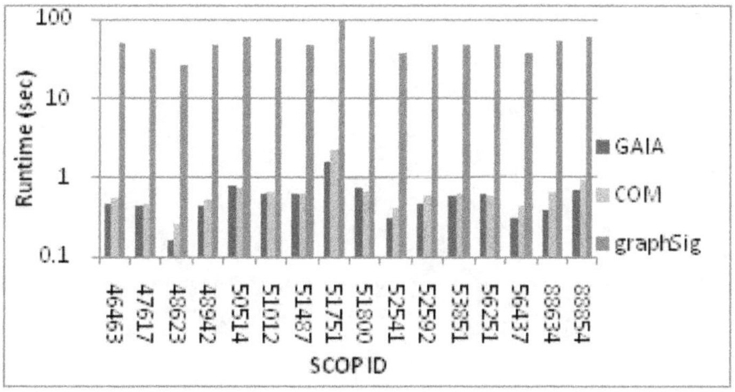

Figure 4. Normalized accuracy comparison for unbalanced protein datasets between GAIA and COM

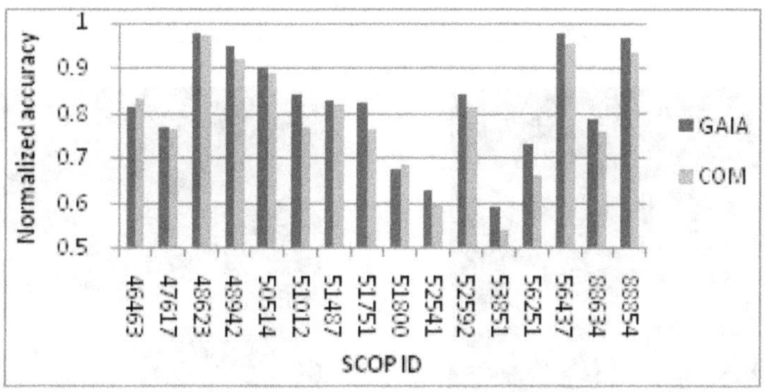

labels and 7 edge labels in protein graphs. In chemical compound graphs, although many different atom types appear in the datasets, the majority of them are C, H, N and O. In addition, most edges in chemical compound graphs are labeled as either "single bond" or "double bond" with very few exceptions. As a result, the number of subgraphs in protein graphs is much larger than the number of subgraphs in chemical compound graphs, which also leads to the much lower performance processing protein graphs than chemical compound graphs for graphSig.

We use the unbalanced protein datasets described at the beginning of this section to provide further comparison between COM and GAIA. In most real applications there are a lot of negative graphs than positive graphs. Figure 4 and Figure 5 compare the normalized accuracy and runtime performance between GAIA and COM, respectively. Compared with using balanced protein datasets, using unbalanced protein datasets leads to 3.47% higher accuracy for GAIA and 5.47% higher accuracy for COM. The difference between accuracies of balanced datasets and unbalanced datasets is statistically significant at the 0.01 level. It demonstrates that having more negative graphs as the background simplifies the graph classification problem and leads to higher accuracy. For unbalanced protein datasets, the average normalized accuracy of GAIA is 2.7% higher than that of COM and GAIA is 1.5 times faster than COM. It can be seen that the gap between the accuracy of GAIA and COM is closing as the graph classification problem becomes easier when more negative graphs are given.

Figure 5. Runtime comparison for unbalanced protein datasets between GAIA and COM

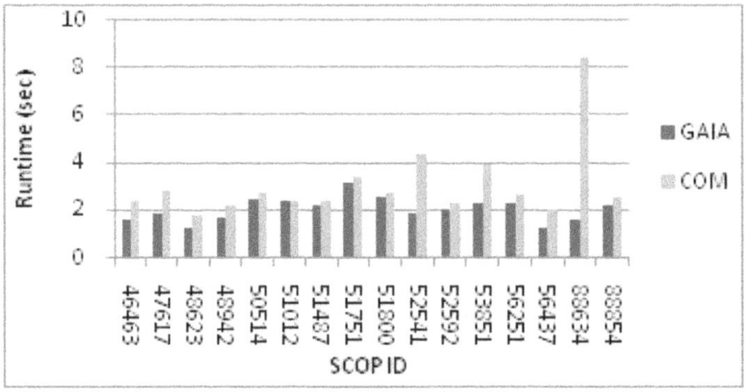

Figure 6. PDB 1a5iA (discriminative pattern: HIS-57, ILE-103, CYS-191, TRP-215; active site: HIS-57, ASP-102, GLY-193, SER-195, GLY-196)

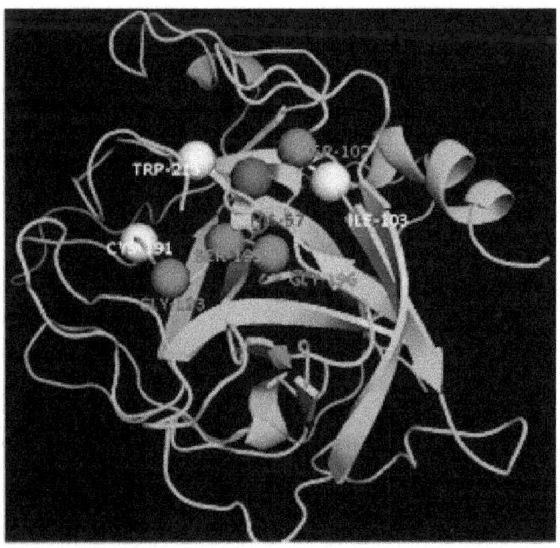

Figure 7. PDB 1aq0A (discriminative pattern: TRP-235, ASN-279, VAL-287, ASN-290; active site: GLU-232, GLU-280, LYS-283, GLU-288)

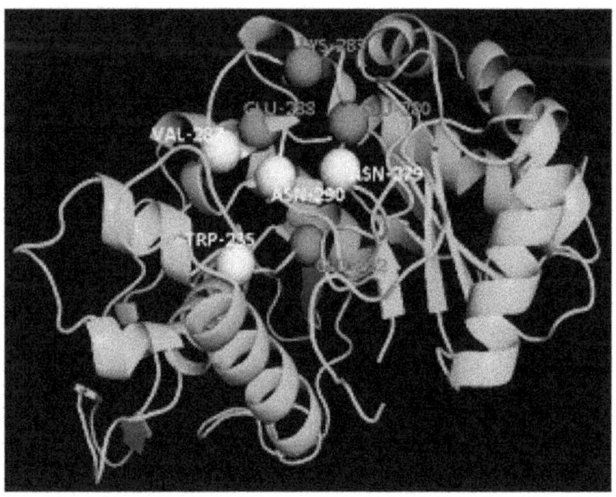

Discriminative Subgraph Patterns vs. Active Sites

In this subsection, we compare the positions of discriminative subgraph patterns and the positions of active sites. We collect the active site information from Catalytic Site Atlas (http://www.ebi.ac.uk/thornton-srv/databases/CSA/), whose active site entries are either literature-based or calculated with PSI-BLAST alignment to literature-based entries using an e value cut-off of 0.0005. We visualize the amino acids composing the discriminative subgraph patterns and active sites in protein structures by PyMOL (http://www.pymol.org/). Figures 6, Figure 7, Figure 8, and Figure 9 show 4 examples. All protein structures are from RCSB Protein Data Bank (http://www.rcsb.org/pdb/). In these figures, the yellow spheres are the amino acids composing the discriminative subgraph patterns and the red sphere are the amino acids composing the active sites. The purple spheres are where discriminative subgraph patterns and active sites overlap.

It can be seen that the patterns found by discriminative subgraph mining algorithms without any information about active sites are close to

Figure 8. PDB 1ijlA (discriminative pattern: TYR-21, CYS-43, CYS-44, HIS-47; active site: GLY-29, HIS-47, ASP-89)

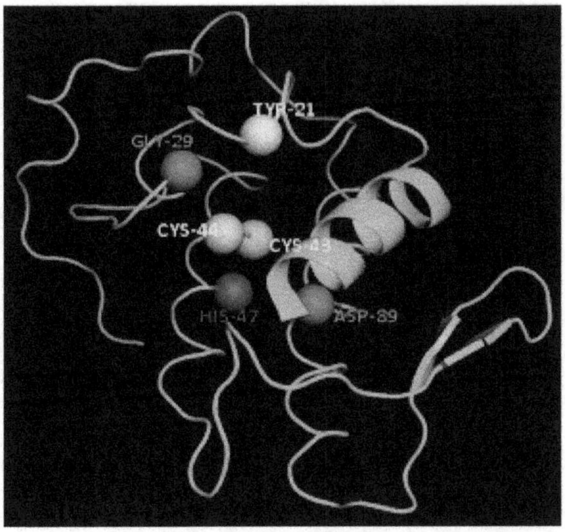

Figure 9. PDB 1ir3A (discriminative pattern: HIS-1130, ARG-1131, ASP-1132, ASP-1150; active site: ASP-1132, ARG-1136)

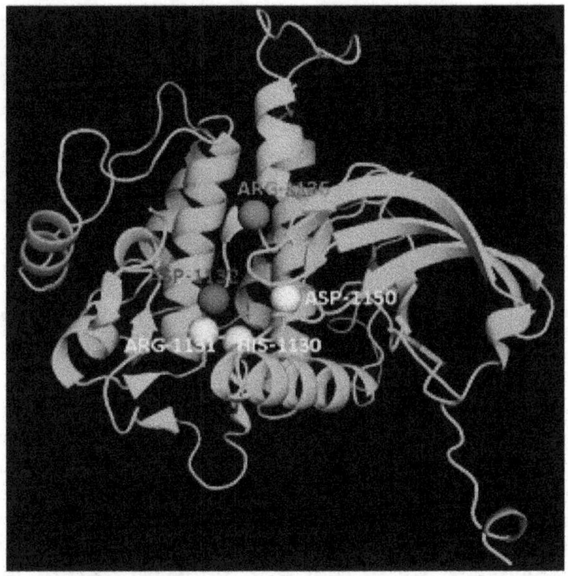

active sites and sometimes even overlap with active sites, which demonstrates that the discriminative patterns are not only of use in protein classification but biologically meaningful as well. Discriminative patterns are not exactly the same as active sites because Catalytic Site Atlas only includes residues that are thought to be directly involved in some aspect of the reaction catalyzed by a protein while discriminative pattern mining algorithms search for residues that can discriminate certain proteins from others. In addition, GAIA and COM compute a set of patterns that can lead to an accurate classifier rather than enumerate all discriminative patterns, which causes them to miss some residues in the active sites. Another reason that discriminative patterns do not completely overlap with active sites is that some of the active sites are too small to be discriminative (e.g., the active sites in Figures 8 and Figure 9) and thereby not considered as discriminative patterns.

CONCLUSION

In this paper, we investigate the problem of graph classification for protein structures and introduce two discriminative subgraph mining algorithms, namely COM and GAIA, for efficient protein classification. COM uses a heuristic subgraph pattern exploration order to search the pattern space such that discriminative patterns are more likely to be detected early. In addition, it reduces the amount of pattern extension operations by approximating large subgraph patterns as co-occurrences of small patterns. GAIA utilizes evolutionary computation in discriminative subgraph pattern mining. During the pattern evolution performed by GAIA, candidate subgraph patterns compete for storage space in memory and CPU time for pattern extensions. Candidate subgraph patterns with high discrimination scores survive the competition and are used to generate graph classifiers. Experiments show that both COM and GAIA outperform graphSig, one of the state-of-the-art graph classification methods, in terms of classification accuracy and runtime efficiency for protein classification. We also demonstrate that discriminative subgraph patterns are close to active sites in 3-D protein structures.

REFERENCES

Bandyopadhyay, D., Huan, J., Liu, J., Prins, J., Snoeyink, J., Wang, W., & Tropsha, A. (2006). Structure-based function inference using protein family-specific fingerprints. *Protein Science*, *15*, 1537–1543. doi:10.1110/ps.062189906

Bandyopadhyay, D., & Snoeyink, J. (2004). Almost Delaunay Simplices: Nearest Neighbor Relations for Imprecise Points. In *Proceedings of SODA*, *2004*, 403–412.

Chang, C.-C., & Lin, C.-J. (2001). *LIBSVM: a library for support vector machines*. Retrieved from http://www.csie.ntu.edu.tw/~cjlin/libsvm

Fei, H., & Huan, J. (2008). Structure feature selection for graph classification. In *Proceedings of the CIKM* (pp. 991-1000).

Fei, H., & Huan, J. (2009). L2 Norm Regularized Feature Kernel Regression for Graph Data. In *Proceedings of the CIKM* (pp. 593-600).

Fröhlich, H., Wegner, J. K., Sieker, F., & Zell, A. (2005). Optimal assignment kernels for attributed molecular graphs. In *Proceedings of the ICML* (pp. 225-232).

Gärtner, T., Flach, P. A., & Wrobel, S. (2003). On graph kernels: hardness results and efficient alternatives. In *Proceedings of the COLT* (pp. 129-143).

Gonzalez, J. A., Holder, L. B., & Hook, D. J. (2002). Graph-based relational concept learning. In *Proceedings of the ICML* (pp. 219-226).

Huan, J., Wang, W., Bandyopadhyay, D., Snoeyink, J., Prins, J., & Tropsha, A. (2004). Mining spatial motifs from protein structure graphs. In *Proceedings of the RECOMB* (pp. 308-315).

Jin, N., Young, C., & Wang, W. (2009). Graph Classification Based on Pattern Co-occurrence. In *Proceedings of the CIKM* (pp. 573-582).

Jin, N., Young, C., & Wang, W. (2010). GAIA: graph classification using evolutionary computation. In *Proceedings of the SIGMOD* (pp. 879-890).

Kashima, H., & Koyanagi, T. (2002). SVM kernels for semi-structured data. In *Proceedings of the ICML* (pp. 291-298).

Kashima, H., Tsuda, K., & Inokuchi, A. (2003). Marginalized kernels between labeled graphs. In *Proceedings of the ICML* (pp. 321-328).

Kudo, T., Maeda, E., & Matsumoto, Y. (2004). An application of boosting to graph classification. In *Proceedings of the NIPS*.

Murzin, A. G., Brenner, S. E., Hubbard, T., & Chothia, C. (1995). SCOP: a structural classification of proteins database for the investigation of sequences and structures. *Journal of Molecular Biology*, *247*, 536–540. doi:10.1016/S0022-2836(05)80134-2

Ranu, S., & Singh, A. K. (2009). GraphSig: A Scalable Approach to Mining Significant Subgraphs in Large Graph Databases. In *Proceedings of the ICDE* (pp. 844-855).

Saigo, H., Kraemer, N., & Tsuda, K. (2008). Partial Least Squares Regression for Graph Mining. In *Proceedings of the SIGKDD* (pp. 578-586).

Smalter, A., Huan, J., & Lushington, G. (2008). A Graph Pattern Diffusion Kernel for Chemical Compound Classification. In *Proceedings of BIBE'08*.

Thoma, M., Cheng, H., Gretton, A., Han, J., Kriegel, H., Smola, A., et al. (2009). Near-optimal supervised feature selection among frequent subgraphs. In *Proceedings of the SDM.*

Wang, G., & Dunbrack, R. L. Jr. (2003). PISCES: a protein sequence culling server. *Bioinformatics (Oxford, England)*, *19*, 1589–1591. doi:10.1093/bioinformatics/btg224

Yan, X., Cheng, H., Han, J., & Yu, P. S. (2008). Mining significant graph patterns by leap search. In *Proceedings of the SIGMOD* (pp. 433-444).

Yan, X., & Han, J. (2002). gSpan: graph-based substructure pattern mining. In *Proceedings of the ICDM* (pp. 721-724).

This work was previously published in the International Journal of Knowledge Discovery in Bioinformatics, Volume 1, Issue 3, edited by Xiao-Li Li and See-Kiong Ng, pp. 36-52, copyright 2010 by IGI Publishing (an imprint of IGI Global).

Chapter 17
Infer Species Phylogenies Using Self-Organizing Maps

Xiaoxu Han
Eastern Michigan University, USA

ABSTRACT

With rapid advances in genomics, phylogenetics has turned to phylogenomics due to the availability of large amounts of sequence and genome data. However, incongruence between species trees and gene trees remains a challenge in molecular phylogenetics for its biological and algorithmic complexities. A state-of-the-art gene concatenation approach was proposed to resolve this problem by inferring the species phylogeny using a random combination of widely distributed orthologous genes screened from genomes. However, such an approach may not be a robust solution to this problem because it ignores the fact that some genes are more informative than others in species inference. This paper presents a self-organizing map (SOM) based phylogeny inference method to overcome its weakness. The author's proposed algorithm not only demonstrates its superiority to the original gene concatenation method by using same datasets, but also shows its advantages in generalization. This paper illustrates that data missing may not play a negative role in phylogeny inferring. This study presents a method to cluster multispecies genes, estimate multispecies gene entropy and visualize the species patterns through the self-organizing map mining.

DOI: 10.4018/978-1-4666-1785-8.ch017

INTRODUCTION

With the surge of genomics, the availability of large amounts of sequence and genome data has revolutionarily changed traditional phylogenetics (Delsuc, Brinkmann, & Philippe, 2005). However, the gene tree and species tree problem still remains a fundamental problem in phylogenetics for its biological and algorithmic complexities (Ma, Li, & Zhang, 2000; Oliver, 2008; Rasmussen & Kellis, 2007). It means that the gene trees, which are phylogenies obtained from individual genes, are incongruent with the species tree. The incongruence between species trees and gene trees simply refers to the fact that they have different branch patterns in the tree topologies (Rasmussen & Kellis, 2007). The species phylogeny reflects the true evolution history of species where all lineages are results of speciation and divergence. Alternatively, a gene tree represents the evolutionary history of an individual gene for a set of organisms, which may not be the evolutionary history of species. Theoretically, it can be viewed as a hypothesis about how a gene evolves through gene duplication, gene loss and nucleotide substitution.

Although the incongruence between gene trees and species trees can be observed from all levels of taxa, it occurs more often for closely related species. For a group of species, the probability of incongruence between a gene trees τ_g and species tree τ_s can be computed as $\Pr(\tau_g \neq \tau_s) = (2e^{-(t/2n)} / 3)$, where t is the speciation time and n is the effective population size. For a fixed effective population size, the less related species have a relatively larger speciation time interval than the closely related species. Thus, the probability of incongruence between the gene trees and species tree is usually high for a set of closely related species (Nei & Kumar, 2000).

Such an incongruence may originate from biological and technical factors. Biologically, the factors like gene duplication, gene loss, lineage sorting or horizontal gene transfer would affect the evolutionary signature of a gene and cause a gene tree different from a species phylogeny (Nei & Kumar, 2000; Whitaker, McConkey, & Westhead, 2009). Technically, the factors like inappropriate substitution models, insufficient data sampling or artifacts in phylogenetic reconstruction methods also play important roles in inferring an incongruent gene tree (Huelsenbeck, 1995; Hahn, 2007). Due to its biological and algorithmic complexities, successful resolving incongruence between the gene trees and species trees may have fundamental impacts on molecular evolution and phylogenetics.

A few reconciliation models built on sequence and genome data have been proposed to resolve the gene tree and species tree problem (Liu et al., 2008; Thomson et al., 2008; Moret, Tang, & Warnow, 2005; Snel, Huynen, & Dutilh, 2005; Rokas, Williams, King, & Carroll, 2003). Two genome-based approaches among them have drawn investigators' attentions. One is called a genome-tree approach that uses the complete genome data to build a phylogenetic tree and guarantees the final evolutionary tree is a species tree (Snel, Huynen, & Dutilh, 2005); Another is called a genome-based gene concatenation method that concatenates a set of widely distributed orthologous genes from the genomes of species interested (Rokas, Williams, King, & Carroll, 2003). Although both of them are promising in the problem-solving, they have their own limitations. The genome-tree method suffers from the relatively immature phylogenetic reconstruction models because classic evolution and phylogenetic models are build on sequence data instead of heterogeneous genome data. Moreover, although thousands of genomes are being sequenced, relatively a small number of relevant genomes are currently available for genome tree reconstruction (Delsuc, Brinkmann, & Philippe, 2005; Snel, Huynen, & Dutilh, 2005). Although the genome based gene concatenation method is welcomed by biologists for the sake of its "simplicity", it has following limitations. Firstly, the gene concatenation number must be empirically determined for each dataset. Since building a phylogenetic tree is actually a NP-hard problem (Felsentein, 2004; Chor & Tuller, 2006), the computational complexity for corresponding phylogenetic analysis would be prohibitive to determine the gene concatenation number, under the situation where large numbers of taxa are existed in the species phylogeny inference. Secondly, the random gene concatenation does not consider that some genes are more informative than others in the species phylogeny inference. It assumes all selected genes are equally informative in inferring a species tree. Thirdly, if the orthology criterion is not robust in data collection, then the genes used in phylogenetic reconstruction may have different evolutionary histories. The random gene concatenation under such a situation will bring noise in the phylogenetic tree reconstruction and the final tree inferred may not be robust or represent the true evolutionary relationships between organisms.

In this work, we propose a self-organizing map (SOM) (Haykin, 1999; Kohonen, 2001) based phylogeny inference algorithm, i.e., phylogenetic mining, to overcome the weakness of the genome based gene concatenation method. The idea of our algorithm is to employ a SOM mining to cluster the widely-distributed orthologous multi-species genes in the genomes interested to gain relationships between them. Then phylogenetically informative genes are selected from these clusters based on the maximum entropy principle to infer a final species tree. We will show that this SOM mining based approach can achieve better results for an eight yeast species tree inference problem than the original gene concatenation method by using the same datasets. To the best of our knowledge, our algorithm first provides a method to cluster multispecies genes, estimate multi-species gene entropy, and visualize species patterns. Moreover, our method can be easily extended to infer other species trees by overcoming the ad-hoc mechanism of the original method. The paper is structured as follows. Section 2 introduces the SOM-based phylogenetic mining method, in addition to simulating random gene concatenation via Bayesian phylogeny analysis. Section 3 shows the main results of our method in species pattern visualization, multispecies gene clustering, multispecies gene entropy estimation and species tree inference through maximum entropy gene concatenation; Section 4 challenges the traditional concepts in phylogenetic species tree reconstruction by illustrating that a robust phylogeny can be obtained under data missing cases. Section 5 concludes this paper.

METHODS

This section first addresses the importance of phylogenetically informative genes in the species tree inference by simulating random gene concatenation via Bayesian phylogeny analysis. We employ software *MrBayes 3.1* to conduct Bayesian phylogeny inference (Huelsenbeck et al., 2002). Then a self-organizing map mining based phylogeny inference is presented. For convenience, we use the same dataset used by the genome based random gene concatenation method, which consists of 106 widely distributed othologous genes from eight yeast genomes: *S.cerevisiae, S. paradoxus, S. mikatae, S. kudriavavzevii, S. bayanus, S. castellii, S. kluyveri* and *Candida albicans* (Rokas, Williams, King, & Carroll, 2003). The whole dataset can be found online in the supplemental data of Rokas, Williams, King, and Carroll (2003).

RANDOM GENE CONCATENATION SIMULATIONS

As we pointed before, the phylogenetically informative degrees of genes may play an important role in resolving the gene tree and species tree problem, because some genes may contain more informative speciation signals than others. To address this concern, we simulate the species tree inference by using Bayesian phylogeny analysis under three different gene concatenation cases: worst-case, random-case and best-case concatenations. For each case, we simulate 10 trials for each sub-case where gene concatenation number is from 2 to 31. For each trial in the three simulations, genes are randomly selected from the 45 congruent gene set, 61 incongruent gene set or the total 106 genes respectively. The whole simulation took about two week running on an 8-node cluster (16 CPUs) under the Sun Grid Engine (SGE), each node with 1.0 GB memory and 1.5 Ghz CPU.

We classify these 106 multi-species orthologous genes into two categories using Bayesian phylogeny analysis: congruent (good) genes and incongruent (bad) genes. A gene is congruent if its gene tree is the species tree under Bayesian phylogeny inference. Otherwise, the gene is incongruent. We have found that the 106 genes can be classified into 45 congruent and 61 incongruent genes. The worst-case or best-case concatenations combine only congruent or incongruent genes in Bayesian phylogeny analysis under the GTR+Γ model (Felsentein, 2004). Alternatively, the random-case concatenation randomly combines genes under the same conduction. We have conducted 10 simulation trials for each combination case. For example, 10 trials of two randomly selected genes are concatenated for the two-gene combination case under the worst-case, best-case or random-case simulations in the Bayesian phylogeny analysis. Although it would be attractive to have more than 10 trails, e.g., 50 trials for each combination case simulation, it would lead to an extremely prohibitive computing demand due to the high complexity nature of such a simulation.

We introduce a tree credibility measure to evaluate the robustness of the inferred phylogenetic tree. The tree credibility is defined as the product of all posterior probabilities of its inferred branches: $t_p = \prod_{i=1}^{|B_I|} b_p^i$, where the b_p^i is the posterior probability in the i_{th} inferred branch and B_I is the set of all inferred branches. The tree credibility t_p of the phylogenetic tree of a congruent gene is in the interval $(0,1]$. The tree credibility of the phylogenetic tree of an incongruent gene is zero. The satisfactory result in the simulation is that the corresponding average phylogenetic tree credibility reaches the 100%. Correspondingly, the final tree is considered as a species tree with the maximum support, i.e., each branch with 100% posterior probability.

We have the following interesting findings. It needs at least 28 incongruent gene combinations to achieve the satisfactory result. However, it needs only 4 congruent gene and 15 random gene combinations to achieve a species tree with the 100% average tree credibility respectively. Figure 1 shows the species phylogeny with the tree credibility 100% for the eight yeast species.

Our simulation results not only support that different genes have unequal phylogenetic information strongly, but also indicate that the equal-phylogenetic information assumption for each gene in the random gene concatenation method is improper or at least inefficient in the species phylogeny inference. Figure 2 shows the average tree credibility for gene concatenation from 2 to 31 genes in the three simulation cases. It is clear to see that the congruent (good) and random gene concatenations not only use few genes to reach a species tree with the maximum support, but also demonstrate stability to reach the satisfactory result. On the other hand, the incongruent (bad) gene concatenation appears to have large oscillations in the average tree credibility. The standard deviation analysis of the average tree credibility values of the three simulation cases also supports this observation.

Figure 3 illustrates the tree credibility values for each trial from 2 to 31 incongruent gene concatenations, where each horizontal axis represents the number of trials and each vertical axis represents the tree credibility values. Obviously, at least 28 incongruent (bad) genes are needed to

Figure 1. The species phylogeny with 100% tree credibility of eight yeast species

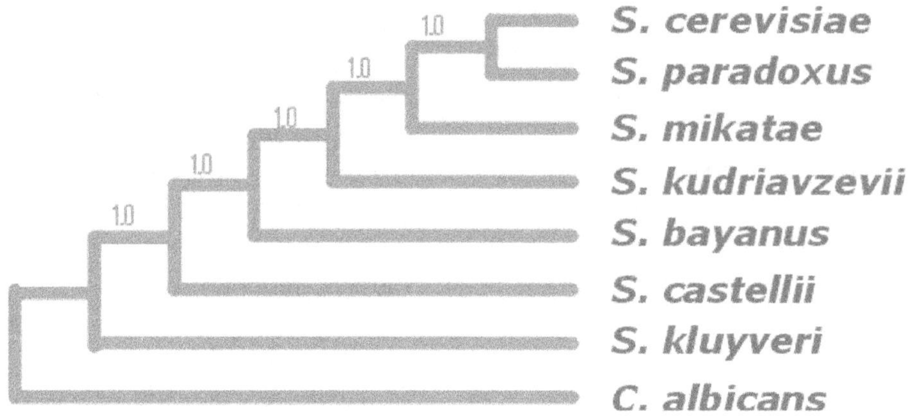

Figure 2. Random gene concatenation simulations for congruent, incongruent and random genes under the Bayesian phylogeny analysis

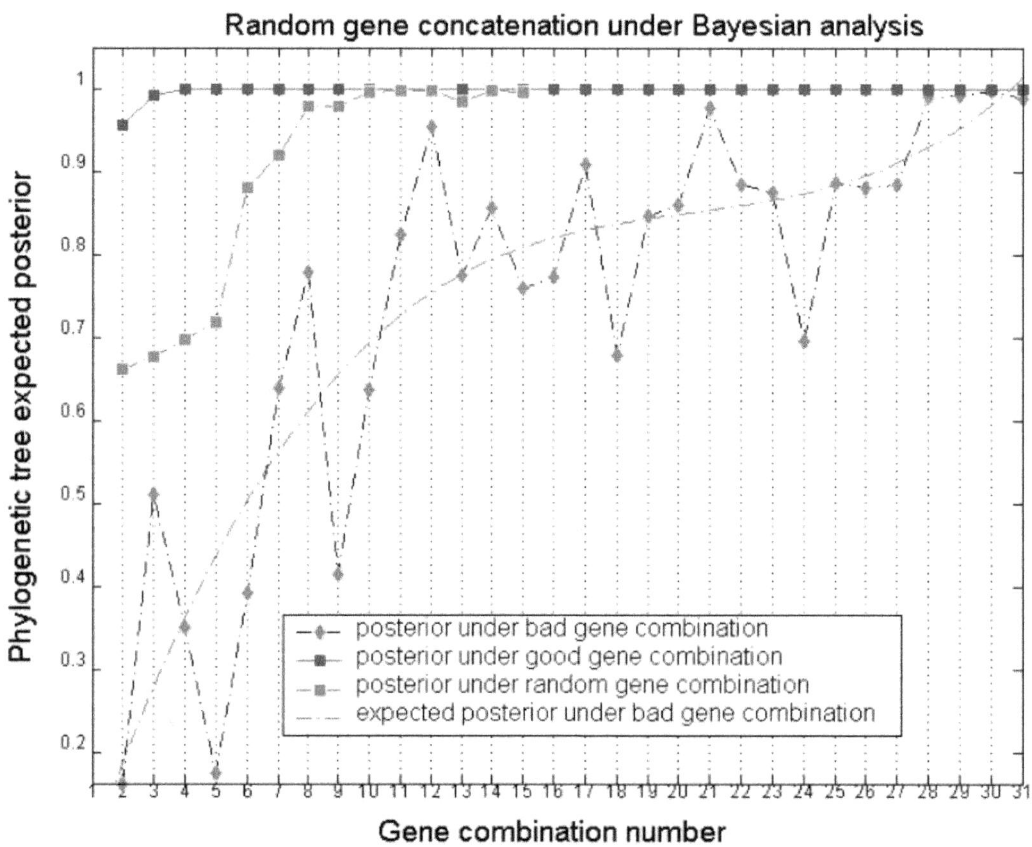

Figure 3. Random gene concatenation simulations for 2 to 31 incongruent genes

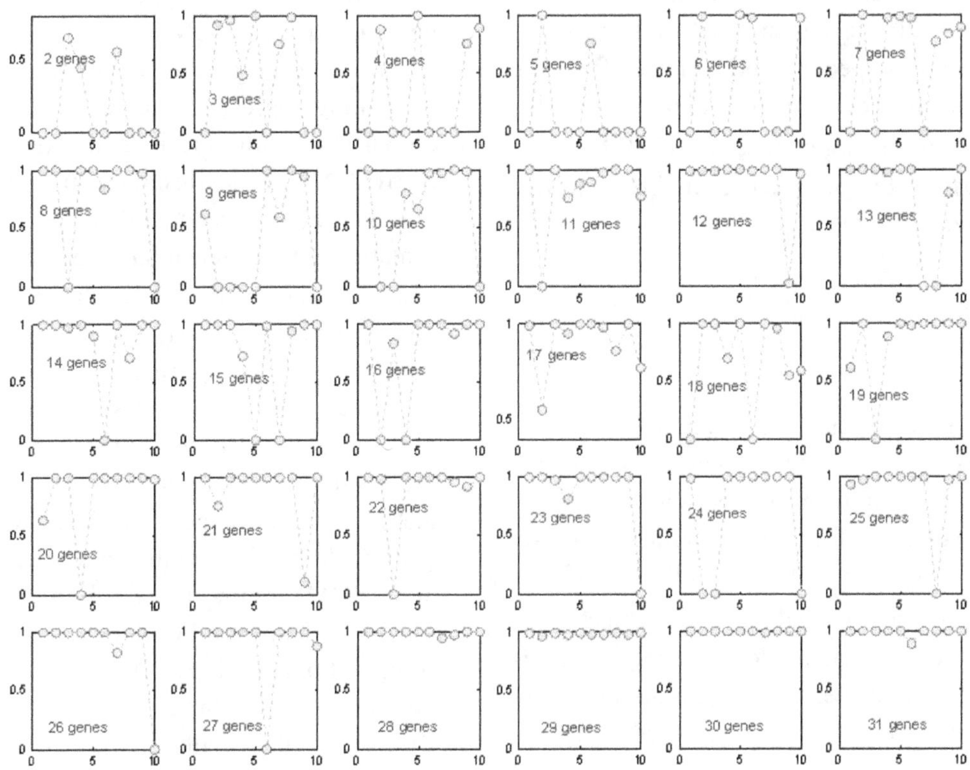

reach a species tree with the maximum support on each branch.

It is clear that this random gene concatenation method reaches its "*worst scenario*", when all concatenation genes are incongruent genes. Since the gene and species tree problem is actually a black box for biologists and investigators have little knowledge about congruence of genes in molecular evaluation, the worst gene combination case is unavoidable in this ad-hoc approach. Although the average species tree credibility increases with the increase of gene concatenation number, a relatively large gene concatenation number demands huge, even prohibitive computing. Furthermore, such an approach will be theoretically infeasible with the increase of number of species, because of the complexities of phylogenetic tree reconstruction (Chor & Tuller, 2006). Thus, we need a robust and systematic approach to resolve the gene tree and species tree problem.

SELF-ORGANIZING MAP (SOM) BASED PHYLOGENETIC INFERENCE

The random gene concatenation method treats all genes uniformly in the phylogeny inference. It would be desirable to use good (congruent) genes or "potentially good" genes, i.e., phylogenetically informative genes in the species tree inference. With this motivation, we propose a self-organizing map (SOM) based phylogeny inference algorithm, i.e., phylogenetic mining, to resolve the gene tree and species tree problem.

The basic idea of this algorithm can be described as follows. Given a set of aligned orthologous multi-species genes selected from the genomes interested: $G = \bigcup g_i$, generally, $|G| \geq 10^2$, a self-organizing map (SOM) is employed to mine the gene set G such that the multi-species genes are clustered as

$$G = \bigcup_{i=1}^{k} S_i, S_i \cap S_j = \varnothing, i \neq j \quad (1)$$

Then phylogenetically or potentially informative genes are selected from each cluster according to the maximum entropy principle to build a phylogenetically informative candidate gene set G_p where $|G_p| = k \ll |G|$. Finally, a phylogenetic reconstruction method $f(\bullet)$, which is selected as the Bayesian phylogeny analysis in this study (Huelsenbeck et al., 2002), is applied to set G_p to infer the final species phylogeny: $f(G_p) \rightarrow T$. We will demonstrate that the final evolutionary tree from this selected set converges to the species tree with the maximum support.

CAN WE CLUSTER MULTISPECIES GENES DIRECTLY?

Multispecies gene clustering is essential in our phylogenetic mining algorithm. However, different genes have different lengths after alignments, it is almost impossible to cluster multispecies genes directly. To cluster genes directly, we have to encode genes into uniform column vectors, which means zeros have to be filled into the encoded vectors for the short genes. The filled zeros will act as missing data in the clustering and it is not advisable to do clustering under an ill-posed condition (Hastie, Tibshirani, & Friedman, 2009).

We present an alternative way to cluster multispecies genes using self-organizing maps (SOM), i.e., cluster the prototypes of the multi-species genes obtained in SOM mining. The prototype of a multi-species gene is a low-dimensional data containing representative features of the high-dimensional multi-species gene. A self organizing map (SOM) is a classic data mining approach that maps a high dimensional data set D to its representative low-dimensional prototype $W \in \mathbb{R}^2$, i.e., $SOM: D \rightarrow W \in \mathbb{R}^2$. It takes a competitive unsupervised learning (self-organization) to partition original data space into a set of corresponding representative prototypes. With the assumption of no prior knowledge about the data to be mined, the unsupervised learning in SOM is a process of feature extraction and data dimension reduction (Ritter, 1995). The details on the self-organizing map (SOM) mining can be found in Appendix. The more details on SOM and its applications in computational biology can be found in (Nikkila, Toronen, Kaski, Venna, Castren, & Wong, 2002; Draghici, 2003; Yanikoglu & Erman, 2002).

SELF-ORGANIZING MAP (SOM) BASED PHYLOGENY INFERENCE ALGORITHMS (PHYLOGENETIC MINING)

The self-organizing map (SOM) based phylogeny inference algorithm, i.e., phylogenetic mining, to resolving the gene tree and species tree problem can be described as following steps. (1) The orthologous multispecies gene selected from genomes of n species $G = \bigcup_i g_i$ is encoded as a numeric matrix $D \in \mathbb{R}^n$, where each column represents corresponding species and a multispecies gene is represented by many rows. (2) The self-organizing map (SOM) mining is conducted on the numeric matrix D to compute the prototype for each species. (3) The prototype of each multi-species, which is denoted as the gene distribution on the SOM plane, is then computed by collecting its "hit-ratios" on the SOM plane. (4) The multi-species genes are clustered by conducting hierarchical clustering for their prototypes. (5) A phylogenetically informative candidate gene set G_p is built by selecting maximum entropy genes from each cluster. (6) A phylogenetic reconstruction algorithm, which is selected as Bayesian phylogeny analysis in this study, is applied to the gene set G_p to infer the final species phylogeny.

In this work, we still use the same dataset used by Rokas, Williams, King, and Carroll (2003) in their random gene concatenation method (Rokas, Williams, King, & Carroll, 2003).The relevant algorithmic implementation details on data encoding, multispecies gene prototype calculation, multispecies gene clustering and multispecies gene entropy calculations are described in the following sections.

Encoding

The orthologous multi-species gene set $G = \bigcup_i g_i$ is a 127026×8 character matrix, which consists of 106 genes of the eight yeast species from genomes of eight yeast species: *Scer*, *Spar*, *Smik*, *Skud*, *Sbay*, *Scas*, *Sklu*, *Calb*. Each column represents a taxon and each row represents a site of a multispecies gene, which usually occupies certain number of rows according to its length. The four orthogonal bases are used to encode '*A*', '*T*',' *C*', '*G*' respectively. We employ orthogonal vectors in the cncoding rather than general integers, e.g., ('1', '2', '3', '4') to represent four nucleotides, because there are no overlapping information in such an encoding format.

$$A = (1,0,0,0)^t, \; T = (0,1,0,0)^t \\ C = (0,0,1,0)^t, \; G = (0,0,0,1)^t \tag{2}$$

The missing nucleotides and gaps in the character matrix are encoded as a vector with four zeros entries. It is worth to note that there are only several missing nucleotides and gaps in the 127026×8 character matrix. After encoding, the input character matrix is encoded as a 508104×8 numeric matrix *D*.

Self-Organizing Map Mining

A self-organization map (SOM) mining is conducted on the numeric matrix *D* to compute the prototype for each species. We employ a sequence learning algorithm in the SOM mining. Although another batch learning algorithm is faster than the sequence learning algorithm employed, it is more likely to encounter local minimum than the sequence learning (Haykin, 1999; Kohonen, 2001). We use principle component analysis (PCA) to initialize all the reference vectors in reference vector matrix *W* in SOM instead of a random initialization for its efficiency (Kohonen, 2001). The neighborhood kernel function is a selected as Gaussian kernel function for its biologically meaningful and fast convergent properties (Kohonen, 2001). Alternatively, we select a SOM plane consisting of 20×20 map units (neurons), because it is desirable to set the number of map units proportional to the square root of the number of data samples (508104), such that the complexity of the SOM learning scales linearly with the number of data sample (Kohonen, 2001). The complexity of each training epoch is $O(nmk)$ where the sample number $n = 508104$, species number $m = 8$ and the number of neurons $k = 400$. The prototypes of species are obtained after the SOM mining with more than 2000 epochs, which are column vectors of the final reference vector matrix $W \in \mathbb{R}^{400 \times 8}$ in SOM. For convenience, we denote the prototypes of eight species as species patterns because they are pattern signatures of eight species on the SOM plane.

Multispecies Gene Distribution Estimation

Just as we pointed out before, the corresponding prototype of a multispecies gene is just its gene distribution on the SOM plane. However, unlike the prototype of a species that can be obtained directly, the multispecies gene distribution on the SOM is computed through collecting the hit-ratios of each multi-species gene.

Traditional methods have encountered hard time in estimating multi-species gene distribution in the sequence space. Given an aligned multispecies gene $x = x_1x_2...x_l$ of *n* species with gene length $|x| = l$ (which a $n \times l$ character matrix), where x_i represents the i_{th} site of the gene, its gene distribution is a collection of probabilities of each site in the sequence space $\sum$, which is the set of all possible combinations of *n-block* nucleotides. The n-block nucleotides are a segment with *n* nucleotides and the total size of the sequence space is $|\sum| = 4^n$. In other words, the gene distribution of a multispecies gene *x* is the probability mass function $p(x) = \{p_1, p_2, ...p_l\}$ of each site in the sequence space $\sum$. The probability of the i_{th} site can be approximated by its relative frequency $p_i = \frac{fr(x_i)}{l}$ where $fr(x_i)$ is the number of occurrence of the i_{th} site in the multispecies gene *x*. However, such an approximation is not reasonable due to the finite sample problem (Durbin, Eddy, Krogh, & Mitchison, 1998), where the sequence space is rather huge and the total aligned gene has relatively short length.

The multi-species gene distribution can be estimated by collecting the hit-rates of the multi-species gene on the SOM plane. For convenience, we define the "hit" and "hit-rate" for a multi-species gene at first.

Hit. If a neuron *j* on the SOM plane is acknowledged as the *best match unit* (BMU) (Haykin, 1999) of a site x_i in a multi-species gene $x = x_1x_2...x_l$ after the SOM mining, we say that x_i hits the neuron *j*; that is, the reference vector on the neuron *j* has the nearest distance to x_i.

Hit-rate. For a neuron *j* on the SOM plane hit by n_j number of sites in a multispecies gene *x* with $|x| = l$ sites after the SOM mining, the *hit-rate* for neuron *j* from the gene *x* is the ratio between n_j and total site number *l*.

The sites of a multispecies gene hit different neurons on the SOM plane in SOM mining according to its nucleotide patterns. Hence, the distribution of the neurons hit by all sites from the multispecies gene on the SOM plane would be a faithful approximation of the underlying distribution of the gene in the sequence space according to the SOM learning mechanism (Haykin, 1999). Alternatively, the distribution of the neurons hit by a multispecies gene can be represented as the hit-rates of all the map units according to the definitions of the hit-rate. Thus, given a multispecies gene $x = x_1x_2...x_l$, its gene distribution on the SOM plane is the probability mass function of each site on the SOM plane that is a low-dimensional feature space, i.e., $p'(y) = \{y_1, y_2...y_k\}$, which is a vector of hit-rates of neurons hit by the multispecies gene. The parameter *k* is the number of neurons on the SOM plane, and y_i is the hit-rate on the ith neuron:

$$y_i = h_i / \sum_{i=1}^{k} h_i,\ i = 1, 2...k \tag{3}$$

The parameter h_i is the cardinality of set s_i, which consists of the sites in the multispecies gene *x* acknowledging the neuron *i* as their best match unit (BMU): $s_i = \{e \mid \arg\min_j \| e - w_j \| = i, j = 1, 2...k\}$, where *e* is a site of the multispecies gene *x* and w_j is the reference vector associative with the jth neuron on the SOM plane. The total number of sites from the multispecies gene *x* is just the total hitting number. The gene distribution on the SOM plane is the gene distribution of the multispecies gene in the low-dimensional feature space induced by SOM mining.

Multispecies Gene Clustering

To cluster the multispecies genes, standard hierarchical clustering is conducted on the gene distributions on the SOM plane, which are the corresponding prototypes of the multispecies genes. Compared with the fuzzy clustering of SOM itself, the hierarchical clustering is more informative because a complete binary tree structure is built between different gene patterns (Hastie, Tibshirani, & Friedman, 1999; Nikkila, Toronen, Kaski, Venna, Castren, & Wong, 2002). The cutoff in the hierarchical clustering can be decided by subjectively specifying the final cluster numbers (e.g., 2 clusters) or using the unified distance matrix (U-matrix) approach. The U-matrix is to visualize the cluster structure on the SOM plane by computing the average distance between the reference vectors of neighboring neurons (Kohonen, 2001). We use both approaches in the multispecies gene clustering.

Multispecies Gene Entropy Estimation and Maximum Entropy Principle Based Gene Selection

The multispecies gene entropy can be viewed as a special case in the block entropy of a DNA sequence where each site can be viewed as a DNA "block" (Durbin, Eddy, Krogh, & Mitchison, 1998). Given a multispecies gene with *n* species and aligned length $l : x = x_1 x_2 ... x_l$, where x_i is the ith site of the gene, its entropy can be defined as $H(x) = -\sum_{i=1}^{|\Sigma|} p(x_i) \log p(x_i)$. The $p(x_i) = p_i$ is the gene distribution of the ith site of the gene in the sequence space Σ ($|\Sigma| = 4^n$). Just as we pointed out before, it would be difficult to estimate p_i in the sequence space due to the finite sample effect (Durbin, Eddy, Krogh, & Mitchison, 1998; Herzel, Ebeling, & Schmitt, 1994; Vinga & Almeida, 2004). However, the multispecies gene entropy value can be estimated by the following formula: $H(x) = -\sum_{i=1}^{k} y_i \log y_i$ because the gene distribution in the sequence space can be approximated faithfully by the gene distribution $p'(y) = \{y_1, y_2 ... y_k\}$ on the SOM plane with *k* neurons. In other words, the multispecies gene entropy can be estimated by using its gene distribution on low-dimensional feature space induced by the SOM mining.

We employ the maximum entropy principle (Haykin, 1999; Hastie, Tibshirani, & Friedman, 2009) to select informative genes from each cluster. Since genes in a same cluster share similar gene distribution on the SOM plane, it is reasonable to believe they may share same or similar phylogenetic properties. Thus, we only need pick one gene from each cluster to build the phylogenetically informative candidate gene set G_p, i.e., concatenate the locally largest entropy genes in each cluster to build G_p. On the other hand, we predict that a gene with large entropy value will contain more phylogenetic information, because a gene with high entropy will contain more information than a gene with low entropy value according to the maximum entropy principle. In other words, a higher entropy gene is more likely to be a phylogenetically informative gene whose gene tree is close or congruent to the final species tree with good probability support on each branch.

RESULTS

Since species patterns (the prototypes of species) are corresponding column vectors in the final reference vector matrix $W \in \mathbb{R}^{400 \times 8}$, we can visualize the species patterns by drawing each final reference vector associated with each neuron on the SOM plane by taking advantage of the power-

Figure 4. Visualizations of eight species patterns on a 20×20 SOM plane

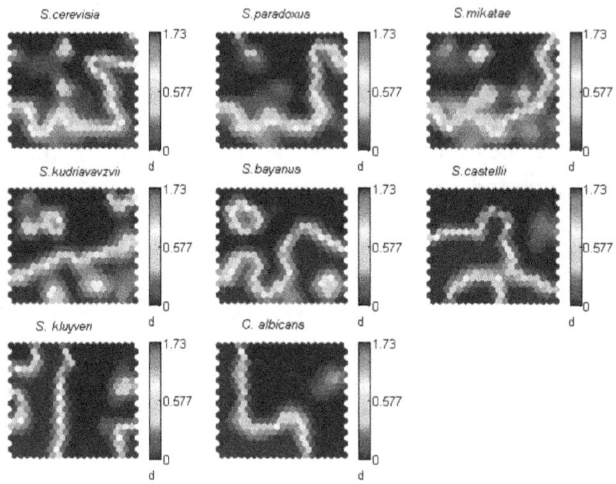

ful visualization capability of SOM. Figure 4 illustrates the eight species patterns on the SOM plane with 20×20 neurons. To the best of our knowledge, this is the first work in species pattern visualizations.

It is interesting to discover some species evolutionary relationships from the species patterns visualization. The first two species patterns share more mutual features than others indicating the first two species have closer evolutionary relationships than other taxa. The 3rd species is the taxon nearest to the 1st and 2nd species since its pattern is most similar to the first and second species pattern. The 4th species pattern is nearest to the first three species patterns and the 5th species pattern is mostly similar to the 4th species pattern. It is interesting to see that the 6th and 7th species patterns both consist of two disjoint parts. So it is reasonable to conclude that the 6th and 7th species have closer evolutionary distances among the eight species. The last species pattern apears to be much different from the others (its corresponding histogram plot is also much different from those of others). It is reasonable to think the last species is an out-group species. Actually, *C. albcians* is the outgroup species in the species phylogeny (Rokas, Williams, King, & Carroll, 2003).

Can the species prototypes be used to build a phylogenetic tree for the species by clustering the species prototypes? It is possible to take advantage of the species prototypes to build a phylogenetic tree. However, the clustering of the species prototypes is equivalent to employing the UPGMA distance method, which has a molecular clock and additivity assumptions (Nei & Kumar, 2000). Thus, there is no guarantee for such an approach to produce a robust tree, although this approach seems to be a simple and free of evolutionary models. In other words, the species prototypes can help us to "sketch" the evolutionary relationships between species. However, we still need the robust phylogenetic reconstruction methods to build an exact phylogeny.

MULTISPECIES GENE CLUSTERING ANALYSIS AND VALIDITY

We have found that the unified distance matrix (U-matrix) indicated 18 naturally grouped clusters for the prototypes of 106 genes. Figure 5 shows

Figure 5. The U-matrix indicating 18 naturally grouped clusters

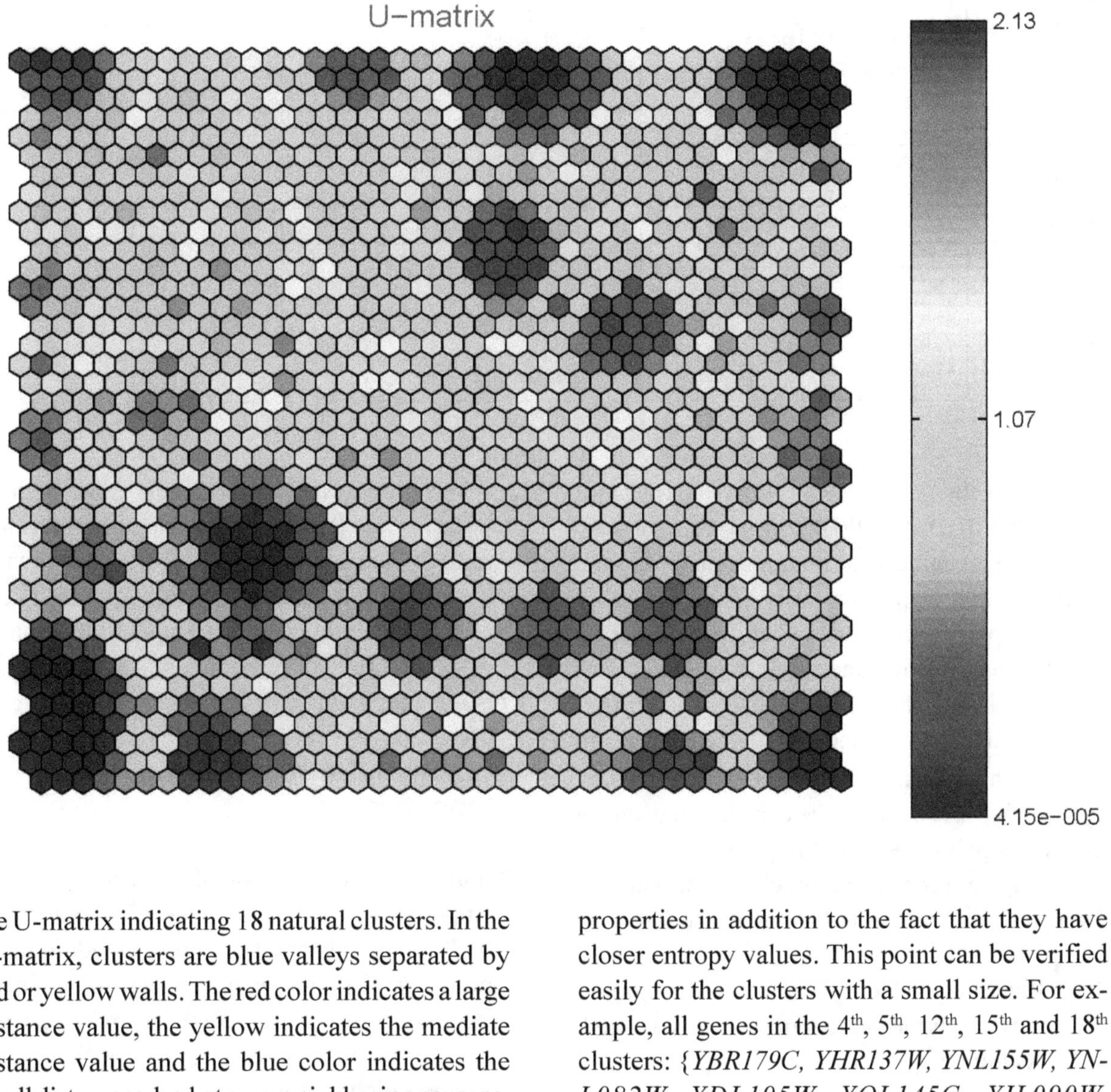

the U-matrix indicating 18 natural clusters. In the U-matrix, clusters are blue valleys separated by red or yellow walls. The red color indicates a large distance value, the yellow indicates the mediate distance value and the blue color indicates the small distance value between neighboring neurons.

Table 1 shows the gene entries in the 18 naturally grouped clusters obtained from the hierarchical clustering. It is possible that genes belonging to a same cluster share similar biological or phylogenetic properties. Hence, clustering multispecies genes before phylogenetic analysis helps us to discriminate relationships between the genes and overcome the "blindness" in the phylogeny reconstruction. Moreover, it is also interesting to see that each cluster shares similar phylogenetic properties in addition to the fact that they have closer entropy values. This point can be verified easily for the clusters with a small size. For example, all genes in the 4th, 5th, 12th, 15th and 18th clusters: {*YBR179C, YHR137W, YNL155W, YNL082W, YDL195W, YOL145C, YIL090W, YPL210C, YKL034W, YER005W, YGR194C, YJL100W, YLR389C*} are actually incongruent genes whose gene trees are not congruent to the species phylogeny.

In addition to clustering multispecies genes into 18 naturally grouped clusters, we grouped the multispecies genes into clusters according to subjectively set cutoff (e.g., cutoff=2,3). To verify such a SOM-based clustering is also meaningful, we have designed a hypothesis test to

Table 1. 18 gene clusters of 106 multispecies genes

Cluster	1	2	3	4	5	6
Genes	3,9,13,14,49, 55,61 65,72, 78,89, 93	25,42, 43 45,62, 80 84, 104	6,23, 36,71 99	11,50 66,102	21,73 88	81
Cluster	**7**	**8**	**9**	**10**	**11**	**12**
Genes	7, 17,30 32,59,98	90	18	63	75	83, 94
Cluster	**13**	**14**	**15**	**16**	**17**	**18**
Genes	8,26,28 48,57,64 67,70,74 79,82,85 87,91,96 101,105	1,2,19,20 22,27,29 41,52,54 56,68,69 76,77,97 103,106	35,46 53,58	51	4,5,10 12,5,16, 24,31,33 34,37,38 39,40,86 92,95,100	44, 47

verify the validity of the gene clustering. The null hypothesis H_0 is that the SOM-based clustering is as same meaningful as random gene prototype selection, that is, the similarity of same number of genes picked from one cluster should not be higher than that of same number of genes randomly picked from the total gene set. The alternative hypothesis H_1 is the complement of H_0. In the hypothesis test, we define a measure *similarity*, which is the mean of the correlation distance between a pair of gene prototypes to measure the similarity between genes. We choose the similarity as the test statistic and follow these steps.

1. Randomly select a cluster C and a number δ which is the number of genes to be tested. A counter η is set as zero.
2. Randomly pick two gene sets s_1 from C and s_2 from the total gene set G. The number of genes in s_1, s_2 is δ.
3. Compute test statistic similarity t_1, t_2 for gene sets s_1, s_2.
4. Compare two similarities: if $t_1 > t_2$ holds, increase the counter η by 1.
5. Repeat 1-4 at least 10,000 iterations and compute the p-value by dividing the counter η by total iterations.

We compute the *p-values* for both hierarchical clustering and K-means clustering for 106 genes. We found there is no difference between the two clustering approaches after SOM mining. The p-value for hierarchical clustering and k-means is 0.9566 and 0.9548 respectively. Our hypothesis testing supports the validity of the clustering.

Maximum Entropy Gene Concatenations

Figure 6 shows the plot of the entropy values computed for 106 genes on the SOM plane with 20×20 map units and their corresponding histogram plot. It is interesting to see that gene entropy values are not subject to a standard normal distribution. Just as we pointed out before, we believe a high entropy gene is more likely to be a phylogenetically informative gene. Thus, we concatenate the local maximum entropy genes from each cluster to build the phylogenetically informative candidate gene set G_p and conduct Bayesian phylogeny analysis to the concatenated genes. It is worth to note that the maximum entropy gene is a "local maximum" gene whose entropy is the maximum for all genes in a particular cluster instead of among the total gene set. Compared with the general random gene concat-

Figure 6. The gene entropy values of 106 genes and its histogram plot

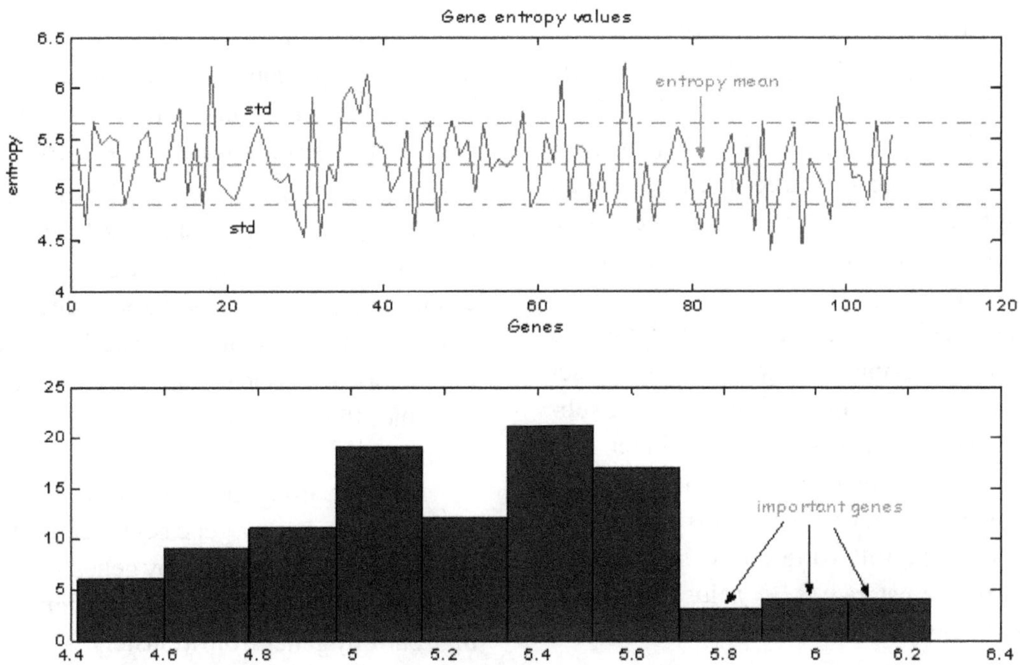

enation, there is no experimental gene concatenation number computing needed if this approach is applied to a different data set. Furthermore, the maximum entropy gene concatenation based on clustering prevents from the "data bias" problem if the total gene set obtained has over-representation of one or several types of genes due to data acquisition issues. This property may ameliorate over-support for some branches in the final consensus tree. We have the following experimental cases for this SOM-based species phylogeny inference algorithm.

EXPERIMENTS

According to our SOM-based phylogeny inference method, we build the gene set G_p by selecting the 'local' maximum entropy gene from each cluster: {89 104 71 102 21 81 17 90 18 63 75 83 91 58 106, 51 38 47} ({*YIL125W, YNL220W, YOL049W, YNL082W, YDL195W, YPL169C, YDL116W, YJL085W, YDL126C, YMR186W, YOR158W, YPL210C, YJL087C, YLR253W, YNR008W, YIL109C, YFR044C, YGR285C*}) among the 18 clusters. The mean entropy of this gene set is 5.2325 bits, which is higher than the average gene entropy: 5.2111. The gene tree inferred from the Bayesian phylogeny analysis under GTR+Γ model is the species tree where the posterior probability of 100% is on each inferred branch (Figure 1).

The result may not be surprising for us because $|G_p| = 18$ is near to the experimental number obtained in the original random gene concatenation method, and it is also > 15, the least gene number (15) to reach a species phylogeny with the maximum support in the random-case Bayesian phylogeny simulation. To response such a concern, we decrease G_p by only selecting the maximum entropy genes from 8 clusters which have higher mean entropy than the mean entropy of all gene clusters, i.e., G_p consists of genes {89, 18, 104, 71, 51, 106, 38, 58} ({*YIL125W, YDL126C, YNL220W, YOL049W, YIL109C, YNR008W, YFR044C, YLR253W*}) with average entropy 5.8218. The final tree inferred through the eight gene concatenation is still a species tree where each inferred branch has 100% posterior probability support.

Just as we said before, we have employed the hierarchical clustering to cluster the total gene set G into an arbitrary number of clusters with subjectively specified cut-off values. Under this situation, we also pick these local maximum entropy genes from each cluster. This experiment would answer the following query: suppose the gene set G is clustered as $\# j$ clusters, which genes are the most informative to build a robust species tree?

If we treat the whole gene set as a cluster (i.e., no clustering work), the maximum entropy gene: gene 18 (*YDL126C*) will be the only gene in G_p and its corresponding tree credibility $t_p = 100\%$ after Bayesian phylogeny analysis. Interestingly, genes with large K-L distances to the global maximum gene (*YDL126C*) are more likely to be incongruent genes. For example, among the four genes: {51,81,90,75} ({*YIL109C, YPL169C, YJL085W, YOR158W*}) with the largest K-L distances to the maximum entropy gene, only gene 90 (*YJL085W*) among them is a congruent gene. On the other hand, there are 8 genes: {*YIL125W, YMR186W, YNL220W, YHR019C, YCR017C, YBL015W, YPR074C, YGL253W*} are congruent genes among the 10 genes with the smallest K-L distances to the global maximum gene. For example, gene 89 and gene 63 (*YIL125W, YMR186W*) are two genes nearest to the maximum entropy gene in the K-L distance. Both of them are congruent genes with high entropy values, and their corresponding tree credibility values are *100%* and *98%* respectively.

Similarly, these phylogenetically informative candidate gene set G_p consists of $\# j$ local maximum entropy genes selected from $\# j$ clusters, if the total gene set G is clustered into $\# j$ clusters. We have found that the corresponding phylogeny inferred from each G_p through Bayesian phylogeny analysis is a species tree with 100% tree credibility. Table 2 shows all detailed information about the maximum entropy gene selection when the input 106 orthologous genes are partitioned from 1 to 14 clusters. The 100% tree credibility for each concatenation case strongly supports the effectiveness of our phylogenetic mining method.

However, it is worthy pointing out that such a maximum entropy gene selection requires that the 'local' maximum entropy genes selected from different clusters. Randomly selecting several high entropy genes from a cluster may not produce good results because more genes selected from a same cluster will make the sampling bias occurrence, which may lead to "over support" for specific branch patterns in the final tree inference.

For convenience, we call our local maximum entropy gene concatenation as a phylogenetically informative gene concatenation. We compared performance of this concatenation with those of the best-case and random-case gene concatenations under Bayesian phylogeny analysis. Just as before, for each gene combination case in the best-case and random-case simulations, we generated 10 trials of randomly selected congruent genes or general genes for them respectively. We have found that the performance of the phylogentically informative gene concatenation is even better than that of the good gene combination. Figure 7 shows the performance comparisons on the three gene concatenation cases. It is obviously that our phylogenetically informative gene concatenation method outperforms the congruent (good) and random gene concatenations from its tree credibility values. Since the species phylogeny is generally unknown to investigators, we suggest that our

Table 2. Maximum entropy gene selection for an arbitrary number of clusters

Cluster#	Genes in G_p	Gene length of G_p	Average entropy of G_p	Tree credibility
1	18	2136	6.1918	1.0
2	71 18	3729	6.1612	1.0
3	71 18 38	4548	6.1474	1.0
4	71 17 38 18	6027	5.8146	1.0
5	89 71 17 38 18	8853	5.7792	1.0
6	89 58 71 17 38 18	10140	5.7637	1.0
7	17 75 89 38 58 71 18	10896	5.6071	1.0
8	91 17 75 89 38 58 71 18	12201	5.5248	1.0
9	91 17 75 89 38 58 71 18 47	13437	5.4267	1.0
10	91 17 75 89 38 58 71 18 47 51	15327	5.4299	1.0
11	91 17 75 89 38 58 71 18 47 51 106	16503	5.4362	1.0
12	91 17 75 89 38 58 71 18 47 51 106 21	17757	5.3866	1.0
13	91 17 75 89 38 58 71 18 47 51 106 21 83	19008	5.3190	1.0
14	91 17 75 89 38 58 71 18 47 51 106 21 83 63	21066	5.3697	1.0

method can provide a systematic solution to the gene tree and species tree problem.

MORE DATA OR LESS DATA?

The general belief for most biologists is that more genes will bring more information contributing to resolving the gene tree and species tree problem (Snel, Huynen, & Dutilh, 2005; Rokas, Williams, King, & Carroll, 2003). However, we believe that phylogenetically informative genes would be more important in inferring a robust species phylogeny. We give a counter example to the Roaks *et al* 's random gene concatenation method (Rokas, Williams, King, & Carroll, 2003). We conducted Bayesian phylogeny analysis for 20 'random' genes selected from the 106 widely distributed orthogous genes: {7 15 20 25 29 33 34 35 53 60 66 68 96 97 13 19 34 38 39 42} ({*YBR070C, YDL006W, YDL166C, YDR054C, YDR361C, YDR531W, YEL037C, YER005W, YJL100W, YML021C, YNL155W, YNL248C, YKR071C, YKR099W, YCL054W, YDL148C, YEL037C, YFR044C, YGL001C, YGL225W*}). The mean entropy of this gene set is 5.2113, which is slightly smaller than the mean entropy of total genes: 5.2122. There are 8 genes with entropy values greater than the average entropy and 12 genes with entropy values less than the average entropy. The tree credibility of the phylogeny inferred from these 20 genes is only 67.22%, which is a rather weak support for a species tree. On the other hand, our maximum entropy gene selection method can reach a species phylogeny with 100% tree credibility using only one gene

Figure 7. Comparing the performance of random gene concatenation and phylogenetically informative gene concatenation approaches

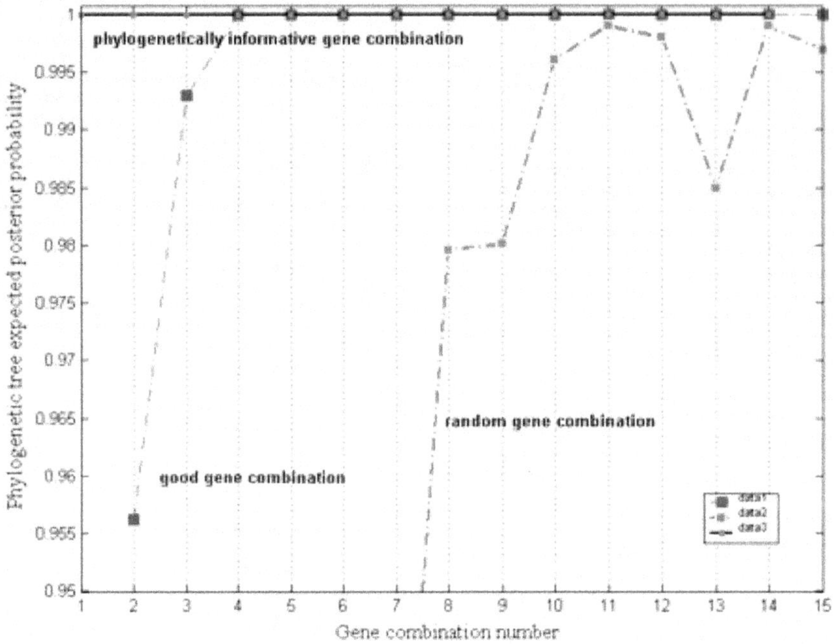

(*YDL126C*). This counter example strongly supports our belief that more genes do not always mean a robust species phylogeny reconstruction.

Another widely accepted belief is that missing data will play a negative role in the phylogenetic tree reconstruction. We believe that data missing may not always affect phylogeny inference negatively, if the gene dataset is sufficiently informative. In some special case, it even may contribute to improving phylogeny inference because data missing increases the gene data entropy value.

We did following experiments to demonstrate this point. In the first experiment, we show that data missing can even achieve better phylogeny inference results. We concatenate gene 1 (*YAL053W*) and gene 2 (*YAR007C*) to infer a species phylogeny. The gene 1 (*YAL053W*) is a congruent gene with 1701 sites (entropy value: 5.3399), whose specie tree has 75% tree credibility. The gene 2 (*YAR007C*) is an incongruent gene with 1458 sites (entropy value: 4.6394). The phylogenetic tree of the concatenation of the two genes is a species tree with tree credibility 54%. Except that the branch connecting *S. kluyveri* and *S. castellii* has 54% posterior probability, all the other inferred branches have 100% posterior probability support. Then, we made the dataset concatenated by the two genes uniformly data missing 15% and 25% percents and conducted Bayesian phylogeny analysis under the same parameter setting and evolution model selection. It is surprising to find that the branch connecting *S. kluyveri* and *S. castellii* under 15% data missing has 99% posterior probability support. Similarly, this branch has 96% posterior probability support under 25% data missing (Just as before, all other branches have 100% posterior probability support).

In the second experiment, we show a robust phylogeny inference can even be achieved under data missing cases, i.e., they would not affect the resolution of the final tree inferred. First, we show that the phylogenetic tree inferred by concatenating gene 1 (*YAL053W*) and gene 18 (*YDL126C*) is a species phylogeny with 100% tree credibility. Similarly, the phylogenetic tree inferred by these following genes: {1,3,4,7,8,15,16,18} is also a species phylogeny with 100% tree credibility. Then, we make the 1st gene concatenation dataset 15% and the 2nd gene concatenation dataset 50% data missing uniformly and build a phylogenetic tree using the two datasets with missing data through the Bayesian phylogeny analysis. It is also surprising to see that both of them attain a species tree with 100% tree credibility. This is maybe because that the two original gene sets are sufficiently informative and the 15% and 25% data missing are not enough to affect their phylogeny inference.

CONCLUSION

In this paper, we provide an analytical solution to resolving gene tree and species tree problem from the data mining viewpoint. Our results show that this approach is more robust than the random gene concatenation approach. This study not only first presents a species pattern visualization approach, but also provides a multispecies gene clustering and entropy estimation approach using SOM mining. Moreover, we demonstrate that missing data may not be a negative factor in the phylogenetic tree building and more genes does not always lead to a better phylogenetic resolution.

Some interesting issues, worthy of additional investigation, arise from these results. Even though our analyses demonstrate the utility of maximum entropy genes in phylogenetic reconstruction, we have not yet found any clear biological property of these genes that is correlated with this measure. From a data-gathering perspective, it is unknown if this method will be useful in specifying additional genes and/or taxa to be targeted for acquisition of new sequence data. That is, the *predictive* utility of this approach still needs further testing. If this method allowed the transfer of predictions of phylogenetic utility of particular genes from one (e.g., well-sampled) group to another, this feature alone would provide considerable benefit to phylogenomic analyses. If generalizable, we plan to take advantage of the powerful feature extract capability of SOM mining and entropy theory to address paralogy and orthology detection problem. The detection of paralogy and orthology is a key problem in phylogenomics but still in its naïve stage (Koonin, 2005; Fitch, 2000). We expect the detection of the paralogous and orthologous genes can be conducted in their corresponding feature spaces and by means of entropy estimations. We are also interested in integrating our phylogentic mining approach to the Bayesian phylogeny analysis to explore more powerful tree reconstruction algorithms. Moreover, we plan to investigate novel evolutionary models based on the entropy theory to provide a theoretical support for our simulation results.

ACKNOWLEDGMENT

Author sincerely thanks Dr. John M. Logsdon Jr. and Dr. Andrew Schurko at the University of Iowa for their computing facility support and valuable discussions. The author also wants to thank Dr. Ying Liu at the University of Texas, Dallas, and Dr. Xiao-li Li, the editor-in-chief, for their valuable suggestions to improve this paper.

REFERENCES

Chor, B., & Tuller, T. (2006). Finding a maximum likelihood tree is hard. *Journal of the ACM, 53*(5), 722–744. doi:10.1145/1183907.1183909

Delsuc, F., Brinkmann, H., & Philippe, H. (2005). Phylogenomics and the reconstruction of the tree of life. *Nature Reviews. Genetics*, *6*, 361–375. doi:10.1038/nrg1603

Draghici, S. (2003). *Data Analysis Tools for DNA Microarrays*. Virginia Beach, VA: Chapman & Hall/CRC.

Durbin, R., Eddy, S., Krogh, A., & Mitchison, G. (1998). *Probabilistic models of proteins and nucleic acids*. Cambridge, UK: Cambridge University Press.

Felsentein, J. (2004). *Inferring Phylogenies*. Sunderland, MA: Sinauer Associates.

Fitch, W. (2000). Homology a personal view on some of the problems. *Trends in Genetics, 16*, 227–231. doi:10.1016/S0168-9525(00)02005-9

Hahn, M. (2007). Bias in phylogenetic tree reconciliation methods: Implications for vertebrate genome evolution. *Genome Biology, 8*, R141. doi:10.1186/gb-2007-8-7-r141

Hastie, T., Tibshirani, R., & Friedman, J. (2009). *The elements of statistical learning: data mining, inference and prediction* (2nd ed.). Berlin: Springer Verlag.

Haykin, S. (1999). *Neural Networks: A Comprehensive Foundation* (2nd ed.). Upper Saddle River, NJ: Prentice-Hall.

Herzel, H., Ebeling, W., & Schmitt, A. O. (1994). Entropies of biosequences: the role of repeats. *Physical Review, 50*, 5061–5071.

Huelsenbeck, J. (1995). The robustness of two phylogenetic methods: Four-taxon simulations reveal a slight superiority of maximum likelihood over neighbor joining. *Molecular Biology and Evolution, 12*, 843–849.

Huelsenbeck, J. (2002). Potential applications and pitfalls of Bayesian inference of phylogeny. *Systematic Biology, 51*, 673–688. doi:10.1080/10635150290102366

Huelsenbeck, J. P., & Ronquist, F. (2001). MRBAYES: Bayesian inference of phylogeny. *Bioinformatics (Oxford, England), 17*, 754–755. doi:10.1093/bioinformatics/17.8.754

Kohonen, T. (2001). *Self-Organizing Maps* (3rd ed.). Berlin: Springer Verlag.

Koonin, E. (2005). Orthologs, paralogs, and evolutionary genomics. *Annual Review of Genetics, 39*, 309–338. doi:10.1146/annurev.genet.39.073003.114725

Liu, L. (2008). Estimating species trees using multiple-allele DNA sequence data. *Evolution; International Journal of Organic Evolution, 62*, 2080–2091. doi:10.1111/j.1558-5646.2008.00414.x

Ma, B., Li, M., & Zhang, L. (2000). From Gene Trees to Species Trees. *SIAM Journal on Computing, 30*, 729–752. doi:10.1137/S0097539798343362

Moret, B., Tang, J., & Warnow, T. (2005). Reconstructing phylogenies from gene-content and gene-order data. In Gascuel, O. (Ed.), *Mathematics of Evolution and Phylogeny* (pp. 321–352). Oxford, UK: Oxford University Press.

Nei, M., & Kumar, S. (2000). *Molecular Evolution and Phylogenetics* (2nd ed.). London: Oxford University Press.

Nikkila, J., Toronen, P., Kaski, S., Venna, J., Castren, E., & Wong, G. (2002). Analysis and visualization of gene expression data using self-organizing maps. *Neural Networks, 15*, 9530–9966. doi:10.1016/S0893-6080(02)00070-9

Oliver, J. C. (2008). AUGIST: inferring species trees while accommodating gene tree uncertainty. *Bioinformatics (Oxford, England), 24*, 2932–2933. doi:10.1093/bioinformatics/btn556

Rasmussen, M., & Kellis, M. (2007). Accurate gene-tree reconstruction by learning gene- and species-specific substitution rates across multiple complete genomes. *Genome Research, 17*, 1932–1942. doi:10.1101/gr.7105007

Ritter, H. (1995). *Self-organizing feature maps: Kohnnen maps. The handbook of Brain Theory and Neural Networks*. Cambridge, MA: MIT Press.

Rokas, A., Williams, B., King, N., & Carroll, S. (2003). Genome-scale approaches to resolving incongruence in molecular phylogenies. *Nature, 425*, 798–804. doi:10.1038/nature02053

Snel, B., Huynen, M., & Dutilh, B. (2005). Genome trees and the nature of genome evolution. *Annual Review of Microbiology, 59*, 191–209. doi:10.1146/annurev.micro.59.030804.121233

Thomson, R. (2008). Developing markers for multilocus phylogenetics in non-model organisms: A test case with turtles. *Molecular Phylogenetics and Evolution, 49*, 514–525. doi:10.1016/j.ympev.2008.08.006

Vesanto, J., & Alhoniemi, E. (2000). Clustering of the Self-Organizing Map. *IEEE Transactions on Neural Networks, 11*, 586–600. doi:10.1109/72.846731

Vinga, S., & Almeida, J. (2004). Renyi continuous entropy of DNA sequences. *Journal of Theoretical Biology, 231*, 377–388. doi:10.1016/j.jtbi.2004.06.030

Whitaker, J., McConkey, G., & Westhead, D. (2009). The transferome of metabolic genes explored: analysis of the horizontal transfer of enzyme encoding genes in unicellular eukaryotes. *Genome Biology, 10*, R36. doi:10.1186/gb-2009-10-4-r36

Yanikoglu, B., & Erman, B. (2002). Minimum Energy Configurations of the 2-Dimensional HP-Model of Proteins by Self-Organizing Networks. *Journal of Computational Biology, 9*(4), 613–620. doi:10.1089/10665270276027734 5

APPENDIX: PRINCIPLE OF SELF-ORGANIZING MAPS

A self-organing map (SOM) consists of an input dataset X , a SOM plane P , which is the input and also the output plane, and an unsupervised learning algorithm $l : SOM = (X, P, l)$. The input data set X is a $n \times m$ matrix. Each column data in X is called a variable or component and each row data is referred as a sample.

The SOM plane P is generally a set of neurons (map units) placed on a two dimensional lattice. For each neuron i ($i = 1, 2...k$) on the SOM plane, there is a reference vector w_i with same dimensions as each sample. A reference vector matrix W can be obtained by collecting all reference vectors w_i ($i = 1, 2...k$) on the SOM plane. Obviously, the reference matrix W is a $k \times m$ dimensional matrix. After the SOM finishes learning, the reference vector matrix W stores prototypes of the original input data matrix X that is the extracted feature-data of the input dataset.

As a nonlinear feature extraction algorithm, the SOM maps a high dimensional dataset X in the input space Ω, into its prototype W in the feature space $\Lambda \subset R^2$ by an unsupervised competitive learning mechanism $SOM : X \mapsto W \in \Lambda$. The self-organizing learning consists of loops of three stages: competition, cooperation and adjusting, in addition to the initialization stage at the beginning of the unsupervised learning.

In the initialization stage, the reference vector in each neuron on the SOM plane P is initialized randomly or by principal component analysis (PCA). The PCA method assumes we have prior knowledge about the reference vector matrix W and picks the first two maximum principal components of training dataset X to initialize it. The PCA based initialization leads better performance in the SOM learning than the general random initialization method (Haykin, 1999).

In the competition stage, the SOM queries all neurons on the SOM plane to find a candidate j^* with closest reference vector W_{j^*} to a sample $x \in X$ according to a specific distance measure, where default is the Euclidean measure: $d(x - W_{j^*}) = \min_i\{d(x, W_i)\}$, Then the candidate j^* , the winning neuron in the competition, will be identified as the best match unit (BMU) for the sample x .

$$W_i(t+1) = W_i(t) + \alpha(t) h_{j^* i}(t)(x - W_i(t)) \quad (4)$$

In the adjusting stage, the learning rate is adjusted and the neighborhood radius decreases with respect to time exponentially to localize reference vector matrix iterations: $\sigma(t) = \sigma(0) e^{\frac{-t}{\tau_1}}$, $\alpha(t) = \alpha(0) e^{\frac{-t}{\tau_2}}$ at *t=0,1,2...* The initial learning rate and neighborhood size are denoted as $\alpha(0)$ and $\sigma(0)$ respectively, where τ_1 and τ_2 are preselected time constants.

These three stages are repeated in the unsupervised learning until the reference vector matrix W converges. Actually, the SOM learning is equivalent to a gradient based optimization algorithm. If the objective function is defined as the local energy function at time *t* for a neighborhood centered in the winning neuron j : $e(t) = \sum_{i=1}^{k} h_{ji}(t) \, || \, x - W_i(t) \, ||^2$, where the distance measure is any specified distance metric. Thus the iteration scheme in the cooperation stage is equivalent to finding the optimal reference vector in a neuron to minimize the specified local energy function specified by the following gradient optimization $W_i(t) = W_i(t) - \alpha(t)I \frac{\partial e(t)}{\partial W_i(t)}$, where $\alpha(t)I$ is the learning rate matrix and I is an identity matrix. Similarly, the global energy function $E(t)$ can be defined as the sum of the all local energy for all samples in the training data set:

$$E(t) = \sum_{x \in X} \left(\sum_{i=1}^{k} h_{ji}(t) ||x - W_i(t)||^2 \right) \tag{5}$$

The computing complexity of one epoch of training to minimize the global energy function is $O(nmk)$. The self-organizing learning algorithm described in SOM mining is called the sequence learning algorithm. There is an acceleration learning algorithm called the batch learning proposed by Kohonen (2001) to speed up the sequence learning. In the batch learning, total data samples are involved in each update of the reference vector matrix W. Compared with the sequential learning, the batch learning algorithm converges fast and is free from learning rate adjustments but is more likely to trap in a local minimum (Kohonen, 2001).

This work was previously published in the International Journal of Knowledge Discovery in Bioinformatics, Volume 1, Issue 2, edited by Xiao-Li Li and See-Kiong Ng, pp. 29-49, copyright 2010 by IGI Publishing (an imprint of IGI Global).

Section 5
Biological Data Mining for Healthcare

Chapter 18

$SPCC_{TDM}$, a Catalogue for Analysis of Therapeutic Drug Monitoring Related Contents in the Drug Prescription Information

Sven Ulrich
Pharmaceutical Consultant, Germany

Pierre Baumann
University of Lausanne, Switzerland

Andreas Conca
Regional Hospital of Bolzano, Italy

Hans-Joachim Kuss
University of Munich, Germany

Viktoria Stieffenhofer
University of Mainz, Germany

Christoph Hiemke
University of Mainz, Germany

ABSTRACT

Therapeutic drug monitoring (TDM) has consistently been shown to be useful for optimization of drug therapy. For the first time, a method has been developed for the text analysis of TDM in SPCs in that a catalogue SPC-Content$_{TDM}$ (SPCC$_{TDM}$) provides a codification of the content of TDM in SPCs. It consists of six structure-related items (dose, adverse drug reactions, drug interactions, overdose, pregnancy/breast feeding, and pharmacokinetics) according to implicit or explicit references to TDM in paragraphs of the SPC, and four theory-guided items according to the information about ranges of plasma concentrations and a recommendation of TDM in the SPC. The catalogue is regarded as valid for the text analysis of SPCs with respect to TDM. It can be used in the comparison of SPCs, in the comparison with medico-scientific evidence and for the estimation of the perception of TDM in SPCs by the reader. Regarding the approach as a model of text mining, it may be extended for evaluation of other aspects reported in SPCs.

DOI: 10.4018/978-1-4666-1785-8.ch018

INTRODUCTION

Therapeutic drug monitoring (TDM), i.e., the assay of plasma concentration of drugs and metabolites during drug therapy, is a practical application of pharmacokinetics. It is helpful when alternative methods of prediction of therapeutic response, adverse reactions and toxicity are limited and if a certain minimal duration of treatment is expected, mainly long-term treatment. The indication of TDM for an individual patient and in a special clinical situation is given in case of non-response to treatment, adverse drug reactions or symptoms of intoxication at normal dose, suspicion of compliance problems, pharmacokinetic drug interactions or pharmacogenetic deviations. TDM is commonly used today for anticonvulsive drugs, immunosuppressant drugs, psychopharmaceuticals, aminoglycoside antibiotic drugs and cardiac glycosides (Baumann et al., 2004; Ensom et al., 1998; Shenfield, 2001).

The prescription information (summary of products characteristics, SPC) of a drug preparation is developed by pharmaceutical companies according to requirements of regulatory authorities. SPCs are prepared for the information of doctors and pharmacists and they are part of the marketing authorisation of a drug. The content of SPCs consists of the results of pre-clinical and clinical studies, post-marketing pharmacovigilance data and information derived from the medico-scientific literature. Pharmacologic and clinical data also include basic pharmacokinetic information. Information on TDM should be included at least for drugs with established TDM. It has to be recognised in this regard that SPCs strongly influence drug therapy in clinical practice. Deviations of SPC-content from medico-scientific evidence may result in sub-optimal treatment. Package inserts of drug preparations are very similar to SPCs, however, they are prepared for patients.

The systematic evaluation of the content of SPCs or package inserts is a new approach and has been applied to problems such as treatment of geriatric patients (Steinmetz et al., 2005), pharmacogenetics (Zineh et al., 2006), and drug interactions (Bergk et al., 2005). A recent study analysed German SPCs of 48 psycho-pharmaceuticals with respect to TDM-content. Considerable disagreement was found in comparison with medico-scientific evidence of TDM of these drugs (Ulrich et al., 2007; Ulrich et al., 2008). A lack of information on TDM was found for well established TDM-drugs such as amitriptyline and clozapine, for example.

A method to assay the content of written communication with respect to a special attribute is regarded as text analysis or content analysis, originating mainly from empirical social sciences. Thereby different definitions, categories, aims and methods of text analysis are used (Kolbe & Burnett, 1991; Krippendorff, 2004; Neuendorf, 2001). In a simple quantitative approach the frequency of relevant key words is assessed. Text analysis may also be performed according to a catalogue of codification instructions, i.e., a theory-guided analysis. This was applied in the catalogue $SPCC_{TDM}$ (SPC-Content with references to TDM) (Ulrich, 2009) for the analysis of German SPCs of psycho-pharmaceuticals (Ulrich et al., 2007; Ulrich et al., 2008). Qualitative text analyses may include a valuation of the content, e.g. pro/con or correct/false. Accordingly, the reduction of the complexity of a text by means of text analysis allows for a better comparison of content, a test of hypotheses, and discussion of perception by the readers.

The aim of this study was to estimate the inter-coder reliability of the catalogue $SPCC_{TDM}$ (Ulrich, 2009) as a measure for the validity of $SPCC_{TDM}$ in the analysis of SPCs.

METHODS AND RESULTS

SPC analysis was done by a recently established algorithm as described in the $SPCC_{TDM}$-catalogue

(Ulrich, 2009). Briefly, the SPC-paragraphs "dose" (item 1), "adverse drug reactions" (item 2), "drug interactions" (item 3), "overdose" (item 4), "pregnancy/breast feeding" (item 5), and "pharmacokinetics" (item 6) are assessed in a manual process for the occurrence of implicit (1 point) or explicit (2 points) references to TDM (0 points for no references to TDM). Thus, items 1 to 6 are analysed strictly according to the paragraphs of the SPC.

Explicit TDM-content was rated when the following information was given:

- Occurrence of words/ word groups such as "therapeutic drug monitoring", "control of plasma levels", "assay of plasma levels", or "analysis of toxic blood levels"
- Recommendations to measure plasma levels
- Recommendation to use a distinct analytical method for drug analysis
- Indication of therapeutic and/or toxic plasma levels or therapeutic range of plasma levels
- Statements on a relationship between plasma levels and therapeutic effect, adverse effects, risk of malformation or severity of intoxication
- Statements on the relationship between dose and steady-state trough plasma levels

Implicit TDM-content contains the way of thinking and the knowledge of TDM in the description and evaluation of too low, normal or too high plasma levels, however, the activity of the assay of plasma levels is not included. Therefore, implicit TDM-content was identified and coded when the text gave rise to the idea to measure plasma levels, e.g.:

- Statements such as "too low", "normal", or "too high" plasma levels without specifications to analyse plasma concentrations
- Statements such as "increased plasma levels" in the elderly or other specific patients (with renal impairment etc.)
- Statements such as "increased/ decreased plasma levels because of drug interactions"
- Statements on induction or inhibition of drug-metabolizing enzymes
- Statements on adverse effects at increased plasma levels
- Recommendations to use lower doses in case of liver or renal dysfunction

Items 7 to 10 are analysed in the entire SPC with search for dose related plasma concentrations (item 7, 3 points), toxic range of plasma concentrations (item 8, 3 points), therapeutic range of plasma concentrations (item 9, 4 points), and recommendation of TDM (item 10, 3 points for recommendation in a special clinical situation, 5 points for general recommendation). The scores of items 1 to 10 are added in an empirical summary score SPC-Content$_{TDM}$ ($SPCC_{TDM}$) and regarded as a crude measure of the information on TDM in SPCs.

Inter-coder reliability of $SPCC_{TDM}$ was estimated by coding 8 German SPCs with 5 coders (Table 1). For comparison, SPCs in English language can be retrieved from internet (Electronic Medicines Compendium, 2009a,b). It should be recognised that the structure of SPCs is equal for Germany, UK, France and other European countries because of standard European regulatory requirements. American SPCs have a very similar structure according to American regulatory requirements. Thus, $SPCC_{TDM}$ is applicable to these SPCs without change. A coder-training using three other German SPCs (Table 1) was conducted prior to the reliability test. The training included presentation and explanation of the $SPCC_{TDM}$-catalogue to the coders, coding of the three training SPCs by each coder, analysis, and discussion of errors. The organisation of this training process as well as reliability test has been performed by one author (SU) who was also a

*Table 1. Overview of Summary of Product Characteristics (SPCs) used in the reliability test and in the training of SPCC$_{TDM}$-coding, SPCs retrieved from Rote Liste (2008), *estimated according to expert opinion by authors*

Drug substance	Drug brand (Germany)	Drug group	Year of SPC-update	Recommended level of clinical TDM-application*
SPCs of reliability test:				
Digitoxin	Digimerck	Cardiac glycoside	2001	Selected cases
Digoxin	Lanicor	Cardiac glycoside	2000	Selected cases
Aripiprazole	Abilify	Antipsychotic	2005	Selected cases
Nitrazepam	Radedorm 5	Anxiolytic	2001	No
Gentamicin	Refobacin	Antibiotic	2005	Routine
Sirolimus	Rapamune	Immunosuppressant	2006	Routine
Gabapentin	Neurontin	Anticonvulsive	2005	Selected cases
Duloxetine	Cymbalta	Antidepressant	2007	Selected cases
SPCs of training:				
Opipramol	Insidon	Antidepressant	2005	No
Imipramine	Pryleugan 25	Antidepressant	2004	Routine
Phenytoin	Phenhydan	Anticonvulsive	1999	Routine

coder. SPCs of reliability test and training were chosen according to different drug groups and different levels of clinical application of TDM as derived from expert knowledge and medico-scientific evidence, e.g. (Baumann et al., 2004). Thus, the entire range of SPCC$_{TDM}$ should be covered. Inter-coder reliability was analysed by means of a free-marginal variant of Fleiss' kappa (Brennan & Prediger, 1981; Randolph, 2008) for the items 1 to 10 (eqn. 1) and by Kendall's coefficient of concordance W for the summary score. Kendall's W was calculated according to eqn. (2) using the mean r_s of the pair-wise Spearman correlations r_s of n = 5 coders (Legendre, 2005). The level of agreement between coders was evaluated according to literature (Landis & Koch, 1977).

$$kappa_{free} = \frac{\left(\frac{1}{(Nn(n-1)}\left(\sum_{i=1}^{N}\sum_{j=1}^{k} n_{ij}^2 - Nn\right)\right) - \frac{1}{k}}{1 - \frac{1}{k}} \tag{1}$$

(N = number of cases, n = number of raters, k = number of rating categories)

$$W = \frac{(n-1)\overline{r}_s + 1}{n} \tag{2}$$

The entire set of data obtained in the coding for 8 SPCs of reliability test and for 3 SPCs of the training is presented in Table 2. The mean values SPCC$_{TDM}$ of the 11 SPCs and of 5 coders are shown in Figure 1. Values were low (< 8 points) for the central nervous system (CNS)-drugs duloxetine, opipramol, aripiprazole, gabapentin and nitrazepam. Medium values (9 to 17 points) were found for the tricyclic antidepressant imipramine and the cardiac glycoside digitoxin. Values of almost 20 and higher were coded for the immunosuppressant sirolimus, the aminoglycoside antibiotic gentamicin, the cardiac glycoside digoxin and the anticonvulsive drug phenytoin.

The results of Fleiss' kappa of SPCC$_{TDM}$-items are presented in Table 2. Values of 0.4 were found in item 3 (paragraph drug interactions of SPC)

Table 2. Overview of structure of SPCC$_{TDM}$, results of SPCC$_{TDM}$-coding of 11 SPCs by 5 coders (presented in the order of coders for the 10 items and the summary score) and reliability calculated (items 1 to 10: free-marginal variant of Fleiss' kappa, SPCC$_{TDM}$: Kendall's coefficient of concordance W)

Drug substance	Items of Summary of Product Characteristics-Content of Therapeutic Drug Monitoring (SPCC$_{TDM}$)										SPCC$_{TDM}$
	Item 1	Item 2	Item 3	Item 4	Item 5	Item 6	Item 7	Item 8	Item 9	Item 10	
	Content in paragraphs of SPC						Content of entire SPC				
	Dose	Adverse drug reaction	Drug interaction	Overdose	Pregnancy/ Breast feeding	Pharmacokinetics	Dose related plasma concentration	Toxic range of plasma concentration	Therapeutic range of plasma concentration	Recommendation of TDM	
	Codings possible										
	0-1-2	0-1-2	0-1-2	0-1-2	0-1-2	0-1-2	0-3	0-3	0-4	0-3-5	0 to 27
SPCs of reliability test:											
Digitoxin	22122	00000	11011	12212	10010	11211	00300	33000	44444	53533	18,16,17,13,13
Digoxin	22222	00000	22212	22202	00010	22222	00000	33333	44444	55555	20,20,20,18,20
Aripiprazole	11111	00000	01111	00010	00000	01111	00000	00000	00000	00000	1,3,3,4,3
Nitrazepam	10101	00000	00111	00000	00000	02122	30000	00000	44044	00030	8,6,3,10,8
Gentamicin	22222	11012	00000	22002	00000	22222	00033	33333	44444	55555	19,19,16,20,23
Sirolimus	22222	21122	12222	00000	00000	12222	33333	03000	04444	55555	14,22,19,20,20
Duloxetine	01111	00000	01111	00000	00000	01011	00000	00000	00000	00000	0,3,2,3,3
Gabapentin	10011	00000	01011	00001	00000	22012	00033	00000	00000	50000	8,3,0,6,8
Reliability	0.63	0.76	0.40	0.59	0.81	0.40	0.60	0.75	0.80	0.74	0.877
SPCs of training:											
Opipramol	00000	00000	11111	00000	00000	01111	00000	00000	00000	00000	1,2,2,2,2
Imipramine	00000	01020	21111	22212	00000	01111	00000	00000	04044	05055	4,14,4,14,13
Phenytoin	22222	22222	21111	22202	22202	22222	30330	33333	44444	55555	27,23,26,22,23

and item 6 (paragraph pharmacokinetics of SPC). This represents a level of fair to moderate agreement. Kappa-values of about 0.6 (items 1, 4 and 7, i.e., paragraphs dose and overdose as well as existence of dose-related plasma concentrations) designate moderate to substantial agreement. Substantial to almost perfect agreement was found in five items, i.e., item 2 (paragraph adverse effects), item 5 (pregnancy/ breast feeding), and items 8 to 10 (toxic and therapeutic ranges of plasma concentration, recommendation of TDM). Kendall's coefficient of concordance W was calculated 0.877 for the summary score SPCC$_{TDM}$ (Table 2).

DISCUSSION

The SPCC$_{TDM}$ was found to be a valid method for the analysis of the content of TDM-related information in SPCs. This was shown by good inter-coder reliability as expressed by kappa statistics for single items and Kendall's coefficient of concordance W = 0.877 for the summary score. According to literature (Landis & Koch, 1977),

Figure 1. Mean (±SD) $SPCC_{TDM}$-values of 5 coders for German SPCs of 11 drugs obtained during training and inter-coder reliability test. The maximum score is 27.

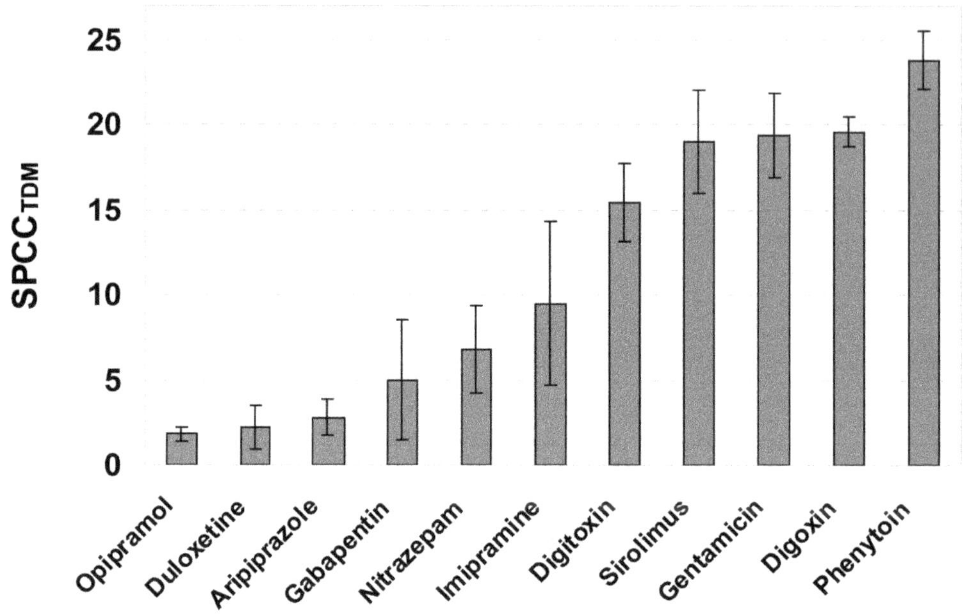

kappa-values of about 0.2 to 0.4 represent a fair agreement, values of 0.4 to 0.6 moderate agreement, 0.6 to 0.8 substantial agreement, and 0.8 to 1.0 almost perfect agreement. Analyses by different coders who were experts in TDM and trained in the application of the catalogue provided these concordant results. Accordingly, users of the $SPCC_{TDM}$ should be familiar with TDM and also perform training with a minimum of three SPCs before a real coding. Kappa-values of < 0 (poor agreement) or 0 to 0.2 (slight agreement) have not been found. The lower agreement in item 6 (paragraph pharmacokinetics) is explained by the difficulty to differentiate between TDM-content and pharmacokinetic content, i.e., pharmacokinetic information is not always identical to TDM-information. Moreover, the decision whether pharmacokinetic information is simultaneously regarded as TDM-information may depend on the drug investigated, e.g. information on maximum (t_{max})-plasma levels is regarded as TDM-information only in case t_{max}-values are used in TDM of a selected drug. The lower agreement in item 3 (paragraph drug interactions) may be explained by the difficulty to find out a single implicit TDM-information among a considerable number of other drug interactions not considered as an implicit TDM-reference, e.g., pharmacodynamic interactions, in two SPCs. It is expected that better training of coders will improve agreement also in these two items in future. Better kappa-values in items 1, 2, 4, and 5 (paragraphs dose, adverse effects, overdose, pregnancy/ breast feeding) can be explained by two points: (i) information on TDM was lower in these paragraphs in general, and (ii) less subtle differences in terms of implicit/ explicit TDM-information. Finally, it is not unexpected that better kappa-values have been achieved for items 7 to 10. These items include the crucial and most obvious information on TDM and there is no longer the difficulty of the discrimination of implicit and explicit TDM-references.

Content analysis of the prescription information of pharmaceuticals is increasingly published (Bergk et al., 2005; Gebran & Al Haidari, 2006; Reggi et al., 2003; Steinmetz et al., 2005; Ulrich et al., 2007; Zineh et al., 2006). However, a test of inter-coder reliability has been found in only one study of drug interactions (Bergk et al., 2005). A mean kappa of 0.92 was reported. This excellent agreement can be explained by the application of a checklist of drug interactions which is regarded as easier to perform than a catalogue of codification instructions such as $SPCC_{TDM}$. The application of an algorithm according to a catalogue such as $SPCC_{TDM}$ includes subjective decisions and is therefore prone to differences in coding by different coders. The application of a checklist, for instance in case of a computerised method, can almost completely exclude subjective differences in coding. A considerably higher reliability is expected. On the other hand, content validity may be more a problem with checklists if relevant content is not included in the list, e.g. not taking into account special and unexpected drug interactions in a check list of drug interactions. Coding by experts according to general codification instructions would not miss such new and unexpected details.

Briefly, content analyses of SPCs or of package inserts published without validation (Gebran & Al Haidari, 2006; Reggi et al., 2003; Steinmetz et al., 2005; Zineh et al., 2006) are described here to show the different approaches. The electronic Physician's Desk Reference of prescription inserts was queried with nine key words in a content analysis of pharmacogenetics, e.g., "genetic", "genomic", and "genotype". The data extracted electronically was compared manually with medico-scientific evidence of a manual PubMed-analysis. The content analysis was completed with a decision on the sufficiency of the content using six criteria, e.g., dosing recommendations based on genotype, and an indication for a drug based on a molecular classification of the patient. One criterion used implicit or explicit evidence of a special content (to suggest that no dose changes are necessary based on metabolising phenotype) which is similar to explicit and implicit content of items 1 to 6 of $SPCC_{TDM}$ (Zineh et al., 2006). In another study, prescription inserts of the electronic Physician's Desk Reference were searched for geriatric information taking into account all statements with reference to older people. Two investigators then independently reviewed the information as classified by three types of information, i.e., pharmacokinetics, precautions and warnings, and dosing. The quantity of geriatric information in the package insert was estimated according to the occurrence of one, two or three such types of information (Steinmetz et al., 2005). This classification of information is very similar to the analysis according to paragraphs of SPC in the $SPCC_{TDM}$. In contrast to these analyses of a special content of SPCs or package inserts, a complete content analysis of package inserts was described using checklists of 32 to 88 indicators per drug (Gebran & Al Haidari, 2006). The indicators were derived from data which was regarded as standard and complete, i.e., the British National Formulary (BNF). Completeness was described as percentage of correct indicators compared with BNF. A similar approach with BNF as standard was used in the comparison of package inserts of different countries. However, only the information on indications, contraindications, adverse effects and dosing regimen was included. The number of indicators was 22 to 48 per drug (Reggi et al., 2003). It is assumed that content analysis was conducted manually in these two studies because no other information is provided on the method.

The $SPCC_{TDM}$ does not count precisely the TDM-content in per cent of words of an SPC or similar. The TDM-content is also not counted in per cent of a defined reference information. In contrast, the $SPCC_{TDM}$ estimates the impression of TDM which gets a reader of an SPC. Two groups of hypothetical categories are applied based on expert knowledge, i.e. structure-based items (items 1 to 6, according to the expectation of any TDM-information in these TDM-relevant parts of the SPC) and theory-based items (items 7 to 10, according to the expectation of crucial TDM-data according to TDM-theory). These categories of $SPCC_{TDM}$ are regarded as sufficient with respect to content-validity of the catalogue. However, the $SPCC_{TDM}$ was developed first for psycho-pharmaceuticals. In the extension of the method to other drug groups, the experts felt that an additional theory-based item may be reasonable, i.e. taking into account information on the chemical assay of the drug in plasma. A considerable and very detailed information on methods of drug analysis was found in some SPCs, e.g., sirolimus (Rapamune). No other theory-based category to be considered in the $SPCC_{TDM}$ emerged during this study.

The SPC-paragraph "Special warnings and precautions" was excluded from the structure-based items of $SPCC_{TDM}$ because a considerable amount of redundancy with respect to the other paragraphs was encountered in German SPCs, and with the aim to create a simple method. In French SPCs of antipsychotic drugs, some implicit TDM-information was exclusively found in this paragraph and the $SPCC_{TDM}$ did, therefore, slightly underestimate the TDM-content (Rougemont et al., in press). It should therefore be considered that codification catalogues in content analysis are no stringent and invariable methods (Kolbe & Burnett, 1991). Rather, an adaptation to a change in the underlying construction of the text is recommended for the catalogue. This must of course be taken into account in a comparison of the results. Moreover, the role of redundancy has to be discussed for the reader's perception of TDM-content in the SPC.

The summary score of $SPCC_{TDM}$ assumes that the scores of the single items represent relative quantities in integer steps with respect to the TDM-impression of an SPC. For example, giving the therapeutic window four points in item 9 is regarded as equal to an explicit TDM-content in the paragraph dose (two points, item 1) and implicit TDM-contents in the paragraphs drug interactions and overdose (one point in each case, items 3 and 4) (2+1+1 = 4). This is of course a highly hypothetical assumption of an ordinal scaled behaviour and based only on expert knowledge on TDM and SPCs so far. However, it appears reasonable to provide the crucial theory-based categories of TDM with higher values than the structure-based categories and also explicit content should obviously be higher rated than only implicit content. It is also assumed in the $SPCC_{TDM}$ that dependence between items does not impair this quantitative estimation of the summary score. Dependence of items is expected to occur between items 1 to 6 on the one hand and items 7 to 10 on the other hand. It may be for example that the more information is provided on ranges of plasma levels and recommendation of TDM (items 7 to 10) it is also more likely to encounter other TDM-information in the paragraphs of items 1 to 6.

The results of $SPCC_{TDM}$ as shown in Figure 1 should roughly represent the actual role of TDM of these drugs, i.e., agreement between SPC and clinical practice. However, deviations are obvious taking into account the expert opinion. For example, routine TDM is recommended for imipramine (Baumann et al., 2004), gentamicin, sirolimus and phenytoin, but the $SPCC_{TDM}$ is considerably lower for the antidepressant imipramine. Ranges of the $SPCC_{TDM}$ to be achieved by an SPC may be defined with respect to the clinical application of TDM of a drug. Thus, $SPCC_{TDM} \leq 2$ should be regarded only as a noise of information in SPCs which may have been arisen by chance and cannot be attributed to an effective TDM-information. $SPCC_{TDM} = 3$ to 8 may indicate absence of TDM or almost no application of TDM. Drugs with recommendation of TDM in selected cases should achieve about $SPCC_{TDM} = 9$ to 17, and ca. $SPCC_{TDM} \geq 18$ should indicate a routine application of TDM.

CONCLUSION

$SPCC_{TDM}$ based text analysis was found to be suitable to estimate TDM related information in SPCs with sufficient validity. In a compromise of time needed and informative value generated, the catalogue $SPCC_{TDM}$ can be used in the text analysis of SPCs with respect to TDM. This allows for a comparison between SPCs or a comparison of content with medico-scientific evidence, for example. It may be also possible to generate hypotheses on the perception of TDM by the reader of the SPCs. Thus, the catalogue $SPCC_{TDM}$ is an approach to the principle question of text analysis "Who says what, to whom, why, to what extent and with what effect?" for TDM in the prescription information. The catalogue $SPCC_{TDM}$ can also be regarded as a basis to create similar catalogues by adapting it to special questions and circumstances. Other catalogues for the content analysis of SPCs may be developed.

The analysis of SPCs by $SPCC_{TDM}$ is a first step in improving SPCs which will finally result in improved patient care.

ACKNOWLEDGMENT

The authors wish to thank Philippe Vincent, Professor of Clinical Psychiatry at the Faculty of Pharmacy at University of Montreal, Canada, for careful reading of the manuscript and correcting the English.

REFERENCES

Baumann, P., Hiemke, C., Ulrich, S., Eckermann, G., Gaertner, I., & Gerlach, M. (2004). The AGNP-TDM expert group consensus guidelines: therapeutic drug monitoring in psychiatry. *Pharmacopsychiatry*, *37*(6), 243–265. doi:10.1055/s-2004-832687

Bergk, V., Haefeli, W. E., Gasse, C., Brenner, H., & Martin-Facklam, M. (2005). Information deficits in the summary of product characteristics preclude an optimal management of drug interactions: a comparison with evidence from the literature. *European Journal of Clinical Pharmacology*, *61*(5-6), 327–335. doi:10.1007/s00228-005-0943-4

Brennan, R. L., & Prediger, D. J. (1981). Coefficient Kappa: Some uses, misuses, and alternatives. *Educational and Psychological Measurement*, *41*(3), 687–699. doi:10.1177/001316448104100307

Electronic Medicines Compendium. (2009a). *SPC Rapamune*. Retrieved September 2009, from http://emc.medicines.org.uk/document.aspx?documentId=5747

Electronic Medicines Compendium. (2009b). *SPC Cymbalta*. Retrieved September 2009, from http://emc.medicines.org.uk/medicine/15694

Ensom, M. H., Davis, G. A., Cropp, C. D., & Ensom, R. J. (1998). Clinical pharmacokinetics in the 21st century: does the evidence support definitive outcomes? *Clinical Pharmacokinetics*, *34*(4), 265–279. doi:10.2165/00003088-199834040-00001

Gebran, N., & Al Haidari, K. (2006). Assessment of prescribing information for generic drugs manufactured in the Middle East and marketed in Saudi Arabia. *Annals of Saudi Medicine*, *26*(3), 192–199.

Kolbe, R., & Burnett, M. (1991). Content-analysis research: an examination of applications with directives for improving research reliability and objectivity. *The Journal of Consumer Research*, *18*(2), 243–250. doi:10.1086/209256

Krippendorff, K. (2004). *Content Analysis: An introduction to its methodology* (2nd ed.). Thousand Oaks, CA: Sage Publications.

Landis, J. R., & Koch, G. G. (1977). The measurement of observer agreement for categorical data. *Biometrics*, *33*(1), 159–174. doi:10.2307/2529310

Legendre, P. (2005). Species associations: The Kendall coefficient of concordance revisited. *Journal of Agricultural Biological & Environmental Statistics*, *10*(2), 226–245. doi:10.1198/108571105X46642

Neuendorf, K. A. (2001). *The content analysis guidebook*. Thousand Oaks, CA: Sage Publications.

Randolph, J. J. (2008). *Online Kappa Calculator.* Retrieved September 2008, from http://justus.randolph.name/kappa

Reggi, V., Balocco-Mattavelli, R., Bonati, M., Breton, I., Figueras, A., & Jambert, E. (2003). Prescribing information in 26 countries: a comparative study. *European Journal of Clinical Pharmacology*, *59*(4), 263–270. doi:10.1007/s00228-003-0607-1

Rote Liste. (2008). Retrieved January 2008, from http://www.fachinfo.de/index_html

Rougemont, M., Ulrich, S., Hiemke, C., Corruble, E., & Baumann, P. (in press). French summaries of product characteristics (SPC): Content in relation to therapeutic monitoring of psychotropic drugs. *Fundamental & Clinical Pharmacology*.

Shenfield, G. M. (2001). Therapeutic drug monitoring beyond 2000. *British Journal of Clinical Pharmacology*, *52*(1), 3S–4S. doi:10.1046/j.1365-2125.2001.0520s1003.x

Steinmetz, K. L., Coley, K. C., & Pollock, B. G. (2005). Assessment of geriatric information on the drug label for commonly prescribed drugs in older people. *Journal of the American Geriatrics Society*, *53*(5), 891–894. doi:10.1111/j.1532-5415.2005.53273.x

Ulrich, S. (2009). *SPCC(TDM): Catalogue for the content analysis of summary of product characteristics (SPCs) with reference to therapeutic drug monitoring (TDM).* Retrieved May 2009, from http://www.agnp.de

Ulrich, S., Hiemke, C., Laux, G., Mueller-Oerlinghausen, B., Havemann-Reinecke, U., & Riederer, P. (2007). Value and actuality of the prescription information for therapeutic drug monitoring of psychopharmaceuticals: A comparison with the medico-scientific evidence. *Pharmacopsychiatry*, *40*(3), 121–127. doi:10.1055/s-2007-977712

Ulrich, S., Laux, G., Mueller-Oerlinghausen, B., Havemann-Reinecke, U., Riederer, P., & Zernig, G. (2008). Nutzen und Aktualität der Fachinformationen für das therapeutische Drug Monitoring von Psychopharmaka. *Psychopharmakotherapie*, *15*(2), 67–74.

Zineh, I., Pebanco, G. D., Aquilante, C. L., Gerhard, T., Beitelshees, A. L., Beasley, B. N., & Hartzema, A. G. (2006). Discordance between availability of pharmacogenetic studies and pharmacogenetics-based prescribing information for the top 200 drugs. *The Annals of Pharmacotherapy*, *40*(4), 639–644. doi:10.1345/aph.1G464

This work was previously published in the International Journal of Knowledge Discovery in Bioinformatics, Volume 1, Issue 2, edited by Xiao-Li Li and See-Kiong Ng, pp. 1-11, copyright 2010 by IGI Publishing (an imprint of IGI Global).

Chapter 19
Hierarchical Density-Based Clustering of White Matter Tracts in the Human Brain

Junming Shao
University of Munich, Germany

Afra Wohlschläeger
Technical University of Munich, Germany

Klaus Hahn
HMGU Helmholtz Center Munich, Germany

Christian Boehm
University of Munich, Germany

Qinli Yang
University of Edinburgh, UK

Nicholas Myers
Technical University of Munich, Germany

Claudia Plant
Florida State University, USA

ABSTRACT

Diffusion tensor magnetic resonance imaging (DTI) provides a promising way of estimating the neural fiber pathways in the human brain non-invasively via white matter tractography. However, it is difficult to analyze the vast number of resulting tracts quantitatively. Automatic tract clustering would be useful for the neuroscience community, as it can contribute to accurate neurosurgical planning, tract-based analysis, or white matter atlas creation. In this paper, the authors propose a new framework for automatic white matter tract clustering using a hierarchical density-based approach. A novel fiber similarity measure based on dynamic time warping allows for an effective and efficient evaluation of fiber similarity. A lower bounding technique is used to further speed up the computation. Then the algorithm OPTICS is applied, to sort the data into a reachability plot, visualizing the clustering structure of the data. Interactive and automatic clustering algorithms are finally introduced to obtain the clusters. Extensive experiments on synthetic data and real data demonstrate the effectiveness and efficiency of our fiber similarity measure and show that the hierarchical density-based clustering method can group these tracts into meaningful bundles on multiple scales as well as eliminating noisy fibers.

DOI: 10.4018/978-1-4666-1785-8.ch019

INTRODUCTION

Diffusion Tensor imaging (DTI) can explore the organization and integrity of human white matter tracts in vivo, using water diffusion properties as a probe (Mori, 2007). It measures for every voxel the diffusivity of water molecules within the tissue, and thus gives valuable insight into the orientation of fiber tracts, since water diffusion is strongest along the direction of fibers and restricted in the directions perpendicular to them. Potential pathways of fiber tracts in the white matter of the human brain can be reconstructed by deterministic and stochastic tractography based on the measured diffusion weighted images. These techniques have attracted attention in the study of anatomical connectivity (Hagmann et al., 2008), brain changes (Huang et al., 2006), and various pathologies related to white matter atrophy, such as schizophrenia (Park et al., 2004), Alzheimer's disease (Damoiseaux et al., 2009), or multiple sclerosis (Law & Grossman, 2005).

Performing fiber tracking in the human brain usually results in large sets of tracks. A problem in clinical research is how to segment and interpret this vast amount of information. A frequently used method is to select fiber groups of interest on the basis of expert knowledge by *virtual dissection*. Experts first specify some regions of interest (ROIs) and select then all fibers that pass through these pre-defined ROIs (Catani et al., 2002). This process tends to be inefficient, since the manual handling of ROIs is time-consuming and also limited by the availability of experts. Moreover, manual specification of ROIs may be biased in patient populations. Therefore, automatic clustering of fiber tracts into bundles of similar fibers is preferable for many applications. Two fibers are considered as similar if they have comparable length, similar shape and similar location (Ding, Gore, & Anderson, 2003). A number of approaches have been proposed to automatically cluster white matter tracts. However, before grouping fibers into clusters, it is necessary to specify a fiber similarity measure.

A fiber similarity measure is a function that quantifies the similarity between pairs of fibers. In the early work by Brun et al. (2003), it is assumed that two fiber tracts with similar end points should be considered as similar. The Euclidean distance between the starting and ending points of the two fibers is used to calculate fiber similarity. However, this assumption is not sufficient in some cases, as not all fibers in a bundle start and end in the same region. It also ignores the shape information contained in all points on the fibers. Ding, Gore, and Anderson (2003) propose a similarity measure by cutting each fiber into corresponding fiber segments and use then the mean Euclidean distance between the segments to define piecewise similarity. This similarity method is efficient, but may lack effectivity since this measure loses the point-by-point information. Several authors acknowledge that the point-by-point correspondence of the trajectories should be included to define a convenient fiber similarity measure. For instance, in (Zhang, Demiralp, & Laidlaw, 2003; Zhang & Laidlaw, 2005) the distance is defined between two fibers as the average distance from any point on the shorter fiber to the closest point on the longer fiber, where only distances above a certain threshold contribute to this average. Another possibility is proposed by Klein et al. (2007). They introduce a grid-based similarity measure. A box around the fibers is partitioned into identical cells. Each point of a fiber is then assigned to the cells with weights. The similarity between a pair of fibers is calculated by the pairwise weights, summed over all cells including the fibers. The side length of the cell is an adjustable parameter that controls the scale of comparison. Corouge, Gouttard, and Gerig (2004) form point pairs, mapping each point of one fiber to the closest point on the other fiber. The resulting point pairs are then used to define the distance between fiber pairs. They define three similarity distances: closest point distance, mean of closest point distance (MCP) and Hausdorff distance (HDD).

Currently, MCP (Corouge, Gouttard, & Gerig, 2004; O'Donnell et al., 2006; O'Donnell & Westin, 2007; Moberts, Vilanova, & Wijk, 2005; Zhang & Laidlaw, 2005) and HDD (Corouge, Gouttard, & Gerig, 2004; Gerig, Gouttard, & Corouge, 2004; Moberts, Vilanova, & Wijk, 2005; Xia, Turken, Whitfield-Gabrieli, & Gabrieli, 2005) are widely used fiber similarity measures. Their formal definition is the following:

Suppose there are two fibers P and Q, then MCP is defined as:

$$d_{MCP}(P,Q) = avg\left(d_m(P,Q), d_m(Q,P)\right) \quad (1)$$

with $d_m(P,Q) = avg_{p_i \in P}\{min_{q_j \in Q} \mid\mid p_i - q_j \mid\mid\}$

While HDD is defined as:

$$d_{HDD}(P,Q) = avg\left(d_h(P,Q), d_h(Q,P)\right) \quad (2)$$

with $d_h(P,Q) = \max_{p_i \in P}\{\min_{q_j \in Q} \mid\mid p_i - q_j \mid\mid\}$

For the point-by-point fiber similarity measures, the computational cost or inefficiency is high as the time complexity is O($m \cdot n$), where m and n are the lengths of any two fibers respectively.

Having defined a convenient similarity measure, the fibers can be clustered. In the following we shortly review such algorithms.

Ding, Gore, and Anderson (2003) establish a k most similar-fibers algorithm to find automatically the different bundles. Based on a similarity measure for fiber segments, a cluster around fiber F is constructed by neighbouring fibers whose similarity to F is greatest. With similar clustering approach, Corouge, Gouttard, and Gerig (2004) use point pair distances for the similarity measure. Clusters of low cardinality are treated as outliers. The number of clusters is controlled by setting a threshold on the accepted maximum distance.

Another frequently used clustering method is spectral clustering (Brun, Park, Knutsson, & Westin, 2003; O'Donnell1 et al., 2006), which refers to techniques relying on the eigenstructure of the similarity matrix to partition fibers into disjoint clusters. Brun et al. (2003) use a spectral embedding technique called Laplacian eigenmaps to map the fibers to a Euclidean feature space and then they use a Gaussian kernel to compare the fibers in this new space, assigning similar fibers to similar colors. O'Donnell et al. (2006) decompose the fiber similarity matrix into their eigenvalues and eigenvectors. The top eigenvectors are used to represent the most important similarity information for each fiber, while removing noise. Then, a specific clustering called k-way normal cut is used to obtain fiber bundles. For this type of clustering the number of expected clusters must be defined a priori by the user. Klein et al. (2007) use an extension of multiple eigenvector clustering, a special variant of spectral clustering. This method includes an automatic detection of the number of clusters.

Zhang and Laidlaw (2005) use an agglomerative hierarchical clustering algorithm. This type of algorithm decomposes a fiber set into several levels of partitions, represented by a dendrogram. Starting with the clustering obtained by placing every object in a cluster, in every step the two closest clusters are merged until all objects are in one cluster. A drawback of hierarchical clustering is that it is hard to find the best partition of the dendrogram.

Maddah et al. (2005, 2006) incorporate anatomical expert knowledge into clustering. They use an atlas of fiber tracts, labeled by the number of their bundle. To build such an atlas, they start with a set of labeled ROIs specified by an expert and calculate the corresponding fiber bundles seeded from those ROIs. Affine registration is then used to map extracted fibers of some subject to the atlas. To make the comparison between fibers of the subject and of the atlas efficient, B-spline representations of the tracts are used. Similar subject and atlas fibers are labeled identically. O'Donnell and Westin (2007) also present an atlas based segmentation of fiber tracts. Their segmentation is performed in two steps. First, an atlas, labeled by experts, is learned from a population of subjects using spectral clustering. Based on this model, fibers of a novel subject are clustered by an extension of the spectral clustering solution stored in the atlas, using the Nystrom method. This segmentation across subjects enables testing of neuroscientific hypotheses with respect to group differences.

Though appreciable progress has been achieved in fiber clustering, the following problems are, in our opinion, still challenging and may justify a novel approach:

1) Fiber Similarity Measure: Effective point-by-point similarity measures for pair-wise fiber comparison are needed for clustering. These similarity measures should effectively capture similarities, but also efficiently, as big sets of fibers need to be analyzed.
2) Flexibility: The fiber clustering approach should be flexible enough to cluster fibers on multiple scales. Hierarchical cluster structures are prevalent in nature and can be found, e.g., for the fibers of the Corpus Callosum.
3) Transparency: A frequently used visualization of a hierarchical clustering result is the dendrogram. This graph lacks of transparency for large data sets and is hard to analyze. More informative visual presentations of a clustering structure would be useful.
4) Outlier-detection and -robustness: Due to experimental limitations, like thermal noise or partial volume effects, the set of fibers produced by tractography contains also imperfect fibers or outliers. It is beneficial for further tract-based analysis to separate out those fibers during clustering to obtain compact clusters. Another aspect is the impact of such outliers on the results of some clustering algorithms. Algorithms which are robust to outliers are necessary.

In view of these issues, we propose in this paper a new fiber similarity measure which is adapted from dynamic time warping, to calculate a point-by-point similarity distance among fibers. Further, a lower bounding technique for this similarity measure is introduced, to reduce computational cost. Considering the different clustering substructures of fiber data and inherent outliers, we apply - to the best of our knowledge - for the first time hierarchical density-based clustering to group fiber tracts. The outlier-robustness of density-based clustering is demonstrated. The resulting clustering structure of the fiber data can be visualized in a transparent manner by a so-called reachability plot. From this plot, the clusters are subsequently calculated by interactive or automatic fiber clustering algorithms on multiple scales. Imperfect fibers or outliers are eliminated during the clustering process. The approach is tested on several synthetic as well as real fiber data sets.

The remainder of the paper is organized as follows: we first present the novel fiber similarity measure and the hierarchical density-based fiber clustering in detail. A series of clustering experiments on synthetic and real data and the relevant results are then described. A discussion of our results follows. Finally, we conclude our work.

FIBER SIMILARITY MEASURE

After deterministic tractography, a fiber is represented as an ordered set of points in space. The steps of the arc length, defined by two successive points of a fiber, are not necessarily identical, if e.g., numerical Runge Kutta methods with dynamical stepsizes are used for tracking. In addition, two fibers may have different lengths and consequently different numbers of points in space. To quantify similarity between two fibers, we adapt for this work the Dynamic Time Warping (DTW) method (Itakura, 1975; Sakoe & Chiba, 1978; Sankoff & Kruskal, 1983). DTW is introduced for the

Figure 1. Cost matrix (a) and optimal warping path for time series (b)

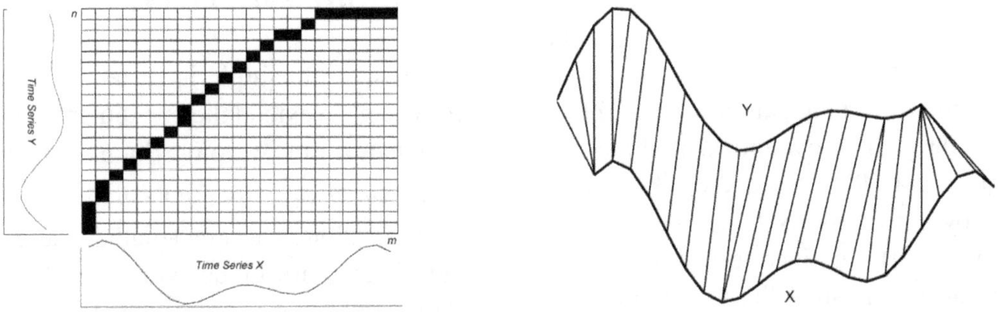

(a) Cost Matrix (b) Optimal warping path (time series)

comparison of time series which are out of phase and is applied in fields as diverse as speech recognition (Sakoe & Chiba, 1978), bioinformatics (Aach & Church, 2001) or data mining (Keogh & Pazzani, 2001).

In the following section, we will first briefly review DTW for time series and then extend it to a similarity measure for space curves. Finally, we present a convenient lower bounding technique to save computational cost.

DYNAMIC TIME WARPING

Dynamic time warping (DTW) is a technique that looks for the optimal alignment of two time series. To achieve this goal, the time series are "warped" together non-linearly, by stretching or shrinking them along the time axes (Salvador & Chan, 2004).

Suppose we have two time series X and Y, of lengths *m* and *n* respectively, where

$$X = (x_1, x_2, \ldots, x_i, \ldots, x_m) \tag{3}$$

$$Y = (y_1, y_2, \ldots, y_j, \ldots, y_m) \tag{4}$$

The objective is to optimize a warping path W:

$$W = (w_1, w_2, \ldots, w_k, \ldots, w_K) \tag{5}$$

where K is the length of W, with $max(m,n) < K < m+n - 1$. The kth element of W is a pair of indices indicating a connection of time points in X and Y and is written as $w_k = (i, j)$, see Figure 1. A warping path follows the constraints (Keogh, 2002):

1) Boundary conditions: $w_1 = (1, 1)$ and $w_K = (m, n)$. This requires the warping path to start and finish in the first and last points of the series respectively;
2) Monotony: Given $w_k = (i, j)$, then $w_{k+1} = (i', j')$, with $i' - i >= 0$ and $j' - j >= 0$. This forces the points in W to be monotonically spaced in time;
3) Continuity: Given $w_k = (i, j)$, then $w_{k+1} = (i', j')$, with $i' - i <= 1$ and $j' - j <= 1$. This restricts the admissible steps in the warping path to adjacent points of the series.

There are many warping paths satisfying the above conditions. In order to find a best match between two time series, we look for that path which minimizes the cumulative distance between them. The distance *dtw* for this optimum path is defined as:

$$dtw(X,Y) = \min\left(\sum_{k=1}^{K} d(w_k)\right) \tag{6}$$

where $d(\cdot)$ is a distance function. We define it as

$$d(w_k) \equiv d(i, j) \equiv |x_i - y_j| \tag{7}$$

The optimum warping path for $dtw(X, Y)$ can be obtained through the dynamical programming approach (Itakura, 1975). It proceeds like follows: First, a m by n cost matrix D is constructed. A component D (i, j) is defined recursively as sum of the distance d (i,j) and the minimum of the cumulative distances in the adjacent elements:

$$D(i, j) = d(i, j) + \min\{D(i-1, j-1), D(i-1, j), D(i, j-1)\} \tag{8}$$

After the entire cost matrix D is filled, starting from D (1, 1), the minimum-distance warping path can be found in reverse order, starting from D (m, n). For this purpose a greedy search is performed to evaluate cells to the left, down, and diagonally to the bottom-left. Whichever of these three adjacent cells has the minimum value is added to the beginning of the warping path found so far, and the search continues from that cell. The search stops if D (1, 1) is reached. Figure 1 (a) shows an example of two time series with their cost matrix and a minimum-distance warping path between them. The warping path is W = {(1, 1), (1, 2), (1, 3), (2, 4), (2, 5), (3, 6), (4, 7), (5, 8), (6,9), (7, 10), (8, 11), (8, 12), (9, 13), (10, 14), (11, 15), (12, 16), (13, 17), (14, 18), (15, 19), (16, 19), (17, 20), (18, 21), (19, 21), (20, 21), (21, 21), (22, 21), (23, 21)}. If the warping path passes through a cell D (i, j) in the cost matrix, the ith point in time series X is warped to the jth point in time series Y. Since a single point may map to multiple points in the other time series, dynamic time warping can handle time series with different lengths. However, we should note that the dynamic time warping distance function doesn't satisfy triangle inequality and thus not a metric. An illustration of the optimum warping path between two time series can be found in Figure 1 (b).

FIBER SIMILARITY MEASURE WITH DTW

To calculate the similarity distance between fibers in space, we extend the one dimensional concept of dynamic time warping. Suppose $p_i(p_i^1, p_i^2, p_i^3)$ and $q_j(q_j^1, q_j^2, q_j^3)$ are points of the fibers P and Q, indexed along the arc length, where the three coordinates are given within the brackets. We define the distance between the two points as:

$$d(p_i, q_j) = |p_i^1 - q_j^1| + |p_i^2 - q_j^2| + |p_i^3 - q_j^3| \tag{9}$$

Using this distance function, we can treat the spatial fiber problem like a time series problem. The optimal warping path can be obtained through dynamical programming and is represented as $W = (w_1, \ldots w_k, \ldots w_K)$, where K is the length of the path and

$$d(w_k) \equiv d(p_i, q_j) \tag{10}$$

To reduce the effect of different lengths of fibers for similarity calculation, we define the similarity distance $DTW(P, Q)$ between the fibers P and Q as the averaged distance for the optimal warping path:

$$DTW(P,Q) = min(\frac{\sum_{k=1}^{K} d(w_k)}{K}) \tag{11}$$

LOWER BOUNDING DISTANCE

The time complexity of our fiber similarity measure DTW is O ($m \cdot n$), which is demanding in terms of CPU time. To deal with this problem, we introduce an easily computed lower bounding distance LB. For two fibers we have $LB \leq DTW$, where an efficient LB should be a tight lower bound to DTW.

For two given spatial fibers $P(P^1, P^2, P^3)$ and $Q(Q^1, Q^2, Q^3)$, we rewrite them as three sequences of point pairs (P^1, Q^1), (P^2, Q^2) and (P^3, Q^3) respectively. Let us first consider the pair-wise sequence (P^1, Q^1). $Max(P^1)$ and $max(Q^1)$ denote the maximum values in P^1 and Q^1, respectively. Min(P1) and min(Q1) define the minimum values. A pair $(min(P^1), max(P^1))$ defines the range R_p of P^1. Without loss of generality, we assume $max(P^1)$ $max(Q^1)$. There are then three possible arrangements for the two ranges R_p and R_Q, see Figure 2.

The lower bounding distance between the two sequences is defined as follows (Yi, Jagadish, & Faloutsos, 1998):

$$lb(P^1,Q^1) = \begin{cases} \sum_{p_i^1 > max(Q^1)} | p_i^1 - max(Q^1) | + \sum_{q_j^1 < min(P^1)} | q_j^1 - min(P^1) | \\ \quad if\ P^1\ and Q^1\ overlap \\ \sum_{p_i^1 > max(Q^1)} | p_i^1 - max(Q^1) | + \sum_{p_i^1 < min(Q^1)} | p_i^1 - min(Q^1) | \\ \quad if\ P^1\ and Q^1\ enclose \\ max\left(\sum_{i=1}^{|P^1|} | p_i^1 - max(Q^1) |, \sum_{j=1}^{|Q^1|} | q_j^1 - min(P^1) | \right) \\ \quad if\ P^1\ and Q^1\ disjoint \end{cases} \tag{12}$$

An analogous definition is used for the other pair-wise sequences (P^2, Q^2) and (P^3, Q^3). Finally, the lower bounding distance between fibers P and Q is defined as:

$$LB(P,Q) = \frac{lb(P^1,Q^1) + lb(P^2,Q^2) + lb(P^3,Q^3)}{m + n - 1} \tag{13}$$

We prove in Appendix A the property: $LB(P,Q) \leq DTW(P,Q)$ for spatial fibers.

This concept is especially useful for density-based clustering, see the following Section. For clustering, it is necessary to perform a so-called range search. That means, it is necessary to query the data set for the most similar fibers with respect to P, i.e. for fibers Q with *DTW*(P, Q) below a given threshold ε. First, *LB*(P, Q) is computed for these fibers. *DTW*(P, Q) must then only be computed for those fibers with *LB*(P, Q) $< \varepsilon$. The Pseudocode of ε-range search with lower bounding distance is provided in Figure 3.

Figure 2. Illustration of possible arrangements of R_p and R_Q

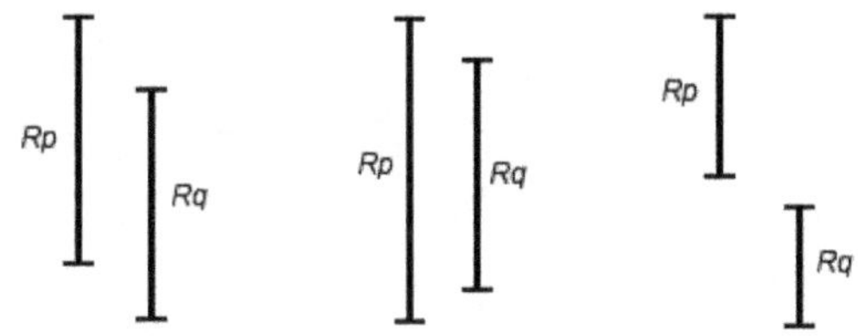

(a) overlap (b) enclose (c) disjoint

HIERARCHICAL DENSITY-BASED FIBER CLUSTERING

After calculation of the fiber similarities, each fiber is regarded as an object in metric space, and a clustering approach can be applied to group fibers. Clustering has attracted much attention during the last decades, which led to the publication of a vast number of research papers, books, and surveys (e.g., Ankerst, Breunig, Kriegel, & Sander, 1999; Dempster, Laird, & Rubin, 1977; Ester, Kriegel, Sander, & Xu, 1996; Ng, Jordan, &Weiss, 2001; Jain & Dubes, 1988). One interesting branch of research considers the clustering problem from a density-based point of view: clusters are regarded as areas of high object density which are separated by areas of low density. This notion has several attractive benefits: In contrast to other methods, the users do not have to specify the number of clusters they want to find. Density-based clustering is able to detect clusters of arbitrary shape. It is robust to noise and outliers. Finally, density-based clustering methods are not restricted to vector data, but

Figure 3. Pseudocode of ε -range query using lower bounding distance

```
algorithm rangeLBQuery (P, ϵ, D)
    Neigbors = {};
    For each object Q ∈ D
        LB_dist = LB(Q, P);
        if(LB_dist < ϵ)
            DTW_dist = DTW(Q, P);
            if(DTW_dist < ϵ)
                Neigbors.add(Q)
            EndIf
        EndIf
    EndFor
Return Neigbors;
```

are applicable to general metric spaces. A number of approaches have recently been proposed, including DBSCAN (Ester, Kriegel, Sander, & Xu, 1996), DBCLASD (Xu, Ester, Kriegel, & Sander, 1998), DENCLUE (Hinneburg & Keim, 1998), and OPTICS (Ankerst, Breunig, Kriegel, & Sander, 1999).

DENSITY-BASED CLUSTERING

The idea behind density-based clustering is, that the object density in neighbourhood of each cluster object P is sufficiently high. The basic algorithm for density-based clustering is DBSCAN (Ester, Kriegel, Sander, & Xu, 1996). In DBSCAN a flat clustering of the data is computed. A hierarchical structure, containing subclusters embedded in bigger clusters, cannot be detected. On the other hand, DBSCAN does not require the user to know the number of clusters a priori and can find clusters of arbitrary shape. Clustering with DBSCAN relies on two parameters: *MinPts*, specifying a minimum number of objects in a neighbourhood, and ε, specifying the radius of the neighbourhood.

These two parameters determine a density threshold for clustering. DBSCAN discriminates three types of objects. An object is called a core object, if it is in the interior of a cluster. In this case at least *MinPts* objects are in its ε-neighbourhood. A border object is in the neighbourhood of a core object, but contains fewer objects than *MinPts* in its eps-neighbourhood. A noise object is neither a core nor a border object. Any two core objects, which are close enough-within the distance ε - are put into the same cluster. Any border object that is close enough to a core object is put into the same cluster as the core object. Noise objects are stored elsewhere (cf., Ester, Kriegel, Sander, & Xu, 1996) for a formal description of DBSCAN), see (Shao et al., 2010) for a first application of DBSCAN to the problem of fiber clustering.

OPTICS (Ankerst, Breunig, Kriegel, & Sander, 1999) is a hierarchical density-based clustering algorithm, which emerges from DBSCAN. In contrast to DBSCAN, OPTICS does not assign cluster membership, but orders objects of a dataset D into a hierarchical cluster structure. This structure contains clustering information, which could also be achieved by all possible DBSCAN clusterings with respect to distances ε' that are smaller than the generating distance ε. To describe the hierarchical cluster structure, each object is characterized by two measures, its core-distance and its reachability-distance. Formally this is captured by the following definitions:

Definition 1: (Core-distance of an object P) Let $P \in D$, $MinPts \in N$, $\varepsilon \in R$, $N_\varepsilon(P)$ be the ε-neighbourhood of P, $MinPts_{dist}(P)$ be the distance from P to its *MinPts*-nearest neighbour. The core-distance of object P is defined as:

$$CoDist(P) = \begin{cases} UNDEFINED, |N_{\varepsilon}(P)| < MinPts \\ MinPts_{dist}(P), otherwise. \end{cases}$$

The core-distance of an object P measures its local density. It is defined as the *MinPts*-nearest neighbour distance of P if P is a core object. Otherwise it is UNDEFINED.

Definition 2: (Reachability-distance of O with respect to object P) Let P, O $\in$ D, *MinPts* $\in$ N, $\varepsilon \in$ R, $N_{\varepsilon}(P)$ be the ε-neighbourhood of P, the reachability distance of object O w.r.t P is defined as:

$$ReDist(O, P) = \begin{cases} UNDEFINED, |N_{\varepsilon}(P)| < MinPts \\ max(CoDist(P), dist(O, P)), otherwise \end{cases}$$

Figure 4 illustrates the notions of core-distance and reachability-distance. The reachability-distance of Q w.r.t. P is the core-distance of P while the reachability-distance of R w.r.t. P is the distance between P and R.

The result of the algorithm OPTICS is an ordering of the data set reflecting the hierarchical cluster structure. More precisely, during the run of the algorithm, OPTICS computes for each object the core distance and a suitable reachability distance and writes this information to an output file. OPTICS starts with an arbitrary unprocessed object O, sets its reachability distance to UNDEFINED and determines its core distance. The object O is now processed and written with its core and reachability distance to the output. If O is a core object, all objects in the ε-neighbourhood of O are retrieved and inserted into a data structure called *seed list*. The seed list is a priority queue and the objects in the seed list are sorted according to the minimum reachability distance with respect to any of the objects processed before. As next object, OPTICS always selects the top object of the seed list, or in case the seed list is empty, an arbitrary unprocessed object from the data set. Consider the case that the start object O is a core object and object P is the top object in the seed list which has minimum reachability distance w.r.t. O. Object P is processed now, which implies that P is written with its core distance and reachability distance to the output. If P is a core object, all objects in the ε neighbourhood are added to the seed list. If necessary, the seed list is updated: Some objects already in the seed list may have a smaller reachability distance from P than from O. The algorithm terminates as soon as all objects have been processed. In this ordering, for every object the smallest reachability distance w.r.t. the preceding objects is determined.

Figure 4. Illustration of core-distance and reachability-distance. The radius of the solid circle indicates the generating distance ε while the radius of the dashed circle is the MinPts-nearest neighbour distance of P (core-distance).

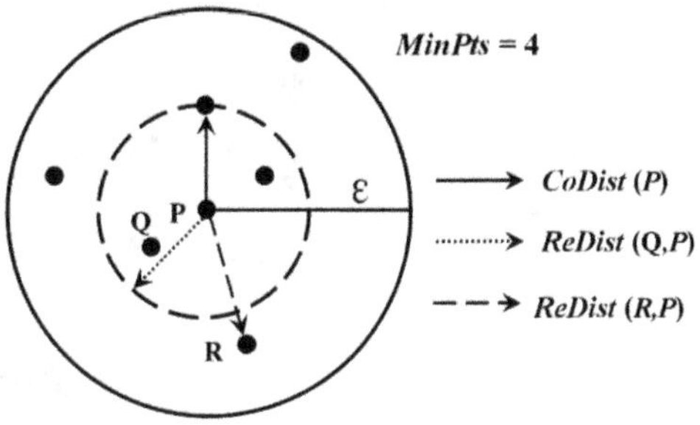

Finally, the structure of the data ordering can be visualized in a reachability plot. This reachability plot is a 2D plot showing the objects' position index on the *x*-axis and the objects' reachability-distance on the *y*-axis. The reachability plot provides a visual representation of the cluster structure of the data from the density-based point of view and enables an intuitive way to find clusters. Valleys

Figure 5. Illustration of a reachability plot. Left: 2D sample data set, where different capital letters indicate hierarchical clusters with different density, shape and size. Right: the reachability plot of the sample data set, the different valleys of this plot indicate the clusterse.

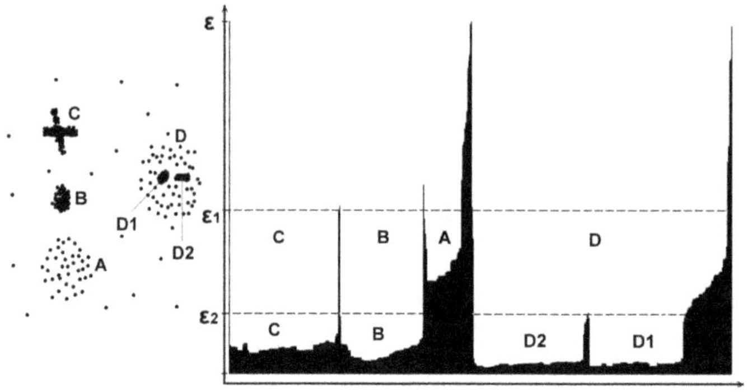

in this plot indicate clusters, the deeper the valley, the denser the cluster. Clusters can be obtained by cutting the reachability plot with a horizontal line. An example of a reachability plot for a 2D data set is presented in Figure 5. The reachablity plot provides information about the general number of clusters, the densities of different clusters, and the hierarchical structure of the data. Another visual exploration paradigm which can simplify navigation and clustering of complex fiber structures was introduced by (Jianu, Demiralp, & Laidlaw, 2009). They use dendrograms from hierarchical clustering together with a two dimensional embedding of tract similarity based on a spring model following Hook's law.

Like DBSCAN, OPTICS requires the two parameters: *MinPts* and ε. For DBSCAN it is not trivial to find the optimal parameters for clustering (Ester, Kriegel, Sander, & Xu, 1996). OPTICS, however, is rather insensitive to the input parameters (Ankerst, Breunig, Kriegel, &

Sander, 1999). For all experiments in this paper which cover very different data structures *MinPts*=10 and ε=30 have been used. See the section of Discussion for the parameter selection.

FIBER CLUSTER EXTRACTION

The simplest way to obtain clusters in a reachability plot is to select a suitable parameter $0< \varepsilon_i < \varepsilon$ and to cut the reachability plot with the corresponding horizontal line. Intuitively, the valleys below the line indicate the clusters. Also their left borders are part of the clusters, if they are not noise. Noise or outliers of clusters are defined by the condition that their core- and reachability distances are larger than ε_i .See for an illustration Figure 5. With parameter ε_1, we obtain four clusters A, B, C, and D. If the line is moved down to ε_2, we obtain four new clusters B, C, D1, and D2, and some data which are interpreted as outliers (e.g., the cluster A and the group on the right side of D1). This demonstrates that a global threshold may not be sufficient, to extract all meaningful clusters (e.g., A with ε_2), since clusters can have varying densities. Therefore, two improved methods are proposed for fiber clustering.

Interactive fiber cluster extraction. To extract fiber bundles with different densities, it is desirable to use a series of localized parameters. We have implemented a novel software to enable such interactive fiber clustering. The parameters can be selected from a screen-graph of the reachability plot in several steps.

Figure 6. Illustration of the interactive clustering approach on a real dataset. (1a): interactive fiber cluster extraction at the first level based on the reachability plot. (1b-1c): The corresponding segmented fiber bundles in Axial and Sagittal perspectives. Different colors indicate different fiber bundles. (2a): Fiber cluster extraction at the second level. (2b-2c): The segmented fiber bundles at the second level, where the fiber bundles with red color are further split into two new fiber bundles: one with cyan color ending at A and another with the red color. Three fibers with black color ending at B are interpreted as noise.

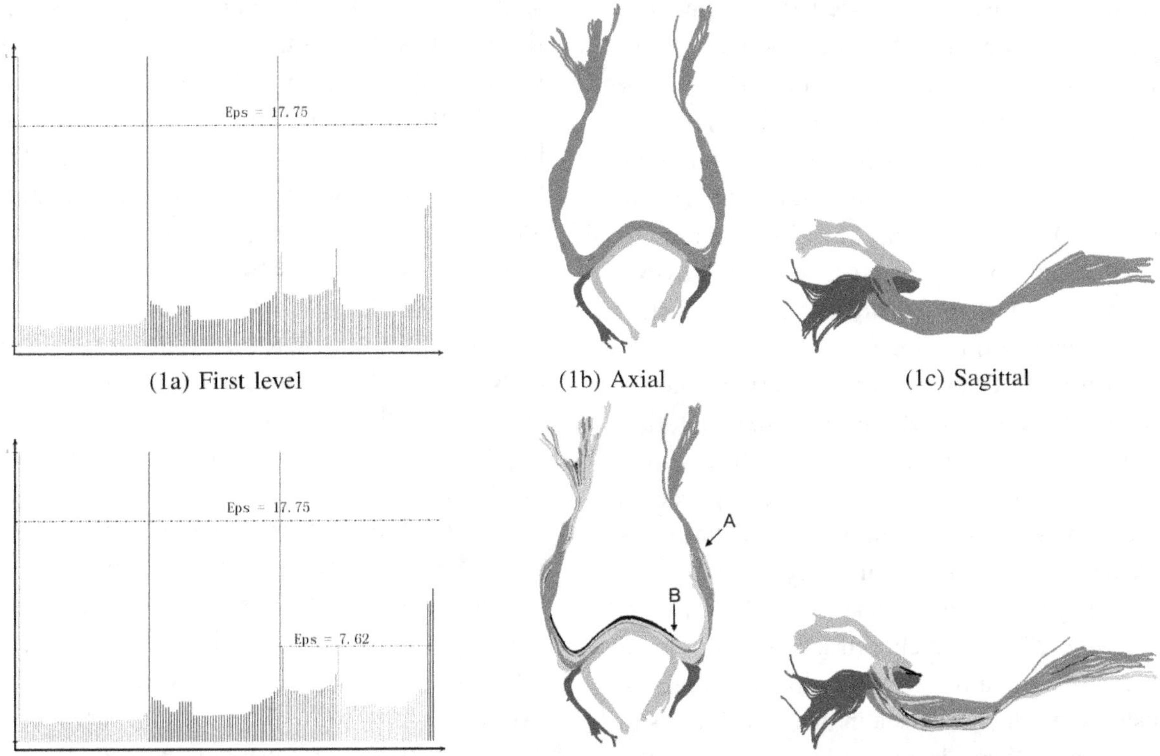

(1a) First level (1b) Axial (1c) Sagittal

(2a) Second level (2b) Axial (2c) Sagittal

1) For a first segmentation of the fibers, we select a global ε_1 (first level). This may create several large clusters $(C_1^1, ..., C_i^1, ..., C_n^1)$ and one cluster O for noise, where the core- and reachability distances are above ε_1. Quantification of noise follows the definition proposed for DBSCAN.
2) For each cluster C_i^l found at level l, there may exist several valleys inside. To obtain those smaller subclusters, we go to the next level l=l+1, selecting the parameter ε_i^l only within C_i^{l-1} and proceed like in step 1). This creates new subclusters and in some cases noise.
3) To find clusters at higher levels, we repeat step 2) until the fibers are appropriately segmented.

To select the parameter values ε on the screen-graph, users only need to "double-click" one position (x,y) in the reachability plot. The y-coordinate of the position indicates the ε, the x-coordinate is used to indicate its range. The first range is automatically global, while further ranges are localized to the cluster chosen. Figure 6 illustrates this approach on a real data set. Here,

250 fibers are seeded in the Splenium of Corpus Callosum (see the Experimental Section for details of the data and of the tractography). For the first segmentation into three clusters ε_1=17.75 is chosen (see Figure 6 (1a-1c) for the segmented reachability plot and for the fiber clusters in identical coloring). At the second level, we use ε_2=7.62, to separate the two obvious valleys within the red cluster. See Figure 6 (2a-2c) for the resulting clusters and for the three noise fibers. The arrows in Figure 6 2b indicate the endpoints of the cluster colored in cyan (A) and of the noise cluster (B), colored black. This fast and easy to use software provides users with a flexible method to quantify a visually meaningful clustering of a reachability plot.

Automatic fiber cluster extraction. The interactive method is useful to get a segmentation of data sets with a simple cluster structure, which can be inspected visually. For more complex data sets, however, an automatic segmentation is needed. To achieve this goal, we adapted the Tree Clustering algorithm (Sander, Qin, Lu, Niu, & Kovarsky, 2003), which extracts a hierarchical clustering from a reachability plot and creates a cluster tree. The basic idea in this algorithm is to identify significant local maxima separating the valleys in the reachability plot and to use these points for cluster extraction. A local maximum is regarded as significant if its height is well above the level of the valleys to its right and left side.

To determine the clusters between the local maxima, two parameters are introduced: *MinSize*, specifying the minimum number of fibers in a cluster or the minimum scale of segmentation, and *Ratio*, which determines if the local maximum is sufficiently above the neighbouring valleys to use it as a split point between two clusters.

We give a short description of our adaption of the Tree Clustering algorithm, which can now also segment highly irregular reachability plots. First, we collect all points P whose reachability distances (RD) are local maxima of the right and left neighbourhoods that enclose at least *MinSize* points. This set of points P is then sorted in descending order according to their RD P.r. Clusters are determined from this list by recursively removing the point P with largest RD, possibly separating clusters, until the list is empty. P is regarded as split point SP of two neighbouring clusters, if the median values of the RDs in both valleys are significantly lower than *SP.r* (*median*/*SP.r* < *Ratio* < 1). Noise points N within a cluster are detected by the condition *N.r* > *SP.r* and *CoDist*(N) > *SP.r*. Clusters without noise are listed in a tree graph, and noise is collected separately. There are two places in the tree where new nodes can be added: they can become children of the current node, or children of the parent node, replacing the current node. If the RDs of the current and parent split points are close, the newly created nodes are attached to the parent node instead of the current node itself.

Figure 7 illustrates the application of this method to the same data set which was used for interactive clustering. The reachability plot (Figure 7(a)) and the cluster tree (Fig. 7(b)) indicate the same fiber clustering and noise, which is presented in Figure 6. *MinSize*=10 and *Ratio*=0.7 are applied. See section of Discussion for the parameter setting.

EXPERIMENTAL RESULT AND ANALYSIS

In this section, we present a series of numerical experiments on synthetic and on real data to explore the efficiency and effectiveness of our fiber clustering approach. All algorithms are implemented in Java and the fiber visualization functions are programmed in MatLab. The calculations have been performed on a workstation with 2.4 GHz CPU and 2.0 GB RAM. We first present a comparison of fiber similarity measures and evaluate then different clustering algorithms with respect to outliers. Finally, an application

Figure 7. Illustration of automatic fiber clustering. (a): Automatic fiber clustering based on the reachability plot with ratio= 0.7, minSize = 10. (b): The cluster structure of the data set.

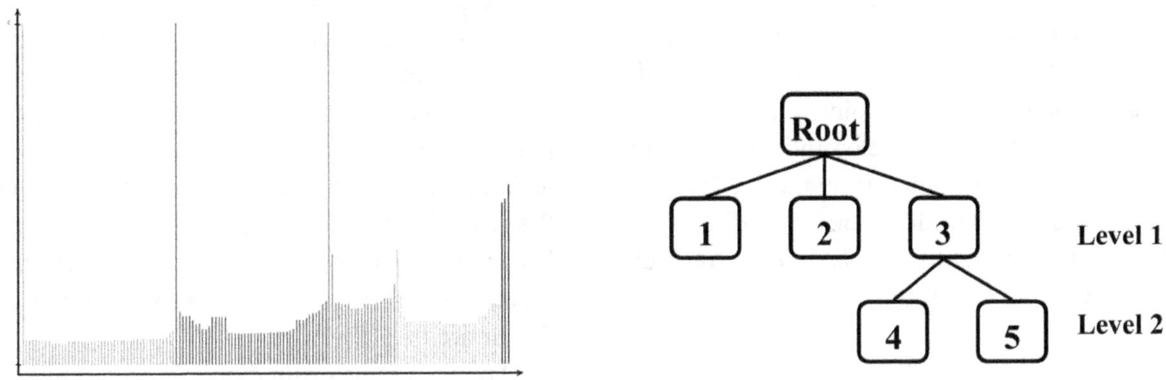

(a) Reachablity Plot

(b) Fiber Cluster Structure

of hierarchical density-based fiber clustering to special real fiber sets is presented.

CLUSTER VALIDATION MEASURE

To compare clustering results for different approaches with a ground truth, an information-theoretic external cluster-validity measure (Dom, 2001) is used. We give a short description of this measure: A clustering of a set of fibers can be described by a set of cluster labels, mapping every fiber to a cluster. The evaluation scheme measures how useful the calculated cluster labels are as predictors of the ground truth cluster labels. In contrast to measures like the Rand Index (Halkidi, Batistakis, & Vazirgiannis, 2001) or the Cluster Purity (Wang, Wirth, & Wang, 2007), different numbers of clusters can be compared. The clustering quality measure is defined as the entire encoding cost, Q(C, K), which is the sum of the empirical conditional entropy H(C|K) and of the code length for the number of clusters CL(C|K):

$$Q(C, K) = H(C|K) + CL(C|K) \quad (14)$$

Where C is the set of ground truth labels and K equals the set of calculated labels.

$$H(C \mid K) = -\sum_{c=1}^{|C|}\sum_{k=1}^{|K|} \frac{h(c,k)}{n} log \frac{h(c,k)}{h(k)}$$

$$CL(C \mid K) = \frac{1}{n}\sum_{k=1}^{|K|} log \binom{h(k)+ \mid C \mid -1}{|C|-1}$$

where h(c, k) is the number of fibers labeled within class C with label c and within class K with label $h(k) \equiv \sum_c h(c,k)$. $|K|$ and $|C|$ is the number of calculated clusters and the number of ground truth clusters respectively. The smaller the value of the entire encoding cost, the better is the quality of the clustering results.

FIBER DATA

To evaluate our approach, realistic human brain data are taken from the open source software "Slicer3-3.4", folder "surgery case" (http://www.slicer.org/). This measurement comprises 55 non-collinear directions for the diffusion weighted images, with a *b-factor*=1000 s/mm^2, and 5 acquisitions for the reference, with a *b-factor* =0 s/mm^2. The brain volume contains 256×256×70 voxels with size 1 × 1×2.6 mm^3. For fiber tracking the numerical Runge Kutta method (4-th order) is applied. The seed points are given in two ways: First, small fiducial seed regions are defined by specifying their central points within the map of Fractional Anisotropy (FA); second, ROI seeding is defined by all seed points within a larger constrained volume. The deterministic tracks for fiber clustering are started from the seed points, following the principal diffusion in both directions. They are stopped for too low FA and for too high curvature of the tracks.

EXPERIMENTS ON FIBER SIMILARITY MEASURE

To explore different similarity measures, a realistic set of fibers is obtained by specifying 6 fiducial seed regions manually (see Figure 8 (a) for the seed regions). They are located in the internal and external Capsules and in the Corpus Callosum. After fiber tractography, gold standard fiber clusters (ground truth) are created with the help of an experienced physician (See Figure 8 (b)). This gold standard includes 372 fibers and shows 6 anatomically meaningful fiber bundles assigned by C1-C6 with different colors and one group of noise or outliers (3 fibers, black color). These fibers *f1*,*f2*,*f3* differ from their neighbours by shape and length.

Based on theses data, we compare our similarity measure (DTW) with two frequently used measures: Mean of closest point distance (MCP) and Hausdorff distance (HDD). For all three measures the reachability plots are calculated and presented in Figure 8 (1a-1c). The interactive clustering method is applied to achieve the result which is closest to the ground truth. For this purpose a global threshold is sufficient, see Figure 8 (2a-2c).

It is clear from the reachability plots that DTW and MCP separate the data into the 6 gold standard clusters (6 big valleys) of the ground truth, whereas HDD already segments the data for a global threshold into 10 valleys. The reachability plot of DTW shows more local maxima than that of MCP, indicating more information about possible cluster separations or noise. For further comparison between DTW and MCP, we focus on the noise fibers indicated in Figure 8 (1a-1b) in black color. Whereas MCP does not isolate *f1*,*f2* from the neighbours, DTW matches the ground truth. As MCP averages the minimum distances of point pairs from one fiber to another, it seems to ``smooth out" the information in the data more than DTW.

Apart from the evaluation of effectiveness of a fiber similarity measure, the efficiency (computation cost) should also be considered. For this purpose we perform experiments on a range search (search the $\varepsilon -$ neighbourhood fibers for one fixed fiber), which is the most time consuming step in fiber clustering. The involved similarity measures include: DTW with lower-bounding distance LB, DTW, MCP and HDD. The number of test fibers in the data set range from 1000 to 5000. Figure 9 presents the computational cost of range search for increasing numbers of fibers.

In Figure 9 we see that the efficiency of DTW outperforms that of MCP and HDD. LB further improves the efficiency of DTW. The larger the size of the data, the more the advantage of the LB technique becomes apparent. For example, LB costs about 60% and 35% of the computation time needed for MCP, when they are applied to 1000 and 5000 fibers respectively. Compared with DTW, LB improves efficiency by about 50% for 5000 fibers.

EXPERIMENTS ON FIBER CLUSTERING

In this section, we perform experiments on synthetic data as well as on real brain data using DTW with LB. The clustering method OPTICS is first compared to two different approaches: Spectral clustering (SC) (Ng, Jordan & Weiss, 2001), and

Figure 8. Experimental results of different fiber similarity measures. (a): Seed points, (b): the gold standard, (1a-1c): reachablity plots for DTW, MCP and HDD, (2a-2c): the corresponding fibers in identical coloring.

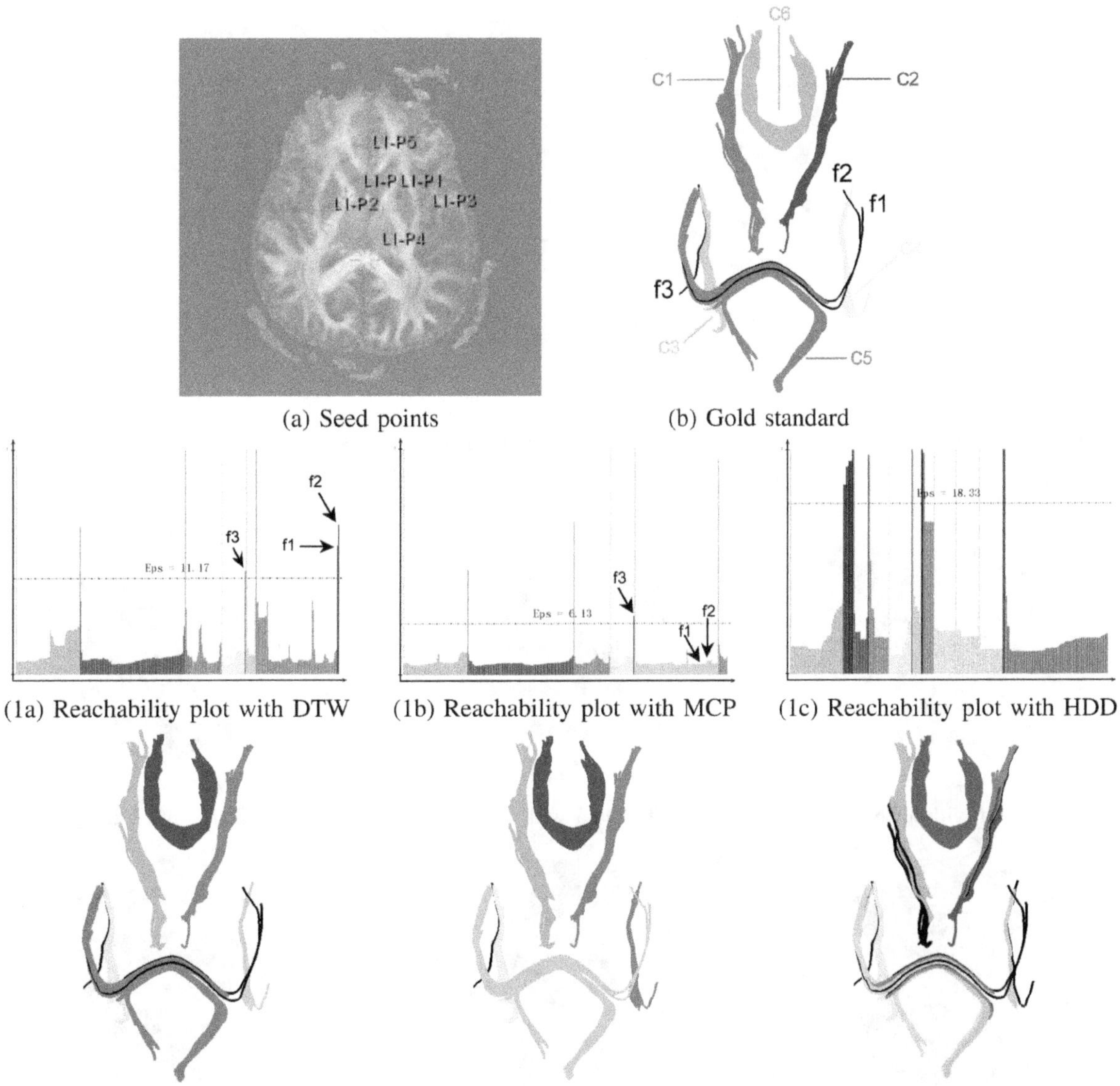

(a) Seed points (b) Gold standard

(1a) Reachability plot with DTW (1b) Reachability plot with MCP (1c) Reachability plot with HDD

(2a) Clustering Results with DTW (2b) Clustering Results with MCP (2c) Clustering Results with HDD

Hierarchical clustering (Single Link, SL) (Zhang, Demiralp, & Laidlaw, 2003).

Synthetic Data. The ground truth of our synthetic data in three dimensions includes five clusters of straight lines and two clusters of helices (See Figure 10 (a)), these clusters are composed of 410 individual fibers. To test the sensitivity to variations within the fibers we add gaussian noise to the fibers. For every point $(x(t), y(t), z(t))$ of a synthetic fiber, we modify the coordinates by $(x(t)+\varepsilon(t), y(t)+\varepsilon(t), z(t)+\varepsilon(t))$, where $\varepsilon(t)$ is Gaussian noise (Figure 10 (b)). Moreover, due to limitations of the experiment and of the tracking method, outliers may appear in fiber data.

Figure 9. Comparison of efficiency with different fiber similarity measures

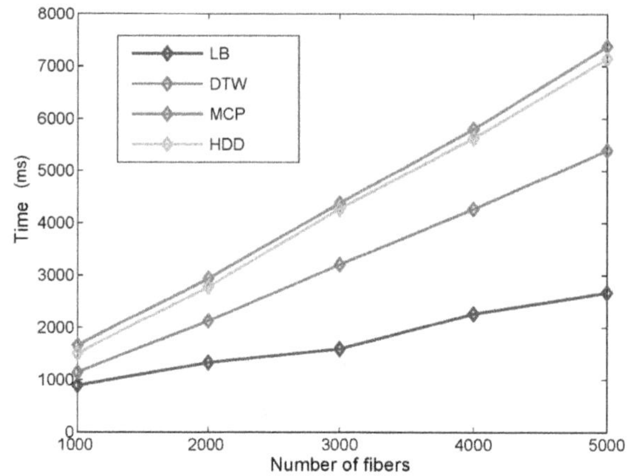

Figure 10. Different fiber clustering results based on synthetic data

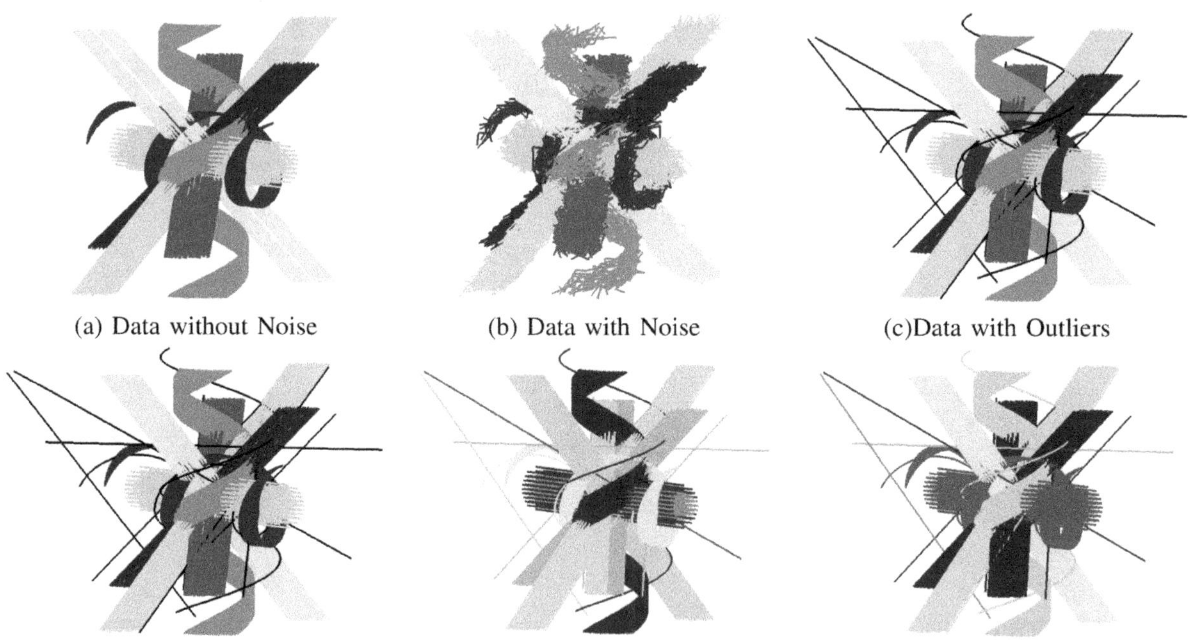

(a) Data without Noise (b) Data with Noise (c)Data with Outliers

(d) Density-based Clustering (e) Spectral Clustering (f) Single Link

To evaluate the three clustering methods with respect to such outliers, we add 8 outlier lines and 2 outlier helices to the synthetic data (Figure 10 (c)), creating a total of 420 synthetic fibers.

For all three clustering methods the ground truth presented in Figure 10 (a-b) is reproduced perfectly. The situation is different for Figure 10 (c). Only OPTICS can reproduce the ground truth, if the outliers are included, see (Figure 10 (d-f)). For SC the user has to specify the desired number of clusters (8 clusters are applied in (Figure 10 (e)). For SL either the perfect split of the dendro-

Figure 11. Dendrograms and reachability Plots for the synthetic data

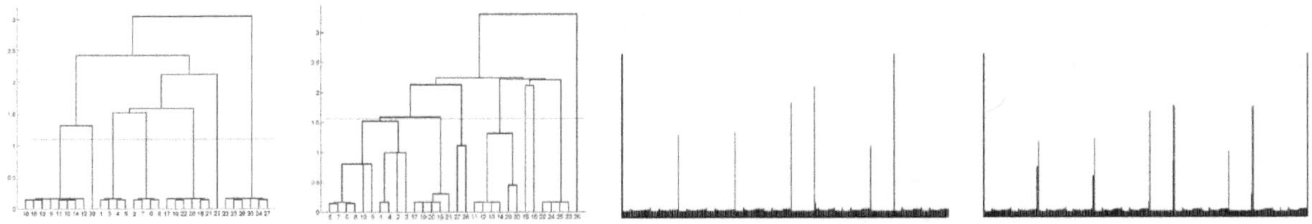

Data without outliers | Data with outliers | Data without outliers | Data with outliers

(a) Dendrogram Tree | (b) Reachability Plot

Table 1. Comparison of different fiber clustering without and with outliers

Measures	Performance without and with Outliers		
	Conditional-entropy	*Code-length*	*Encoding-cost*
OPTICS	0.0|0.0	0.304|0.358	**0.304|0.358**
SL	0.0|0.515	0.304|0.311	**0.304|0.826**
SC	0.0|0.775	0.304|0.232	**0.304|1.001**

gram has to be chosen or, like in SC, the number of clusters must be given (8 clusters are chosen). For the perfect split some information theoretic methods, like Minimal Description Length (Hansen & Yu, 2001), can be used.

Figure 11(a-b) shows the dendrograms and reachability plots for the synthetic fibers without (410 fibers) and with the 10 additional noisy fibers. Only the top 30 nodes in the dendrogram are visualized: the total number of layers is 419. We see that the dendrogram is appreciably complicated by adding a few outliers. The reachability plots in both cases, however, are robust to the outliers and already indicate visually the 7 clusters and the outliers.

These results are quantified in Table 1, where the conditional-entropy, the code-length and the Encoding-cost are given for the synthetic data without and with outliers. The strong impact of only few outliers on the clustering results is apparent for SL and SC.

Applications of hierarchical density-based clustering to real data. In this section, we apply OPTICS and automatic clustering on two real data sets with very different clustering structure, to demonstrate the method's flexibility.

Real Data 1: For Figure 12, we seeded 1768 fibers from 5 fiducial seed regions, chosen according to (Mori, Wakana, Nagae-Poetscher, & Zijl, 2005). Six big fiber groups are thus created, comprising two branches of the Corticospinal Tracts (red, magenta), part of the middle Cerebellar Peduncle (cyan), two parts of the Cingulum (yellow, blue) and a smaller section of the Corpus Callosum (green). Figure 12 (a) shows the reachability plot and its colored segmentation, which corresponds to the cluster structure in Figure12 (b). A low *Ratio*=0.2 and *MinSize*=30 are used to achieve a coarse clustering. Finer details of the structures within the clusters should not be detected. Figures 12 (1a-1c) present the result in different perspectives. The six clusters are well detected, but some noise is found as well. Noise is presented in Figure 12 (1d).

Real Data 2: Figure 13 presents a segmentation of the Corpus Callosum. Starting from the mid-sagittal slice of an FA-map within the Corpus Callosum, 1100 fiber tracts are calculated by ROI

Figure 12. Automatic fiber clustering on the real data 1. (a): Automatic clustering of the fibers with parameter ratio=0.2, MinSize=30 based on the reachablity plot. (b): The fiber cluster structure of the data, where the data are split into 6 fiber bundles; (1a-1c): Fiber clustering results, the different color indicates various fiber bundles and black color means outliers. (1d): The outliers in axial perspective.

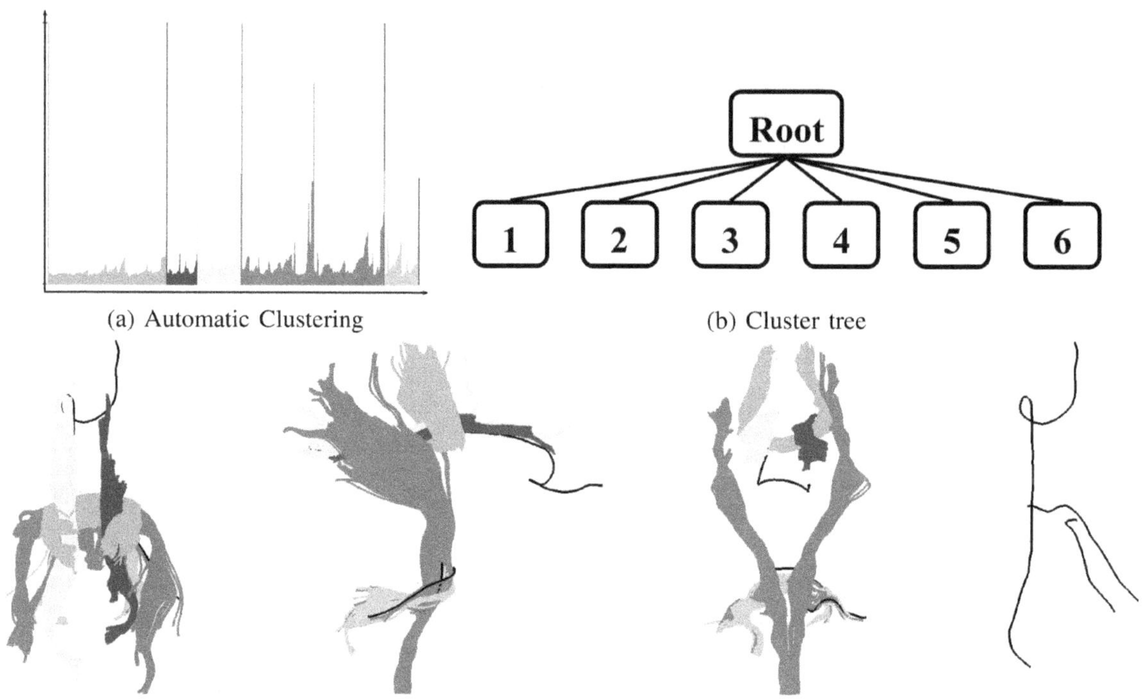

(a) Automatic Clustering (b) Cluster tree

(1a) Axial (1b) Sagittal (1c) Coronal (1d) Axial (Noise)

seeding of the slicer data. The main part of the fibers connects cortical areas in approximate mirror-image sites. A smaller subset is built of commissural trajectories to the Temporal Lobes (Mori, Wakana, Nagae-Poetscher, & Zijl, 2005). Automatic fiber clustering is applied to the reachability plot of these data with *Ratio*=0.7 and *MinSize*=30 to detect also subclusters within any hierarchy. The final reachability plot with colored segmentation and the cluster tree are given in Figure 13 (a-b). Automatic Tree Clustering creates three levels, as illustrated in Figure 13 (1a-1c). Noise fibers are presented in black color. At the first level, only three big clusters are detected. This flat clustering result can also be achieved by an application of DBSCAN, see (Shao et al., 2010). Now, the clustering hierarchy is detected. The second level splits up cluster number three (blue color) into four clusters. At the third level, a finer segmentation into nine clusters of similar fibers is performed. Noise is given separately in Figures 13 (2a-2c). The final compact clustering without noise is given in Figures 13 (3a-3c) in three perspectives.

DISCUSSION

We present in this paper a novel framework for clustering of white matter tracts. First, we introduce a novel similarity measure DTW between two fibers which is based on dynamical time warping. This method, developed for time series which are out of phase, is adapted to comparisons

Figure 13. The automatic fiber clustering result for Corpus Callosum. (a):Automatic clustering of fibers with parameters ratio=0.7, MinSize=30. (b): The Cluster tree of Corpus Callosum. (1a-1c): Clustering results at three levels respectively. (2a-2c): The noise fibers at different levels. (3a-3c): The different views of clustering results at level 3 (without noise fibers).

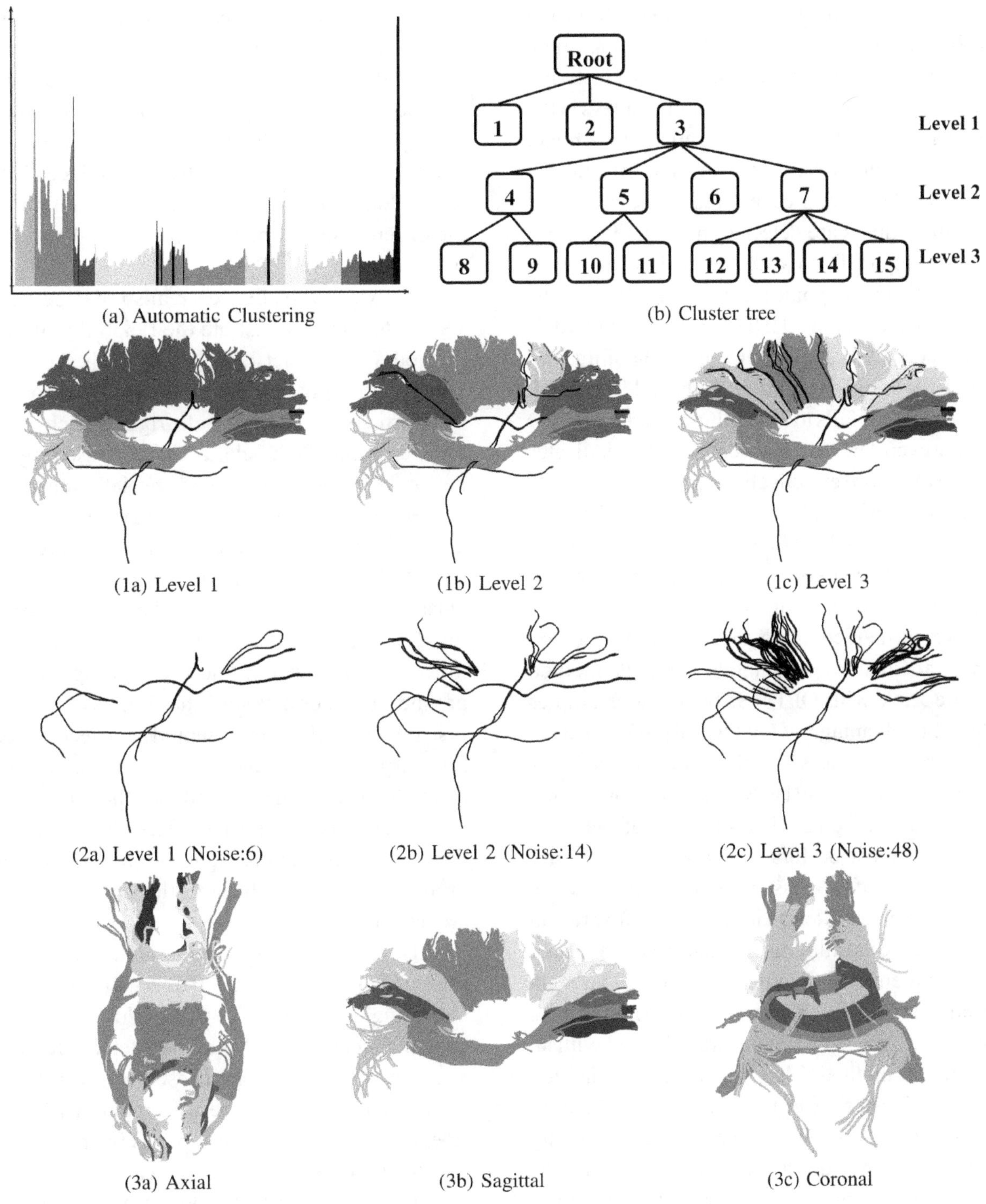

between space curves. Then, a convenient lower bound is introduced to speed up the calculation of this measure. Second, the robust hierarchical density-based clustering method OPTICS is applied. To the best of our knowledge, this is the first application of OPTICS to the fiber clustering problem. The outcome is a reachabiltiy plot, which gives a transparent visualization of the clustering structure of the fibers. The clusters or valleys in the reachability plot are then determined by a novel interactive method and by a new modification of an automatic method, called Tree Clustering. Noise or outlier fibers are detected in a local fashion.

To compare DTW with the standard measures, mean of closest point distance (MCP) and the Hausdorff distance (HDD), a gold standard is introduced. We show that the reachability plot for DTW perfectly reproduces the gold standard, whereas MCP fails to detect the outliers and HDD cannot even detect the main clusters. This indicates that DTW more effectively maps the information about clustering into the reachability plot than MCP and HDD do. Also efficiency is highest for DTW, which can even be improved by the use of a lower bound technique.

We further compare the clustering method Spectral Clustering and the hierarchical method Single Link with OPTICS. Synthetic examples show no advantage of OPTICS for regular data, but more robustness if outliers are included.

The result of OPTICS is a reachability plot visualizing the clustering structure of the data. This plot must be further analyzed to quantify the location of the clusters. For this purpose we introduce an easy to use interactive software and demonstrate its utility on a realistic fiber set seeded in the Splenium of Corpus Callosum. The main bundles are reasonably detected, also noise fibers are found. The same results are achieved with an automatic method. This method is an adaption of Tree Clustering to the irregular reachability plots based on DTW. To make Tree Clustering more robust, we use the median to calculate the significance of the split points and eliminate noise fibers at every recursive clustering step. Noise or outliers are identified by a criterion defined in density based clustering (Ankerst, Breunig, Kriegel, & Sander, 1999). Our method detects noise in a local fashion, as at every level of the cluster hierarchy noise can be detected. A fiber in a cluster is defined as noise, if its core and reachability distances are above the reachability distance of the corresponding split point. Other concepts of local outliers or noise in density based clustering can be found in (Breunig, Kriegel, Ng, & Sander, 2000). A comparative exploration of the different noise concepts is beyond the scope of this work.

Finally two very different realistic data sets are segmented, to demonstrate the flexibility of our method. For data set 1, including partly the Cingulum (left, right branch), the Corpus Callsoum, the Corticospinal Tracts (left, right branch) and the Cerebellar Peduncle, a coarse clustering is performed. This is achieved by special parameters in Tree Clustering. For *Ratio*=0.2 the finer details of any hierarchy in the clusters are suppressed and *MinSize*=30 prevented too small clusters. The cluster tree shows just one level of a flat clustering. Ignoring any hierarchy, the fiber clusters fit to a segmentation into the 6 big groups of the ground truth. For data set 2 the Corpus Callosum is segmented, a hierarchy should be detected. For this purpose the parameters *Ratio*=0.7 and *MinSize*=30 are applied and produced three levels in the hierarchical cluster tree. The clustering fits well to the anatomical expectation, looking for groups with similar fibers. At the end 11 clusters are detected.

Also, at every level, noise is detected in data set2, which comprising 1100 fibers. The first level clustering detects 6 noise fibers; this adds up to 14 at the second level. Finally we detect a total of 48 noise fibers or outliers. These numbers reflect, on the one hand, the degree of irregularity in the fiber data set, indicating erroneous fiber tracking due to wrong seed points, experimental noise, or partial volume effects. On the other hand, as mentioned, these numbers may also be influenced by the special definition of noise applied.

Our approach of automatic fiber clustering needs 4 parameters to be specified. For the algorithm OPTICS *MinPts* and *eps* must be given. For Tree Clustering, *Ratio* and *MinSize* are needed.

All calculations in this paper with OPTICS are performed with the same parametrization: *MinPts*=10 and *eps*=30. This insensitivity of OPTICS to its parameters is discussed by Ankerst et al. (1999), and greatly simplifies the application of OPTICS. Ankers et al. show that there is a broad range of possible parameters for which we can see the same clustering structure of the data in the reachability plots. Important is that *eps* is sufficiently large. The smaller we define *eps*, the more objects have an UNDEFINED reachability distance. Consequently, clusters with low density may disappear.

For Tree Clustering, the two parameters are more versatile and their selection should be guided by the user's intention: *MinSize* constrains the resolution or scale of clustering, preventing very small clusters. *Ratio* determines whether a flat or more hierarchical clustering should be detected. The larger *Ratio*, the more valleys in the reachability plot is accepted as clusters, enabling the detection of the hierarchy within the reachability plot.

CONCLUSION

We present a first application of hierarchical density-based clustering to the fiber clustering problem. A novel similarity measure is introduced, and its efficiency is increased by a lower bounding technique. The algorithm OPTICS is applied to calculate reachability plots. These plots present the cluster structure of the fiber data in a transparent manner. They are the basis for final cluster detection. We introduce for this purpose a novel interactive and an automatic cluster extraction method. Our methods compare well with existing techniques and show advantages with respect to transparency, flexibility, effectivity, efficiency and outlier robustness. The approach is tested successfully on several synthetic and real data sets.

REFERENCES

Aach, J., & Church, G. (2001). Aligning gene expression time series with time warping algorithms. *Bioinformatics (Oxford, England)*, *17*, 495–508. doi:10.1093/bioinformatics/17.6.495

Ankerst, M., Breunig, M. M., Kriegel, H. P., & Sander, J. (1999). OPTICS: Ordering Points to Identify the Clustering Structure. In *Proceedings of the ACM SIGMOD*, Philadelphia (pp. 49-60).

Breunig, M. M., Kriegel, H. P., Ng, R. T., & Sander, J. (2000). LOF: Identifying Density-Based Local Outliers. In *Proceedings of the ACM SIGMOD 2000 Int. Conf. Management of Data*, Dallas, TX (Vol. 29, No. 2, pp. 93-104).

Brun, A., Park, H. J., Knutsson, H., & Westin, C. F. (2003). Coloring of DTMRI fiber traces using Laplacian eigenmaps. In *Proceedings of the Ninth Int. Conf. on Computer Aided Systems Theory* (Vol. 2809, pp. 564-572). Berlin: Springer Verlag.

Catani, M., Howard, R. J., Pajevic, S., & Jones, D. K. (2002). Virtual in vivo interactive dissection of white matter fasciculi in the human brain. *NeuroImage*, *17*, 77–94. doi:10.1006/nimg.2002.1136

Corouge, I., Gouttard, S., & Gerig, G. (2004). Towards a shape model of white matter fiber bundles using diffusion tensor MRI. In *Proceedings of the IEEE Int. Symposium on Biomedical Imaging*, Arlington, VA (pp. 344-347).

Damoiseaux, J. S., Smith, S. M., Witter, M. P., Sanz-Arigita, E. J., Barkhof, F., & Scheltens, P. (2009). White Matter Tract Integrity in Aging and Alzheimer's Disease. *Human Brain Mapping*, *30*, 1051–1059. doi:10.1002/hbm.20563

Dempster, A. P., Laird, N. M., & Rubin, D. B. (1977). Maximum Likelihood from Incomplete Data via the EM Algorithm. *Journal of the Royal Statistical Society. Series A (General)*, *39*(1), 1–31.

Ding, Z., Gore, J., & Anderson, A. (2003). Classification and quantification of neuronal fiber pathways using diffusion tensor MRI. *Magnetic Resonance in Medicine*, *49*, 716–721. doi:10.1002/mrm.10415

Dom, B. E. (2001). *An information-theoretic external cluster-validity measure* (Tech. Rep. No. RJ10219). IBM.

Ester, M., Kriegel, H. P., Sander, J., & Xu, X. (1996). A density-based algorithm for discovering clusters in large spatial databases with noise. In *Proceedings of the Second Int. Conf. on Knowledge Discovery and Data Mining* (pp. 226-231). Portland, OR: AAAI Press.

Gerig, G., Gouttard, S., & Corouge, I. (2004). Analysis of brain white matter via fiber tract modelling. In *Proceedings of the 26th Ann. Int. Conf. IEEE Engineering in Medicine and Biology Society (EMBS)*, San Francisco (pp. 4421-4424).

Hagmann, P., Cammoun, L., Gigandet, X., Meuli, R., Honey, C. J., Wedeen, V. J., & Sporns, O. (2008). Mapping the structural core of the human cerebral cortex. *PLoS Biology*, *6*(7), e159. doi:10.1371/journal.pbio.0060159

Halkidi, M., Batistakis, Y., & Vazirgiannis, M. (2001). On clustering validation techniques. *Journal of Intelligent Information Systems*, *17*(2-3), 107–145. doi:10.1023/A:1012801612483

Hansen, M., & Yu, B. (2001). Model selection and the principle of minimum description Length. *JASA*, *96*(454), 746–774.

Hinneburg, A., & Keim, D. A. (1998). An Efficient Approach to Clustering in Large Multimedia Databases with Noise. In *Proceedings of the 4th Int. Conf. on Knowledge Discovery and Data Mining*, New York (pp. 58-65).

Huang, H., Zhang, J., Wakana, S., Zhang, W., Ren, T., & Richards, L. J. (2006). White and gray matter development in human fetal, newborn and pediatric brains. *NeuroImage*, *33*, 27–38. doi:10.1016/j.neuroimage.2006.06.009

Itakura, F. (1975). Minimum Prediction Residual Principle Applied to Speech Recognition. In *Proceedings of the IEEE Trans. Acoustics, Speech, and Signal Proc.* (Vol. ASSP-23, pp. 52-72).

Jagadish, B., Yi, H., & Faloutsos, C. (1998). Efficient retrieval of similar time sequences under time warping. In *Proceedings of the 14th Int. Conf. on Data Engineering*, Orlando, FL (pp. 201-220).

Jain, A. K., & Dubes, R. C. (1988). *Algorithms for Clustering Data*. Upper Saddle River, NJ: Prentice-Hall.

Jianu, R., Demiralp, C., & Laidlaw, D. H. (2009). Exploring 3D DTI fiber tracts with linked 2D representations. *IEEE Transactions on Visualization and Computer Graphics*, *15*(6), 1449–1456. doi:10.1109/TVCG.2009.141

Keogh, E. (2002). Exact Indexing of Dynamic Time Warping. In *Proceedings of the 28th Int. Conf. on Very Large Data Bases*, Hong Kong (pp. 406-417).

Keogh, E., & Pazzani, M. (2001). Derivative Dynamic Time Warping. *In Proceedings of the First SIAM Int. Conf. on Data Mining*, Chicago (pp. 5-7).

Klein, J., Bittihn, P., Ledochowitsch, P., Hahn, H. K., Konrad, O., Rexilius, J., & Peitgen, H. O. (2007). Grid-Based Spectral Fiber Clustering. In *Proceedings of the SPIE 6509* (p. 65091E). doi:10.1117/12.706242

Law, G. Y., & Grossman, R. (2005). Application of diffusion tensor MR imaging in multiple sclerosis. *Annals of the New York Academy of Sciences*, *1064*, 202–219. doi:10.1196/annals.1340.039

Maddah, M., Grimson, W., & Warfield, S. (2006). Statistical modeling and EM clustering of white matter fiber tracts. In *Proceedings of the IEEE Int. Symp. on Biomedical Imaging*, Arlington, VA (pp. 53-56).

Maddah, M., Mewes, A., Haker, S., Grimson, W. E. L., & Warfield, S. (2005). Automated atlas-based clustering of white matter fiber tracts from DTMRI. In *Proceedings of the Int. Conf. on Medical Image Computing and Computer Assisted Intervention*, Palm Springs, CA (pp. 188-195).

Moberts, B., Vilanova, A., & van Wijk, J. J. (2005). Evaluation of Fiber Clustering Methods for diffusion tensor imaging. In *Proceedings of the IEEE Visualization*, Minneapolis, MN (pp. 65-72).

Mori, S. (2007). *Introduction to Diffusion Tensor Imaging*. New York: Elsevier.

Mori, S., Wakana, S., Nagae-Poetscher, L. M., & van Zijl, P. C. M. (2005). *MRI Atlas of Human White Matter*. Amsterdam, The Netherlands: Elsevier.

Ng, A. Y., Jordan, M. I., & Weiss, Y. (2001). On spectral clustering:Analysis and an algorithm. In *Advances in Neural Information Proc* (*Vol. 14*, pp. 849–856). Vancouver, BC, Canada: Systems.

O'Donnell, L. J., Kubicki, M., Shenton, M. E., Grimson, W. E. L., & Westin, C. F. (2006). A method for clustering white matter fiber tracts. *AJNR. American Journal of Neuroradiology*, *27*(5), 1032–1036.

O'Donnell, L. J., & Westin, C. F. (2007). Automatic Tractography Segmentation Using a High-dimensional White Matter Atlas. *IEEE Transactions on Medical Imaging*, *26*(11), 1562–1575. doi:10.1109/TMI.2007.906785

Park, H. J., Westin, C. F., Kubicki, M., Maier, S. E., Niznikiewicz, M., & Baer, A. (2004). White matter hemisphere asymmetries in healthy subjects and in schizophrenia: A diffusion tensor MRI study. *NeuroImage*, *24*, 213–223. doi:10.1016/j.neuroimage.2004.04.036

Sakoe, H., & Chiba, S. (1978). Dynamic Programming Optimization for Spoken Word Recognition. *IEEE Transactions on Acoustics, Speech, and Signal Processing*, *26*, 623–625. doi:10.1109/TASSP.1978.1163055

Salvador, S., & Chan, P. (2004). FastDTW: Toward accurate dynamic time warping in linear time and space. In *Proceedings of the 3rd Workshop On Mining Temporal and Sequential Data (ACM KDD)*, Seattle, WA (pp. 22-25).

Sander, J., Qin, X., Lu, Z., Niu, N., & Kovarsky, A. (2003). Automatic Extraction of Clusters from Hierarchical Clustering Representations. In *Proceedings of the 7th Pacific-Asia Conf. on Knowledge Discovery and Data Mining*, South Korea (pp. 75-87).

Sankoff, D., & Kruskal, J. B. (1983). *Time Warps, String Edits, and Macromolecules: The Theory and Practice of Sequence Comparison*. Reading, MA: Addison-Wesley.

Shao, J., Hahn, K., Yang, Q., Boehm, C., Wohlschlaeger, A., Myers, N., & Plant, C. (2010). Combining Time Series Similarity with Density-based Clustering to Identify Fiber Bundles in the Human Brain. In *Proceedings of the Int. Conf. on Data Mining (ICDM), Workshop on Biological Data Mining and its Applications in Healthcare*.

Wang, X., Wirth, A., & Wang, L. (2007). Structure-based statistical features and multivariate time series clustering. In *Proceedings of the 7th IEEE ICDM*, Omaha, NE (pp. 351-360).

Xia, Y., Turken, U., Whitfield-Gabrieli, S. L., & Gabrieli, J. D. (2005). Knowledge-based classification of neuronal fibers in entire brain. In *Proceedings of the Int. Conf. on Medical Image Computing and Computer Assisted Intervention* (pp. 205-212). New York: Springer.

Xu, X., Ester, M., Kriegel, H. P., & Sander, J. (1998). A Distribution-Based Clustering Algorithm for Mining in Large Spatial Databases. In *Proceedings of the 14th Int. Conf. on Data Engineering*, Los Alamitos, CA (pp. 324-331). Washington, DC: IEEE Computer Society Press.

Zhang, S., Demiralp, C., & Laidlaw, D. H. (2003). Visualizing diffusion tensor MR images using streamtubes and streamsurfaces. *IEEE Transactions on Visualization and Computer Graphics*, *9*(4), 454–462. doi:10.1109/TVCG.2003.1260740

Zhang, S., & Laidlaw, D. H. (2005). DTI fiber clustering and cross-subject cluster analysis. In: International Society for Magnetic Resonance in Medicine (Ed.), *Proceedings of the 13th Scientific Meeting ISMRM*, Miami, FL (pp. 2727).

APPENDIX

Remark: LB is a lower bound to DTW for spatial fibers.

For two fibers $P = (p_1, \ldots, p_m)$ and $Q = (q_1, \ldots, q_n)$, for each dimension, we have $lb(p^d, q^d) < dtw(p^d, q^d)$, $d = \{1, 2, 3\}$ (Yi, Jagadish, & Faloutsos, 1998). We want to prove

$DTW(P, Q) \geq LB(P, Q)$

This is shown in the following.

$$DTW(P,Q) = \frac{\sum_{k=1}^{K}(|p_i^1 - q_j^1| + |p_i^2 - q_j^2| + |p_i^3 - q_j^3|)}{K}$$

$$\geq \frac{\sum_{k=1}^{K}(|p_i^1 - q_j^1| + |p_i^2 - q_j^2| + |p_i^3 - q_j^3|)}{m+n-1}$$

$$= \frac{\sum_{k=1}^{K}(|p_i^1 - q_j^1|)}{m+n-1} + \frac{\sum_{k=1}^{K}(|p_i^2 - q_j^2|)}{m+n-1} + \frac{\sum_{k=1}^{K}(|p_i^3 - q_j^3|)}{m+n-1}$$

$$= \frac{dtw(p^1, q^1) + dtw(p^2, q^2) + dtw(p^3, q^3)}{m+n-1}$$

$$\geq \frac{lb(p^1, q^1) + lb(p^2, q^2) + lb(p^3, q^3)}{m+n-1}$$

$$= LB(P,Q)$$

This work was previously published in the International Journal of Knowledge Discovery in Bioinformatics, Volume 1, Issue 4, edited by Xiao-Li Li and See-Kiong Ng, pp. 1-25, copyright 2010 by IGI Publishing (an imprint of IGI Global).

Chapter 20
Revealing the Origin and Nature of Drug Resistance of Dynamic Tumour Systems

Ricardo Santiago-Mozos
National University of Ireland Galway, Ireland

Imtiaz A. Khan
Cardiff University, UK

Michael G. Madden
National University of Ireland Galway, Ireland

ABSTRACT

In this paper, the authors identify the strategies that resistant subpopulations of cancer cells undertake to overcome the effect of the anticancer drug Topotecan. For the analyses of cell lineage data encoded from timelapse microscopy, data mining tools are chosen that generate interpretable models of the data, addressing their statistical significance. By interpreting the short-term and long-term cytotoxic effect of Topotecan through these data models, the authors reveal the strategies that resistant subpopulations of cells undertake to maximize their clonal expansion potential. In this context, this paper identifies a pattern of cell death independent of cytotoxic effect. Finally, it is observed that cells exposed to Topotecan have higher movement over time, indicating a putative relationship between cytotoxic effect and cell motility.

A central challenge in cancer biology is to understand of the strategies followed by cells to overcome the effect of anticancer drugs (Smith, Khan, & Errington, 2009). Cancer, a highly complex and heterogeneous disease (Heppner, 1984; Rubin, 1990), can be described as an evolving system, that can be best illustrated by cell lineages. In experimental terms, a cell lineage reflects the relationship between descendents from a common progenitor that was exposed to a given influence, such as a bioactive drug, for a time period. The behaviour of both the progenitor and the evolving progeny reveals the time-integrated response to this influence (i.e., the pharmacodynamic response). The study of cell lineages has been, and remains, of importance in developmental biology

DOI: 10.4018/978-1-4666-1785-8.ch020

(Stern & Fraser, 2001; Alvarez-Buylla, García-Verdugo, & Tramontin, 2001; Anderson, Gage, & Weissman, 2001; Ardavín et al., 2001; Dor, Brown, Martinez, & Melton, 2004; Kim & Shibata, 2002; Noctor, Flint, Weissman, Dammerman, & Kriegstein, 2001) and medicine (Bernards & Weinberg, 2002; Hope, Jin, & Dick, 2004; Tang et al., 2003; Weigelt et al., 2003; Yamamoto et al., 2003).

Here we select and apply appropriate data mining techniques that provide interpretable models on previously encoded cell lineage data (Khan et al., 2007), in order to to reveal the degree of heterogeneity of a tumour system in response to therapy, as well as the strategies (or patterns) the resistant fraction incorporates in order to overcome the effect of the anticancer drug and thus maximize their clonal expansion potential.

DATA

Biological Sample Preparation

The human osteosarcoma cell line [U-2 OS (ATCC HTB 96)23], derived from a 15-year-old Caucasian female and transfected with a fluorescent reporter cyclin B1 GFP, is selected. The cells are maintained at 37 °C and 5% CO_2 using standard tissue culture techniques. Media used is McCoys 5A modified (Sigma) supplemented with 2 mM glutamine, 100 units/ml penicillin, 100 mg/ml streptomycin, 10% fetal calf serum and 1000 mg/ml geneticin.

Cells are treated with 1 μM and 10 μM bolus dose of anticancer agent Topotecan (TPT) (Bailly, 2000). TPT is a water soluble derivative of the alkaloid Camptothecin and act as a topoisomerase I, a nuclear enzyme involved in DNA replication and repair, inhibitor (Wang, 1996). TPT is used for the treatment of a wide range of cancers, including lung (Ichinose et al., 2010), breast (Cheung et al., 2008), ovarian (Lorusso et al., 2010) and bone (Seibel et al., 2007), both in experimental and clinical contexts.

An hour post treatment, the cultured dishes are placed onto a time lapse instrument designed to capture transmission phase images from multi well plates. Image sequences are taken for 115 hours at 15-minute time intervals. The cell lineage data is encoded by ProgeniTRAK (Khan et al., 2006) and is retrieved from a cell lineage database ProgeniDB (Khan et al., 2007).

Cell Lineage Data

Information from 253 lineages is available: 168 are Control, 37 are 1μM and 48 are 10μM lineages. Each lineage is characterised as a tree where the genealogy of the Progenitor cell and its offspring is represented.

Figure 1 shows an example of a Control lineage along the experiment; we observe that Sixth Generation is reached in the 115 hours. Figure 2 shows the encoding, along with the labeling convention (refer to Table 1 for notation): an arc between two successive nodes, where both nodes represents dividing (mitosis) cells, indicates the cell cycle time or inter-mitotic-time (IMT). In labeling terms, for example, $Cell_{3\_1}$ and $Cell_{3\_2}$ in Figure 2 are sisters; both are daughters of $Cell_{2\_1}$, which is a daughter of $Cell_{1\_1}$, which is again a daughter of a Progenitor cell. The daughters of a Progenitor are termed First Generation cells ($Cell_{1\_1}$ and $Cell_{1\_2}$ in Figure 2). The daughters of the First Generation cells are termed Second Generation cells ($Cell_{2\_1}$,..., $Cell_{2\_4}$ in Figure 2), and so on. For this work, we consider nodes or cells up to the Fourth Generation, with the assumption that any living cells after Third Generation can be considered to be part of the resistant population. By the Third Generation, due to the amount of time elapsed since the administration of the drug and the cell proliferation rate, the cytotoxic effect of the drug is diminished.

We record the status of each cell in the lineage; see Table 2 for possible statuses and their defini-

Figure 1. A lineage encoded from a real progenitor cell using ProgeniTRAK, spanning up to six generations, where each green node represents cell division, red node cell death

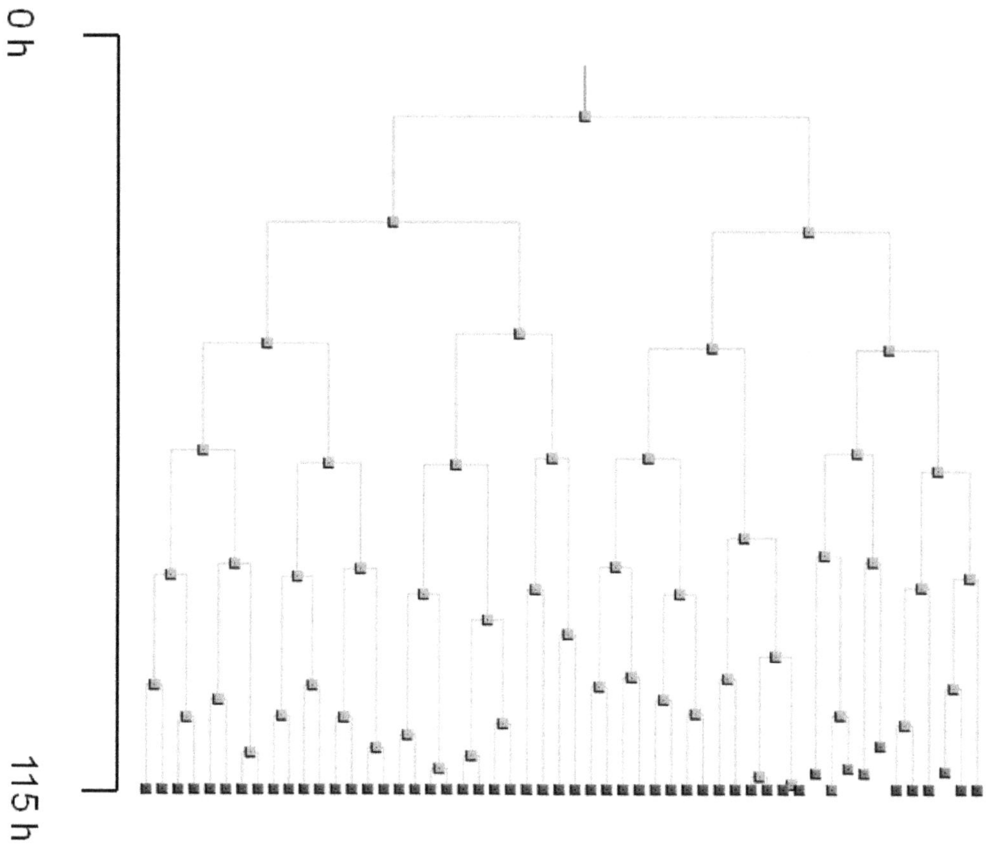

Figure 2. Generalized labeling convention for transforming real lineages into machine learning terms. For this work we included lineage data up to the third generation.

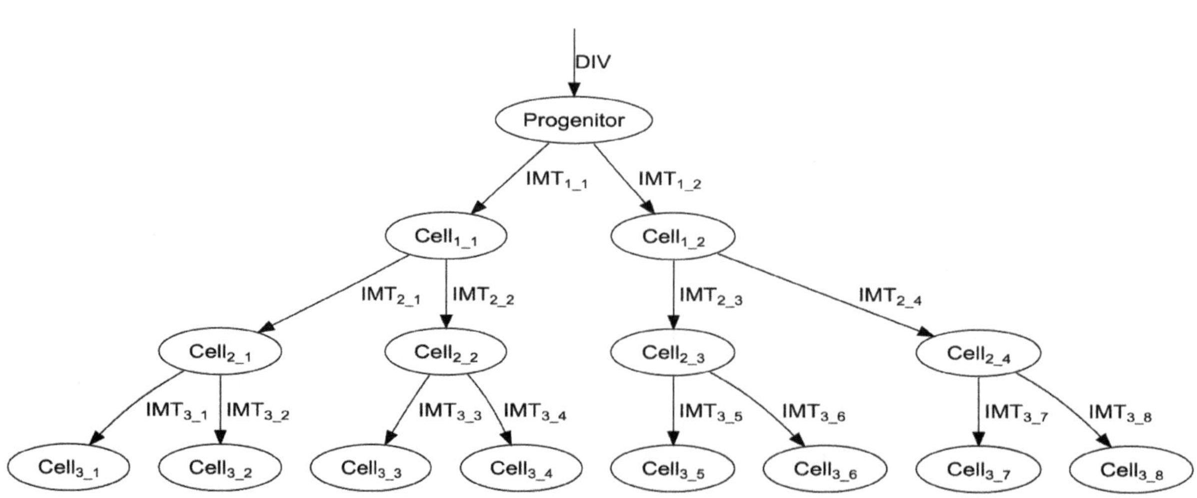

Table 1. Notation and definitions

Variable	Definition
$Cell_{I\_J}$	Cell J in Generation I.
IMT	Inter-Mitotic Time. The time needed by a cell to divide.
IMD	Inter-Mitotic Displacement. The distance travelled by a cell until its division.
DIV	The elapsed time from the start of the experiment until division of a Progenitor cell.
DIV_D	The distance travelled by a Progenitor cell from the start of the experiment until division.
$Status_{I\_J}$	Status of Cell J in Generation I.
IMT_{I_J}	The IMT for cell J in Generation I.
IMD_{I_J}	The IMD for cell J in Generation I.
Resistant	Cells whose descendants reach Fourth Generation. When applied to lineages, lineages that reach Fourth Generation.
Stressed	The opposite to resistant.
Control	Cells (lineages) that have not been exposed to drug.
Perturbed	Cells (lineages) that have been exposed to drug.
Drug	A variable that carries information to distinguish the type of lineage.

Table 2. Possible statuses for cells

Cell status	Definition
M4	The cell divides into four daughters.
M3	The cell divides into three daughters.
Unresolved	The cell status could not be measured.
Polyploidy	The cell divides into one cell.
Re–fused	The cell divides but its daughters cells fuse together.
Arrested	The cell survives without dividing until the end of the experiment.
Lost	The cell is lost from the field of view of the microscope.
Never–Born	The mother of the cell does not divide.
Death	The cell dies due to apoptosis (programmed cell death) or necrosis (uncontrolled cell death).
Divided	Normal division. The cell divides into two daughters.

tions. When a cell divides, both the IMT and its inter-mitotic displacement (IMD) are recorded. IMT refers to the time taken by a cell to divide, while IMD refers to its displacement during that time. The time needed by the Progenitor cell to divide is recorded as DIV and its displacement from the start of the experiment until it divides is recorded as DIV_D.

Data Transformation. Sister cells are sorted according their statuses (following the order given in Table 2) and when both cells divide, the one with shorter IMT goes first. For example, if both First Generation cells divide, the one with shorter IMT is labeled as $Cell_{1\_1}$ and the other as $Cell_{1\_2}$. The daughters of $Cell_{1\_1}$ are called $Cell_{2\_1}$ and $Cell_{2\_2}$, and are sorted according their statuses and IMTs as before. IMT and IMD numbering follow their cell numbering; for example, IMT_{3_5} refers to the IMT of $Cell_{3\_5}$ (see Table 1). Lineage information is codified in a vector of 44 variables:

1. DIV, DIV_D: all progenitor cells divide in our dataset.
2. $Status_{I\_J}$ (14 variables, one for each cell apart from the Progenitor cell; see Table 2 for their possible values).
3. IMT_{I_J} and IMD_{I_J} (28 variables): these are numeric values except if a cell does not divide, in which case the *missing* value is recorded.

Implications. The sorting of the data have some implications: since *Divided* is the last status in the ordering, if the first sister divides, the second sister also divides, and has a larger IMT. Of course, if the second sister does not divide, neither does the first and their statuses are dependent according the ordering in Table 2. For example, If $Status_{3\_5}$ equals *Divided* that implies that $Cell_{3\_6}$ also divided; and it implies that $Cell_{2\_3}$ divided (its mother) and this last fact implies that $Cell_{2\_4}$ divided; and finally than $Cell_{1\_2}$ divided. Therefore, when we say that $Cell_{3\_5}$ divided we are implicitly saying that $Cell_{1\_2}$, $Cell_{2\_3}$, $Cell_{2\_4}$ and $Cell_{3\_6}$ also divided, which makes this a very informative event.

STUDIES

The data are analysed to understand the behaviour of Control lineages, and we explore any deviations from this behaviour that arise when the drug is applied. We have selected four main studies, as described below: (1) cell death in Control lineages; (2) cytotoxic effect of the drug; (3) characterization of Stressed lineages; and (4) classification of Control versus Perturbed lineages.

Study of Cell Death in Control Lineages

The goal of this study is to explore *Death* events in Control lineages. The question considered is: '*is there any discernible pattern in the* Death *events*?' For this analysis, we divide the cells into two groups: the first contains the mothers of cell sisters where at least one sister dies, while in the second contains the mothers of cell sisters where both of the sisters survive. For each generation, the cell *death factor*, defined as the ratio of *Death* cells to *Divided* cells, and the dynamical behaviour of the mothers of *Death* and *Non-Death* cells (those with any status except *Death*, *Never–Born* or *Lost*) are measured and compared. Also, we construct classifiers to predict whether a cell pair is going to have at least one *Death*, given their ancestors' information. Note that it is not possible in this classification task to distinguish between sisters, as they share the same ancestors' information, even though they can be different: for example, one might divide while the other dies. Accordingly, the task is inherently noisy.

Study of the Cytotoxic Effect of Topotecan

The goal of this study is to determine the effectiveness of the drug. We consider the difference of ratios of cell events in Control, 1μM and 10μM lineages, by focusing on the following measures:

1. The ratio of *Death Cells* to *Non-Death Cells* measures the catastrophic cytotoxic effect of the drug.
2. The ratio of *Stressed Cells* to *Resistant Cells*. Stressed cells are those that manage to divide but whose offspring does not have Third Generation dividing cells. Resistant cells have an offspring with at least one Third Generation dividing cell. This ratio measures the cytotoxic effect on the current generation based on the outcome in the Third Generation, which can be interpreted as a moderated cytotoxic effect.
3. The ratio of *Full Clonal Potential Cells* (FCP) to *Partial Clonal Potential Cells* (PCP). FCP cells are those whose offspring in the Third Generation are all dividing cells. For cells

at the First Generation, full clonal potential means that they have four descendants that divide in Third Generation, and obviously for cells in the Second Generation, full clonal potential means they have two descendants that divide in the Third Generation. Partial clonal potential cells are those whose offspring have *at least one* Third Generation dividing cell but are *not* full clonal potential. This proportion measures the cytotoxic effect on the current generation based on the outcome in Third Generation resistant cells; in other words, it measures the cytotoxic effect on the cells' clonal potential.

In addition, the effect of the drug on the cell dynamics (IMT and IMD) is investigated.

Study of Stressed Lineages

The goal of this study is to identify the common strategies followed by the cells to reach the Fourth Generation. Lineages are divided into a *Stressed* group, which do not reach this generation, and a *Resistant* group (the others). Then, classification rules that identify both groups are induced.

Study of Control versus Perturbed Lineages

When the drug is applied, cell lineages are expected to be perturbed. The goal of this study is to identify what lineages have been perturbed. That is, what differences can be identified between Control and Perturbed lineages. For this study, the data is separated into two classes, one containing the Control lineages and the other containing both 1μM and 10μM lineages.

METHODS

To compare proportions of events among lineage types, the events are counted and arranged in contingency tables, and the Fisher test is used to asses if the difference between the proportions is statistically significant (Campbell, 2007). The null hypothesis is: *both lineage types have the same proportion on these events*, and the alternative hypothesis is: *the proportions are different*. As several tests are performed, the false discovery rate (FDR) must be controlled (Benjamini & Hochberg, 1995; Benjamini & Yekutieli, 2001). The FDR is the expected number of false rejections among all rejections of the null hypothesis. To control the FDR with Fisher tests, we use the method proposed in Carlson, Heckerman, and Shani (2009), with an FDR value of 0.1.

To compare the dynamic behaviours of lineages we select the empirical cumulative distribution function (ECDF), which approximates the probability that a variable is equal to or lower than any threshold. The ECDF converges to the actual cumulative distribution function as the number of samples tends to infinity (Vapnik, 1998). To compare ECDF distributions, the Kolmogorov-Smirnov two-sample test (Smirnov, 1939) is used. The null hypothesis is: *both samples come from a common distribution*, and we choose as the alternative hypothesis: *both samples come from different distributions*. To control the FDR of these tests, the general method proposed in (Benjamini & Yekutieli, 2001, Theorem 1.3) is used. The reason for using two different methods to control the FDR (one for the Fisher tests and other for the Kolmogorov-Smirnov test) is because the method proposed in Carlson et al. (2009) is more appropriate (and has more power) for contingency tables than the general procedure. To learn patterns from the data, we use RIPPER (Cohen, 1995), which is a rule inducer, and C4.5 (Quinlan, 1996), which is an algorithm to build tree classifiers, because these algorithms can provide interpretable representations of their classification decisions.

Data is also explored using Bayesian networks (Charniak, 1991; Pearl, 1988; Howard & Matheson, 2005). Continuous models are avoided

Table 3. Cell death factor for Control lineages

	Gen. 1	Gen. 2	Gen. 3
Status=*Death*	13	51	40
Status=*Divided*	234	363	591
Cell death factor	5.5%	14%	6.7%

because the distribution of the data is unknown. Instead, we consider the following methods for data discretization: equal-width bins, equal-frequency bins and the supervised discretization proposed in (Fayyad & Irani, 1993). After learning the Bayesian network from the data, the probability distributions of the variables given the values of their mothers provide valuable information about the data.

The performance of the classification algorithms is measured by leave-one-out cross-validation, and all processing steps in the data are done inside the leave-one-out loop. Finally, models that are presented for discussion in this paper are induced from all the data.

RESULTS

In this section, we show the results of applying the different methods to the data, with the aim of addressing the research questions identified above.

Study of Cell Death in Control Lineages

Table 3 summarises the results for cell death in Control lineages. There is a clear increase in the cell death factor in the Second Generation. However, the most important observation here is that cell death occurs spontaneously as part of the regulation of the population growth in the Control group. Given that there are 10 possible statuses and *Death* is not a particularly frequent one, an important pattern is found in death cells: *Death-Death* sisters pairs account for almost half of deaths in the First Generation (6 out of 13), more than half in the Second Generation (28 out of 51) and half in the Third Generation (20 out of 40).

The next analysis focuses on cell dynamics. IMT_1 (both IMT_{1_1} and IMT_{1_2}) is divided into two groups:

1. IMT_1 of cells where at least one daughter dies.
2. IMT_1 of cells where neither of the two daughters die.

Figure 3 shows the ECDFs of both groups. When comparing both curves with the Kolmogorov-Smirnov test, the p-value is 0.003, which provides statistical evidence against the null hypothesis: both datasets come from the same distribution. The IMT_1 distribution of cells in Group 1 tends to be delayed with respect to the IMT_1 distribution of Group 2. For example, 78% of cells in Group 2 have an IMT_1 below 22.5 hours, but only 46% of cells in Group 1 have an IMT_1 below this threshold. A similar effect, also statistically significant, is observed in the cells of the Third Generation (Figure 4): 78% of cells in Group 2 have an IMT_2 lower than 21 hours, but only 48% of cells in Group 1 have an IMT_2 below this threshold.

Now moving to the prediction of *Death* events in the Second Generation, we use the RIPPER algorithm with inputs DIV and IMT_1 to discover the following rule set:

1. If (IMT_1 <= 18.75 hours), no daughter cell dies.

Figure 3. Comparison of the ECDFs of IMT_1 for cells that die and cells that do not

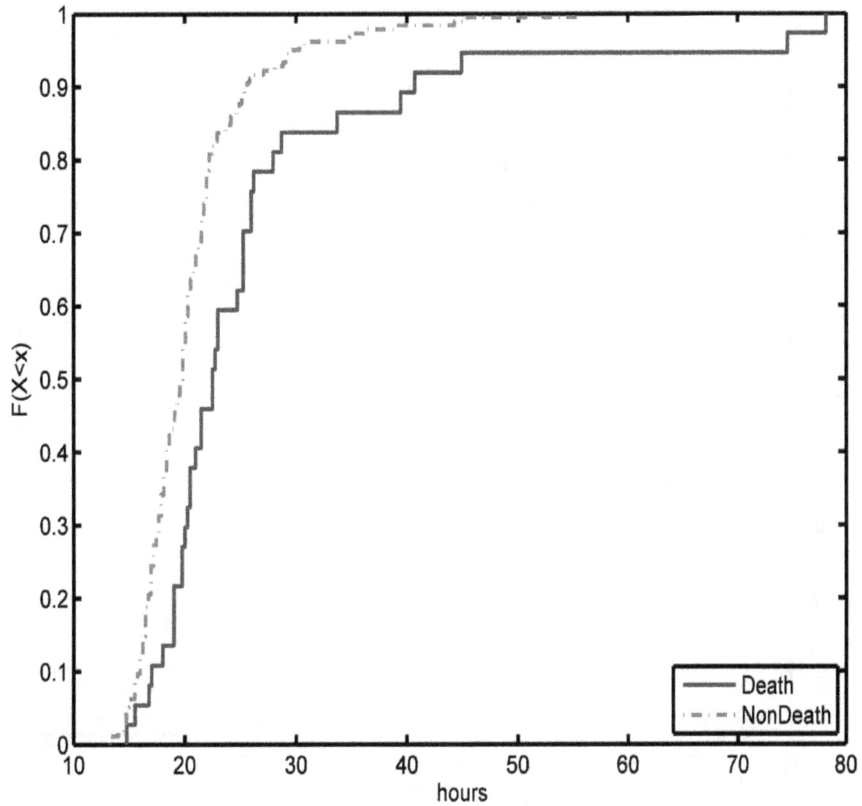

2. Else if (DIV >= 21.5 hours), no daughter cell dies.
3. Else if ($21.75 <= IMT_1 <= 22.25$) hours, no daughter cell dies.
4. Else if ($19.25 <= IMT_1 <= 21.25$) hours and (DIV <= 12.5 hours), no daughter cell dies.
5. Else (DIV <= 4.5 hours), no daughter cell dies.
6. Else at least one daughter cell dies.

(Note that to apply a RIPPER rule set to a database, the prediction of the first rule that fires is taken.) Using this rule set, the overall prediction accuracy is 75% (159 out of 213), where the accuracy for *Non-Death* cells is 78% (135 out of 176) and the accuracy for the other class is 65% (24 out of 37)[1]. We use the WEKA (Hall et al., 2009) implementation of RIPPER (called JRIP) with default parameters[2]. Because of high class imbalance, a cost-sensitive[3] classifier is used, where zero cost is assigned to correct classifications and a cost inversely proportional to the size of the actual class is assigned to errors. Our aim is to require the classifier to learn both classes. We find a similar result when predicting Third Generation *Death* events. The rule set from JRIP is in this case:

1. If ($IMT_1 <= 22$ hours) and ($IMT_2 <= 21.75$ hours), no daughter cell dies.
2. Else if ($22.75 <= IMT_1 <= 26.5$) hours, no daughter cell dies.
3. Else if ($IMT_2 <= 17.75$ hours), no daughter cell dies.
4. Else if ($23.75 <= IMT_2 <= 26$) hours, no daughter cell dies.

Figure 4. Comparison of the ECDFs of IMT_2 for cells that die and cells that do not

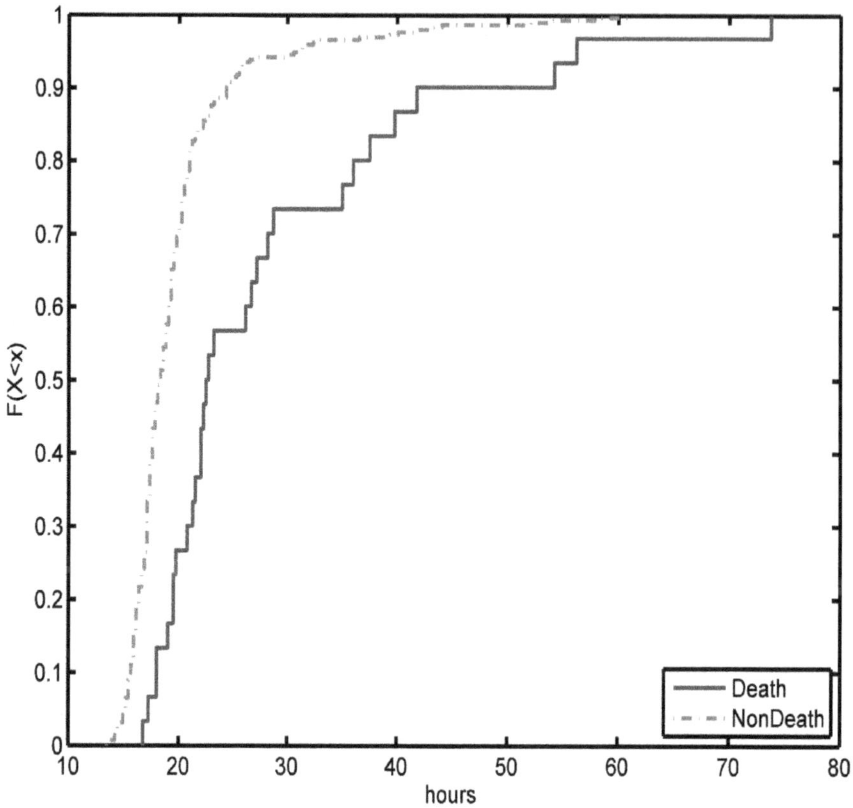

5. Else at least one daughter cell dies.

The overall accuracy in this case is 80% (256 out of 320), while the accuracy for *Death* and *Non-Death* cell classes are respectively 67% (20 out of 30) and 81% (236 out of 290).

Study of the Cytotoxic Effect of Topotecan

The results are summarised in Table 4 where ``Gen.'' stands for Generation. It is statistically evident that the drug is killing more cells in the Perturbed lineages than in Controls in the First Generation, but there is no statistical evidence to say the same in the Second Generation. Surprisingly, the proportion of *Deaths* in the Second Generation for 1µM is lower than that for Control and 10µM. For Stressed versus Resistant and Full Clonal Potential (FCP) versus Partial Clonal Potential (PCP), there is a trend from Control to 10µM in both generations: as more drug is added, there are more Stressed cells and fewer FCP cells. However, possibly due to the small sample size, we found the differences to be statistically significant only when comparing the Control and 10µM groups. The values of the proportions are shown in Table 5.

Cell Dynamic Behaviour. Here, we consider the dynamic behaviour of cells: that is, how the ECDFs of DIV, IMT_1, IMT_2 and IMT_3 change when the system is perturbed with the drug. To compute the ECDF of each lineage type in each generation, the cells that have successfully divided in that generation are considered.

Table 4. Cytotoxic effect in the First and Second Generations for 1µM and 10µM. Those with statistically different proportions with respect to Control are shown in **boldface**. *The values in parentheses are the corresponding p-values*

Proportion	Gen.	Control	1µM	10µM
Death/Non-Death	1st	13/293	**8/63** (0.038)	**9/76** (0.033)
	2nd	51/384	11/99	18/93
Stressed/Resistant	1st	35/178	10/40	15/38
	2nd	29/318	11/65	**19/58** (<0.0002)
FCP/PCP	1st	88/23	13/4	**4/7** (<0.005)
	2nd	196/13	30/3	**11/6** (0.001)

Table 5. The values of proportions from Table 4

Proportion	Gen.	Control	1µM	10µM
Death/Non-Death	1st	0.0444	0.1270	0.1184
	2nd	0.1328	0.1111	0.1935
Stressed/Resistant	1st	0.1966	0.2500	0.3947
	2nd	0.0912	0.1692	0.3276
FCP/PCP	1st	3.8261	3.2500	0.5714
	2nd	15.0769	10.0000	1.8333

Figure 5 shows the ECDF of DIV for Control, 1µM and 10µM lineages. It shows a big discrepancy among the distributions of Control, 1µM and 10µM starting at 10 hours. 1µM is delayed and 10µM is greatly delayed when compared to Control. The Kolmogorov-Smirnov test finds statistical evidence *against* the null hypothesis for both Control versus 1µM (p-value $< 10^{-3}$) and Control versus 10µM (p-value $< 10^{-5}$). Just to illustrate this difference, 91% of the Control group's DIV times are less than to 22 hours, but only 73% of the 1µM group's DIV and 50% of the 10µM group's DIV are below that threshold. The maximum difference between Control and 1µM ECDFs is 35%, which is at 19.25 hours; and between Control and 10µM ECDFs is 41%, which occurs at 22 hours.

Figure 6 shows the ECDF of IMT_1 for Control, 1µM and 10µM lineages. We observe that 1µM is slightly delayed with respect to Control, and 10µM is slightly delayed with respect to 1µM. The Kolmogorov-Smirnov test finds statistical evidence showing that the distribution of IMT_1 is different in Control and 10µM (p-value $<10^{-5}$). 57% of Control IMT_1 are lower than 20.5 hours, but only 44% of 1µM IMT_1 and 18% of 10µM IMT_1 are below that threshold. The maximum difference between Control and 10µM is 39%, which is located at 20.5 hours.

Figure 7 shows the ECDFs of IMT_2. We observe that 1µM is very similar to Control (it almost follows the Control distribution) and 10µM is delayed. The Kolmogorov-Smirnov test finds statistical evidence showing that the distribution of IMT_2 is different in Control group and 10µM group (p-value $<10^{-5}$). There is also statistical evidence showing that the distribution of 1µM and 10µM are different (p-value $<10^{-3}$). 60% of Controls and 58% of 1µM IMT_2 are less than 19.5 hours, but only 27% of 10µM IMT_2 are less than

Figure 5. Comparison of the ECDFs of the Progenitor division time

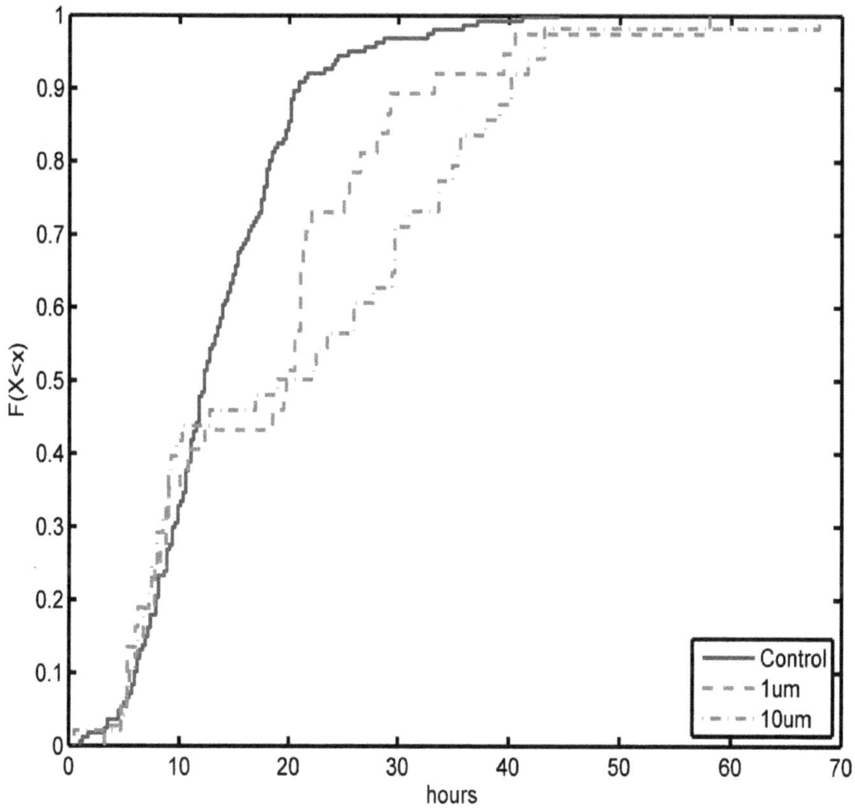

this threshold. The maximum difference between Controls and 10μM is 33%, which occurs at 19.5 hours; and between 1μM and 10μM is 32%, which is at 17.5 hours.

Figure 8 shows the ECDFs of IMT_3. We observe the same trends as in Figure 7: the 1μM group is very close to the Control group and the 10μM group is closer to the Control group than before. The Kolmogorov-Smirnov test finds statistical evidence showing that the distribution of IMT3 is different in Control and 10μM (p-value $<10^{-5}$). There is also evidence showing that the distribution of 1μM and 10μM are different (p-value $<10^{-4}$). 59% of 1μM IMT_3 and 57% of Control IMT_3 are lower than 19.5 hours, but only 29% of 10μM IMT_3 are lower than this threshold. The maximum difference between Control and 10μM is 28%, which is at 19.5 hours; and between 1μM and 10μM is 30%, which is at the same time point.

Study of Stressed Lineages

We now consider patterns that can be used to predict whether a lineage is Stressed or Resistant. We construct a series of classifiers using the following inputs:

1. Progenitor: Drug and DIV are used as input to the classifier.
2. First: Drug, DIV and First Generation information are used.
3. Second: Drug, DIV, First and Second Generations information are used.

Figure 6. Comparison of the ECDFs of the first generation cells IMT

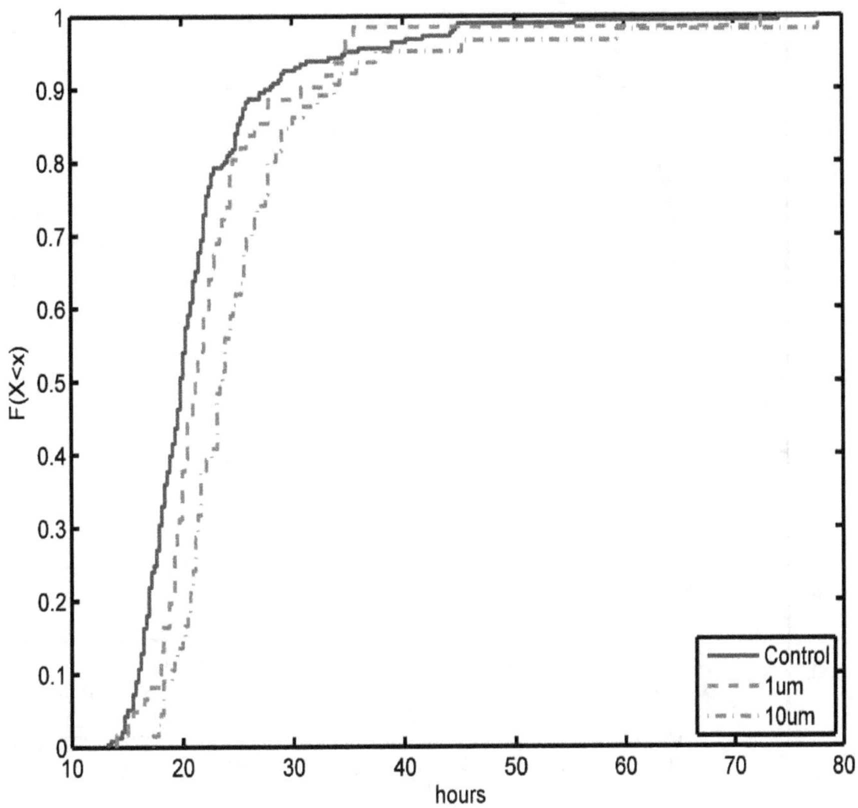

Drug is the lineage type: Control, 1μM or 10μM.

The experiments are summarised in Table 6. The overall accuracy and the accuracy per class (Stressed and Resistant) are provided. The default parameters are used in WEKA's J48[4] and JRIP. In this case, it is not necessary to handle class imbalance. We observe that the classifiers with the best accuracies are those that include Second Generation inputs. However, the classifiers using Second Generation information are not considered because the rules they provide are obvious and do not add more insight into lineages' behaviour. Therefore, the classifiers for the *First* input data set are selected. The final rule set induced using JRIP for this input data is:

1. If ($Status_{1\_1}$ = *Arrested*), the lineage is Stressed.
2. Else if (IMT_{1_2} >= 39.5 hours), the lineage is Stressed.
3. Else if ($Status_{1\_2}$ = *Death*), the lineage is Stressed.
4. Else if ($Status_{1\_1}$ = *Unresolved*), the lineage is Stressed.
5. Else the lineage is Resistant.

The first rule identifies that many of the *Arrested* cells in the First Generation have also an *Arrested* sister and therefore there are neither Second nor Third Generations in those lineages. The second rule shows that Stress lineages have at least one cell in the First Generation with a very long IMT_1 (the threshold is fixed to 39.5 hours). The next rule is obvious: if one cell's status is

Figure 7. Comparison of the ECDFs of the second generation cells IMT

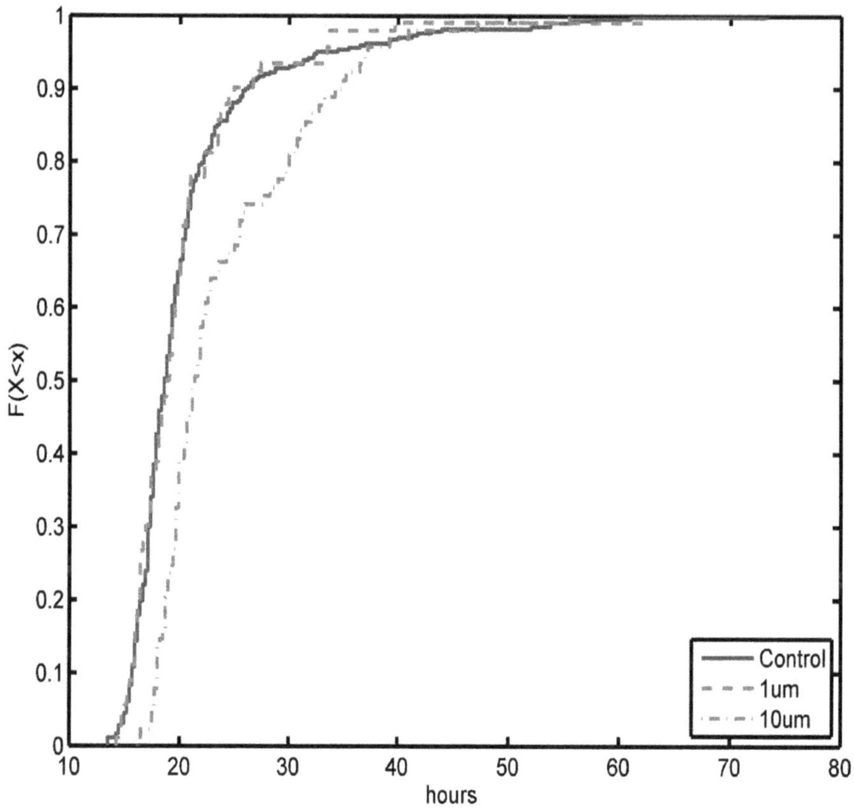

Death, then the other does not divide and the lineage must be stressed. The fourth rule is an artifact of the dataset: it happens in the data that when one cell in the First Generation is *Unresolved* the lineage is stressed.

The classifier resulting from the J48 algorithm is shown in Figure 9. This tree, which is also very simple, complements the JRIP ruleset for the Resistant lineages: they have at least one First Generation cell with an IMT lower than 36.5 hours.

Study of Control versus Perturbed lineages

When we follow the procedure described in the previous section (training a series of classifiers with incremental generation information) to distinguish between Control and Perturbed lineages, the best result is given by J48 using all but Third Generation information. The overall accuracy of this classifier (Figure 10) is 78.4% (196 out of 250), where the accuracies on the Control and Perturbed lineages are 81.8% (135 out of 165) and 71.8% (61 out of 85) respectively. This classifier uses information of the time to divide (DIV) and displacement (DIV_D) of the progenitor as its most discriminative features. For example, if a lineage progenitor cell takes longer than 20.7 hours to divide and moves more than 30.41 microns, the lineage is declared Perturbed. In addition uses $Cell_{1\_2}$, $Cell_{2\_4}$ and the number Second Generation Divisions to classify the lineages.

As it appears that using all the available information for training the classifier does not result in the best possible performance, next we explore

Figure 8. Comparison of the ECDFs of the third generation cells IMT

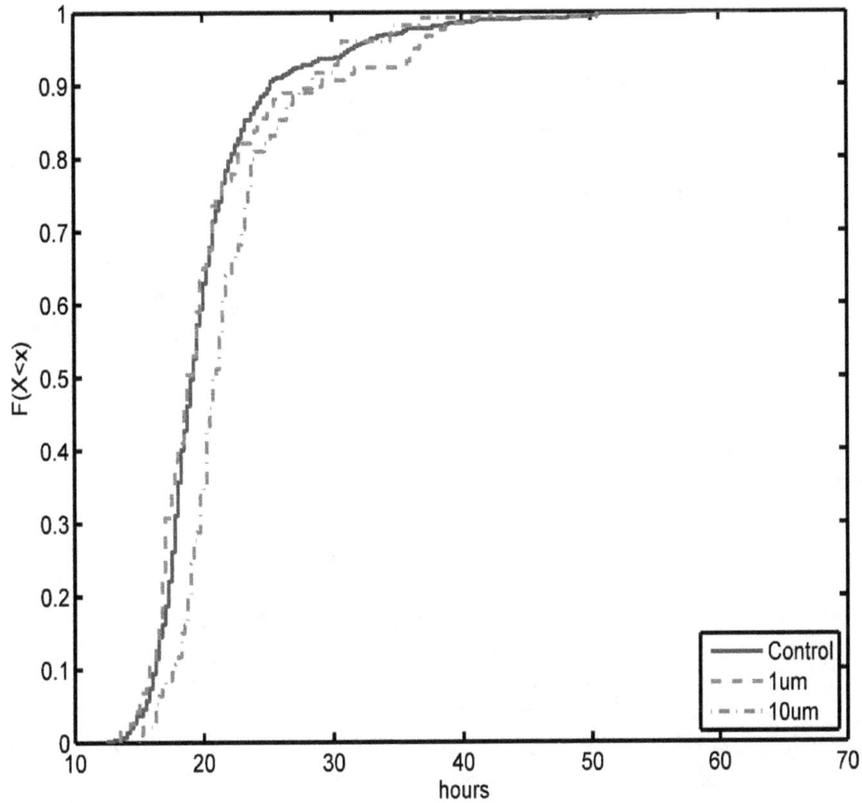

Table 6. Per-class and overall accuracies, leave-one-out result, for JRIP and J48 using inputs until generation "Gen." (inclusive)

Gen.	Method	Overall	Stressed	Resistant
Progenitor	JRIP	76.0%	36.8%	90.0%
	J48	65.9%	0.0%	89.4%
First	JRIP	91.2%	75.4%	96.9%
	J48	92.2%	75.4%	98.1%
Second	JRIP	94.9%	89.5%	96.9%
	J48	95.8%	89.5%	98.1%

Bayesian network modelling to find a suitable subset of variables.

Bayesian Network Modelling. In a Bayesian Network, the Markov Blanket of a variable *T* is the set that consist of its parents, its children and its children's parents (where these refer to the relationships between nodes in the Bayesian network, and are not to be confused with relationships between cells). If the values of the variables in the Markov Blanket are known, the rest of variables in the Bayesian Network become irrelevant for predicting the value of *T*.

Figure 11 shows the Markov Blanket of the best automatically learned network that we have

Figure 9. Tree classifier for the prediction of resistant versus stressed lineages using Progenitor and first generation information

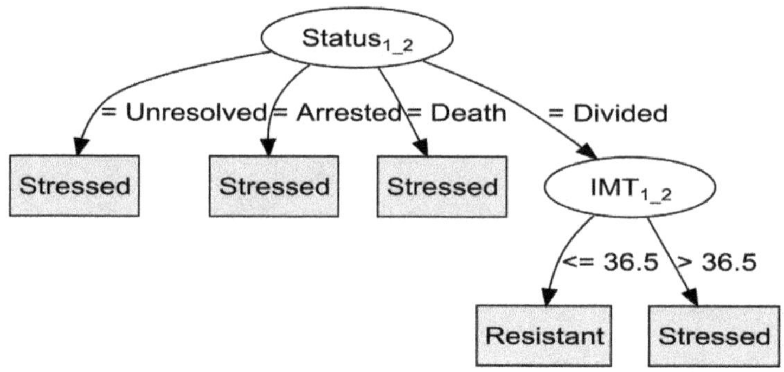

Figure 10. Tree classifier for the prediction of control versus perturbed lineages using Progenitor, first and second generations information

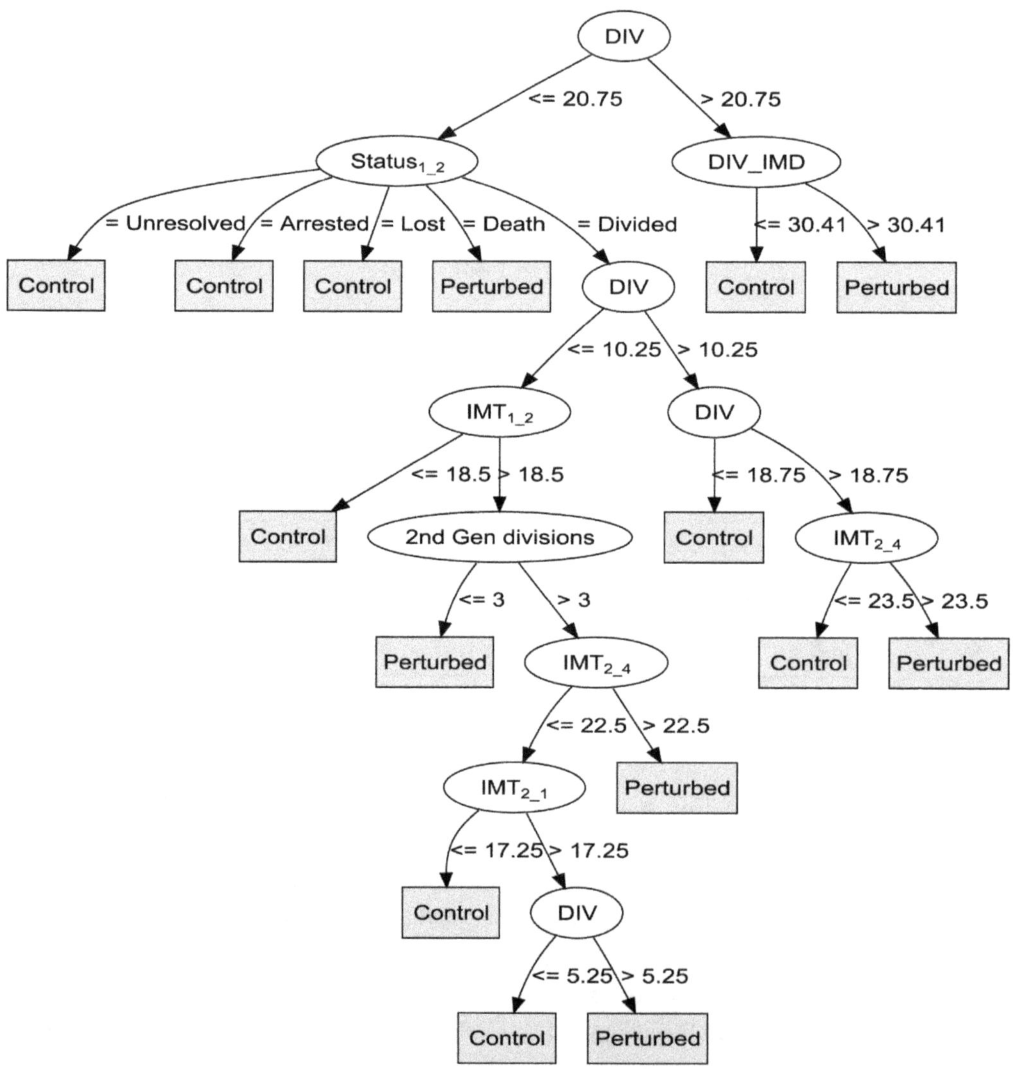

Figure 11. Markov Blanket of the automatically–learned Bayesian network for classifying perturbed versus control lineages (Drug)

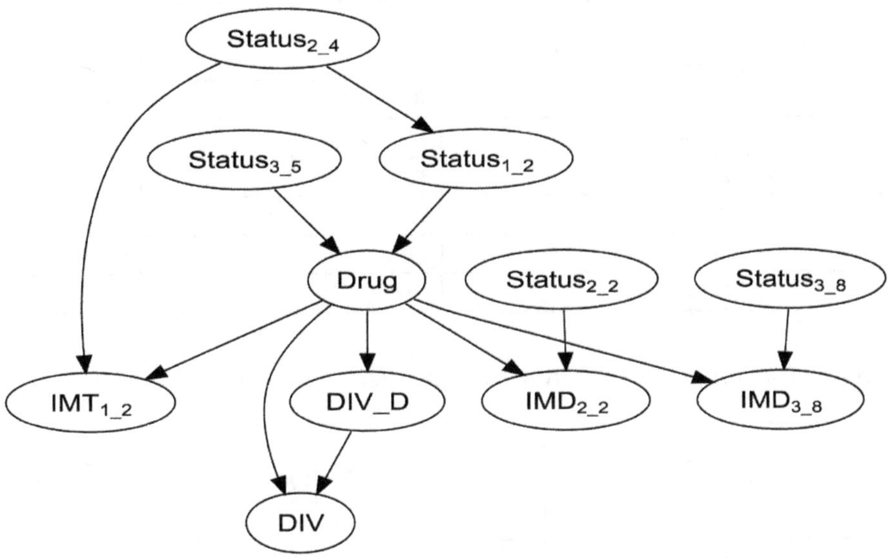

Table 7. Extract of the probability distribution of Drug given $Status_{3\_5}$ and $Status_{1\_2}$ for the network in Figure 11

		Drug	
$Status_{3\_5}$	$Status_{1\_2}$	Control	Perturbed
Refused	Divided	75.0%	25.0%
Arrested	Divided	34.6%	65.4%
Lost	Lost	81.3%	18.8%
Lost	Divided	47.6%	52.4%
Never–Born	Unresolved	75.0%	25.0%
Never–Born	Arrested	87.9%	12.1%
Never–Born	Death	29.2%	70.8%
Never–Born	Divided	63.5%	36.5%
Death	Divided	59.1%	40.9%
Divided	Divided	80.4%	19.6%

induced from the data. This network is the result of applying to the data the supervised discretization proposed in Fayyad and Irani, (1993), followed by a local hill climbing search starting with the naive Bayes network. All those steps were done in WEKA. The overall accuracy of this network is 76% (190 out of 250), and the accuracy for Control and Perturbed classes are respectively 83% (137 out of 165) and 62.3% (53 out of 85).

The probability distribution tables provide further insights on the data. Table 7 shows that the events with a high probability for Control lineages have $Status_{1\_2}$=*Divided* with the exception of $Status_{3\_5}$=*Arrested*. This probability reaches its maximum value (0.8) when both statuses are

Table 8. Extract of the probability distribution of IMT_{1_2} given Drug and $Status_{2\_4}$ for the network in Figure 11

		IMT_{1_2}	
Drug	$Status_{2\_4}$	(0, 19.1]	(19.1, ∞)
Control	Arrested	70.0%	30.0%
Control	Death	10.0%	90.0%
Control	Divided	32.9%	67.1%
Perturbed	Arrested	16.7%	83.3%
Perturbed	Death	5.0%	95.0%
Perturbed	Divided	7.8%	92.2%

Table 9. Probability distribution of DIV given drug and DIV_D for the network in Figure 11

		DIV		
Drug	DIV_D	(0,9.4]	(9.4,20.9]	(20.9,inf)
Control	(0,59.4]	29.8%	60.0%	10.2%
Control	(59.4,inf)	3.2%	67.7%	29.0%
Perturbed	(0,59.4]	50.4%	16.2%	33.3%
Perturbed	(59.4,inf)	11.9%	15.3%	72.9%

Divided. The events that show high probability for Perturbed lineages are $Status_{1\_2}$=*Death.* In addition, the *Arrested* event in the First Generation is more associated with Control lineages but the *Arrested* event in the Third Generation is more associated with Perturbed lineages.

Table 8 shows that the supervised discretization procedure selects a threshold for IMT_{1_2} at 19.1 hours (which suggest that this threshold is informative for this task). We also observe that: (1) Control and Perturbed lineages behave differently when the $Cell_{2\_4}$ cell is *Arrested*, as in the case of Control cells it is more probable that their mothers need less than 19.1 hours to divide but the opposite is more probable for Perturbed cells; (2) Control and Perturbed lineages behave similarly when a $Cell_{2\_4}$ dies, as in both cases is very probable that its mother needs more than 19.1 hours to divide; (3) When the $Cell_{2\_4}$ divides, there is a probability of 1/3 of IMT_{1_2} being lower than 19.1 hours for Control lineages, but the probability of the same event is less than 1/12 for Perturbed lineages.

Table 9 shows that the supervised discretization selects a threshold for DIV at 9.4 hours and another one at 20.9 hours. We observe that most Control, as opposed to Perturbed, cells divide some time between 9.4 and 20.9 hours. We also observe that there is a positive correlation between DIV and DIV_D. Table 10 shows that the selected threshold for DIV_D is 59.4 microns. We observe that when the lineage is Control, the cell moves less than this threshold 91% of the time.

Table 10. Probability distribution of DIV_D given drug for the network in Figure 11

	DIV_D	
Drug	(0,59.4]	(59.4,inf)
Control	91.3%	8.7%
Perturbed	66.9%	33.1%

JRIP Result with Bayesian Variable Selection. If the data is assumed to be generated using a Bayesian network, all the relevant information about a variable is found in its Markov Blanket. Therefore, the Bayesian Network modeling can be used as a variable selection procedure, i.e., only the variables in the Markov Blanket of *Drug* are used to predict its outcome. Accordingly, we combine this variable selection procedure with a JRIP classifier (with both being performed inside the leave-one-out loop). Table 11 shows how many times each variable were selected. This method is consistent in all the leave-one-out iterations: the same variables are selected in almost all the iterations. In some iterations, however, some other variables are included, but this happens less than 2% of the time.

This procedure's overall accuracy is 81% (203 out of 250), and the accuracy on Control and Perturbed classes are respectively 87.3% (144 out of 165) and 69.4% (59 out of 85). In this case, the JRIP rule set is:

1. If (DIV >= 21 hours) and (DIV_D >= 32.25 microns), the lineage is Perturbed.
2. Else if (DIV <= 10 hours) and (IMT_{1_2} >= 22.7 hours), the lineage is Perturbed.
3. Else the lineage is Control.

Performing variable selection using the Markov blanket in this way has yielded an effective and simple model of the data that has better accuracy than previous ones.

Dynamic Behaviour. IMD is the Euclidean distance from point where a cell is born to the point where a cell divides. This is not the same as the total cell displacement, because the cell can move in any direction and doesn't have to maintain one direction. However, we can assume that this movement can be characterised by an unknown probability distribution; as we have samples from this distribution, we can compare perturbed and unperturbed populations with the ECDF. The ECDFs of the 'displacement' of the

Table 11. Bayesian variable selection in each leave-one-out iteration. The number of times a variable was selected by the procedure is recorded

Variable	number of times selected
DIV	250
IMT_{1_1}	5
IMT_{1_2}	250
DIV_D	250
IMD_{2_2}	250
IMD_{3_8}	250
$Status_{1\_2}$	250
$Status_{2\_2}$	249
$Status_{2\_4}$	243
$Status_{3\_4}$	1
$Status_{3\_5}$	250
$Status_{3\_8}$	250

cells (IMD) are now considered for Progenitor to Third Generation. These complement the IMT ECDFs that were shown in Figures 5 to 8.

Figure 12 shows the ECDFs of the displacement of Progenitor cells. We observe a clear difference between the distributions of the movements among Control, 1µM and 10µM groups, with Control cells having the lowest displacements and 10µM cells having the greatest displacements. For example, 77% Controls, 65% 1µM and 44% 10µM are lower than 37 microns. The maximum difference between Control and 1µM is 26%, which is located at 27 microns. The maximum difference between Control and 10µM is 33%, which is located at 37 microns. The differences between Control and 1µM and between Control and 10µM are statistically significant using a Kolmogorov-Smirnov test (p-values $<10^{-3}$). Comparing this figure with Figure 5, we observe that the ECDFs have different profiles; for example, 10µM and Controls are separated from the beginning of the graph in Figure 12.

Figure 13 shows the ECDFs of the First Generation cells IMD, the distributions are closer to each other and no statistical evidence is found for

Figure 12. Comparison of the ECDFs of the Progenitor displacement

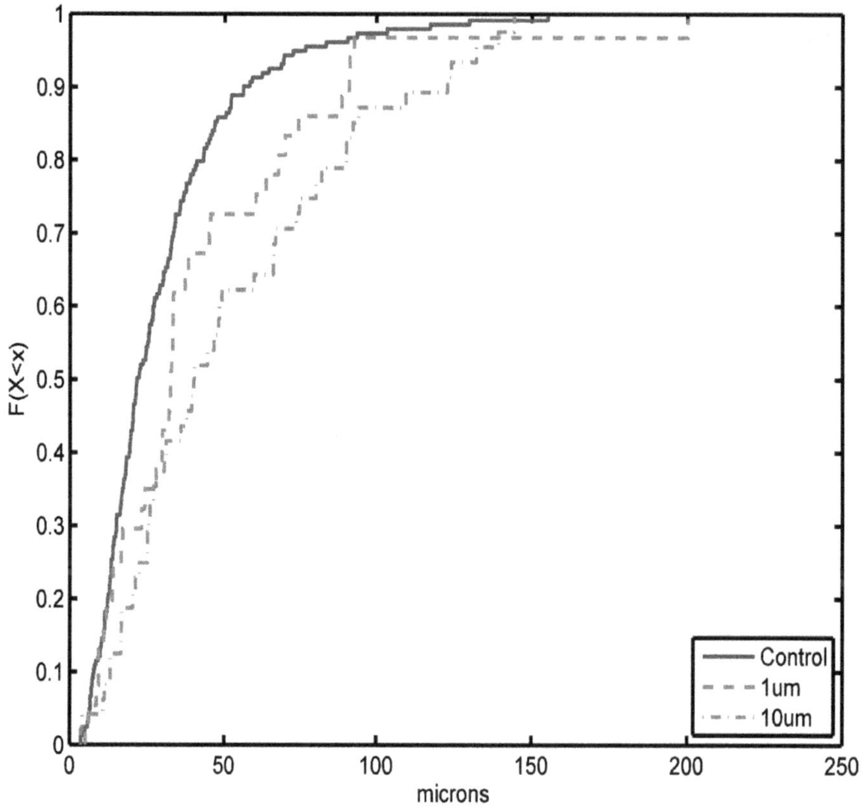

a difference between them. Again, these ECDFs have a different profile to the corresponding IMT ones in Figure 6.

Figure 14 shows the ECDFs of Second Generation cells IMD. The ECDFs are even closer to each other in this generation and 1μM cells are delayed more than 10μM cells over some ranges. We found no statistically significant differences between the curves using a Kolmogorov-Smirnov test. There is a notable difference between these ECDFs and its IMT: in Figure 7 the 1μM distribution is almost overlaid with the Control distribution, but this is not the case in its IMD counterpart.

Figure 15 shows the ECDFs of Third Generation cells IMD: the distributions of 1μM and 10μM IMD are, surprisingly, more delayed than in previous generations with respect to Control, breaking the trend that was observed in the previous generations. For example, 71% Controls, 52% 1μM and 44% 10μM are lower than 51 microns. The maximum difference between Control and 1μM is 20% and it is located at 55 microns. The maximum difference between Control and 10μM is 27% and it is located at 51 microns. Both differences are statistically significant using a Kolmogorov-Smirnov test (p-values $<10^{-3}$). As before, we observe that the behaviour is different to that of the corresponding IMT graph (Figure 8).

Cell 'Speed' and cytotoxic level. The previous subsection shows that the distribution of IMD are different for Control and Perturbed populations. Here we show that the relative 'speed' distributions of the cell are also different. We define the *speed* of any particular cell as the ratio

Figure 13. Comparison of the ECDFs of the first generation cells IMD

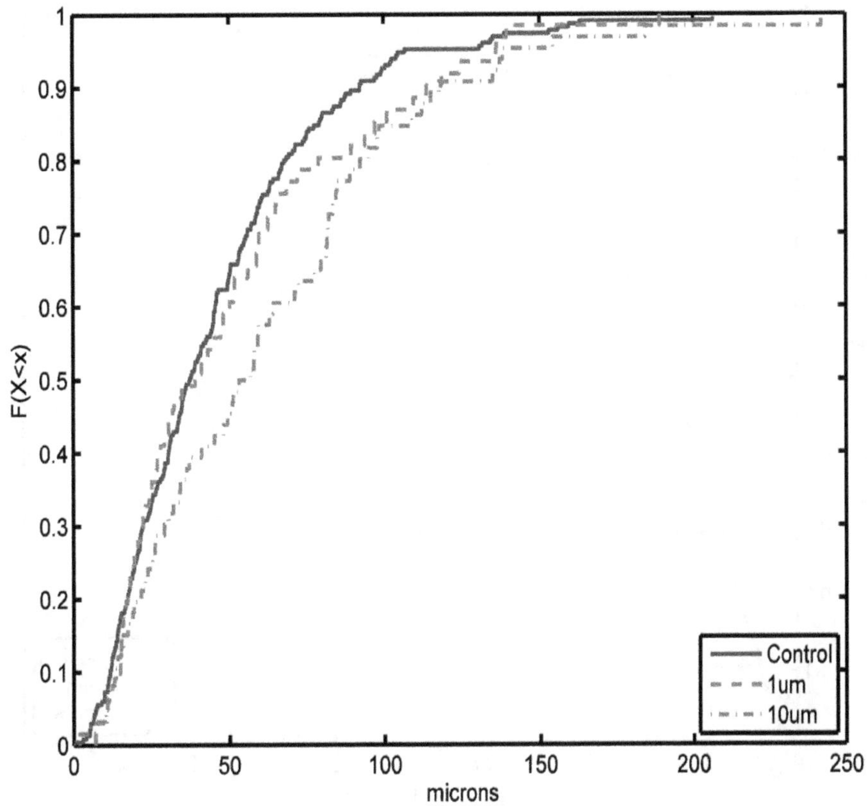

of its IMD to its IMT. Figures 16 to 19 show the empirical distribution of this ratio.

Figure 16 and Figure 17 show this difference for Progenitor and First Generation cells. We observe that 10μM moves slightly faster than Controls, but we do not find this difference to be statistically significant with a Kolmogorov-Smirnov test.

Figure 18 shows this difference for Second Generation cells. Again, we observe that there appear to be slight differences between the distributions, but these are not found to be statistically significant with a Kolmogorov-Smirnov test.

Figure 19 shows this difference for Third Generation cells. We observe that 1μM moves slightly faster than Controls and 10μM moves faster than both. This time, both differences are statistically significant with a Kolmogorov-Smirnov test when compared to Control (p-values $< 3\cdot10^{-4}$). For example, 73% Controls, 54% 1μM and 49% 10μM are lower than 2.7 micron/hour. The maximum difference between Control and 1μM is 21% and it is located at 2.8 micron/hour. The maximum difference between Control and 10μM is 24% and it is located at 2.7 micron/hour.

DISCUSSION

Here we analyse the results and propose some hypotheses that might be worthy to explore further.

Figure 14. Comparison of the ECDFs of the second generation cells IMD

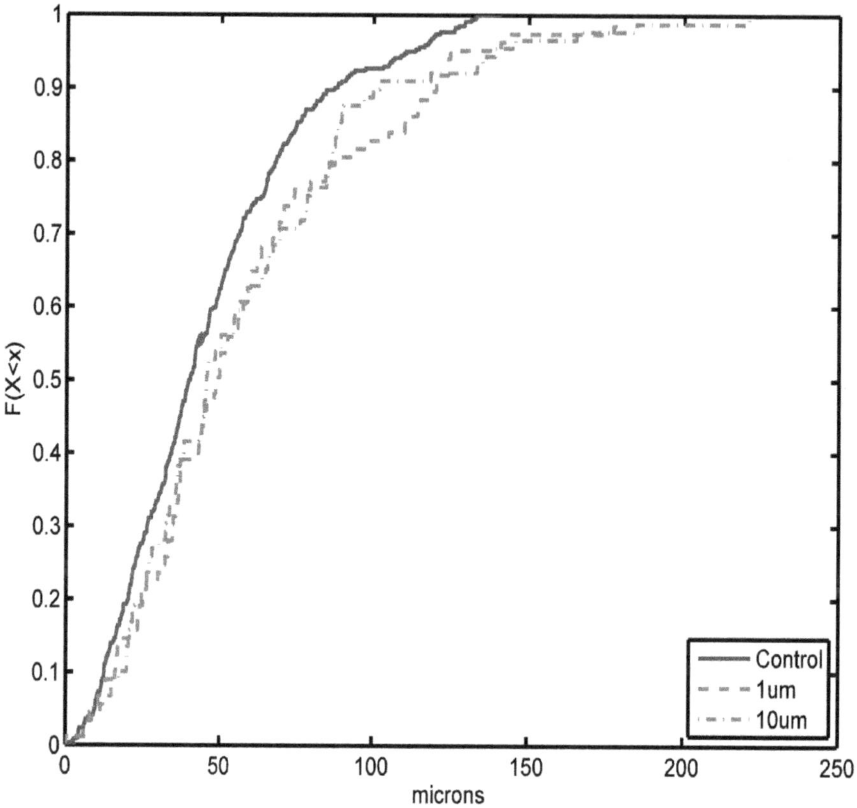

Study of Cell Death in Control Lineages

Cell death, is deemed to be a part of the regulatory mechanism of tumour system, which invokes detailed investigation of such dynamic system. This work has revealed the following patterns for the *Death* event for Control lineages: Firstly, there is a strong pattern of paired sister death, i.e. when one cell dies, it is very probable that its sister also dies; Secondly, cell death in general can be related with longer cell cycle duration of the corresponding mother cell.

Study of the Cytotoxic Effect of Topotecan

The cytotoxic effect of the drug is prominent in 10μM, where statistically significant differences between Control and 10μM lineages are found in almost all tests. On the other hand, the effect of the drug in 1μM has only a statistical significant impact in the *Death* to *Non-Death* proportion and in the ECDFs of DIV and IMT_1. *Death Non-Death* proportion in the Second Generation is lower for 1μM than both Control and 10μM. This might be an artifact of the data or something to be further investigated: *does a low dosage of drug result in fewer cells death the Second Generation after the drug administration is stopped?* The trends in the other cases suggest that in low dose, i.e., 1μM has still some cytotoxic effect, but it is

Figure 15. Comparison of the ECDFs of the third generation cells IMD

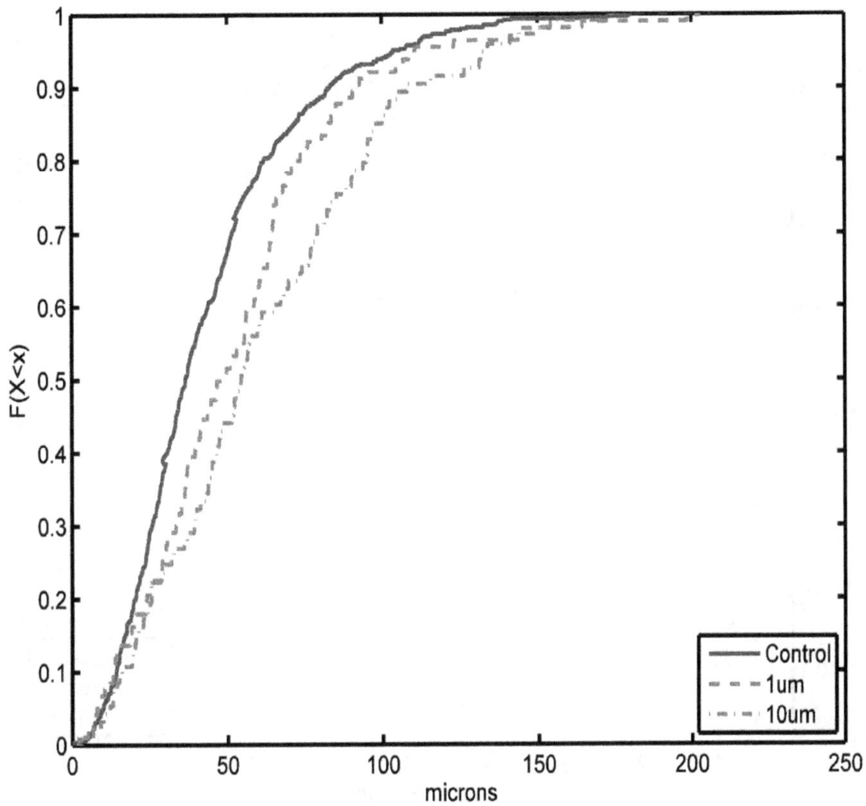

almost lost in the Second Generation, as we see in Figure 7 where 1μM IMT_2 distribution is very close to IMT_2 Control distribution (although its IMD is still affected as shown in Figure 14). The greatest variation between Control and 10μM is found in the proportion of Full Clonal Potential to Partial Clonal Potential. 10μM drug dosage makes this proportion 6.7 times lower than Control for the First Generation and 8.2 times lower for the Second Generation.

There is an interesting relationship found in the ECDF of DIV (Figure 5): For progenitor cells, TPT prevents division between the times T=10 hours and T=20 hours, establishing the S-phase specific cytotoxic effect of TPT (Pommier, 2006). When TPT was administered, the progenitor cells were distributed in all 3 phases (G2, S and G2M) phases of the cell cycle, and from the start of the experiment the cohorts of progenitor cells that delivers to mitosis within the first 0–10 h zone could be considered to be in G2 during the drug treatment, S-phase if time to mitosis occurred between 11–18 h and G1 if delivery occurred after 18 h. (Feeney et al., 2003; Errington, Marquez, Chappell, Wiltshire, & Smith, 2005).

Study of Stressed Lineages

In the algorithms to classify the lineages (Control population also has stressed lineages), the *Drug* variable is not selected, which suggests, as the accuracy of the classifiers is quite high, that we do not learn much from the *Drug* variable after considering the other data. The key factor that explains whether a lineage will be stressed is the behaviour of the lineage in First Generation, for

Figure 16. Comparison of the ECDFs of the Progenitor displacement – time to divide ratio

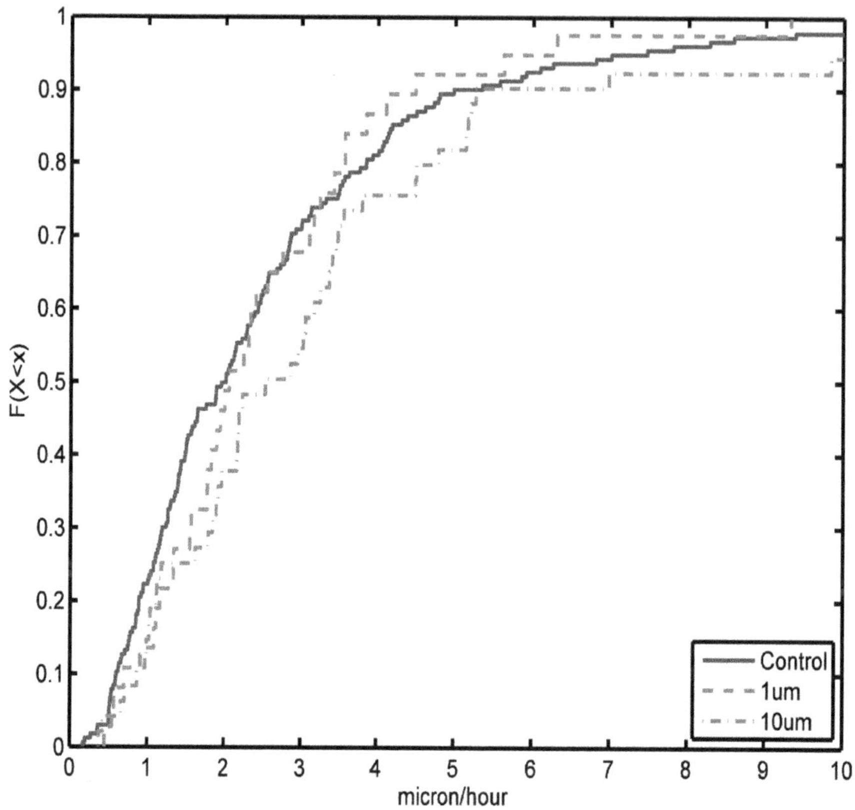

all lineage types. In addition, *Arrested* cells tend to appear as sister pairs in the First Generation. For example, for Control lineages, 48 out of 49 *Arrested* cells come from *Arrested* sister pairs.

Study of Control versus Perturbed Lineages

When we analyse a data set with a small number of examples relative to the number of variables, we might expect to encounter the well-known *curse of dimensionality*. In general, as the number of variables increases, the number of possible hypotheses to explain the data increases (assuming a sufficiently rich representation language), which increases the risk of discovering hypotheses that are based just on artefacts of the data, which in turn leads to models having lower accuracies. In addition, correlated variables, noisy variables and partially redundant variables all increase the hypotheses search space, making the search harder, without providing new information. We suggest that this is the reason why we get models with better accuracy when we follow the common practice of feature selection, rather than using all the variables.

We hypothesise that the extra displacement of 1μM and 10μM lineages, when compared to Control lineages, may be caused by some internal cell mechanism that compels the cell to move away from stressful environments. If we look at Figure 12, we see that the greater the concentration of drug, the greater the displacement, which gives some support to this hypothesis. If we look at the movement of the cells, the general trend is that Control cells move less than 1μM cells, which

Figure 17. Comparison of the ECDFs of the first generation cells IMD – IMT ratio

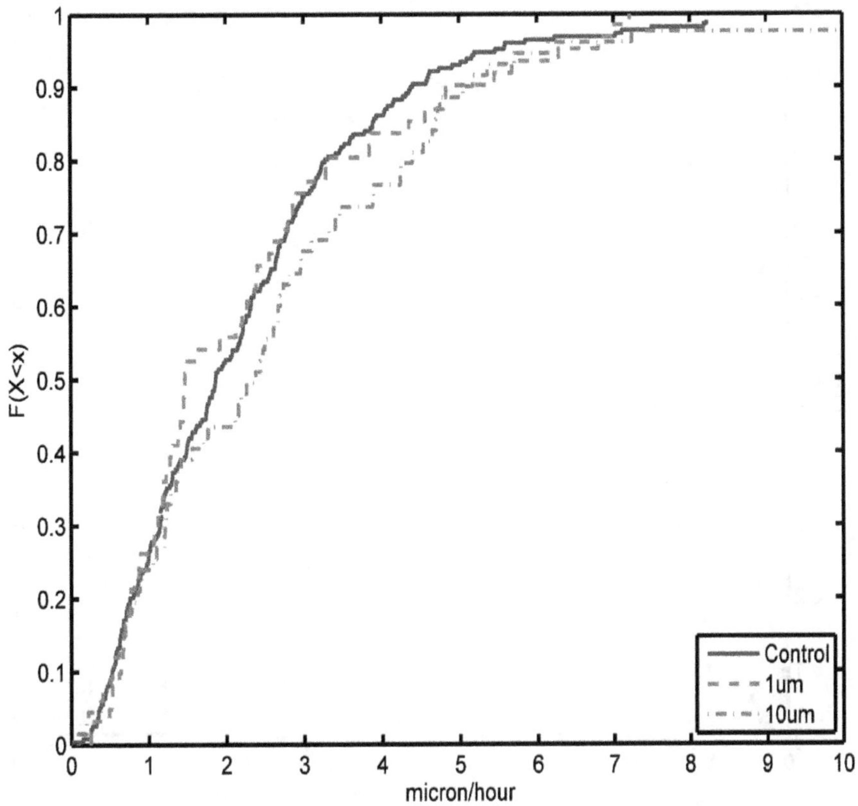

in turn move less than 10μM cells. This does not happen in the Second Generation, where 1μM and 10μM distributions are interlaced, and over some ranges the 10μM distribution is closer to the Control distribution (see Figure 14).

When comparing IMD and IMT distributions, we observe that in the Second Generation, the 1μM IMT ECDF is very close to Control distribution, but the same does not happen with the 1μM IMD ECDF. This is also found in the Third Generation. The greatest differences among IMD ECDFs are observed in Progenitor and Third Generation cells.

When comparing the cell *speed* distributions we observe that the difference among populations is not only in the distance they travel but also in the speed at they move, which indicates that perturbed populations not only move more but also faster. As a general trend, 10μM is the fastest population with the exception of Second Generation (Figure 18) where 1μM is the fastest one. This is related to what we already noted above (see Figure 14) where 1μM and 10μM IMD distributions are similar, but we learn from Figure 18 that 1μM cells are moving faster in a shorter IMT. Also surprising is that the behaviour of 1μM departs from that of the Control group in Second Generation. The behaviour of both 1μM and 10μM is clearly different (statistically significant) from Control in Third Generation (Figure 19).

From these observations, we conclude that there is something more than a simple correlation between IMT and IMD. This leads to the following new questions and hypothesis to explore in the future: Why do Perturbed cells move more? Does the drug in small dosages make cancer cells expand more rapidly?

Figure 18. Comparison of the ECDFs of the second generation cells IMD – IMT ratio

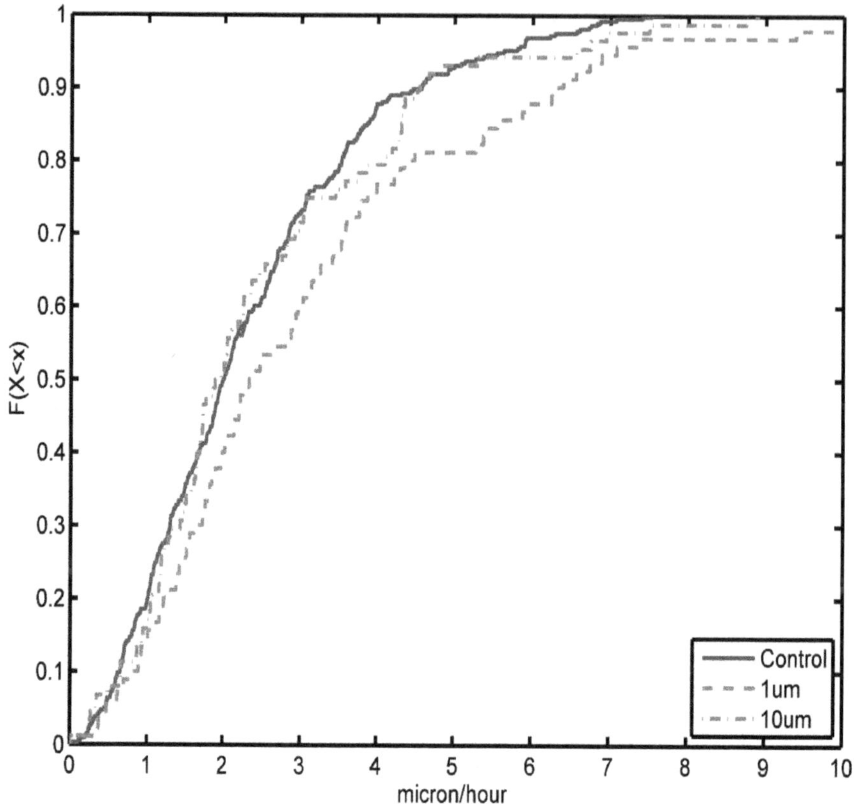

The informative events (probabilities close to zero or one) in Bayesian networks probability tables help us to consider interesting expected and unexpected behaviours. In this case we focus on the *Arrested* event: cells exhibit different behaviours in Control and Perturbed lineages as expected. However, it is surprising how the probability of the event changes from the First to the Third Generation and how it changes depending on the time taken by the mother to divide. We also observed that the division time of the ancestors has an impact on the probability of some future events of their offspring.

By modelling the data with a Bayesian Network, we can perform a variable selection that allows us to subsequently discover simpler and better models using rule-based learners. The JRIP model gets to 81% overall accuracy, it is simple and uses the displacement information. Again, cell movement is important and increases accuracy.

CONCLUSION

Through this work, we have demonstrated how data mining procedures can reveal important patterns of a dynamic cell system. By our analyses, we have found the cytotoxic effect of Topotecan on both the dynamics and event frequencies of the system. The results reiterate the finding that initial cell cycle positioning is an important factor for immediate cytotoxic response. The results also suggest that by distributing the cytotoxic effect asymmetrically within the progeny, the system adopts strategies that facilitate the generation of

Figure 19. Comparison of the ECDFs of the third generation cells IMD – IMT ratio

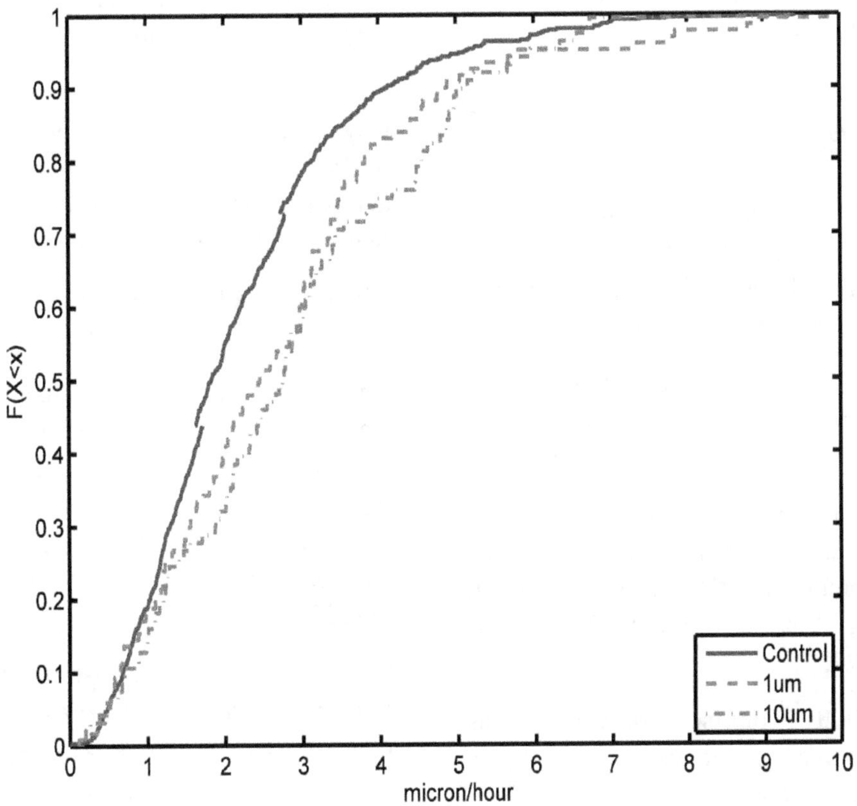

drug resistance progenies, thus maximizing its clonal expansion potential.

We also observe cellular deaths in un-perturbed condition with well defined patterns, indicating the existence of an innate regulation mechanism of the tumour system. Investigation on such a mechanism requires incorporation of gene and protein network data. In addition, we have found simple models that can predict death in cells in an unperturbed condition, and also how inheritance information like the cell cycle duration of a mother cell can influence the fate of daughters. Our models are successful at classifying between perturbed and unperturbed lineages, thus revealing the population level cytotoxic effect. More importantly, by predicting the clonal potential, we are not only able to identify the resistant subpopulation in early generations, but we are also able to identify strategies that the following generations adopt in order to maximize their clonal potential. In the spatial context, we have observed a putative relationship between cytotoxic effect and cellular movement; such results can be exploited to understand the metastasis potential of the tumour (Cavanna, Pokorna, Vesely, Gray, & Zicha, 2007), which is another important aspect of cancer research.

ACKNOWLEDGMENT

This research has been supported by a Marie Curie Transfer of Knowledge Fellowship of the EU 6th Framework Programme, contract CT-2005-029611. It has also been supported in part by the Science Foundation Ireland under Grant 08/RFP/CMS1254. Ricardo Santiago-Mozos has been partly supported by Fundación Española para la Ciencia y la Tecnología (Ministerio de Educación de España).

REFERENCES

Alvarez-Buylla, A., García-Verdugo, J. M., & Tramontin, A. D. (2001). A unified hypothesis on the lineage of neural stem cells. *Nature Reviews. Neuroscience*, *2*, 287–293. doi:10.1038/35067582

Anderson, D. J., Gage, F. H., & Weissman, I. L. (2001). Can stem cells cross lineage boundaries? *Nature Medicine*, *7*, 393–395. doi:10.1038/86439

Ardavín, C., Martínez del Hoyo, G., Martín, P., Anjuère, F., Arias, C. F., & Marín, A. R. (2001). Origin and differentiation of dendritic cells. *Trends in Immunology*, *22*, 691–700. doi:10.1016/S1471-4906(01)02059-2

Bailly, C. (2000). Topoisomerase I poisons and suppressors as anticancer drugs. *Current Medicinal Chemistry*, *7*, 39–58. doi:. doi:10.2174/0929867003375489

Benjamini, Y., & Hochberg, Y. (1995). Controlling the false discovery rate: a practical and powerful approach to multiple testing. *Journal of the Royal Statistical Society. Series B. Methodological*, *57*, 289–300.

Benjamini, Y., & Yekutieli, D. (2001). The control of the false discovery rate in multiple testing under dependency. *Annals of Statistics*, *29*, 1165–1188. doi:.doi:10.1214/aos/1013699998

Bernards, R., & Weinberg, R. A. (2002). Metastasis genes: A progression puzzle. *Nature*, *418*, 823. doi:10.1038/418823a

Campbell, I. (2007). Chi-squared and Fisher-Irwin tests of two-by-two tables with small sample recommendations. *Statistics in Medicine*, *26*, 3661–3675. doi:10.1002/sim.2832

Carlson, J. M., Heckerman, D., & Shani, G. (2009). *Estimating false discovery rates for contingency tables* (Tech Rep. No. 2009-53). Microsoft Research.

Cavanna, T., Pokorna, E., Vesely, P., Gray, C., & Zicha, D. (2007). Evidence for protein 4.1 B acting as a metastasis suppressor. *Journal of Cell Science*, *120*, 606–616. doi:10.1242/jcs.000273

Charniak, E. (1991). Bayesian Networks without Tears. *AI Magazine*, *12*, 50–63.

Cheung, S., Evans, N. D., Chappell, M. J., Godfrey, K. R., Smith, P. J., & Errington, R. J. (2008). Exploration of the intercellular heterogeneity of topotecan uptake into human breast cancer cells through compartmental modelling. *Mathematical Biosciences*, *213*, 119–134. doi:10.1016/j.mbs.2008.03.008

Cohen, W. W. (1995). Fast effective rule induction. In *Proceedings of the Twelfth International Conference on Machine Learning* (pp. 115-123).

Dor, Y., Brown, J., Martinez, O. I., & Melton, D. A. (2004). Adult pancreatic beta-cells are formed by self-duplication rather than stem-cell differentiation. *Nature*, *429*, 41–46. doi:10.1038/nature02520

Errington, R. J., Marquez, N., Chappell, S. C., Wiltshire, M., & Smith, P. J. (2005). Time-Lapse Microscopy Approaches to Track Cell Cycle Progression at the Single-Cell Level. In *Current Protocols in Cytometry*. New York: John Wiley & Sons. doi:10.1002/0471142956.cy1204s31

Fayyad, U. M., & Irani, K. B. (1993). Multi-interval discretization of continuous valued attributes for classification learning. In *Proceedings of the Thirteenth International Joint Conference on Articial Intelligence* (Vol. 2, pp. 1022-1027). San Francisco: Morgan Kaufmann Publishers.

Feeney, G. P., Errington, R. J., Wiltshire, M., Marquez, N., Chappell, S. C., & Smith, P. J. (2003). Tracking the cell cycle origins for escape from topotecan action by breast cancer cells. *British Journal of Cancer*, *88*, 1310–1317. doi:10.1038/sj.bjc.6600889

Hall, M., Frank, E., Holmes, G., Pfahringer, B., Reutemann, P., & Witten, I. H. (2009). The WEKA Data Mining Software: An Update. *SIGKDD Explorations*, *11*, 10–18. doi:10.1145/1656274.1656278

Heppner, G. H. (1984). Tumor heterogeneity. *Cancer Research*, *44*, 2259–2265.

Hope, K. J., Jin, L., & Dick, J. E. (2004). Acute myeloid leukemia originates from a hierarchy of leukemic stem cell classes that differ in self-renewal capacity. *Nature Immunology*, *5*, 738–743. doi:10.1038/ni1080

Howard, R. A., & Matheson, J. E. (2005). Influence diagrams. *Decision Analysis*, *2*, 144–147. doi:10.1287/deca.1050.0050

Ichinose, Y., Seto, T., Nishiwaki, Y., Kiura, K., Sakai, H., & Yokoyama, A. (2010). Phase I Study of Topotecan and Cisplatin in Patients with Small Cell Lung Cancer. *Japanese Journal of Clinical Oncology*. doi:10.1093/jjco/hyq177

Khan, I. A., Hedley, C. J., White, N. S., Ali, R., Chappell, M. J., & Evans, N. D. (2006). A novel integrative bioinformatics environment for encoding and interrogating timelapse microscopy images. In *Modelling and Control in Biomedical Systems* (pp. 273–278). Amsterdam, The Netherlands: Elsevier.

Khan, I. A., Husemann, P., Campbell, L., White, N. S., White, R. J., & Smith, P. J. (2007). ProgeniDB: a novel cell lineage database for generation associated phenotypic behavior in cell-based assays. *Cell Cycle (Georgetown, Tex.)*, *6*, 868–874. doi:10.4161/cc.6.7.4045

Kim, K. M., & Shibata, D. (2002). Methylation reveals a niche: stem cell succession in human colon crypts. *Oncogene*, *21*, 5441–5449. doi:10.1038/sj.onc.1205604

Lorusso, D., Pietragalla, A., Mainenti, S., Masciullo, V., Di Vagno, G., & Scambia, G. (2010). Review role of topotecan in gynaecological cancers: current indications and perspectives. *Critical Reviews in Oncology/Hematology*, *74*, 163–174. doi:10.1016/j.critrevonc.2009.08.001

Noctor, S. C., Flint, A. C., Weissman, T. A., Dammerman, R. S., & Kriegstein, A. R. (2001). Neurons derived from radial glial cells establish radial units in neocortex. *Nature*, *409*, 714–720. doi:10.1038/35055553

Pearl, J. (1988). *Probabilistic reasoning in intelligent systems: networks of plausible inference*. San Francisco: Morgan Kaufmann.

Pommier, Y. (2006). Topoisomerase I inhibitors: camptothecins and beyond. *Nature Reviews. Cancer*, *6*, 789–802. doi:10.1038/nrc1977

Quinlan, J. R. (1996). Improved use of continuous attributes in C4.5. *Journal of Artificial Intelligence Research*, *4*, 77–90.

Rubin, H. (1990). The significance of biological heterogeneity. *Cancer and Metastasis Reviews*, *9*, 1–20. doi:10.1007/BF00047585

Seibel, N. L., Krailo, M., Chen, Z., Healey, J., Breitfeld, P. P., & Drachtman, R. (2007). Upfront window trial of topotecan in previously untreated children and adolescents with poor prognosis metastatic osteosarcoma: childrens Cancer Group (CCG) 7943. *Cancer*, *109*, 1646–1653. doi:10.1002/cncr.22553

Smirnov, N. (1939). Estimate of deviation between empirical distribution functions in two independent samples. *Bull. Moscow Univ*, *2*, 3–16.

Smith, P. J., Khan, I. A., & Errington, R. J. (2009). Cytomics and cellular informatics - coping with asymmetry and heterogeneity in biological systems. *Drug Discovery Today*, *14*, 271–277. doi:10.1016/j.drudis.2008.11.012

Stern, C. D., & Fraser, S. E. (2001). Tracing the lineage of tracing cell lineages. *Nature Cell Biology*, *3*, E216–E218. doi:10.1038/ncb0901-e216

Tang, M., Pires, Y., Schultz, M., Duarte, I., Gallegos, M., & Wistuba, I. I. (2003). Microsatellite analysis of synchronous and metachronous tumors: a tool for double primary tumor and metastasis assessment. *Diagnostic Molecular Pathology*, *12*, 151–159.

Vapnik, V. (1998). *Statistical learning theory*. New York: Wiley.

Wang, J. C. (1996). DNA topoisomerases. *Annual Review of Biochemistry*, *65*, 635–692. doi:10.1146/annurev.bi.65.070196.003223

Weigelt, B., Glas, A. M., Wessels, L. F., Witteveen, A. T., Peterse, J. L., & Vant Veer, L. J. (2003). Gene expression profiles of primary breast tumors maintained in distant metastases. In *Proceedings of the National Academy of Sciences of the United States of America* (Vol. 100, pp. 15901-15905). doi:10.1073/pnas.2634067100

Yamamoto, N., Yang, M., Jiang, P., Xu, M., Tsuchiya, H., & Tomita, K. (2003). Determination of clonality of metastasis by cell-specific color-coded fluorescent-protein imaging. *Cancer Research*, *63*, 7785.

ENDNOTES

1 The 51 cell deaths in the Second Generation are comprised of 14 (*Death, Death*) sister pairs and 23 (*Death, non-Death*) sister pairs.

2 Default parameters for JRIP are: 'error rate >= ½' is included in stopping criterion; 1 of 3 folds is used for pruning; the minimum total weight of the instances in a rule is set to 2; and the initial seed is 1. We perform 25 optimization runs to minimise the sensitivity to the initial seed.

3 The cost-sensitive classifier, also implemented in WEKA (CostSensitiveClassifier), uses the resampling method.

4 Default parameters for the J48 are: not binary splits on nominal attributes; confidence factor used for pruning equals 0.25; minimum number of instances per leaf equals 2; 1 of 3 folds using for pruning, the rest for growing the tree; C.4.5 pruning; sub-tree raising operation when pruning.

This work was previously published in the International Journal of Knowledge Discovery in Bioinformatics, Volume 1, Issue 4, edited by Xiao-Li Li and See-Kiong Ng, pp. 26-53, copyright 2010 by IGI Publishing (an imprint of IGI Global).

Compilation of References

Aach, J., & Church, G. (2001). Aligning gene expression time series with time warping algorithms. *Bioinformatics (Oxford, England)*, *17*, 495–508. doi:10.1093/bioinformatics/17.6.495

Acuña, V., Chierichetti, F., Lacroix, V., Marchetti-Spaccamela, A., Sagot, M. F., & Stougie, L. (2009). Modes and cuts in metabolic networks: Complexity and algorithms. *Bio Systems*, *95*(1), 51–60. doi:10.1016/j.biosystems.2008.06.015

Adamcsek, B., Palla, G., Farkas, I. J., Dernyi, I., & Vicsek, T. (2006). Cfinder: locating cliques and overlapping modules in biological networks. *Bioinformatics (Oxford, England)*, *22*(8), 1021–1023. doi:10.1093/bioinformatics/btl039

Aebersold, R., & Mann, M. (2003). Mass spectrometry-based proteomics. *Nature*, *422*, 198–207. doi:10.1038/nature01511

Agrawal, R., Imielinski, T., & Swami, A. N. (1993). Mining association rules between sets of items in large databases. In *Proceedings of the 1993 ACM SIGMOD International Conference on Management of Data* (pp. 207-216).

Agrawal, R., Mannila, H., Srikant, R., Toivonen, H., & Verkamo, A. I. (1996). Fast discovery of association rules. In *Advances in Knowledge Discovery and Data Mining* (pp. 307–328). Menlo Park, CA: AAAI Press.

Akutsu, T., Kuhara, S., Maruyama, O., & Miyano, S. (1998). Identification of gene regulatory networks by strategic gene disruptions and gene overexpressions. In *Proceedings of the ACM-SIAM Symposium on Discrete Algorithms*.

Albert, R., Jeong, H., & Barabasi, A.-L. (2000). Error and attack tolerance of complex networks. *Nature*, *406*, 378–482. doi:10.1038/35019019

Alm, E., & Arkin, A. P. (2002). Biological networks. *Current Opinion in Structural Biology*, *13*, 193–202. doi:10.1016/S0959-440X(03)00031-9

Alon, U. (2007). Network motifs: theory and experimental approaches. *Nature*, *8*, 450–461.

Aloy, P., Ceulemans, H., Stark, A., & Russell, R. B. (2003). The relationship between sequence and interaction divergence in proteins. *Journal of Molecular Biology*, *332*, 989–998. doi:10.1016/j.jmb.2003.07.006

Altschul, S. F., Gish, W., Miller, W., Myers, E. W., & Lipman, D. J. (1990). Basic local alignment search tool. *Journal of Molecular Biology*, *215*(3), 403–410.

Altschul, S. F., Madden, T. L., Schaffer, A. A., Zhang, J., Zhang, Z., & Miller, W. (1997). Gapped BLAST and PSI-BLAST: a new generation of protein database search programs. *Nucleic Acids Research*, *25*, 3389–3402. doi:10.1093/nar/25.17.3389

Alvarez-Buylla, A., García-Verdugo, J. M., & Tramontin, A. D. (2001). A unified hypothesis on the lineage of neural stem cells. *Nature Reviews. Neuroscience*, *2*, 287–293.. doi:10.1038/35067582

Anderson, D. J., Gage, F. H., & Weissman, I. L. (2001). Can stem cells cross lineage boundaries? *Nature Medicine*, *7*, 393–395..doi:10.1038/86439

Anderson, K. D., Morin, M. A., Beckel-Mitchener, A., Mobarak, C. D., Neve, R. L., & Furneaux, H. M. (2000). Overexpression of HuD, but not of its truncated form HuD I+II, promotes GAP-43 gene expression and neurite outgrowth in PC12 cells in the absence of nerve growth factor. *Journal of Neurochemistry*, *75*, 1103–1114. doi:10.1046/j.1471-4159.2000.0751103.x

Andreeva, A., Howorth, D., Chandonia, J. M., Brenner, S. E., Hubbard, T. J., & Chothia, C. (2008). Data growth and its impact on the SCOP database: new developments. *Nucleic Acids Research*, *36*, D419–D425. doi:10.1093/nar/gkm993

Ankerst, M., Breunig, M. M., Kriegel, H. P., & Sander, J. (1999). OPTICS: Ordering Points to Identify the Clustering Structure. In *Proceedings of the ACM SIGMOD*, Philadelphia (pp. 49-60).

Antonie, M.-L., & Zaiane, O. (2004). Mining positive and negative association rules: An approach for confined rules. In *Proceedings of the European PKDD Conf.*

Apostolico, A., Ciriello, G., Guerra, C., Heitsch, C., Hsiao, C., & Williams, L. (2008). Finding 3d motifs in ribosomal rna structures. *Nucleic Acids Reserch.*

Apostolico, A., Bock, M. E., & Lonardi, S. (2003). Monotony of surprise and large-scale quest for unusual words. *Journal of Computational Biology*, *10*(2/3), 283–311. doi:10.1089/10665270360688020

Apostolico, A., Parida, L., & Rombo, S. E. (2008). Motif patterns in 2d. *Theoretical Computer Science*, *390*(1), 40–55. doi:10.1016/j.tcs.2007.10.019

Aranda, B. (2010). The IntAct molecular interaction database in 2010. *Nucleic Acids Research*, *38*, D525–D531. doi:10.1093/nar/gkp878

Ardavín, C., Martínez del Hoyo, G., Martín, P., Anjuère, F., Arias, C. F., & Marín, A. R. (2001). Origin and differentiation of dendritic cells. *Trends in Immunology*, *22*, 691–700..doi:10.1016/S1471-4906(01)02059-2

Arnold, K., Bordoli, L., Kopp, J., & Schwede, T. (2006). The SWISS-MODEL Workspace: A web-based environment for protein structure homology modelling. *Bioinformatics (Oxford, England)*, *22*, 195–201. doi:10.1093/bioinformatics/bti770

Asburner, S. (2000). Gene ontology: tool for the unification of biology. the gene ontology consortium. *Nature Genetics*, *25*, 25–29. doi:10.1038/75556

Ashburner, M., Ball, C. A., Blake, J. A., Botstein, D., Butler, H., & Cherry, J. M. (2000). Gene Ontology: tool for the unification of biology. *Nature Genetics*, *25*, 25–29. doi:10.1038/75556

Asirvatham, A. J., Gregorie, C. J., Hu, Z., Magner, W. J., & Tomasi, T. B. (2008). MicroRNA targets in immune genes and the Dicer/Argonaute and ARE machinery components. *Molecular Immunology*, *45*, 1995–2006. doi:10.1016/j.molimm.2007.10.035

Asirvatham, A. J., Magner, W. J., & Tomasi, T. B. (2009). miRNA regulation of cytokine genes. *Cytokine*, *45*, 58–69. doi:10.1016/j.cyto.2008.11.010

Audic, Y., & Hartley, R. S. (2004). Post-transcriptional regulation in cancer. *Biology of the Cell*, *96*, 479–498. doi:10.1016/j.biolcel.2004.05.002

Bach, F. R., Lanckriet, G. R. G., & Jordan, M. I. (2004). Multiple kernel learning, conic duality, and the SMO algorithm. In *Proceedings of the 21st international conference on machine learning*. New York.

Bader, J. S., Chaudhuri, A., Rothberg, J. M., & Chant, J. (2004). Gaining confidence in high-throughput protein interaction networks. *Nature Biotechnology*, *22*, 78–85. doi:10.1038/nbt924

Bailey, T., & Elkan, C. (1994). Fitting a mixture model by expectation maximization to discover motifs in biopolymers. In *Proceedings of the Second International Conference on Intelligent Systems for Molecular Biology (ISMB-94)* (pp. 28-36). Menlo Park, CA: AAAI Press.

Bailey, T., & Elkan, C. (1995). Unsupervised learning of multiple motifs in biopolymers using expectation maximization. *Machine Learning*, *21*, 51–80. doi:10.1007/BF00993379

Bailly, C. (2000). Topoisomerase I poisons and suppressors as anticancer drugs. *Current Medicinal Chemistry*, *7*, 39–58. doi:.doi:10.2174/0929867003375489

Baker, N. A., Sept, D., Joseph, S., Holst, M. J., & McCammon, J. A. (2001). Electrostatics of nanosystems: application to microtubules and the ribosome. *Proceedings of the National Academy of Sciences of the United States of America, 98*, 10037–10041. doi:10.1073/pnas.181342398

Bakheet, T., Frevel, M., Williams, B. R., Greer, W., & Khabar, K. S. (2001). ARED: human AU-rich element-containing mRNA database reveals an unexpectedly diverse functional repertoire of encoded proteins. *Nucleic Acids Research, 29*, 246–254. doi:10.1093/nar/29.1.246

Bakheet, T., Williams, B. R., & Khabar, K. S. (2003). ARED 2.0: an update of AU-rich element mRNA database. *Nucleic Acids Research, 31*, 421–423. doi:10.1093/nar/gkg023

Bakheet, T., Williams, B. R., & Khabar, K. S. (2006). ARED 3.0: the large and diverse AU-rich transcriptome. *Nucleic Acids Research, 34*, D111–D114. doi:10.1093/nar/gkj052

Bandyopadhyay, D., Huan, J., Liu, J., Prins, J., Snoeyink, J., Wang, W., & Tropsha, A. (2006). Structure-based function inference using protein family-specific fingerprints. *Protein Science, 15*, 1537–1543. doi:10.1110/ps.062189906

Bandyopadhyay, D., & Snoeyink, J. (2004). Almost Delaunay Simplices: Nearest Neighbor Relations for Imprecise Points. In. *Proceedings of SODA, 2004*, 403–412.

Bandyopadhyay, S., Sharan, R., & Ideker, T. (2006). Systematic identification of functional orthologs based on protein network comparison. *Genome Research, 16*(3), 428–435. doi:10.1101/gr.4526006

Barabasi, A.-L., & Albert, R. (1999). Emergence of scaling in random networks. *Science, 286*, 509–511. doi:10.1126/science.286.5439.509

Barabasi, A.-L., & Oltvai, Z. N. (2004). Network biology: understanding the cell's functional organization. *Nature Reviews. Genetics, 5*, 101–113. doi:10.1038/nrg1272

Barber, C. B., Dobkin, D. P., & Huhdanpaa, H. T. (1996). The quickhull algorithm for convex hulls. *ACM Transactions on Mathematical Software, 22*(4), 469–483. doi:10.1145/235815.235821

Barnes, M., Freudenberg, J., Thompson, S., Aronow, B., & Pavlidis, P. (2005). Experimental comparison and cross-validation of the Affymetrix and Illumina gene expression analysis platforms. *Nucleic Acids Research, 33*, 5914–5923. doi:10.1093/nar/gki890

Barreau, C., Paillard, L., & Osborne, H. B. (2005). AU-rich elements and associated factors: are there unifying principles? *Nucleic Acids Research, 33*, 7138–7150. doi:10.1093/nar/gki1012

Bartel, D. P. (2004). MicroRNAs: genomics, biogenesis, mechanism, and function. *Cell, 116*, 281–297. doi:10.1016/S0092-8674(04)00045-5

Bartel, D. P. (2009). MicroRNAs: target recognition and regulatory functions. *Cell, 136*, 215–233. doi:10.1016/j.cell.2009.01.002

Bar-Yehuda, R., Geiger, D., Naor, J., & Roth, R. M. (1998). Approximation algorithms for the feedback vertex set problem with applications to constraint satisfaction and Bayesian inference. *SIAM Journal on Computing, 27*, 942–959. doi:10.1137/S0097539796305109

Bastide, Y., Taouil, R., Pasquier, N., Stumme, G., & Lakhal, L. (2000). Mining frequent patterns with counting inference. *SIGKDD Explorations*, 2.

Basu, S., Banerjee, A., et al. (2002). *Semi-supervised clustering by seeding*. Paper presented at the International Conference on Machine Learning.

Baumann, P., Hiemke, C., Ulrich, S., Eckermann, G., Gaertner, I., & Gerlach, M. (2004). The AGNP-TDM expert group consensus guidelines: therapeutic drug monitoring in psychiatry. *Pharmacopsychiatry, 37*(6), 243–265. doi:10.1055/s-2004-832687

Bayardo, R. J., & Agrawal, R. (1999). Mining the most interesting rules. In *Proceedings of the ACM SIGKDD Conf.*

Becker, S. (1996). Mutual information maximization: Models of cortical self-organization. *Network (Bristol, England), 7*(1), 7–31. doi:10.1088/0954-898X/7/1/003

Becquet, C., Blachon, S., Jeudy, B., Boulicaut, J. F., & Gandrillon, O. (2003). Strong-association-rule mining for large-scale gene-expression data analysis: a case study o human SAGE data. *Genome Biology, 12*, 1–16.

Behre, J., Wilhelm, T., Kamp, A., Ruppin, E., & Schuster, S. (2008). Structural robustness of metabolic networks with respect to multiple knockouts. *Journal of Theoretical Biology*, *252*, 433–441. doi:10.1016/j.jtbi.2007.09.043

Bellman, R. E. (1961). *Adaptive Control Processes*. Princeton, NJ: Princeton University Press.

Ben-Dor, A., Chor, B., Karp, R., & Yakhini, Z. (2003). Discovering local structure in gene expression data: the order-preserving submatrix problem. *Journal of Computational Biology*, *10*, 373–384. doi:10.1089/10665270360688075

Ben-Hur, A., & Noble, W. S. (2005). Kernel methods for predicting protein-protein interactions. *Bioinformatics (Oxford, England)*, *21*(Suppl. 1), i38–i46. doi:10.1093/bioinformatics/bti1016

Benjamini, Y., & Hochberg, Y. (1995). Controlling the false discovery rate: a practical and powerful approach to multiple testing. *Journal of the Royal Statistical Society. Series B. Methodological*, *57*, 289–300.

Benjamini, Y., & Yekutieli, D. (2001). The control of the false discovery rate in multiple testing under dependency. *Annals of Statistics*, *29*, 1165–1188. doi:.doi:10.1214/aos/1013699998

Berg, J., & Lassig, M. (2004). Local graph alignment and motif search in biological networks. *Proceedings of the National Academy of Sciences of the United States of America*, *101*(41), 14689–14694. doi:10.1073/pnas.0305199101

Bergk, V., Haefeli, W. E., Gasse, C., Brenner, H., & Martin-Facklam, M. (2005). Information deficits in the summary of product characteristics preclude an optimal management of drug interactions: a comparison with evidence from the literature. *European Journal of Clinical Pharmacology*, *61*(5-6), 327–335. doi:10.1007/s00228-005-0943-4

Berman, H. M., Westbrook, J., Feng, Z., Gilliland, G., Bhat, T. N., & Weissig, H. (2000). The Protein Data Bank. *Nucleic Acids Research*, *28*(1), 235–242. doi:10.1093/nar/28.1.235

Bernards, R., & Weinberg, R. A. (2002). Metastasis genes: A progression puzzle. *Nature*, *418*, 823.. doi:10.1038/418823a

Bernhart, S. H., Tafer, H., Muckstein, U., Flamm, C., Stadler, P. F., & Hofacker, I. L. (2006). Partition function and base pairing probabilities of RNA heterodimers. *Algorithms for Molecular Biology; AMB*, *1*, 3. doi:10.1186/1748-7188-1-3

Berriz, G. F., & King, O. D. (2003). Characterizing gene sets with FuncAssociate. *Bioinformatics (Oxford, England)*, *19*(18), 2502–2504. doi:10.1093/bioinformatics/btg363

Betel, D., Wilson, M., Gabow, A., Marks, D. S., & Sander, C. (2008). The microRNA.org resource: targets and expression. *Nucleic Acids Research*, *36*, D149–D153. doi:10.1093/nar/gkm995

Bickel, S., & Tobias, S. (2004). Multi-View Clustering. In *Proceedings of the Fourth IEEE International Conference on Data Mining (ICDM'04)*.

Biemar, F., Zinzen, R., Ronshaugen, M., Sementchenko, V., Manak, J. R., & Levine, M. S. (2005). Spatial regulation of microRNA gene expression in the Drosophila embryo. *Proceedings of the National Academy of Sciences of the United States of America*, *102*, 15907–15911. doi:10.1073/pnas.0507817102

Bilenko, M., Basu, S., et al. (2004). Integrating constraints and metric learning in semi-supervised clustering. In *Proceedings of the International Conference on Machine Learning (ICML '04)*.

Binkowski, T. A., & Joachimiak, A. (2008). Protein functional surfaces: global shape matching and local spatial alignments of ligand binding sites. *BMC Structural Biology*, *8*, 45. doi:10.1186/1472-6807-8-45

Blanchette, M. (2001). Algorithms for phylogenetic footprinting. In *Proceedings of the Fifth Annual International Conference on Computational Molecular Biology (RECOMB 01)* (pp. 49-58).

Bleakley, K., Biau, G., & Vert, J.-P. (2007). Supervised reconstruction of biological networks with local models. *Bioinformatics (Oxford, England)*, *23*(13), i57–i65. doi:10.1093/bioinformatics/btm204

Boguski, M. S., Lowe, T. M., & Tolstoshev, C. M. (1993). dbEST--database for "expressed sequence tags". *Nature Genetics*, *4*, 332–333. doi:10.1038/ng0893-332

Bordner, A. J., & Abagyan, R. A. (2004). Large-scale prediction of protein geometry and stability changes for arbitrary single point mutations. *Proteins*, *57*(2), 400–413. doi:10.1002/prot.20185

Bowie, J. U., Luthy, R., & Eisenberg, D. (1991). A method to identify protein sequences that fold into a known three-dimensional structure. *Science*, *253*(5016), 164–170. doi:10.1126/science.1853201

Boyd, S., & Vandenberghe, L. (2004). *Convex optimization*. Cambridge University Press.

Bracewell, R. (1999). The Hilbert Transform. In *The Fourier Transform and Its Applications* (3rd ed., pp. 267–272). New York: McGraw-Hill.

Brauer, M. J., Saldanha, A. J., Dolinski, K., & Botstein, D. (2005). Homeostatic adjustment and metabolic remodeling in glucose-limited yeast cultures. *Molecular Biology of the Cell*, *16*, 2503–2517. doi:10.1091/mbc.E04-11-0968

Braunstein, M., Griffin, T. J. IV, Kriakov, J. I., Friedman, S. T., Grindley, N. D., & Jacobs, W. R. Jr. (2000). Identification of genes encoding exported Mycobacterium tuberculosis proteins using a Tn552'phoA in vitro transposition system. *Journal of Bacteriology*, *182*, 2732–2740. doi:10.1128/JB.182.10.2732-2740.2000

Brazma, A., Jonassen, I., Eidhammer, I., & Gilbert, D. (1998). Approaches to the automatic discovery of patterns in biosequences. *Journal of Computational Biology*, *5*(2), 277–304. doi:10.1089/cmb.1998.5.279

Breitkreutz, B.-J., Stark, C., Reguly, T., Boucher, L., Breitkreutz, A., & Livstone, M. (2008). The BioGRID interaction database: 2008 update. *Nucleic Acids Research*, *36*, D637–D640. doi:10.1093/nar/gkm1001

Brennan, R. L., & Prediger, D. J. (1981). Coefficient Kappa: Some uses, misuses, and alternatives. *Educational and Psychological Measurement*, *41*(3), 687–699. doi:10.1177/001316448104100307

Brennecke, J., Stark, A., Russell, R. B., & Cohen, S. M. (2005). Principles of microRNA-target recognition. *PLoS Biology*, *3*, e85. doi:10.1371/journal.pbio.0030085

Breunig, M. M., Kriegel, H. P., Ng, R. T., & Sander, J. (2000). LOF: Identifying Density-Based Local Outliers. In *Proceedings of the ACM SIGMOD 2000 Int. Conf. Management of Data*, Dallas, TX (Vol. 29, No. 2, pp. 93-104).

Brun, A., Park, H. J., Knutsson, H., & Westin, C. F. (2003). Coloring of DTMRI fiber traces using Laplacian eigenmaps. In *Proceedings of the Ninth Int. Conf. on Computer Aided Systems Theory* (Vol. 2809, pp. 564-572). Berlin: Springer Verlag.

Bshouty, N. (1995). Exact learning boolean functions via the monotone theory. *Information and Computation*, *123*, 146–153. doi:10.1006/inco.1995.1164

Buhler, J., & Tompa, M. (2002). Finding motifs using random projections. *Journal of Computational Biology*, *9*, 225–242. doi:10.1089/10665270252935430

Burgard, A. P., Pharkya, P., & Maranas, D. (2003). OptKnock: a bilevel programming framework for identifying gene knockout strategies for microbial strain optimization. *Biotechnology and Bioengineering*, *84*, 647–657. doi:10.1002/bit.10803

Burkhardt, S., Crauser, A., Ferragina, P., Lenhof, H.-P., Rivals, E., & Vingron, M. (1999). *q*-gram based database searching using a suffix array. In *RECOMB '99* (pp. 77–83).

Burley, S. K. (2000). An overview of structural genomics. *Nature Structural Biology*, *7*(Suppl), 932–934. doi:10.1038/80697

Bushati, N., & Cohen, S. M. (2007). microRNA functions. *Annual Review of Cell and Developmental Biology*, *23*, 175–205. doi:10.1146/annurev.cellbio.23.090506.123406

Butte, A. J., Tamayo, P., Slonim, D., Golub, T. R., & Kohane, I. S. (2000). Discovering functional relationships between RNA expression and chemotherapeutic susceptibility using relevance networks. *Proceedings of the National Academy of Sciences of the United States of America*, *97*(22), 12182–12186. doi:10.1073/pnas.220392197

Cai, D., He, X., & Han, J. (2007, December). *Spectral regression: a unified approach for sparse subspace learning.* Paper presented at the IEEE International Conference on Data Mining, Omaha, NE.

Caldarelli, G. (2007). *Scale-free networks: Complex webs in nature and technology.* Oxford Univ Press.

Calders, T., & Goethals, B. (2003). Minimal k-free representations of frequent sets. In *Proceedings of the European PKDD Conf.*

Calders, T., & Goethals, B. (2005). Quick inclusion-exclusion. In *Proceedings of the ECML-PKDD Workshop on Knowl-edge Discovery in Inductive Databases*.

Campbell, I. (2007). Chi-squared and Fisher-Irwin tests of two-by-two tables with small sample recommendations. *Statistics in Medicine*, *26*, 3661–3675.. doi:10.1002/sim.2832

Cancho, R. F., Janssen, C., & Sole, R. V. (2008). Topology of technology graphs: Small world patterns in electronic circuits. *Physical Review E: Statistical, Nonlinear, and Soft Matter Physics*, *64*(4), 046119. doi:10.1103/PhysRevE.64.046119

Canterakis, N. (1999). 3D Zernike moments and Zernike affine invariants for 3D image analysis and recognition. *Proc. 11th Scandinavian Conference on Image Analysis* (pp. 85-93).

Cao, H., Deterding, L. J., & Blackshear, P. J. (2007). Phosphorylation site analysis of the anti-inflammatory and mRNA-destabilizing protein tristetraprolin. *Expert Review of Proteomics*, *4*, 711–726. doi:10.1586/14789450.4.6.711

Cao, X., Li, S. C., & Tung, A. T. (2005). Indexing DNA sequences using *q*-grams. In *CPM'03. Lecture Notes in Computer Science*, *3453*, 4–16.

Capriotti, E., Fariselli, P., & Casadio, R. (2004). A neural-network-based method for predicting protein stability changes upon single point mutations. *Bioinformatics (Oxford, England)*, *20*(1), I63–I68. doi:10.1093/bioinformatics/bth928

Carlson, J. M., Heckerman, D., & Shani, G. (2009). *Estimating false discovery rates for contingency tables* (Tech Rep. No. 2009-53). Microsoft Research.

Carter, S. L., Eklund, A. C., Mecham, B. H., Kohane, I. S., & Szallasi, Z. (2005). Re- definition of Affymetrix probe sets by sequence overlap with cDNA microarray probes reduces cross-platform inconsistencies in cancer-associated gene expression measurements. *BMC Bioinformatics*, *6*, 107. doi:10.1186/1471-2105-6-107

Catani, M., Howard, R. J., Pajevic, S., & Jones, D. K. (2002). Virtual in vivo interactive dissection of white matter fasciculi in the human brain. *NeuroImage*, *17*, 77–94. doi:10.1006/nimg.2002.1136

Cavanna, T., Pokorna, E., Vesely, P., Gray, C., & Zicha, D. (2007). Evidence for protein 4.1 B acting as a metastasis suppressor. *Journal of Cell Science*, *120*, 606–616.. doi:10.1242/jcs.000273

Chalifa-Caspi, V., Yanai, I., Ophir, R., Rosen, N., Shmoish, M., & Benjamin-Rodrig, H. (2004). GeneAnnot: comprehensive two-way linking between oligonuceotide array probesets and GeneCards genes. *Bioinformatics (Oxford, England)*, *20*, 1457–1458. doi:10.1093/bioinformatics/bth081

Chang, C.-C., & Lin, C.-J. (2001). *LIBSVM: a library for support vector machines*. Retrieved from http://www.csie.ntu.edu.tw/~cjlin/libsvm

Chang, T. C., & Mendell, J. T. (2007). microRNAs in vertebrate physiology and human disease. *Annual Review of Genomics and Human Genetics*, *8*, 215–239. doi:10.1146/annurev.genom.8.080706.092351

Charniak, E. (1991). Bayesian Networks without Tears. *AI Magazine*, *12*, 50–63.

Chatraryamontri, A., Ceol, A., Montecchi-Palazzi, L., Nardelli, G., Schneider, M. V., Castagnoli, L., & Cesareni, G. (2007). MINT: the Molecular INTeraction database. *Nucleic Acids Research*, *35*, D572–D574. doi:10.1093/nar/gkl950

Cheadle, C., Fan, J., Cho-Chung, Y. S., Werner, T., Ray, J., & Do, L. (2005). Control of gene expression during t cell activation: alternate regulation of mrna transcription and mrna stability. *BMC Genomics*, *6*(1), 75. doi:10.1186/1471-2164-6-75

Chen, J., Hsu, W., & Lee, M. L. (2007). Labeling network motifs in protein interactomes for protein function prediction. In *ICDE*.

Chen, J., Hsu, W., Lee, M. L., et al. (2006). NeMoFinder: dissecting genome-wide protein-protein interac-tions with meso-scale network motifs. In *KDD* (pp. 106–115).

Chen, C. Y., Gherzi, R., Ong, S. E., Chan, E. L., Raijmakers, R., & Pruijn, G. J. (2001). AU binding proteins recruit the exosome to degrade ARE-containing mRNAs. *Cell*, *107*, 451–464. doi:10.1016/S0092-8674(01)00578-5

Chen, C. Y., & Shyu, A. B. (1994). Selective degradation of early-response-gene mRNAs: functional analyses of sequence features of the AU-rich elements. *Molecular and Cellular Biology*, *14*, 8471–8482.

Cheng, H., Yan, X., Han, J., & Hsu, C.-W. (2007). Discriminative frequent pattern analysis for effective classification. In *Proceedings of International Conference on Data Engineering* (pp. 716-725).

Cheng, H., Yan, X., Han, J., & Yu, P. (2008). Direct discriminative pattern mining for effective classification. In *Proceedings of International Conference on Data Engineering* (pp. 169-178).

Cheng, Y., & Church, G. M. (2000). Biclustering of Expression Data. In *Proceedings of the 8th Int'l Conf. Intelligent Systems for Molecular Biology (ISMB 00)* (pp. 93-103). New York: ACM Press.

Chen, G., Larsen, P., Almasri, E., & Dai, Y. (2008). Rank-based edge reconstruction for scale-free genetic regulatory networks. *BMC Bioinformatics*, *9*(1), 75. doi:10.1186/1471-2105-9-75

Cheng, C.-Y., Huang, C.-Y., & Sun, C.-T. (2008). Mining bridge and brick motifs from complex biological networks for functionally and statistically significant discovery. *IEEE Transactions on Systems, Man, and Cybernetics–Part B*, *38*(1), 17–24. doi:10.1109/TSMCB.2007.908842

Cheng, J., Randall, A., & Baldi, P. (2006). Prediction of protein stability changes for single-site mutations using support vector machines. *Proteins*, *62*(4), 1125–1132. doi:10.1002/prot.20810

Cheng, K., Law, N., Siu, W., & Liew, A. (2008). Identification of coherent patterns in gene expression data using an efficient biclustering algorithm and parallel coordinate visualization. *BMC Bioinformatics*, *9*, 210. doi:10.1186/1471-2105-9-210

Chen, X. (2009). Curve-Based Clustering of Time Course Gene Expression Data Using Self-Organizing Maps. [PubMed]. *Journal of Bioinformatics and Computational Biology*, *7*, 645–661. doi:10.1142/S0219720009004291

Cheung, S., Evans, N. D., Chappell, M. J., Godfrey, K. R., Smith, P. J., & Errington, R. J. (2008). Exploration of the intercellular heterogeneity of topotecan uptake into human breast cancer cells through compartmental modelling. *Mathematical Biosciences*, *213*, 119–134.. doi:10.1016/j.mbs.2008.03.008

Chiang, J. H., & Yu, H. C. (2003). MeKE: discovering the functions of gene products from biomedical literature via sentence alignment. *Bioinformatics (Oxford, England)*, *19*(11), 1417–1422. doi:10.1093/bioinformatics/btg160

Chikhi, R. & Kihara, D. (2009). *Protein binidng ligand prediction using local surface descriptors*.

Chitale, M., Hawkins, T., & Kihara, D. (2009a). Automated prediction of protein function from sequence. In J. Bujnicki (Ed.), *Prediction of Protein Strucutre, Functions, and Interactions* (pp. 63-86). John Wiley & Sons Ltd.

Chitale, M., Hawkins, T., Park, C., & Kihara, D. (2009b). ESG: Extended similarity group method for automated protein function prediction. *Bioinformatics (Oxford, England)*, *25*, 1739–1745. doi:10.1093/bioinformatics/btp309

Chor, B., & Tuller, T. (2006). Biological networks: Comparison, conservation, and evolutionary trees. In *RECOMB* (pp. 30–44).

Chor, B., & Tuller, T. (2006). Finding a maximum likelihood tree is hard. *Journal of the ACM*, *53*(5), 722–744. doi:10.1145/1183907.1183909

Chothia, C., & Lesk, A. M. (1986). The relation between the divergence of sequence and structure in proteins. *The EMBO Journal*, *5*, 823–826.

Cho, Y.-R., Hwang, W., Ramanathan, M., & Zhang, A. (2007). Semantic integration to identify overlapping functional modules in protein interaction networks. *BMC Bioinformatics*, *8*, 265. doi:10.1186/1471-2105-8-265

Cho, Y.-R., & Zhang, A. (2010). Identification of functional hubs and modules by converting interactome networks into hierarchical ordering of proteins. *BMC Bioinformatics*, *11*(3), S3. doi:10.1186/1471-2105-11-S3-S3

Chua, H. N., Sung, W. K., & Wong, L. (2006). Exploiting indirect neighbours and topological weight to predict protein function from protein-protein interactions. *Bioinformatics (Oxford, England)*, *22*(13), 1623–1630. doi:10.1093/bioinformatics/btl145

Ciriello, G., & Guerra, C. (2008). A review on models and algorithms for motif discovery in protein-protein interaction network. *Briefings in Functional Genomics and Proteomics*.

Clout, N. J., Tisi, D., & Hohenester, E. (2003). Novel fold revealed by the structure of a Fas1 domain pair from the insect cell adhesion molecule Fasciclin I. *Structure (London, England)*, *11*, 197–203. doi:10.1016/S0969-2126(03)00002-9

Coggill, P., Finn, R. D., & Bateman, A. (2008). Identifying protein domains with the Pfam database. *Curr. Protoc. Bioinformatics*.

Cohen, J., Briand, F., & Newman, C. (1990). *Community Food Webs: Data and Theory*. Springer.

Cohen, W. W. (1995). Fast effective rule induction. In *Proceedings of the Twelfth International Conference on Machine Learning* (pp. 115-123).

Cong, G., Tung, A., Xu, X., Pan, F., & Yang, J. (2004). FARMER: Finding Interesting Rule Groups in Microarray Datasets. In *Proceedings of the ACM SIGMOD Int'l Conf. Management of Data* (pp. 143-154).

Cong, G., Tan, K., Tung, A., & Feng, P. (2004). Mining Frequent Closed Patterns in Microarray Data. In *ICDM'04* (pp. 363–366). Washington, DC: IEEE Press.

Corouge, I., Gouttard, S., & Gerig, G. (2004). Towards a shape model of white matter fiber bundles using diffusion tensor MRI. In *Proceedings of the IEEE Int. Symposium on Biomedical Imaging*, Arlington, VA (pp. 344-347).

Creighton, C., & Hanash, S. (2003). Mining gene expression databases for association rules. *Bioinformatics (Oxford, England)*, *19*, 79–86. doi:10.1093/bioinformatics/19.1.79

Cui, T., Zhang, L., Wang, X., & He, Z. G. (2009). Uncovering new signaling proteins and potential drug targets through the interactome analysis of Mycobacterium tuberculosis. *BMC Genomics*, *10*, 118. doi:10.1186/1471-2164-10-118

Dai, M., Wang, P., Boyd, A. D., Kostov, G., Athey, B., & Jones, E. G. (2005). Evolving gene/transcript definitions significantly alter the interpretation of GeneChip data. *Nucleic Acids Research*, *33*, e175. doi:10.1093/nar/gni179

Damoiseaux, J. S., Smith, S. M., Witter, M. P., Sanz-Arigita, E. J., Barkhof, F., & Scheltens, P. (2009). White Matter Tract Integrity in Aging and Alzheimer's Disease. *Human Brain Mapping*, *30*, 1051–1059. doi:10.1002/hbm.20563

Dasika, M. S., Burgard, A., & Maranas, D. (2006). A computational framework for the topological analysis and targeted disruption of signal transduction networks. *Biophysical Journal*, *91*, 382–398. doi:10.1529/biophysj.105.069724

Daubechies, I. (1992). *Ten Lectures on Wavelets*. Philadelphia: SIAM. doi:10.1137/1.9781611970104

Davey, B. A., & Priestley, H. A. (1990). *Introduction to lattices and order*. Cambridge, UK: Cambridge University Press.

Davidson, I., & Wagstaff, K. L. (2006). Measuring constraint-set utility for partitional clustering algorithms. In *Proceedings of Knowledge Discovery in Databases* (*Vol. 4213*, pp. 115–126). Pkdd. doi:10.1007/11871637_15

Davis, F. P., Barkan, D. T., Eswar, N., McKerrow, J. H., & Sali, A. (2007). Host pathogen protein interactions predicted by comparative modeling. *Protein Science*, *16*, 2585–2596. doi:10.1110/ps.073228407

Dayhoff, M. O., Schwartz, R. M., & Orcut, B. C. (1978). A model for evolutionary change in proteins. In Dayhoff, M. O. (Ed.), *Atlas of Protein Sequence and Structure* (*Vol. 5*, pp. 345–352). Washington, DC: National Biomedical Research Foundation.

de Berg, M., Cheong, O., van Kreveld, M., & Overmars, M. (2008). *Computational Geometry: Algorithms and Applications*. Berlin: Springer Verlag.

De la Torre, F., & Black, M. J. (2003). A framework for robust subspace learning. *International Journal of Computer Vision*, *54*(1-3), 117–142. doi:10.1023/A:1023709501986

de Souto, M. C., & Costa, I. G. (2008). Clustering cancer gene expression data: a comparative study. *BMC Bioinformatics*, *9*, 497. doi:10.1186/1471-2105-9-497

Deane, C. M., Salwinski, L., Xenarios, I., & Eisenberg, D. (2002). Protein interactions: two methods for assessment of the reliability of high throughput observations. *Molecular & Cellular Proteomics*, *1*(5), 349–356. doi:10.1074/mcp.M100037-MCP200

Dehouck, Y., Grosfils, A., Folch, B., Gilis, D., Bogaerts, P., & Rooman, M. (2009). Fast and accurate predictions of protein stability changes upon mutations using statistical potentials and neural networks: PoPMuSiC-2.0. *Bioinformatics (Oxford, England)*, *25*(19), 2537–2543. doi:10.1093/bioinformatics/btp445

Delsuc, F., Brinkmann, H., & Philippe, H. (2005). Phylogenomics and the reconstruction of the tree of life. *Nature Reviews. Genetics*, *6*, 361–375. doi:10.1038/nrg1603

Dempster, A. P., Laird, N. M., & Rubin, D. B. (1977). Maximum Likelihood from Incomplete Data via the EM Algorithm. *Journal of the Royal Statistical Society. Series A (General)*, *39*(1), 1–31.

Deng, M., Sun, F., & Chen, T. (2003). Assessment of the reliability of protein protein interactions and protein function prediction. In *Pac. Symp. Biocomput. (PSB2003)* (pp. 140–151).

Deng, X., Li, G., Li, Z., Ma, B., & Wang, L. (2003). Generic Drug Design without Side Effect. *SIAM Journal on Computing*, *32*(4), 1073–1090. doi:10.1137/S0097539701397825

Deshpande, M., Kuramochi, M., Wale, N., & Karypis, G. (2005). Frequent sub-structure based approaches for classifying chemical compounds. *IEEE Transactions on Knowledge and Data Engineering*, *17*(8), 1036–1050. doi:10.1109/TKDE.2005.127

Dharan, A., & Nair, A. S. (2009). Biclustering of gene expression data using reactive greedy randomized adaptive search procedure. *BMC Bioinformatics*, *10*(1), S27. doi:10.1186/1471-2105-10-S1-S27

Ding, Z., Gore, J., & Anderson, A. (2003). Classification and quantification of neuronal fiber pathways using diffusion tensor MRI. *Magnetic Resonance in Medicine*, *49*, 716–721. doi:10.1002/mrm.10415

Dittrich, M. T., Klau, G. W., Rosenwald, A., Dandekar, T., & Muller, T. (2008). Identifying functional modules in protein-protein interaction networks: an integrated exact approach. *Bioinformatics (Oxford, England)*, *24*(13), i223–i231. doi:10.1093/bioinformatics/btn161

Dixon, D. A., Balch, G. C., Kedersha, N., Anderson, P., Zimmerman, G. A., Beauchamp, R. D., & Prescott, S. M. (2003). Regulation of cyclooxygenase-2 expression by the translational silencer TIA-1. *The Journal of Experimental Medicine*, *198*, 475–481. doi:10.1084/jem.20030616

Dobrin, R., Beg, Q. K., Barabasi, A. L., & Oltvai, Z. N. (2004). Aggregation of topological motifs in the escherichia coli transcriptional regulatory network. *BMC Bioinformatics*, *5*(10).

Dom, B. E. (2001). *An information-theoretic external cluster-validity measure* (Tech. Rep. No. RJ10219). IBM.

Dong, G., Jiang, C., Pei, J., Li, J., & Wong, L. (2005). Mining succinct systems of minimal generators of formal concepts. In *Proceedings of the Int'l Conf. Database Systems for Advanced Applications*.

Donoho, D. L., & Johnstone, I. M. (1994). Ideal spatial adaptation by wavelet shrinkage. *Biometrika*, *81*, 425–455. doi:10.1093/biomet/81.3.425

Donoho, D. L., & Johnstone, I. M. (1995). Adapting to unknown smoothness via wavelet shrinkage. *Journal of the American Statistical Association*, *90*, 1200–1224. doi:10.2307/2291512

Dopazo, J., Rodr'ıguez, A., S'aiz, J. C., & Sobrino, F. (1993). Design of primers for PCR amplification of highly variable genomes. *CABIOS*, *9*, 123–125.

Dor, Y., Brown, J., Martinez, O. I., & Melton, D. A. (2004). Adult pancreatic beta-cells are formed by self-duplication rather than stem-cell differentiation. *Nature*, *429*, 41–46.. doi:10.1038/nature02520

Draghici, S. (2003). *Data Analysis Tools for DNA Microarrays*. Virginia Beach, VA: Chapman & Hall/CRC.

Durbin, R., Eddy, S., Krogh, A., & Mitchison, G. (1998). *Probabilistic models of proteins and nucleic acids*. Cambridge, UK: Cambridge University Press.

Duret, L., & Bucher, P. (1997). Searching for regulatory elements in human noncoding sequences. *Current Opinion in Structural Biology*, *7*, 399–406. doi:10.1016/S0959-440X(97)80058-9

Duret, L., Dorkeld, F., & Gautier, C. (1993). Strong conservation of non-coding sequences during vertebrates evolution: potential involvement in post-transcriptional regulation of gene expression. *Nucleic Acids Research*, *21*, 2315–2322. doi:10.1093/nar/21.10.2315

Dyer, M. D., Murali, T. M., & Sobral, B. W. (2007). Computational prediction of host-pathogen protein-protein interactions. *Bioinformatics (Oxford, England)*, *23*, i159–i166. doi:10.1093/bioinformatics/btm208

Dyer, M. D., Murali, T. M., & Sobral, B. W. (2008). The landscape of human proteins interacting with viruses and other pathogens. *PLoS Pathogens*, *4*, e32. doi:10.1371/journal.ppat.0040032

Eckman, B. A., & Kaufmann, A. (2004). Querying BLAST within a data federation. *Bulletin of the IEEE Computer Society Technical Committee on Data Engineering*, *27*, 12–19.

Electronic Medicines Compendium. (2009a). *SPC Rapamune*. Retrieved September 2009, from http://emc.medicines.org.uk/document.aspx?documentId=5747

Electronic Medicines Compendium. (2009b). *SPC Cymbalta*. Retrieved September 2009, from http://emc.medicines.org.uk/medicine/15694

Elo, L. L., Jarvenpaa, H., Oresic, M., Lahesmaa, R., & Aittokallio, T. (2007). Systematic construction of gene coexpression networks with applications to human T helper cell differentiation process. *Bioinformatics (Oxford, England)*, *23*(16), 2096–2103. doi:10.1093/bioinformatics/btm309

Elo, L. L., Lahti, L., Skottman, H., Kyl¨aniemi, M., Lahesmaa, R., & Aittokallio, T. (2005). Integrating probe-level expression changes across generations of Affymetrix arrays. *Nucleic Acids Research*, *33*, e193. doi:10.1093/nar/gni193

Ensom, M. H., Davis, G. A., Cropp, C. D., & Ensom, R. J. (1998). Clinical pharmacokinetics in the 21st century: does the evidence support definitive outcomes? *Clinical Pharmacokinetics*, *34*(4), 265–279. doi:10.2165/00003088-199834040-00001

Erdos, P., & Renyi, A. (1959). On random graphs. *Publicationes Matematicae*, *6*, 290–297.

Erdos, P., & Renyi, A. (1960). On the evolution of random graphs. *Publ. Mat. Inst. Hungarian Acad. Sc.*, *5*, 17–61.

Errington, R. J., Marquez, N., Chappell, S. C., Wiltshire, M., & Smith, P. J. (2005). Time-Lapse Microscopy Approaches to Track Cell Cycle Progression at the Single-Cell Level. In *Current Protocols in Cytometry.* New York: John Wiley & Sons. doi:10.1002/0471142956.cy1204s31

Espel, E. (2005). The role of the AU-rich elements of mRNAs in controlling translation. *Seminars in Cell & Developmental Biology*, *16*, 59–67. doi:10.1016/j.semcdb.2004.11.008

Esquela-Kerscher, A., & Slack, F. J. (2006). Oncomirs - microRNAs with a role in cancer. *Nature Reviews. Cancer*, *6*, 259–269. doi:10.1038/nrc1840

Ester, M., Kriegel, H. P., Sander, J., & Xu, X. (1996). A density-based algorithm for discovering clusters in large spatial databases with noise. In *Proceedings of the Second Int. Conf. on Knowledge Discovery and Data Mining* (pp. 226-231). Portland, OR: AAAI Press.

Evangelista, P. F., Embrechts, M. J., & Szymanski, B. K. (2006). Taming the curse of dimensionality in kernels and novelty detection. In Abraham, A., Baets, B., Koppen, M., & Nickolay, B. (Eds.), *Applied Soft Computing Technologies: The Challenge of Complexity* (pp. 425–440). Berlin: Springer Verlag. doi:10.1007/3-540-31662-0_33

Even, G., Naor, J., Schieber, B., & Sudan, M. (1998). Approximating minimum feedback sets and multicuts in directed graphs. *Algorithmica*, *20*, 151–174. doi:10.1007/PL00009191

Evgeniou, T., Micchelli, C. A., & Pontil, M. (2005). Learning multiple tasks with kernel methods. *Journal of Machine Learning Research*, *6*, 615–637.

Fabian, M. R., Sonenberg, N., & Filipowicz, W. (2010). Regulation of mRNA translation and stability by microRNAs. *Annual Review of Biochemistry*, *79*, 351–379. doi:10.1146/annurev-biochem-060308-103103

Fan, W., Zhang, K., Cheng, H., Gao, J., Yan, X., Han, J., et al. (2008). Direct mining of discriminative and essential graphical and itemset features via model-based search tree. In *Proceedings of the ACM SIGKDD international conference on knowledge discovery in databases* (pp. 230-238).

Fang, G., Kuang, R., Pandey, G., Steinbach, M., Myers, C. L., & Kumar, V. (2010). Subspace Differential Co-expression Analysis: Problem Definition and A General Approach. In *Proceedings of the 15th Pacific Symposium on Biocomputing (PSB)* (Vol. 15, pp. 145-156).

Fayyad, U. M., & Irani, K. B. (1993). Multi-interval discretization of continuous valued attributes for classification learning. In *Proceedings of the Thirteenth International Joint Conference on Articial Intelligence* (Vol. 2, pp. 1022-1027). San Francisco: Morgan Kaufmann Publishers.

Feeney, G. P., Errington, R. J., Wiltshire, M., Marquez, N., Chappell, S. C., & Smith, P. J. (2003). Tracking the cell cycle origins for escape from topotecan action by breast cancer cells. *British Journal of Cancer*, *88*, 1310–1317.. doi:10.1038/sj.bjc.6600889

Fei, H., & Huan, J. (2008). Structure feature selection for graph classification. In *Proceedings of the CIKM* (pp. 991-1000).

Fei, H., & Huan, J. (2009). L2 Norm Regularized Feature Kernel Regression for Graph Data. In *Proceedings of the CIKM* (pp. 593-600).

Felsentein, J. (2004). *Inferring Phylogenies*. Sunderland, MA: Sinauer Associates.

Ferrer-Costa, C., Gelpi, J. L., Zamakola, L., Parraga, I., de la Cruz, X., & Orozco, M. (2005). PMUT: a web-based tool for the annotation of pathological mutations on proteins. *Bioinformatics (Oxford, England)*, *21*(14), 3176–3178. doi:10.1093/bioinformatics/bti486

Filipowicz, W., Bhattacharyya, S. N., & Sonenberg, N. (2008). Mechanisms of post-transcriptional regulation by microRNAs: are the answers in sight? *Nature Reviews. Genetics*, *9*, 102–114. doi:10.1038/nrg2290

Finn, R. D., Marshall, M., & Bateman, A. (2005). iPfam: visualization of protein-protein interactions in PDB at domain and amino acid resolutions. *Bioinformatics (Oxford, England)*, *21*, 410–412. doi:10.1093/bioinformatics/bti011

Finn, R. D., Tate, J., Mistry, J., Coggill, P. C., Sammut, S. J., & Hotz, H. R. (2008). The Pfam protein families database. *Nucleic Acids Research*, *36*, D281–D288. doi:10.1093/nar/gkm960

Fionda, V., Panni, S., Palopoli, L., & Rombo, S. E. (2008). Protein-protein interaction network querying by a "focus and zoom" approach. In *Proceedings of BioInformatics Research and Development (BIRD'08)* (pp. 331-346).

Fionda, V., Panni, S., Palopoli, L., & Rombo, S. E. (n.d.). A technique to search functional similarities in ppi networks. *International Journal of Data Mining and Bioinformatics*.

Fitch, W. (2000). Homology a personal view on some of the problems. *Trends in Genetics*, *16*, 227–231. doi:10.1016/S0168-9525(00)02005-9

Ford, L. P., Watson, J., Keene, J. D., & Wilusz, J. (1999). ELAV proteins stabilize deadenylated intermediates in a novel in vitro mRNA deadenylation/degradation system. *Genes & Development*, *13*, 188–201. doi:10.1101/gad.13.2.188

Francke, C., Siezen, R. J., & Teusink, B. (2005). Reconstructing the metabolic network of a bacterium from its genome. *Trends in Microbiology*, *13*(11), 550–559. doi:10.1016/j.tim.2005.09.001

Frank, E., Hall, M., Trigg, L., Holmes, G., & Witten, I. H. (2004). Data mining in bioinformatics using Weka. *Bioinformatics (Oxford, England)*, *20*(15), 2479–2481. doi:10.1093/bioinformatics/bth261

Frey, B. J., & Dueck, D. (2007). Clustering by passing messages between data points. *Science*, *315*, 972–976. doi:10.1126/science.1136800

Friedman, R. C., Farh, K. K., Burge, C. B., & Bartel, D. P. (2009). Most mammalian mRNAs are conserved targets of microRNAs. *Genome Research*, *19*, 92–105. doi:10.1101/gr.082701.108

Fröhlich, H., Wegner, J. K., Sieker, F., & Zell, A. (2005). Optimal assignment kernels for attributed molecular graphs. In *Proceedings of the ICML* (pp. 225-232).

Galban, S., Martindale, J. L., Mazan-Mamczarz, K., Lopez de Silanes, I., Fan, J., & Wang, W. (2003). Influence of the RNA-binding protein HuR in pVHL-regulated p53 expression in renal carcinoma cells. *Molecular and Cellular Biology*, *23*, 7083–7095. doi:10.1128/MCB.23.20.7083-7095.2003

Ganter, B., & Wille, R. (1999). *Formal concept analysis: Mathematical foundations*. New York: Springer-Verlag.

Gardy, J. L., Laird, M. R., Chen, F., Rey, S., Walsh, C. J., & Ester, M. (2005). PSORTb v.2.0: expanded prediction of bacterial protein subcellular localization and insights gained from comparative proteome analysis. *Bioinformatics (Oxford, England)*, *21*, 617–623. doi:10.1093/bioinformatics/bti057

Garey, M., & Johnson, D. (1979). *Computers and intractability: A guide to the theory of NP-completeness*. New York: Freeman.

Gärtner, T., Flach, P. A., & Wrobel, S. (2003). On graph kernels: hardness results and efficient alternatives. In *Proceedings of the COLT* (pp. 129-143).

Gautier, L., Mooller, M., Friis-Hansen, L., & Knudsen, S. (2004). Alternative mapping of probes to genes for Affymetrix chips. *BMC Bioinformatics*, *5*, 111. doi:10.1186/1471-2105-5-111

Gavin, A.-C. (2002). Functional organization of the yeast proteome by systematic analysis of protein complexes. *Nature*, *415*, 141–147. doi:10.1038/415141a

Gebran, N., & Al Haidari, K. (2006). Assessment of prescribing information for generic drugs manufactured in the Middle East and marketed in Saudi Arabia. *Annals of Saudi Medicine*, *26*(3), 192–199.

Ge, H., Liu, Z., Church, G. M., & Vidal, M. (2001). Correlation between transcriptome and interactome mapping data from Saccharomyces cerevisiae. *Nature Genetics*, *29*, 482–486. doi:10.1038/ng776

Gene Ontology Consortium. (2004). The gene ontology (GO) database and informatics resource. *Nucleic Acids Research*, *32*(Database issue), D258–D261. doi:10.1093/nar/gkh036

Gerig, G., Gouttard, S., & Corouge, I. (2004). Analysis of brain white matter via fiber tract modelling. In *Proceedings of the 26th Ann. Int. Conf. IEEE Engineering in Medicine and Biology Society (EMBS)*, San Francisco (pp. 4421-4424).

Gherzi, R., Lee, K. Y., Briata, P., Wegmuller, D., Moroni, C., Karin, M., & Chen, C. Y. (2004). A KH domain RNA binding protein, KSRP, promotes ARE-directed mRNA turnover by recruiting the degradation machinery. *Molecular Cell*, *14*, 571–583. doi:10.1016/j.molcel.2004.05.002

Gibbons, F. D., & Roth, F. P. (2002). Judging the quality of gene expression-based clustering methods using gene annotation. *Genome Research*, *12*(10), 1574–1581. doi:10.1101/gr.397002

Giot, L. (2003). A protein interaction map of Drosophila melanogaster. *Science*, *302*, 1727–1736. doi:10.1126/science.1090289

Girvan, M., & Newman, M. E. J. (2002). Community structure in social and biological networks. *Proceedings of the National Academy of Sciences of the United States of America*, *99*(12), 7821–7826. doi:10.1073/pnas.122653799

Glass, L., & Mackey, M. C. (1979). A simple model for phase locking of biological oscillators. [PubMed]. *Journal of Mathematical Biology*, *7*, 339–352. doi:10.1007/BF00275153

Glenisson, P., & Antal, P. (2003). Evaluation of the vector space representation in text-based gene clustering. *Pacific Symposium on Biocomputing. Pacific Symposium on Biocomputing*, 391–402.

Glenisson, P., & Mathys, J. (2003). Meta-clustering of gene expression data and literature-based information. *SIGKDD Explor. Newsl.*, *5*(2), 101–112. doi:10.1145/980972.980985

Goethals, B., & Zaki, M. (2003). Advances in frequent itemset mining implementations: report on FIMI'03. *SIGKDD Explorations*, *6*, 109–117. doi:10.1145/1007730.1007744

Gonzalez, J. A., Holder, L. B., & Hook, D. J. (2002). Graph-based relational concept learning. In *Proceedings of the ICML* (pp. 219-226).

Gotoh, O. (1982). An improved algorithm for matching biological sequences. *Journal of Molecular Biology*, *162*(3), 705–708. doi:10.1016/0022-2836(82)90398-9

Gowri, V. S., Krishnadev, O., Swamy, C. S., & Srinivasan, N. (2006). MulPSSM: a database of multiple position-specific scoring matrices of protein domain families. *Nucleic Acids Research*, *34*, D243–D246. doi:10.1093/nar/gkj043

Gowtham, A., Gupta, R., Fang, G., Pandey, G., Steinbach, M., & Kumar, V. (2009). Association Analysis Techniques for Bioinformatics Problems. In *Proceedings of the 1st International Conference on Bioinformatics and Computational Biology (BICoB)* (pp. 1-13).

Griffiths-Jones, S., Saini, H. K., van Dongen, S., & Enright, A. J. (2008). miRBase: tools for microRNA genomics. *Nucleic Acids Research*, *36*, D154–D158. doi:10.1093/nar/gkm952

Grimson, A., Farh, K. K., Johnston, W. K., Garrett-Engele, P., Lim, L. P., & Bartel, D. P. (2007). MicroRNA targeting specificity in mammals: determinants beyond seed pairing. *Molecular Cell*, *27*, 91–105. doi:10.1016/j.molcel.2007.06.017

Grochow, J., & Kellis, M. (2007). Network motif discovery using sub-graph enumeration and symmetry-breaking. *Research in Computational Molecular Biology* (pp. 92-106).

Guerois, R., Nielsen, J. E., & Serrano, L. (2002). Predicting changes in the stability of proteins and protein complexes: a study of more than 1000 mutations. *Journal of Molecular Biology*, *320*(2), 369–387. doi:10.1016/S0022-2836(02)00442-4

Guex, N., & Peitsch, M. C. (1997). SWISS-MODEL and the Swiss-PdbViewer: An environment for comparative protein modelling. *Electrophoresis*, *18*, 2714–2723. doi:10.1002/elps.1150181505

Guimerà, R., Sales-Pardo, M., & Amaral, L. A. N. (2007). A network-based method for target selection in metabolic networks. *Bioinformatics (Oxford, England)*, *23*(13), 1616–1622. doi:10.1093/bioinformatics/btm150

Gunopulos, D., Khardon, R., Mannila, H., Saluja, S., Toivonen, H., & Sharma, R. (2003). Discovering all most specific sentences. *ACM Transactions on Database Systems*, *28*, 140–174. doi:10.1145/777943.777945

Guo, Z., Li, Y., Gong, X., Yao, C., Ma, W., & Wang, D. (2007). Edge-based scoring and searching method for identifying condition-responsive protein protein interaction sub-network. *Bioinformatics (Oxford, England)*, *23*(16), 2121–2128. doi:10.1093/bioinformatics/btm294

Gusfield, D. (1997). *Algorithms on Strings, Trees, and Sequences: Computer Science and Computational Biology*. Cambridge, UK: Cambridge University Press. doi:10.1017/CBO9780511574931

Guy, E. S., & Mallampalli, A. (2008). Managing TB in the 21st century: existing and novel drug therapies. *Therapeutic Advances in Respiratory Disease*, *2*, 401–408. doi:10.1177/1753465808099522

Hagmann, P., Cammoun, L., Gigandet, X., Meuli, R., Honey, C. J., Wedeen, V. J., & Sporns, O. (2008). Mapping the structural core of the human cerebral cortex. *PLoS Biology*, *6*(7), e159. doi:10.1371/journal.pbio.0060159

Hahn, M. (2007). Bias in phylogenetic tree reconciliation methods: Implications for vertebrate genome evolution. *Genome Biology*, *8*, R141. doi:10.1186/gb-2007-8-7-r141

Halees, A. S., El-Badrawi, R., & Khabar, K. S. (2008). ARED Organism: expansion of ARED reveals AU-rich element cluster variations between human and mouse. *Nucleic Acids Research*, *36*, D137–D140. doi:10.1093/nar/gkm959

Halkidi, M., Batistakis, Y., & Vazirgiannis, M. (2001). On clustering validation techniques. *Journal of Intelligent Information Systems*, *17*(2-3), 107–145. doi:10.1023/A:1012801612483

Hall, M., Frank, E., Holmes, G., Pfahringer, B., Reutemann, P., & Witten, I. H. (2009). The WEKA Data Mining Software: An Update. *SIGKDD Explorations*, *11*, 10–18..doi:10.1145/1656274.1656278

Hammell, M., Long, D., Zhang, L., Lee, A., Carmack, C. S., & Han, M. (2008). mirWIP: microRNA target prediction based on microRNA-containing ribonucleoprotein-enriched transcripts. *Nature Methods*, *5*, 813–819. doi:10.1038/nmeth.1247

Handorf, T., Christian, N., Ebenhöh, O., & Kahn, D. (2008). An environmental perspective on metabolism. *Journal of Theoretical Biology*, *252*(3), 530–537. doi:10.1016/j.jtbi.2007.10.036

Handorf, T., Ebenhöh, O., & Heinrich, R. (2005). Expanding metabolic networks: scopes of compounds, robustness, and evolution. *Journal of Molecular Evolution*, *61*, 498–512. doi:10.1007/s00239-005-0027-1

Han, J.-D. J., Bertin, N., Hao, T., Goldberg, D. S., Berriz, G. F., & Zhang, L. V. (2004). Evidence for dynamically organized modularity in the yeast protein-protein interaction network. *Nature*, *430*(6995), 88–93. doi:10.1038/nature02555

Hansen, M., & Yu, B. (2001). Model selection and the principle of minimum description Length. *JASA*, *96*(454), 746–774.

Harbig, J., Sprinkle, R., & Enkemann, S. A. (2005). A sequence-based identification of the genes detected by probesets on the Affymetrix U133 plus 2.0 array. *Nucleic Acids Research*, *33*, e31. doi:10.1093/nar/gni027

Hartwell, L. H., Hopfield, J. J., Leibler, S., & Murray, A. W. (1999). From molecular to modular cell biology. *Nature*, *402*, c47–c52. doi:10.1038/35011540

Hartwell, L., Hopfield, J., Leibler, A., & Murray, A. (1999). From molecular to modular cell biology. *Nature*, *402*, c47–c52. doi:10.1038/35011540

Hastie, T., Tibshirani, R., & Friedman, J. (2009). *The elements of statistical learning: data mining, inference and prediction* (2nd ed.). Berlin: Springer Verlag.

Haus, U. U., Klamt, S., & Stephen, T. (2008). Computing knock-out strategies in metabolic networks. *Journal of Computational Biology*, *15*(3), 259–268. doi:10.1089/cmb.2007.0229

Hawkins, T., Chitale, M., & Kihara, D. (2008). New paradigm in protein function prediction for large scale omics analysis. *Molecular BioSystems*, *4*, 223–231. doi:10.1039/b718229e

Hawkins, T., Chitale, M., Luban, S., & Kihara, D. (2009). PFP: Automated prediction of gene ontology functional annotations with confidence scores using protein sequence data. *Proteins*, *74*, 566–582. doi:10.1002/prot.22172

Hawkins, T., & Kihara, D. (2007). Function prediction of uncharacterized proteins. *Journal of Bioinformatics and Computational Biology*, *5*, 1–30. doi:10.1142/S0219720007002503

Hawkins, T., Luban, S., & Kihara, D. (2006). Enhanced automated function prediction using distantly related sequences and contextual association by PFP. *Protein Science*, *15*, 1550–1556. doi:10.1110/ps.062153506

Haykin, S. (1999). *Neural Networks: A Comprehensive Foundation* (2nd ed.). Upper Saddle River, NJ: Prentice-Hall.

Haykin, S. (2002). *Adaptive Filter Theory*. Englewood Cliffs, NJ: Prentice Hall.

Hendlich, M., Rippmann, F., & Barnickel, G. (1997). LIGSITE: automatic and efficient detection of potential small molecule-binding sites in proteins. *Journal of Molecular Graphics & Modelling*, *15*, 359–363, 389. doi:10.1016/S1093-3263(98)00002-3

Heppner, G. H. (1984). Tumor heterogeneity. *Cancer Research*, *44*, 2259–2265.

Hertz, G., & Stormo, G. (1995). Identification of consensus patterns in unaligned DNA and protein sequences: a large-deviation statistical basis for penalizing gaps. In *Proceedings of the 3rd Int'l Conf. Bioinformatics and Genome Research* (pp. 201-216).

Herzel, H., Ebeling, W., & Schmitt, A. O. (1994). Entropies of biosequences: the role of repeats. *Physical Review*, *50*, 5061–5071.

Hinneburg, A., & Keim, D. A. (1998). An Efficient Approach to Clustering in Large Multimedia Databases with Noise. In *Proceedings of the 4th Int. Conf. on Knowledge Discovery and Data Mining*, New York (pp. 58-65).

Hohmann, S., & Mager, W. H. (Eds.). (2003). *Yeast Stress Responses*. Berlin: Springer. doi:10.1007/3-540-45611-2

Hope, K. J., Jin, L., & Dick, J. E. (2004). Acute myeloid leukemia originates from a hierarchy of leukemic stem cell classes that differ in self-renewal capacity. *Nature Immunology*, *5*, 738–743..doi:10.1038/ni1080

Houben, E. N., Nguyen, L., & Pieters, J. (2006). Interaction of pathogenic mycobacteria with the host immune system. *Current Opinion in Microbiology*, *9*, 76–85. doi:10.1016/j.mib.2005.12.014

Howard, R. A., & Matheson, J. E. (2005). Influence diagrams. *Decision Analysis*, *2*, 144–147..doi:10.1287/deca.1050.0050

Ho, Y. (2002). Systematic identification of protein complexes in Saccharomyces cerevisiae by mass spectrometry. *Nature*, *415*, 180–183. doi:10.1038/415180a

Hsu, P. W., Huang, H. D., Hsu, S. D., Lin, L. Z., Tsou, A. P., & Tseng, C. P. (2006). miRNAMap: genomic maps of microRNA genes and their target genes in mammalian genomes. *Nucleic Acids Research, 34*, D135–D139. doi:10.1093/nar/gkj135

Hsu, S. D., Chu, C. H., Tsou, A. P., Chen, S. J., Chen, H. C., & Hsu, P. W. (2008). miRNAMap 2.0: genomic maps of microRNAs in metazoan genomes. *Nucleic Acids Research, 36*, D165–D169. doi:10.1093/nar/gkm1012

Huan, J., Wang, W., Bandyopadhyay, D., Snoeyink, J., Prins, J., & Tropsha, A. (2004). Mining spatial motifs from protein structure graphs. In *Proceedings of the RECOMB* (pp. 308-315).

Huang, B., & Schroeder, M. (2006). LIGSITEcsc: predicting ligand binding sites using the Connolly surface and degree of conservation. *BMC Structural Biology, 6*, 19. doi:10.1186/1472-6807-6-19

Huang, H., Zhang, J., Wakana, S., Zhang, W., Ren, T., & Richards, L. J. (2006). White and gray matter development in human fetal, newborn and pediatric brains. *NeuroImage, 33*, 27–38. doi:10.1016/j.neuroimage.2006.06.009

Huang, L. T., Gromiha, M. M., & Ho, S. Y. (2007). iPTREE-STAB: interpretable decision tree based method for predicting protein stability changes upon mutations. *Bioinformatics (Oxford, England), 23*(10), 1292–1293. doi:10.1093/bioinformatics/btm100

Huang, T. W., Tien, A. C., Huang, W. S., Lee, Y. C., Peng, C. L., & Tseng, H. H. (2004). POINT: a database for the prediction of protein-protein interactions based on the orthologous interactome. *Bioinformatics (Oxford, England), 20*, 3273–3276. doi:10.1093/bioinformatics/bth366

Hubbard, T. J. P., Aken, B. L., Beal, K., Ballester, B., Caccamo, M., Chen, Y., Clarke, L., Coates, G., Cunningham, F., Cutts, T., et al. (2007). Ensembl 2007. *Nucleic Acids Research, 35(Database-Issue)*, 610–617.

Hubbard, T. J., Aken, B. L., Beal, K., Ballester, B., Caccamo, M., & Chen, Y. (2007). Ensembl 2007. *Nucleic Acids Research, 35*, D610–D617. doi:10.1093/nar/gkl996

Hubert, L., & Arabie, P. (1985). Comparing partitions. *J. Classification, 2*, 193–218. doi:10.1007/BF01908075

Huelsenbeck, J. (1995). The robustness of two phylogenetic methods: Four-taxon simulations reveal a slight superiority of maximum likelihood over neighbor joining. *Molecular Biology and Evolution, 12*, 843–849.

Huelsenbeck, J. (2002). Potential applications and pitfalls of Bayesian inference of phylogeny. *Systematic Biology, 51*, 673–688. doi:10.1080/10635150290102366

Huelsenbeck, J. P., & Ronquist, F. (2001). MRBAYES: Bayesian inference of phylogeny. *Bioinformatics (Oxford, England), 17*, 754–755. doi:10.1093/bioinformatics/17.8.754

Hulo, N., Bairoch, A., Bulliard, V., Cerutti, L., De, C. E., & Langendijk-Genevaux, P. S. (2006). The PROSITE database. *Nucleic Acids Research, 34*, D227–D230. doi:10.1093/nar/gkj063

Hunt, E. (2004). Indexed searching on proteins using a suffix sequoia. *Bulletin of the IEEE Computer Society Technical Committee on Data Engineering, 27*, 24–31.

Hunt, E. (2007). Exhaustive peptide searching using relations. In *BNCOD'07. Lecture Notes in Computer Science, 4587*, 13–24. doi:10.1007/978-3-540-73390-4_3

Huttenhower, C., Flamholz, A., Landis, J., Sahi, S., Myers, C., & Olszewski, K. (2007). Nearest Neighbor Networks: clustering expression data based on gene neighborhoods. *BMC Bioinformatics, 8*(1), 250. doi:10.1186/1471-2105-8-250

Hyyrö, H., & Navarro, G. (2003). A practical index for genome searching. In *SPIRE'03. Lecture Notes in Computer Science, 2857*, 341–349.

Ichinose, Y., Seto, T., Nishiwaki, Y., Kiura, K., Sakai, H., & Yokoyama, A. (2010). Phase I Study of Topotecan and Cisplatin in Patients with Small Cell Lung Cancer. *Japanese Journal of Clinical Oncology*..doi:10.1093/jjco/hyq177

Ideker, T., Ozier, O., Schwikowski, B., & Siegel, A. F. (2002). Discovering regulatory and signalling circuits in molecular interaction networks. *Bioinformatics (Oxford, England), 18*(1), S233–S240.

Ideker, T., & Thorsson, V. (2001). Integrated genomic and proteomic analyses of a systematically perturbed metabolic network. *Science, 292*(5518), 929–934. doi:10.1126/science.292.5518.929

Ideker, T., Thorsson, V., Ranish, J. A., Christmas, R., Buhler, J., & Eng, J. K. (2001). Integrated genomic and proteomic analyses of a systemically perturbed metabolic network. [PubMed]. *Science*, *292*, 929–934. doi:10.1126/science.292.5518.929

INSERT FIGURE 001 Rekha, N., Machado, S. M., Narayanan, C., Krupa, A., & Srinivasan, N. (2005). Interaction interfaces of protein domains are not topologically equivalent across families within superfamilies: Implications for metabolic and signaling pathways. *Proteins*, *58*, 339–353. doi:10.1002/prot.20319

Itakura, F. (1975). Minimum Prediction Residual Principle Applied to Speech Recognition. In *Proceedings of the IEEE Trans. Acoustics, Speech, and Signal Proc.* (Vol. ASSP-23, pp. 52-72).

Ito, T., Chiba, T., Ozawa, R., Yoshida, M., Hattori, M., & Sakaki, Y. (2001). A comprehensive two-hybrid analysis to explore the yeast protein interactome. *Proceedings of the National Academy of Sciences of the United States of America*, *98*(8), 4569–4574. doi:10.1073/pnas.061034498

Jagadish, B., Yi, H., & Faloutsos, C. (1998). Efficient retrieval of similar time sequences under time warping. In *Proceedings of the 14th Int. Conf. on Data Engineering*, Orlando, FL (pp. 201-220).

Jain, A. K., & Dubes, R. C. (1988). *Algorithms for clustering data*. Upper Saddle River, NJ: Prentice Hall.

Jain, R. G., Andrews, L. G., McGowan, K. M., Gao, F., Keene, J. D., & Pekala, P. P. (1995). Hel-N1, an RNA-binding protein, is a ligand for an A + U rich region of the GLUT1 3' UTR. *Nucleic Acids Symposium Series*, 209–211.

Jansen, R., Greenbaum, D., & Gerstein, M. (2002). Relating whole-genome expression data with protein-protein interactions. *Genome Research*, *12*, 37–46. doi:10.1101/gr.205602

Jansen, R., Yu, H., Greenbaum, D., Kluger, Y., Krogan, N. J., & Chung, S. (2003). A Bayesian networks approach for predicting protein-protein interactions from genomic data. *Science*, *302*, 449–453. doi:10.1126/science.1087361

Jaroszewicz, S., & Simovici, D. A. (2002). Support approximations using bonferroni-type inequalities. In *Proceedings of the 6th European Conference on Principles of Data Mining and Knowledge Discovery*.

Jeong, H., Tombor, B., Albert, R., Oltval, Z. N., & Barabási, A. L. (2000). The large-scale organization of metabolic networks. *Nature*, *407*, 651–654. doi:10.1038/35036627

Jiang, D., Zhou, S., & Chen, Y.-P. P. (2009). Compensatory ability to null mutation in metabolic networks. *Biotechnology and Bioengineering*, *103*, 361–369. doi:10.1002/bit.22237

Jianu, R., Demiralp, C., & Laidlaw, D. H. (2009). Exploring 3D DTI fiber tracts with linked 2D representations. *IEEE Transactions on Visualization and Computer Graphics*, *15*(6), 1449–1456. doi:10.1109/TVCG.2009.141

Jin, N., Young, C., & Wang, W. (2009). Graph Classification Based on Pattern Co-occurrence. In *Proceedings of the CIKM* (pp. 573-582).

Jin, N., Young, C., & Wang, W. (2010). GAIA: graph classification using evolutionary computation. In *Proceedings of the SIGMOD* (pp. 879-890).

Jing, Q., Huang, S., Guth, S., Zarubin, T., Motoyama, A., & Chen, J. (2005). Involvement of microRNA in AU-rich element-mediated mRNA instability. *Cell*, *120*, 623–634. doi:10.1016/j.cell.2004.12.038

Ji, W., Zhou, W., Gregg, K., Yu, N., Davis, S., & Davis, S. (2004). A method for cross-species gene expression analysis with high-density oligonucleotide arrays. *Nucleic Acids Research*, *32*, e93. doi:10.1093/nar/gnh084

John, B., Enright, A. J., Aravin, A., Tuschl, T., Sander, C., & Marks, D. S. (2004). Human microRNA targets. *PLoS Biology*, *2*, e363. doi:10.1371/journal.pbio.0020363

John, B., Sander, C., & Marks, D. S. (2006). Prediction of human microRNA targets. *Methods in Molecular Biology (Clifton, N.J.)*, *342*, 101–113.

Józefowski, S., Sobota, A., & Kwiatkowska, K. (2008). How Mycobacterium tuberculosis subverts host immune responses. *Bioessays*, *30*, 943-954.

Kafri, R., Bar-Even, A., & Pilpel, Y. (2005). Transcription control reprogramming in genetic backup circuits. *Nature Genetics*, *37*(3), 295–299. doi:10.1038/ng1523

Kahraman, A., Morris, R. J., Laskowski, R. A., & Thornton, J. M. (2007). Shape variation in protein binding pockets and their ligands. *Journal of Molecular Biology*, *368*, 283–301. doi:10.1016/j.jmb.2007.01.086

Kalidas, Y., & Chandra, N. (2008). PocketDepth: a new depth based algorithm for identification of ligand binding sites in proteins. *Journal of Structural Biology*, *161*, 31–42. doi:10.1016/j.jsb.2007.09.005

Kamal, A., Azeeza, S., Malik, M. S., Shaik, A. A., & Rao, M. V. (2008). Efforts towards the development of new antitubercular agents: potential for thiolactomycin based compounds. *Journal of Pharmacy & Pharmaceutical Sciences*, *11*, 56s–80s.

Kanehisa, M., Araki, M., Goto, S., Hattori, M., Hirakawa, M., & Itoh, M. (2008). KEGG for linking genomes to life and the environment. *Nucleic Acids Research*, *36*, D480–D484. doi:10.1093/nar/gkm882

Karp, P. D., Keseler, I. M., Shearer, A., Latendresse, M., Krummenacker, M., & Paley, S. M. (2007). Multidimensional annotation of the Escherichia coli K-12 genome. *Nucleic Acids Research*, *35*, 7577–7590. doi:10.1093/nar/gkm740

Karp, P. D., Ouzounis, C. A., & Moore-Kochlacs, C. (2005). Expansion of the BioCyc collection of pathway/genome databases to 160 genomes. *Nucleic Acids Research*, *19*, 6083–6089. doi:10.1093/nar/gki892

Karsch-Mizrachi, I., & Ouellette, B. F. (2001). The GenBank sequence database. *Methods of Biochemical Analysis*, *43*, 45–63. doi:10.1002/0471223921.ch3

Kashima, H., & Koyanagi, T. (2002). SVM kernels for semi-structured data. In *Proceedings of the ICML* (pp. 291-298).

Kashima, H., Tsuda, K., & Inokuchi, A. (2003). Marginalized kernels between labeled graphs. In *Proceedings of the ICML* (pp. 321-328).

Kashtan, N., Itzkovitz, S., Milo, R., & Alon, U. (2004). Topological generalizations of network motifs. *Physical Review E: Statistical, Nonlinear, and Soft Matter Physics*, *70*(3), 031909. doi:10.1103/PhysRevE.70.031909

Kasturi, J., & Acharya, R. (2004). Clustering of diverse genomic data using information fusion. In *Proceedings of the 2004 ACM symposium on Applied computing*, Nicosia, Cyprus (pp. 116-120). New York: ACM Press.

Kato, T., Tsuda, K., & Asai, K. (2005). Selective integration of multiple biological data for supervised network inference. *Bioinformatics (Oxford, England)*, *21*(10), 2488–2495. doi:10.1093/bioinformatics/bti339

Katsanou, V., Papadaki, O., Milatos, S., Blackshear, P. J., Anderson, P., Kollias, G., & Kontoyiannis, D. L. (2005). HuR as a negative posttranscriptional modulator in inflammation. *Molecular Cell*, *19*, 777–789. doi:10.1016/j.molcel.2005.08.007

Kawabata, T., & Go, N. (2007). Detection of pockets on protein surfaces using small and large probe spheres to find putative ligand binding sites. *Proteins*, *68*, 516–529. doi:10.1002/prot.21283

Keich, U., & Pevzner, P. (2002a). Finding motifs in the twilight zone. *Bioinformatics (Oxford, England)*, *18*, 1374–1381. doi:10.1093/bioinformatics/18.10.1374

Keich, U., & Pevzner, P. (2002b). Subtle motifs: defining the limits of motif finding algorithms. *Bioinformatics (Oxford, England)*, *18*, 1382–1390. doi:10.1093/bioinformatics/18.10.1382

Keller, G. H., & Manak, M. M. (1989). *DNA Probes* (p. 12). New York: Stockton Press.

Kent, W. (2002). BLAT - The BLAST-like alignment tool. *Genome Research*, *12*, 656–664.

Keogh, E. (2002). Exact Indexing of Dynamic Time Warping. In *Proceedings of the 28th Int. Conf. on Very Large Data Bases*, Hong Kong (pp. 406-417).

Keogh, E., & Pazzani, M. (2001). Derivative Dynamic Time Warping. *In Proceedings of the First SIAM Int. Conf. on Data Mining*, Chicago (pp. 5-7).

Kerr, G., & Ruskin, H. J. (2008). Techniques for clustering gene expression data. *Computers in Biology and Medicine*, *38*(3), 283–293. doi:10.1016/j.compbiomed.2007.11.001

Khabar, K. S. (2005). The AU-rich transcriptome: more than interferons and cytokines, and its role in disease. *Journal of Interferon & Cytokine Research*, *25*, 1–10. doi:10.1089/jir.2005.25.1

Khabar, K. S., Bakheet, T., & Williams, B. R. (2005). AU-rich transient response transcripts in the human genome: expressed sequence tag clustering and gene discovery approach. *Genomics*, *85*, 165–175. doi:10.1016/j.ygeno.2004.10.004

Khan, I. A., Hedley, C. J., White, N. S., Ali, R., Chappell, M. J., & Evans, N. D. (2006). A novel integrative bioinformatics environment for encoding and interrogating timelapse microscopy images. In *Modelling and Control in Biomedical Systems* (pp. 273–278). Amsterdam, The Netherlands: Elsevier.

Khan, I. A., Husemann, P., Campbell, L., White, N. S., White, R. J., & Smith, P. J. (2007). ProgeniDB: a novel cell lineage database for generation associated phenotypic behavior in cell-based assays. *Cell Cycle (Georgetown, Tex.)*, *6*, 868–874. doi:10.4161/cc.6.7.4045

Kihara, D., Sael, L., & Chikhi, R. (2009). Local surface shape-based protein function prediction using Zernike descriptors. *Biophysical Journal*, *96*, 650a. doi:10.1016/j.bpj.2008.12.3435

Kihara, D., & Skolnick, J. (2004). Microbial Genomes have over 72% structure assignment by the threading algorithm PROSPECTOR_Q. *Proteins*, *55*, 464–473. doi:10.1002/prot.20044

Kim, K. M., & Shibata, D. (2002). Methylation reveals a niche: stem cell succession in human colon crypts. *Oncogene*, *21*, 5441–5449..doi:10.1038/sj.onc.1205604

Kim, Y. J., Boyd, A., Athey, B. D., & Patel, J. M. (2005). miBLAST: scalable evaluation of a batch of nucleotide sequence queries with BLAST. *Nucleic Acids Research*, *33*, 4335–4344. doi:10.1093/nar/gki739

Kitano, H. (2004). Cancer as a robust system: implications for anticancer therapy. *Nature Reviews. Cancer*, *4*, 227–235. doi:10.1038/nrc1300

Klamt, S., & Gilles, E. D. (2004). Minimal cut sets in biochemical reaction networks. *Bioinformatics (Oxford, England)*, *20*(2), 226–234. doi:10.1093/bioinformatics/btg395

Klamt, S., & Stelling, J. (2002). Combinatorial complexity of pathway analysis in metabolic networks. *Molecular Biology Reports*, *29*, 233–236. doi:10.1023/A:1020390132244

Klein, D., Kamvar, S. D., et al. (2002). *From instance-level constraints to space- level constraints: Making the most of prior knowledge in data clustering*. Paper presented at the International Conference on Machine Learning (ICML'02).

Klein, J., Bittihn, P., Ledochowitsch, P., Hahn, H. K., Konrad, O., Rexilius, J., & Peitgen, H. O. (2007). Grid-Based Spectral Fiber Clustering. In *Proceedings of the SPIE 6509* (p. 65091E). doi:10.1117/12.706242

Klevecz, R. R., Kauffman, S. A., & Shymko, R. M. (1984). Cellular clocks and oscillators. [PubMed]. *International Review of Cytology*, *86*, 97–128. doi:10.1016/S0074-7696(08)60178-3

Klevecz, R. R., Kros, J., & Gross, S. D. (1978). Phase response versus positive and negative division delay in animal cells. [PubMed]. *Experimental Cell Research*, *116*, 285–290. doi:10.1016/0014-4827(78)90450-0

Kloosterman, W. P., & Plasterk, R. H. (2006). The diverse functions of microRNAs in animal development and disease. *Developmental Cell*, *11*, 441–450. doi:10.1016/j.devcel.2006.09.009

Knijnenburg, T. A., Daran, J.-M. G., Broek, M. A. d., Daran-Lapujade, P. A., Winde, J. H. d., & Pronk, J. T. (2009). Combinatorial effects of environmental parameters ontranscriptional regulation in Saccharomyces cerevisiae: Aquantitative analysis of a compendium of chemostat-basedtranscriptome data. *BMC Genomics*, *10*(53).

Kohonen, T. (2001). *Self-Organizing Maps* (3rd ed.). Berlin: Springer Verlag.

Kolbe, R., & Burnett, M. (1991). Content-analysis research: an examination of applications with directives for improving research reliability and objectivity. *The Journal of Consumer Research*, *18*(2), 243–250. doi:10.1086/209256

Kollman, P. A., Massova, I., Reyes, C., Kuhn, B., Huo, S., & Chong, L. (2000). Calculating structures and free energies of complex molecules: combining molecular mechanics and continuum models. *Accounts of Chemical Research*, *33*(12), 889–897. doi:10.1021/ar000033j

Koonin, E. (2005). Orthologs, paralogs, and evolutionary genomics. *Annual Review of Genetics*, *39*, 309–338. doi:10.1146/annurev.genet.39.073003.114725

Korf, I., Yandell, M., & Bedell, J. (2003). *BLAST.* Sebastopol, CA, USA: O'Reilly & Associates, Inc.

Koyuturk, M., Kim, Y., Topkara, U., Subramaniam, S., Szpankowski, W., & Grama, A. (2006). Pairwise align-ment of protein interaction networks. *Journal of Computational Biology*, *13*(2), 182–199. doi:10.1089/cmb.2006.13.182

Krek, A., Grun, D., Poy, M. N., Wolf, R., Rosenberg, L., & Epstein, E. J. (2005). Combinatorial microRNA target predictions. *Nature Genetics*, *37*, 495–500. doi:10.1038/ng1536

Krippendorff, K. (2004). *Content Analysis: An introduction to its methodology* (2nd ed.). Thousand Oaks, CA: Sage Publications.

Krishnadev, O., & Srinivasan, N. (2008). A data integration approach to predict host-pathogen protein-protein interactions: application to recognize protein interactions between human and a malarial parasite. *In Silico Biology*, *8*, 235–250.

Krogan, N. J., & Cagney, G. (2006). Global landscape of protein complexes in the yeast saccharomyces cerevisiae. *Nature*, *440*(7084), 637–643. doi:10.1038/nature04670

Kryszkiewicz, M. (2001). Concise representation of frequent patterns based on disjunction-free generators. In *Proceedings of the Int'l Conf. on Data Mining*.

Kryszkiewicz, M. (2005). Generalized disjunction-free representation of frequent patterns with negation. *Journal of Experimental & Theoretical Artificial Intelligence*, *17*, 63–82. doi:10.1080/09528130512331315882

Kudo, T., Maeda, E., & Matsumoto, Y. (2004). An application of boosting to graph classification. In *Proceedings of the NIPS.*

Kumar, P., Henikoff, S., & Ng, P. C. (2009). Predicting the effects of coding non-synonymous variants on protein function using the SIFT algorithm. *Nature Protocols*, *4*(7), 1073–1081. doi:10.1038/nprot.2009.86

Kwasigroch, J. M., Gilis, D., Dehouck, Y., & Rooman, M. (2002). PoPMuSiC, rationally designing point mutations in protein structures. *Bioinformatics (Oxford, England)*, *18*(12), 1701–1702. doi:10.1093/bioinformatics/18.12.1701

Lacroix, V., Fernandes, C. G., & Sagot, M.-F. (2006). Motif search in graphs: Application to metabolic networks. *IEEE* [TCBB]. *ACM Transactions on Computational Biology and Bioinformatics*, *3*(4), 360–368. doi:10.1109/TCBB.2006.55

Lagus, K., Kaski, S., & Kohonen, T. (2004). Mining massive document collections by the WEBSOM method. *Information Sciences*, *163*, 135–156. doi:10.1016/j.ins.2003.03.017

Lal, A., Mazan-Mamczarz, K., Kawai, T., Yang, X., Martindale, J. L., & Gorospe, M. (2004). Concurrent versus individual binding of HuR and AUF1 to common labile target mRNAs. *The EMBO Journal*, *23*, 3092–3102. doi:10.1038/sj.emboj.7600305

Lall, S., Grun, D., Krek, A., Chen, K., Wang, Y. L., & Dewey, C. N. (2006). A genome-wide map of conserved microRNA targets in C. elegans. *Current Biology*, *16*, 460–471. doi:10.1016/j.cub.2006.01.050

Lalonde, S., & Ehrhardt, D.W., Loqu_, D., Chen, J., Rhee, S.Y., & Frommer, W.B. (2008). Molecular and cellular approaches for the detection of protein-protein interactions: latest techniques and current limitations. *The Plant Journal*, *53*, 610–635. doi:10.1111/j.1365-313X.2007.03332.x

Lanckriet, G. R. G., Bie, T. D., Cristianini, N., Jordan, M., & Noble, W. (2004). A statistical framework for genomic data fusion. *Bioinformatics (Oxford, England)*, *20*(16), 2626–2635. doi:10.1093/bioinformatics/bth294

Lanctot, K., Li, M., Ma, B., Wang, S., & Zhang, L. (1999). Distinguishing string selection problems. In *Proceedings of the 10th ACM-SIAM Symp. on Discrete Algorithms* (pp. 633-642).

Landis, J. R., & Koch, G. G. (1977). The measurement of observer agreement for categorical data. *Biometrics*, *33*(1), 159–174. doi:10.2307/2529310

Lascaris, R. F., Mager, W. H., & Planta, R. J. (1999). DNA-binding requirements of the yeast protein Rap1p as selected in silico from ribosomal protein gene promoter sequences. *Bioinformatics (Oxford, England), 15*, 267–277. doi:10.1093/bioinformatics/15.4.267

Laskowski, R. A. (1995). SURFNET: a program for visualizing molecular surfaces, cavities, and intermolecular interactions. *Journal of Molecular Graphics, 13*, 323–328. doi:10.1016/0263-7855(95)00073-9

Laskowski, R. A., Luscombe, N. M., Swindells, M. B., & Thornton, J. M. (1996). Protein clefts in molecular recognition and function. *Protein Science, 5*, 2438–2452.

Laskowski, R. A., Watson, J. D., & Thornton, J. M. (2005). Protein function prediction using local 3D templates. *Journal of Molecular Biology, 351*, 614–626. doi:10.1016/j.jmb.2005.05.067

Law, G. Y., & Grossman, R. (2005). Application of diffusion tensor MR imaging in multiple sclerosis. *Annals of the New York Academy of Sciences, 1064*, 202–219. doi:10.1196/annals.1340.039

Lawrence, C., & Reilly, A. (1990). An expectation maximization (EM) algorithm for the identification and characterization of common sites in unaligned biopolymer sequences. *Proteins, 7*, 41–51. doi:10.1002/prot.340070105

Lee, S. A., Chan, C. H., Tsai, C. H., Lai, J. M., Wang, F. S., & Kao, C. Y. (2008). Ortholog-based protein-protein interaction prediction and its application to inter-species interactions. *BMC Bioinformatics, 9*(Suppl 12), S11. doi:10.1186/1471-2105-9-S12-S11

Legendre, P. (2005). Species associations: The Kendall coefficient of concordance revisited. *Journal of Agricultural Biological & Environmental Statistics, 10*(2), 226–245. doi:10.1198/108571105X46642

Lehmann, M., Pasamontes, L., Lassen, S. F., & Wyss, M. (2000). The consensus concept for thermostability engineering of proteins. *Biochimica et Biophysica Acta, 1543*(2), 408–415.

Leong, H. S., Yates, T., Wilson, C., & Miller, C. J. (2005). ADAPT: a database of Affymetrix probesets and transcripts. *Bioinformatics (Oxford, England), 21*, 2552–2553. doi:10.1093/bioinformatics/bti359

Levine, M., & Davidson, E. H. (2005). Gene regulatory networks for development. *Proceedings of the National Academy of Sciences of the United States of America, 102*, 4936–4942. doi:10.1073/pnas.0408031102

Levitt, D. G., & Banaszak, L. J. (1992). POCKET: a computer graphics method for identifying and displaying protein cavities and their surrounding amino acids. *Journal of Molecular Graphics, 10*, 229–234. doi:10.1016/0263-7855(92)80074-N

Levy, E. D., & Pereira-Leal, J. B. (2008). Evolution and dynamics of protein interactions and networks. *Current Opinion in Structural Biology, 18*(3), 349–357. doi:10.1016/j.sbi.2008.03.003

Lewis, B. P., Burge, C. B., & Bartel, D. P. (2005). Conserved seed pairing, often flanked by adenosines, indicates that thousands of human genes are microRNA targets. *Cell, 120*, 15–20. doi:10.1016/j.cell.2004.12.035

Lewis, B. P., Shih, I. H., Jones-Rhoades, M. W., Bartel, D. P., & Burge, C. B. (2003). Prediction of mammalian microRNA targets. *Cell, 115*, 787–798. doi:10.1016/S0092-8674(03)01018-3

Li, M., Ma, B., & Wang, L. (1999). Finding Similar Regions in Many Strings. In *Proceedings of the Thirty-first Annual ACM Symposium on Theory of Computing,* Atlanta (pp. 473-482).

Li, B., Turuvekere, S., Agrawal, M., La, D., Ramani, K., & Kihara, D. (2007). Characterization of local geometry of protein surfaces with the visibility criterion. *Proteins, 71*, 670–683. doi:10.1002/prot.21732

Li, F., & Stormo, G. D. (2001). Selection of optimal DNA oligos for gene expression arrays. *Bioinformatics (Oxford, England), 17*, 1067–1076. doi:10.1093/bioinformatics/17.11.1067

Li, M., Ma, B., & Wang, L. (2002a). Finding Similar Regions in Many Sequences. *Journal of Computer and System Sciences, 65*, 73–96. doi:10.1006/jcss.2002.1823

Li, M., Ma, B., & Wang, L. (2002b). On the closest string and substring problems. *Journal of the ACM, 49*(2), 157–171. doi:10.1145/506147.506150

Li, S. (2004). A map of the interactome network of the metazoan C. elegans. *Science, 303*, 540–543. doi:10.1126/science.1091403

Lise, S., Archambeau, C., Pontil, M., & Jones, D. T. (2009). Prediction of hot spot residues at protein-protein interfaces by combining machine learning and energy-based methods. *BMC Bioinformatics*, *10*, 365. doi:10.1186/1471-2105-10-365

Liu, G., Loraine, A. E., Shigeta, R., Cline, M., Cheng, J., & Valmeekam, V. (2003). NetAffx: Affymetrix probesets and annotations. *Nucleic Acids Research*, *31*, 82–86. doi:10.1093/nar/gkg121

Liu, L. (2008). Estimating species trees using multiple-allele DNA sequence data. *Evolution; International Journal of Organic Evolution*, *62*, 2080–2091. doi:10.1111/j.1558-5646.2008.00414.x

Liu, X., & Wang, L. (2007). Computing the maxim um similarity bi-clusters of gene expression data. *Bioinformatics (Oxford, England)*, *23*(1), 50–56. doi:10.1093/bioinformatics/btl560

Lo, D., Cheng, H., Han, J., Khoo, S., & Sun, C. (2009). Classification of software behaviors for failure detection: a discriminative pattern mining approach. In *Proceedings of the 15th ACM SIGKDD international conference on Knowledge discovery and data mining* (pp. 557-566).

LoBue, P. (2009). Extensively drug-resistant tuberculosis. *Current Opinion in Infectious Diseases*, *22*, 167–173. doi:10.1097/QCO.0b013e3283229fab

Lorusso, D., Pietragalla, A., Mainenti, S., Masciullo, V., Di Vagno, G., & Scambia, G. (2010). Review role of topotecan in gynaecological cancers: current indications and perspectives. *Critical Reviews in Oncology/Hematology*, *74*, 163–174..doi:10.1016/j.critrevonc.2009.08.001

Lucas, K., Busch, M., M¨ossinger, S., & Thompson, J. A. (1991). An improved microcom-puter program for finding gene- or gene family-specific oligonucleotides suitable as primers for polymerase chain reactions or as probes. *CABIOS*, *7*, 525–529.

Lu, X., & Zhang, X. (2006). The effect of GeneChip gene definitions on the microarray study of cancers. *BioEssays*, *28*, 739–746. doi:10.1002/bies.20433

Ma, B., Li, M., & Zhang, L. (2000). From Gene Trees to Species Trees. *SIAM Journal on Computing*, *30*, 729–752. doi:10.1137/S0097539798343362

Maddah, M., Grimson, W., & Warfield, S. (2006). Statistical modeling and EM clustering of white matter fiber tracts. In *Proceedings of the IEEE Int. Symp. on Biomedical Imaging*, Arlington, VA (pp. 53-56).

Maddah, M., Mewes, A., Haker, S., Grimson, W. E. L., & Warfield, S. (2005). Automated atlas-based clustering of white matter fiber tracts from DTMRI. In *Proceedings of the Int. Conf. on Medical Image Computing and Computer Assisted Intervention*, Palm Springs, CA (pp. 188-195).

Madeira, S. C., & Oliveira, A. L. (2004). Biclustering algorithms for biological data analysis: a survey. *IEEE/ACM TCBB*, *1*(1), 24-45.

Ma, F., Liu, X., Li, D., Wang, P., Li, N., Lu, L., & Cao, X. (2010). MicroRNA-466l upregulates IL-10 expression in TLR-triggered macrophages by antagonizing RNA-binding protein tristetraprolin-mediated IL-10 mRNA degradation. *Journal of Immunology (Baltimore, MD.: 1950)*, *184*, 6053–6059. doi:10.4049/jimmunol.0902308

Mangan, S., & Alon, U. (2003). Structure and function of the feed-forward loop network motif. *Proceedings of the National Academy of Sciences of the United States of America*, *100*(21), 11980–11985. doi:10.1073/pnas.2133841100

Mangan, S., Itzkovitz, S., Zaslaver, A., & Alon, U. (2005). The incoherent feed-forward loop accelerates the response-time of the *gal* system of *escherichia coli*. *Journal of Molecular Biology*, *356*(5), 1073–1081. doi:10.1016/j.jmb.2005.12.003

Mannila, H., & Toivonen, H. (1996). Multiple uses of frequent sets and condensed representations. In *Proceedings of the International Conference on Knowledge Discovery and Data Mining*.

Maragkakis, M., Alexiou, P., Papadopoulos, G. L., Reczko, M., Dalamagas, T., & Giannopoulos, G. (2009). Accurate microRNA target prediction correlates with protein repression levels. *BMC Bioinformatics*, *10*, 295. doi:10.1186/1471-2105-10-295

Margolin, A., Nemenman, I., Basso, K., Wiggins, C., Stolovitzky, G., & Favera, R. (2006). ARACNE: An Algorithm for the Reconstruction of Gene Regulatory Networks in a Mammalian Cellular Context. *BMC Bioinformatics*, *7*(1), S7. doi:10.1186/1471-2105-7-S1-S7

Marple, S. L. Jr. (1999). Computing the discrete-time 'analytic' signal via FFT. *IEEE Transactions on Signal Processing, 47*, 2600–2603. doi:10.1109/78.782222

Masso, M., Lu, Z., & Vaisman, I. I. (2006). Computational mutagenesis studies of protein structure-function correlations. *Proteins, 64*, 234–245. doi:10.1002/prot.20968

Masso, M., & Vaisman, I. I. (2007). Accurate prediction of enzyme mutant activity based on a multibody statistical potential. *Bioinformatics (Oxford, England), 23*, 3155–3161. doi:10.1093/bioinformatics/btm509

Masso, M., & Vaisman, I. I. (2008). Accurate prediction of stability changes in protein mutants by combining machine learning with structure based computational mutagenesis. *Bioinformatics (Oxford, England), 24*(18), 2002–2009. doi:10.1093/bioinformatics/btn353

Matsumura, M., Wozniak, J. A., Sun, D. P., & Matthews, B. W. (1989). Structural studies of mutants of T4 lysozyme that alter hydrophobic stabilization. *The Journal of Biological Chemistry, 264*(27), 16059–16066.

Matys, V., & Fricke, E. (2003). TRANSFAC(R): transcriptional regulation, from patterns to profiles. *Nucleic Acids Research, 31*(1), 374–378. doi:10.1093/nar/gkg108

Matys, V., Kel-Margoulis, O. V., Fricke, E., Liebich, I., Land, S., & Barre-Dirrie, A. (2006). Transfac and its module transcompel: transcriptional gene regulation in eukaryotes. *Nucleic Acids Research, 34*(Database issue), D108–D110. doi:10.1093/nar/gkj143

Mazurie, A., Bottani, S., & Vergassola, M. (2005). An evolutionary and functional assessment of regulatory network motifs. *Genome Biology, 6*, R35. doi:10.1186/gb-2005-6-4-r35

McIntosh, T., & Chawla, S. (2007). High confidence rule mining for microarray analysis. *IEEE/ACM TCBB, 4*(4), 611-623.

McKay, B. (1978). Computing automorphisms and canonical labelling of graphs. *LNM, 686*, 223–232.

McPearson, M. J., Quirke, P., & Taylor, G. R. (1991). *PCR A Practical Approach* (p. 8). New York: Oxford University Press.

Mecham, B. H., Klus, G. T., Strovel, J., Augustus, M., Byrne, D., & Bozso, P. (2004a). Sequence-matched probes produce increased cross-platform consistency and more reproducible biological results in microarray-based gene expression measurements. *Nucleic Acids Research, 32*, e74. doi:10.1093/nar/gnh071

Mecham, B. H., Wetmore, D. Z., Szallasi, Z., Sadovsky, Y., Kohane, I., & Mariani, T. J. (2004b). Increased measurement accuracy for sequence-verified microarray probes. *Physiological Genomics, 18*, 308–315. doi:10.1152/physiolgenomics.00066.2004

Mendes, N. D., Freitas, A. T., & Sagot, M. F. (2009). Current tools for the identification of miRNA genes and their targets. *Nucleic Acids Research, 37*, 2419–2433. doi:10.1093/nar/gkp145

Messmer, B. T., & Bunke, H. (1998). A new algorithm for error-tolerant sub-graph isomorphism detection. *IEEE Transactions on Pattern Analysis and Machine Intelligence, 20*(5), 493–504. doi:10.1109/34.682179

Mewes, H. W., Albermann, K., Heumann, K., Liebl, S., & Pfeier, F. (1997). MIPS -a database for protein sequences, homology data and yeast genome information. *Nucleic Acids Research, 25*, 28–30. doi:10.1093/nar/25.1.28

Mewes, H. W., Dietmann, S., Frishman, D., Gregory, R., Mannhaupt, G., & Mayer, K. F. X. (2008). MIPS: analysis and annotation of genome information in 2007. *Nucleic Acids Research, 36*, D196–D201. doi:10.1093/nar/gkm980

Mielikainen, T. (2003). Intersecting data to closed sets with constraints. In *Proceedings of the Workshop on Frequent Itemset Mining Implementations*.

Miller, J.P., Lo, R.S., Ben-Hur, A., Desmarais, C., Stagljar, I., & Stafford, W., Noble, & Fields, S. (2005). Large-scale identification of yeast integral membrane protein interactions. *Proceedings of the National Academy of Sciences of the United States of America, 102*(34), 12123–121. doi:10.1073/pnas.0505482102

Milo, R., Shen-Orr, S., Itzkovitz, S., Kashtan, N., & Chklovskii, D. (2002). Network motifs: Simple building blocks of complex networks. *Science, 298*(5594), 824–828. doi:10.1126/science.298.5594.824

Milo, R., Shen-Orr, S., Itzkovitz, S., Kashtan, N., Chklovskii, D., & Alon, U. (2002). Network motifs: simple building blocks of complex networks. *Science*, *298*, 824–827. doi:10.1126/science.298.5594.824

Mitchell, T. (1982). Generalization as search. *Artificial Intelligence*, *18*, 203–226. doi:10.1016/0004-3702(82)90040-6

Mitchison, J. M. (1971). *The Biology of the Cell Cycle*. London: Cambridge University Press.

Mizuno, A., Tabei, H., & Iwahuti, M. (2006). Characterization of low-acetic-acid-producing yeast isolated from 2-deoxyglucose-resistant mutants and its application to high-gravity brewing. [PubMed]. *Journal of Bioscience and Bioengineering*, *101*, 31–37. doi:10.1263/jbb.101.31

Mobarak, C. D., Anderson, K. D., Morin, M., Beckel-Mitchener, A., Rogers, S. L., & Furneaux, H. (2000). The RNA-binding protein HuD is required for GAP-43 mRNA stability, GAP-43 gene expression, and PKC-dependent neurite outgrowth in PC12 cells. *Molecular Biology of the Cell*, *11*, 3191–3203.

Moberts, B., Vilanova, A., & van Wijk, J. J. (2005). Evaluation of Fiber Clustering Methods for diffusion tensor imaging. In *Proceedings of the IEEE Visualization*, Minneapolis, MN (pp. 65-72).

Moon, H. S., Bhak, J., Lee, H. K., & Lee, D. (2005). Architecture of basic building blocks in protein and domain structural interaction networks. *Bioinformatics (Oxford, England)*, *21*(8), 1479–1486. doi:10.1093/bioinformatics/bti240

Moret, B., Tang, J., & Warnow, T. (2005). Reconstructing phylogenies from gene-content and gene-order data. In Gascuel, O. (Ed.), *Mathematics of Evolution and Phylogeny* (pp. 321–352). Oxford, UK: Oxford University Press.

Mori, S. (2007). *Introduction to Diffusion Tensor Imaging*. New York: Elsevier.

Mori, S., Wakana, S., Nagae-Poetscher, L. M., & van Zijl, P. C. M. (2005). *MRI Atlas of Human White Matter*. Amsterdam, The Netherlands: Elsevier.

Mukherjee, D., Gao, M., O'Connor, J. P., Raijmakers, R., Pruijn, G., Lutz, C. S., & Wilusz, J. (2002). The mammalian exosome mediates the efficient degradation of mRNAs that contain AU-rich elements. *The EMBO Journal*, *21*, 165–174. doi:10.1093/emboj/21.1.165

Murali, T., & Kasif, S. (2003a). Extracting conserved gene expression motifs from gene expression data. In *Proceedings of the Pac Symp Biocomput* (pp. 77-88).

Murali, T., & Kasif, S. (2003 b). RankGene: identification of diagnostic genes based on expression data. In *Proceedings of the Pacific Symposium on Biocomputing*, *8*, 77–88.

Murzin, A. G., Brenner, S. E., Hubbard, T., & Chothia, C. (1995). SCOP: a structural classification of proteins database for the investigation of sequences and structures. *Journal of Molecular Biology*, *247*, 536–540. doi:10.1016/S0022-2836(05)80134-2

Myers, E. W. (1994). A sublinear algorithm for approximate keyword searching. *Algorithmica*, *12*, 345–374. doi:10.1007/BF01185432

Naef, F., Lim, D. A., Patil, N., & Magnasco, M. (2002). DNA hybridization to mismatched templates: A chip study. *Physical Review E: Statistical, Nonlinear, and Soft Matter Physics*, *65*, 040902. doi:10.1103/PhysRevE.65.040902

Nam, S., Kim, B., Shin, S., & Lee, S. (2008). miRGator: an integrated system for functional annotation of microRNAs. *Nucleic Acids Research*, *36*, D159–D164. doi:10.1093/nar/gkm829

Nanavati, A., Chitrapura, K., Joshi, S., & Krishnapuram, R. (2001). Association rule mining: Mining generalised disjunctive association rules. In *Proceedings of the ACM CIKM Conf.*

Navarro, G., & Baeza-Yates, R. (2000). A hybrid indexing method for approximate string matching. *Journal of Discrete Algorithms*, *1*, 205–239.

Nei, M., & Kumar, S. (2000). *Molecular Evolution and Phylogenetics* (2nd ed.). London: Oxford University Press.

Neuendorf, K. A. (2001). *The content analysis guidebook*. Thousand Oaks, CA: Sage Publications.

Newman, M. A., & Hammond, S. M. (2010). Emerging paradigms of regulated microRNA processing. *Genes & Development*, *24*, 1086–1092. doi:10.1101/gad.1919710

Newman, M. E. J. (2003). The structure and function of complex networks. *SIAM Review*, *45*(2), 167–2. doi:10.1137/S003614450342480

Ng, A. Y., Jordan, M. I., & Weiss, Y. (2001). On spectral clustering:Analysis and an algorithm. In *Advances in Neural Information Proc* (*Vol. 14*, pp. 849–856). Vancouver, BC, Canada: Systems.

Ng, P. C., & Henikoff, S. (2006). Predicting the effects of amino acid substitutions on protein function. *Annual Review of Genomics and Human Genetics*, *7*, 61–80. doi:10.1146/annurev.genom.7.080505.115630

Nicholas, G., & Prigogine, I. (1977). *Self-Organization in Non-Equilibrium Systems*. New York: Wiley.

Nikkila, J., Toronen, P., Kaski, S., Venna, J., Castren, E., & Wong, G. (2002). Analysis and visualization of gene expression data using self-organizing maps. *Neural Networks*, *15*, 9530–9966. doi:10.1016/S0893-6080(02)00070-9

Ning, Z., Cox, A., & Mullikin, J. (2001). SSAHA: A fast search method for large DNA databases. *Genome Research*, *11*, 1725–1729. doi:10.1101/gr.194201

Noctor, S. C., Flint, A. C., Weissman, T. A., Dammerman, R. S., & Kriegstein, A. R. (2001). Neurons derived from radial glial cells establish radial units in neocortex. *Nature*, *409*, 714–720..doi:10.1038/35055553

Noon, L. A., & Lloyd, A. C. (2007). Traeting leprosy:an Erb-al remedy? *Trends in Pharmacological Sciences*, *28*, 103–105. doi:10.1016/j.tips.2007.01.004

Novotni, M., & Klein, R. (2003). 3D Zernike descriptors for content based shape retrieval. *ACM Symposium on Solid and Physical Modeling, Proceedings of the eighth ACM symposium on Solid modeling and applications* (pp. 216-225).

O'Donnell, L. J., Kubicki, M., Shenton, M. E., Grimson, W. E. L., & Westin, C. F. (2006). A method for clustering white matter fiber tracts. *AJNR. American Journal of Neuroradiology*, *27*(5), 1032–1036.

O'Donnell, L. J., & Westin, C. F. (2007). Automatic Tractography Segmentation Using a Highdimensional White Matter Atlas. *IEEE Transactions on Medical Imaging*, *26*(11), 1562–1575. doi:10.1109/TMI.2007.906785

Ogden, R. T. (1997). *Essential Wavelets for Statistical Applications and Data Analysis*. Boston: Birkhäuser. doi:10.1007/978-1-4612-0709-2

O'Hare, H., & Juillerat, A., Dianiskov_, P., & Johnsson, K. (2008). A split-protein sensor for studying protein-protein interaction in mycobacteria. *Journal of Microbiological Methods*, *73*, 79–84. doi:10.1016/j.mimet.2008.02.008

Okada, K., Kanaya, S., & Asai, K. (2005). Accurate extraction of functional associations between proteins based on common interaction partners and common domains. *Bioinformatics (Oxford, England)*, *21*(9), 2043–2048. doi:10.1093/bioinformatics/bti305

Olhede, S. C., & Walden, A. T. (2004). Analytic wavelet thresholding. *Biometrika*, *91*, 955–973. doi:10.1093/biomet/91.4.955

Oliver, J. C. (2008). AUGIST: inferring species trees while accommodating gene tree uncertainty. *Bioinformatics (Oxford, England)*, *24*, 2932–2933. doi:10.1093/bioinformatics/btn556

Oliver, S. (2000). Guilt-by-association goes global. *Nature*, *403*, 601–603. doi:10.1038/35001165

Oppenheimer, A. V., & Schafer, R. W. (1975). *Digital Signal Processing*. Englewood Cliffs, NJ: Prentice-Hall.

Pan, F., Cong, G., Tung, K., Yang, J., & Zaki, M. (2004). Carpenter: Finding closed patterns in long biological datasets. In *Proceedings of the ACM SIGKDD Intl. Conf. Knowledge Discovery and Data Mining (KDD)* (pp. 637-642).

Pandey, G., Atluri, G., Steinbach, M., Myers, C. L., & Kumar, V. (2009). An association analysis approach to biclusting. In *Proceedings of the ACM Conf. on Kownlege Discovery and Data Mining* (pp. 677-686).

Papadopoulos, G. L., Reczko, M., Simossis, V. A., Sethupathy, P., & Hatzigeorgiou, A. G. (2009). The database of experimentally supported targets: a functional update of TarBase. *Nucleic Acids Research*, *37*, D155–D158. doi:10.1093/nar/gkn809

Papin, J. A., Price, N. D., Wiback, S. J., Fell, D. A., & Palsson, B. O. (2003). Metabolic pathways in the post-genome era. *Trends in Biochemical Sciences*, *28*, 250–258. doi:10.1016/S0968-0004(03)00064-1

Papoulis, A. (1991). *Probability, Random Variables, and Stochastic Processes* (3rd ed.). New York: McGraw-Hill.

Parida, L. (2008). *Pattern Discovery in Bioinformatics, Theory and Algorithms*. Chapman and HAll/CRC.

Parida, L. (2007). Discovering topological motifs using a compact notation. *Journal of Computational Biology*, *14*(3), 46. doi:10.1089/cmb.2006.0142

Park, H. J., Westin, C. F., Kubicki, M., Maier, S. E., Niznikiewicz, M., & Baer, A. (2004). White matter hemisphere asymmetries in healthy subjects and in schizophrenia: A diffusion tensor MRI study. *NeuroImage*, *24*, 213–223. doi:10.1016/j.neuroimage.2004.04.036

Parrish, J. R., Gulyas, K. D., & Finley, R. L. (2006). Yeast two-hybrid contributions to interactome mapping. *Current Opinion in Biotechnology*, *17*, 387–393. doi:10.1016/j.copbio.2006.06.006

Parthiban, V., Gromiha, M. M., Hoppe, C., & Schomburg, D. (2007). Structural analysis and prediction of protein mutant stability using distance and torsion potentials: role of secondary structure and solvent accessibility. *Proteins*, *66*(1), 41–52. doi:10.1002/prot.21115

Pasquier, N., Bastide, Y., Taouil, R., & Lakhal, L. (1999). Discovering frequent closed itemsets for association rules. In *Proceedings of the 7th Intl. Conf. on Database Theory*.

Pavlidis, P., Weston, J., Cai, J., & Noble, W. S. (2002). Learning gene functional classifications from multiple data types. *Journal of Computational Biology*, *9*(2), 401–411. doi:10.1089/10665270252935539

Pearl, J. (1988). *Probabilistic reasoning in intelligent systems: networks of plausible inference*. San Francisco: Morgan Kaufmann.

Pearson, W. R., & Lipman, D. J. (1988). Improved tools for biological sequence comparison. *Proceedings of the National Academy of Sciences of the United States of America*, *85*, 2444–2448. doi:10.1073/pnas.85.8.2444

Peng, S. S., Chen, C. Y., Xu, N., & Shyu, A. B. (1998). RNA stabilization by the AU-rich element binding protein, HuR, an ELAV protein. *The EMBO Journal*, *17*, 3461–3470. doi:10.1093/emboj/17.12.3461

Percival, D. B., & Walden, A. T. (2000). *Wavelet Methods for Time Series Analysis*. London: Cambridge University Press.

Pereira, J. B., Enright, A.J., & Ouzounis, C.A. (2004). Detection of functional modules from protein interac-tion networks. *Proteins: Structure, Fuctions, and Bioinformatics*, (20), 49.

Perez-Iratxeta, C., & Andrade, M. (2005). Inconsistencies over time in 5% of NetAffx probe-to-gene annotations. *BMC Bioinformatics*, *6*(1), 183. doi:10.1186/1471-2105-6-183

Persico, M., Ceol, A., Gavrila, C., Hoffmann, R., Florio, A., & Cesareni, G. (2005). HomoMINT: an inferred human network based on orthology mapping of protein interactions discovered in model organisms. *BMC Bioinformatics*, *6*(Suppl 4), S21. doi:10.1186/1471-2105-6-S4-S21

Pevzner, P., & Sze, S. (2000). Combinatorial approaches to finding subtle signals in DNA sequences. In *Proceedings of the 8th International Conference on Intelligent Sys-tems for Molecular Biology* (pp. 269-278).

Pfaltz, J., & Jamison, R. (2001). Closure systems and their structure. *Information Sciences*, *139*, 275–286. doi:10.1016/S0020-0255(01)00169-4

Piecyk, M., Wax, S., Beck, A. R., Kedersha, N., Gupta, M., & Maritim, B. (2000). TIA-1 is a translational silencer that selectively regulates the expression of TNF-alpha. *The EMBO Journal*, *19*, 4154–4163. doi:10.1093/emboj/19.15.4154

Pieters, J. (2008). Mycobacterium tuberculosis and the macrophage: Maintaining a balance. *Cell Host & Microbe*, *3*, 399–407. doi:10.1016/j.chom.2008.05.006

Pitera, J. W., & Kollman, P. A. (2000). Exhaustive mutagenesis in silico: multicoordinate free energy calculations on proteins and peptides. *Proteins*, *41*(3), 385–397. doi:10.1002/1097-0134(20001115)41:3<385::AID-PROT100>3.0.CO;2-R

Pizzuti, C., & Rombo, S. E. (2009). Multi-functional protein clustering in PPI networks. *Biological Data Mining in Protein Interaction Networks*. Hershey, PA, USA: IGI Global.

Pommier, Y. (2006). Topoisomerase I inhibitors: camptothecins and beyond. *Nature Reviews. Cancer*, *6*, 789–802.. doi:10.1038/nrc1977

Poteete, A. R., & Hardy, L. W. (1994). Genetic analysis of bacteriophage T4 lysozyme structure and function. *Journal of Bacteriology*, *176*(22), 6783–6788.

Prasad, T. S. K. (2009). Human protein reference database - 2009 update. *Nucleic Acids Research*, *37*, D767–D772. doi:10.1093/nar/gkn892

Price, A., Ramabhadran, S., & Pevzner, P. (2003). Finding Subtle Motifs by Branching from Sample Strings. *Bioinformatics (Oxford, England)*, *19*, 149–155. doi:10.1093/bioinformatics/btg1072

Prill, R. J., Iglesias, P. A., & Levchenko, A. (2005). Dynamic properties of network motifs contribute to biological network organization. *PLoS Biology*, *3*(11), e343. doi:10.1371/journal.pbio.0030343

Proutski, V., & Holme, E. C. (1996). Primer Master: a new program for the design and analysis of PCR primers. *CABIOS*, *12*, 253–255.

Pruitt, K. D., Tatusova, T., & Maglott, D. R. (2005). NCBI Reference Sequence (Ref-Seq): a curated non-redundant sequence database of genomes, transcripts and proteins. *Nucleic Acids Research*, *33*, D501–D504. doi:10.1093/nar/gki025

Pruitt, K. D., Tatusova, T., & Maglott, D. R. (2007). NCBI reference sequences (RefSeq): a curated non-redundant sequence database of genomes, transcripts and proteins. *Nucleic Acids Research*, *35*, D61–D65. doi:10.1093/nar/gkl842

Qamra, R., Mande, S. C., Coates, A. R., & Henderson, B. (2005). The unusual chaperonins of Mycobacterium tuberculosis. [edinb]. *Tuberculosis (Edinburgh, Scotland)*, *85*, 385–394. doi:10.1016/j.tube.2005.08.014

Quinlan, J. R. (1996). Improved use of continuous attributes in C4.5. *Journal of Artificial Intelligence Research*, *4*, 77–90.

Raghavan, V. V., & Wong, S. K. M. (1986). A critical analysis of vector space model for information retrieval. *Journal of the American Society for Information Science American Society for Information Science*, *37*(5), 279–287.

Raineri, I., Wegmueller, D., Gross, B., Certa, U., & Moroni, C. (2004). Roles of AUF1 isoforms, HuR and BRF1 in ARE-dependent mRNA turnover studied by RNA interference. *Nucleic Acids Research*, *32*, 1279–1288. doi:10.1093/nar/gkh282

Rakotomamonjy, A., Bach, F., Canu, S., & Grandvalet, Y. (2008). SimpleMKL. *Journal of Machine Learning Research*, *9*, 2491–2521.

Ramakrishnan, N., Kumar, D., Mishra, B., Potts, M., & Helm, R. (2004). Turning cartwheels: An alternating algorithm for mining redescriptions. In *Proceedings of the ACM SIGKDD Conf.*

Randolph, J. J. (2008). *Online Kappa Calculator.* Retrieved September 2008, from http://justus.randolph.name/kappa

Rand, W. M. (1971). Objective Criteria for Evaluation of Clustering Methods. *Journal of the American Statistical Association*, *66*(336), 846. doi:10.2307/2284239

Ranu, S., & Singh, A. K. (2009). GraphSig: A Scalable Approach to Mining Significant Subgraphs in Large Graph Databases. In *Proceedings of the ICDE* (pp. 844-855).

Rasmussen, M., & Kellis, M. (2007). Accurate gene-tree reconstruction by learning gene- and species-specific substitution rates across multiple complete genomes. *Genome Research*, *17*, 1932–1942. doi:10.1101/gr.7105007

Rat Genome Sequencing Project Consortium. (2004). Genome sequence of the brown Norway rat yields insights into mammalian evolution. *Nature*, *428*, 493–521. doi:10.1038/nature02426

Ravasz, E., Somera, A. L., Mongru, D. A., Oltvai, Z. N., & Barabasi, A.-L. (2002). Hierarchical organization of modularity in metabolic networks. *Science*, *297*, 1551–1555. doi:10.1126/science.1073374

Reggi, V., Balocco-Mattavelli, R., Bonati, M., Breton, I., Figueras, A., & Jambert, E. (2003). Prescribing information in 26 countries: a comparative study. *European Journal of Clinical Pharmacology*, *59*(4), 263–270. doi:10.1007/s00228-003-0607-1

Rehmsmeier, M., Steffen, P., Hochsmann, M., & Giegerich, R. (2004). Fast and effective prediction of microRNA/target duplexes. *RNA (New York, N.Y.)*, *10*, 1507–1517. doi:10.1261/rna.5248604

Rennell, D., Bouvier, S. E., Hardy, L. W., & Poteete, A. R. (1991). Systematic mutation of bacteriophage T4 lysozyme. *Journal of Molecular Biology, 222*(1), 67–88. doi:10.1016/0022-2836(91)90738-R

Rhoades, M. W., Reinhart, B. J., Lim, L. P., Burge, C. B., Bartel, B., & Bartel, D. P. (2002). Prediction of plant microRNA targets. *Cell, 110*, 513–520. doi:10.1016/S0092-8674(02)00863-2

Rhodes, D. R., Tomlins, S. A., Varambally, S., Mahavisno, V., Barrette, T., & Kalyana-Sundaram, S. (2005). Probabilistic model of the human protein-protein interaction network. *Nature Biotechnology, 23*, 951–959. doi:10.1038/nbt1103

Rickman, L., Saldanha, J. W., Hunt, D. M., Hoar, D. N., Colston, M. J., & Millar, J. B. (2004). A two-component signal transduction system with a PAS domain-containing sensor is required for virulence of Mycobacterium tuberculosis in mice. *Biochemical and Biophysical Research Communications, 314*, 259–267. doi:10.1016/j.bbrc.2003.12.082

Ritter, H. (1995). *Self-organizing feature maps: Kohnnen maps. The handbook of Brain Theory and Neural Networks*. Cambridge, MA: MIT Press.

Roberts, R. J. (2001). PubMed Central: The GenBank of the published literature. *Proceedings of the National Academy of Sciences of the United States of America, 98*(2), 381–382. doi:10.1073/pnas.98.2.381

Rodriguez, A., Vigorito, E., Clare, S., Warren, M. V., Couttet, P., & Soond, D. R. (2007). Requirement of bic/microRNA-155 for normal immune function. *Science, 316*, 608–611. doi:10.1126/science.1139253

Rojas, R., & Feldman, J. (1996). *Neural networks: A systematic introduction*. Berlin, New York: Springer-Verlag.

Rokas, A., Williams, B., King, N., & Carroll, S. (2003). Genome-scale approaches to resolving incongruence in molecular phylogenies. *Nature, 425*, 798–804. doi:10.1038/nature02053

Rombo, S. E., & Palopoli, L. (2007). Pattern discovery in biosequences: From simple to complex patterns. *Data Mining Patterns: New Methods and Applications* (pp. 85-105). Hershey, PA, USA: IGI Global.

Rote Liste. (2008). Retrieved January 2008, from http://www.fachinfo.de/index_html

Roth, F. P., & Hughes, J. D. (1998). Finding DNA regulatory motifs within unaligned noncoding sequences clustered by whole-genome mRNA quantitation. *Nature Biotechnology, 16*(10), 939–945. doi:10.1038/nbt1098-939

Rougemont, M., Ulrich, S., Hiemke, C., Corruble, E., & Baumann, P. (in press). French summaries of product characteristics (SPC): Content in relation to therapeutic monitoring of psychotropic drugs. *Fundamental & Clinical Pharmacology*.

Rousseeuw, P. (1987). Silhouettes. a graphical aid to the interpretation and validation of cluster analysis. *Journal of Computational and Applied Mathematics, 20*, 53–65. doi:10.1016/0377-0427(87)90125-7

Rubin, H. (1990). The significance of biological heterogeneity. *Cancer and Metastasis Reviews, 9*, 1–20.. doi:10.1007/BF00047585

Sael, L., & Kihara, D. (2009). Protein surface representation and comparison: New approaches in structural proteomics. In J. Chen & S. Lonardi (Eds.), *Biological data mining*. Boca Raton, Florida, USA: Chapman & Hall/CRC Press.

Sael, L., La, D., Li, B., Rustamov, R., & Kihara, D. (2008a). Rapid comparison of properties on protein surface. *Proteins, 73*, 1–10. doi:10.1002/prot.22141

Sael, L., Li, B., La, D., Fang, Y., Ramani, K., & Rustamov, R. (2008b). Fast protein tertiary structure retrieval based on global surface shape similarity. *Proteins, 72*, 1259–1273. doi:10.1002/prot.22030

Saigo, H., Kraemer, N., & Tsuda, K. (2008). Partial Least Squares Regression for Graph Mining. In *Proceedings of the SIGKDD* (pp. 578-586).

Sakoe, H., & Chiba, S. (1978). Dynamic Programming Optimization for Spoken Word Recognition. *IEEE Transactions on Acoustics, Speech, and Signal Processing, 26*, 623–625. doi:10.1109/TASSP.1978.1163055

Salvador, S., & Chan, P. (2004). FastDTW: Toward accurate dynamic time warping in linear time and space. In *Proceedings of the 3rd Workshop On Mining Temporal and Sequential Data (ACM KDD)*, Seattle, WA (pp. 22-25).

Salwinski, L., Miller, C. S., Smith, A. J., Pettit, F. K., Bowie, J. U., & Eisenberg, D. (2004). The Database of Interacting Proteins: 2004 update. *Nucleic Acids Research, 32*, D449–D451. doi:10.1093/nar/gkh086

Samanta, M. P., & Liang, S. (2003). Predicting protein functions from redundancies in largescale protein interaction networks. *Proceedings of the National Academy of Sciences of the United States of America, 100*(22), 12579–12583. doi:10.1073/pnas.2132527100

Sandberg, R., & Larsson, O. (2007). Improved precision and accuracy for microarrays using updated probe set definitions. *BMC Bioinformatics, 8*, 48. doi:10.1186/1471-2105-8-48

Sander, J., Qin, X., Lu, Z., Niu, N., & Kovarsky, A. (2003). Automatic Extraction of Clusters from Hierarchical Clustering Representations. In *Proceedings of the 7th Pacific-Asia Conf. on Knowledge Discovery and Data Mining*, South Korea (pp. 75-87).

Sankoff, D., & Kruskal, J. B. (1983). *Time Warps, String Edits, and Macromolecules: The Theory and Practice of Sequence Comparison*. Reading, MA: Addison-Wesley.

Sarkar, B., Xi, Q., He, C., & Schneider, R. J. (2003). Selective degradation of AU-rich mRNAs promoted by the p37 AUF1 protein isoform. *Molecular and Cellular Biology, 23*, 6685–6693. doi:10.1128/MCB.23.18.6685-6693.2003

Savasere, A., Omiecinski, E., & Navathe, S. (1998). Mining for strong negative associations in a large database of customer transactions. In *Proceedings of the ICDE Conf.*

Schleif, R. (2000). Regulation of the l-arabinose operon of Escherichia coli. *Trends in Genetics, 16*, 559. doi:10.1016/S0168-9525(00)02153-3

Schölkopf, B., & Smola, A. J. (2002). *Learning with kernels*. MIT Press, Cambridge MA.

Schreiber, F., & Schwbbermeyer, H. (2005). MAVisto: a tool for the exploration of network motifs. *Bioinformatics (Oxford, England), 21*(17), 3572–3574. doi:10.1093/bioinformatics/bti556

Schuster, S., Dandekar, T., & Fell, D. A. (1999). Detection of elementary flux modes in biochemical networks: a promising tool for pathway analysis and metabolic engineering. *Trends in Biotechnology, 17*, 53–60. doi:10.1016/S0167-7799(98)01290-6

Schuster, S., Fell, D. A., & Dandekar, T. (2000). A general definition of metabolic pathways useful for sys-tematic organization and analysis of complex metabolic networks. *Nature Biotechnology, 18*, 326–332. doi:10.1038/73786

Schwede, T., Kopp, J., Guex, N., & Peitsch, M. C. (2003). SWISS-MODEL: an automated protein homology-modeling server. *Nucleic Acids Research, 31*, 3381–3385. doi:10.1093/nar/gkg520

Seibel, N. L., Krailo, M., Chen, Z., Healey, J., Breitfeld, P. P., & Drachtman, R. (2007). Upfront window trial of topotecan in previously untreated children and adolescents with poor prognosis metastatic osteosarcoma: childrens Cancer Group (CCG) 7943. *Cancer, 109*, 1646–1653.. doi:10.1002/cncr.22553

Sempere, L. F., Sokol, N. S., Dubrovsky, E. B., Berger, E. M., & Ambros, V. (2003). Temporal regulation of microRNA expression in Drosophila melanogaster mediated by hormonal signals and broad-Complex gene activity. *Developmental Biology, 259*, 9–18. doi:10.1016/S0012-1606(03)00208-2

Sethupathy, P., Corda, B., & Hatzigeorgiou, A. G. (2006). TarBase: A comprehensive database of experimentally supported animal microRNA targets. *RNA (New York, N.Y.), 12*, 192–197. doi:10.1261/rna.2239606

Severgnini, M., Bicciato, S., Mangano, E., Scarlatti, F., Mezzelani, A., & Mattioli, M. (2006). Strategies for comparing gene expression profiles from different microarray platforms: application to a case-control experiment. *Analytical Biochemistry, 353*, 43–56. doi:10.1016/j.ab.2006.03.023

Shao, J., Hahn, K., Yang, Q., Boehm, C., Wohlschlaeger, A., Myers, N., & Plant, C. (2010). Combining Time Series Similarity with Density-based Clustering to Identify Fiber Bundles in the Human Brain. In *Proceedings of the Int. Conf. on Data Mining (ICDM), Workshop on Biological Data Mining and its Applications in Healthcare*.

Shapira, M., Segal, E., & Botstein, D. (2004). Disruption of yeast forkhead-associated cell cycle transcription by oxidative stress. [PubMed]. *Molecular Biology of the Cell, 15*, 5659–5669. doi:10.1091/mbc.E04-04-0340

Sharan, R., & Ideker, T. (2006). Modeling cellular machinery through biological network comparison. *Nature Biotechnology, 24*(4), 427–433. doi:10.1038/nbt1196

Sharan, R., Ulitsky, I., & Shamir, R. (2007). Network-based prediction of protein function. *Molecular Systems Biology*, *3*, 88. doi:10.1038/msb4100129

Shenfield, G. M. (2001). Therapeutic drug monitoring beyond 2000. *British Journal of Clinical Pharmacology*, *52*(1), 3S–4S. doi:10.1046/j.1365-2125.2001.0520s1003.x

Shen-Orr, S. S., Milo, R., Mangan, S., & Alon, U. (2002). Network motifs in the trascriptional regulation network of *escherichia coli*. *Nature*, *31*, 64.

Shima, Y., Mitsuishi, S., Hirata, K., & Harao, M. (2004). Extracting minimal and closed monotone dnf formulas. In *Proceedings of the Int'l Conf. on Discovery Science*.

Shindyalov, I. N., & Bourne, P. E. (1998). Protein structure alignment by incremental combinatorial extension (CE) of the optimal path. *Protein Engineering*, *11*, 739–747. doi:10.1093/protein/11.9.739

Simonis, N., Gonze, D., Orsi, C., van Helden, J., & Wodak, S. J. (2006). Modularity of the Transcriptional Response of Protein Complexes in Yeast. *Journal of Molecular Biology*, *363*(2), 589. doi:10.1016/j.jmb.2006.06.024

Singh, R., Xu, J., & Berger, B. (2007). Pairwise global alignment of protein interaction networks by matching neighborhood topology. In *RECOMB* (pp. 16–31).

Singh, R., Xu, J., & Berger, B. (2008). IsoRank: Global Alignment of Multiple Protein Interaction Networks with Applications to Functional Orthology Detection. *Proceedings of the National Academy of Sciences of the United States of America*, *105*(35), 12763–12768. doi:10.1073/pnas.0806627105

Siomi, H., & Siomi, M. C. (2010). Posttranscriptional regulation of microRNA biogenesis in animals. *Molecular Cell*, *38*, 323–332. doi:10.1016/j.molcel.2010.03.013

Sippl, M. J. (1993). Boltzmann's principle, knowledge-based mean fields and protein folding. An approach to the computational determination of protein structures. *Journal of Computer-Aided Molecular Design*, *7*(4), 473–501. doi:10.1007/BF02337562

Smalter, A., Huan, J., & Lushington, G. (2008). A Graph Pattern Diffusion Kernel for Chemical Compound Classification. In *Proceedings of BIBE'08*.

Smirnov, N. (1939). Estimate of deviation between empirical distribution functions in two independent samples. *Bull. Moscow Univ*, *2*, 3–16.

Smith, P. J., Khan, I. A., & Errington, R. J. (2009). Cytomics and cellular informatics - coping with asymmetry and heterogeneity in biological systems. *Drug Discovery Today*, *14*, 271–277..doi:10.1016/j.drudis.2008.11.012

Snel, B., Huynen, M., & Dutilh, B. (2005). Genome trees and the nature of genome evolution. *Annual Review of Microbiology*, *59*, 191–209. doi:10.1146/annurev.micro.59.030804.121233

Sonnenburg, S., Rätsch, G., Schäfer, C., & Schölkopf, B. (2006). Large scale multiple kernel learning. *Journal of Machine Learning Research*, *7*, 1531–1565.

Soong, T.-t., Wrzeszczynski, K. O., & Rost, B. (2008). Physical protein-protein interactions predicted from microarrays. *Bioinformatics (Oxford, England)*, *24*(22), 2608–2614. doi:10.1093/bioinformatics/btn498

Spellman, P. T., Gavin, S., Zhang, M. Q., Iyer, V. R., Anders, A., & Eisen, M. B. (1998). Comprehensive Identification of Cell Cycle-regulated Genes of the Yeast Saccharomyces cerevisiae by Microarray Hybridization. [PubMed]. *Molecular Biology of the Cell*, *9*, 3273–3297.

Spellman, P. T., & Sherlock, G. (1998). Identification of cell cycle regulated genes in yeast by DNA microarray hybridization. *Molecular Biology of the Cell*, *9*, 371a–371a.

Spirin, V., & Mirny, L. A. (2003). Protein complexes and functional modules in molecular networks. *Proceedings of the National Academy of Sciences of the United States of America*, *100*(21), 12123–12128. doi:10.1073/pnas.2032324100

Sprinzak, E., Sattath, S., & Margalit, H. (2003). How reliable are experimental protein-protein interaction data? *Journal of Molecular Biology*, *327*(5), 919. doi:10.1016/S0022-2836(03)00239-0

Sridhar, P., Song, B., Kahveci, T., & Ranka, S. (2008). Mining metabolic networks for optimal drug targets. *Proceedings of Pacific Symposium on Biocomputing*, *2008*, 291–302. doi:10.1142/9789812776136_0029

Stalteri, M. A., & Harrison, A. P. (2007). Interpretation of multiple probe sets mapping to the same gene in Affymetrix GeneChips. *BMC Bioinformatics*, *8*, 13. doi:10.1186/1471-2105-8-13

Stark, A., Brennecke, J., Russell, R. B., & Cohen, S. M. (2003). Identification of Drosophila microRNA targets. *PLoS Biology*, *1*, E60. doi:10.1371/journal.pbio.0000060

Steinmetz, K. L., Coley, K. C., & Pollock, B. G. (2005). Assessment of geriatric information on the drug label for commonly prescribed drugs in older people. *Journal of the American Geriatrics Society*, *53*(5), 891–894. doi:10.1111/j.1532-5415.2005.53273.x

Steitz, J. A., & Vasudevan, S. (2009). miRNPs: versatile regulators of gene expression in vertebrate cells. *Biochemical Society Transactions*, *37*, 931–935. doi:10.1042/BST0370931

Stelling, J., Klamt, S., Bettenbrock, K., Schuster, S., & Gilles, E. D. (2002). Metabolic network structure determines key aspects of functionality and regulation. *Nature*, *420*, 190–193. doi:10.1038/nature01166

Stephens, M., Palakal, M., et al. (2001). Detecting gene relations from MEDLINE abstracts. In *Proceedings of the sixth Ann Pac Symp Biocomp (PSB 2001)*.

Stephens, S. M., Chen, J. Y., Davidson, M. G., Thomas, S., & Trute, B. M. (2005). Oracle database 10g: a platform for BLAST search and regular expression pattern matching in life sciences. *Nucleic Acids Research*, *33*, D675–D679. doi:10.1093/nar/gki114

Stern, C. D., & Fraser, S. E. (2001). Tracing the lineage of tracing cell lineages. *Nature Cell Biology*, *3*, E216–E218.. doi:10.1038/ncb0901-e216

Stoecklin, G., Colombi, M., Raineri, I., Leuenberger, S., Mallaun, M., & Schmidlin, M. (2002). Functional cloning of BRF1, a regulator of ARE-dependent mRNA turnover. *The EMBO Journal*, *21*, 4709–4718. doi:10.1093/emboj/cdf444

Stone, E. A., & Sidow, A. (2005). Physicochemical constraint violation by missense substitutions mediates impairment of protein function and disease severity. *Genome Research*, *15*(7), 978–986. doi:10.1101/gr.3804205

Stormo, G. (1990). Consensus patterns in DNA. In Doolittle, R. F. (Ed.), *Molecular evolution: computer analysis of protein and nucleic acid sequences* (*Vol. 183*, pp. 211–221). Methods in Enzymology.

Subramanian, A., Tamayo, P., Mootha, V., Mukherjee, S., Ebert, B., & Gillette, M. (2005). Gene set enrichment analysis: A knowledge-based approach for interpreting genome-wide expression profiles. *Proceedings of the National Academy of Sciences of the United States of America*, *102*(43), 15545–15550. doi:10.1073/pnas.0506580102

Suter, B., Kittanakom, S., & Stagljar, I. (2008). Two-hybrid technologies in proteomics research. *Current Opinion in Biotechnology*, *19*, 316–323. doi:10.1016/j.copbio.2008.06.005

Tailleux, L., Waddell, S. J., Pelizzola, M., Mortellaro, A., Withers, M., & Tanne, A. (2008). Probing host pathogen cross-talk by transcriptional profiling of both Mycobacterium tuberculosis and infected human dendritic cells and macrophages. *PLoS ONE*, *3*, e1403. doi:10.1371/journal.pone.0001403

Tamayo, P., Slonim, D., Mesirov, J., Zhu, Q., Kitareewan, S., & Dmitrovsky, E. (1999). Interpreting patterns of gene expression with self-organizing maps: Methods and application to hematopietic differentiation. [PubMed]. *Proceedings of the National Academy of Sciences of the United States of America*, *96*, 2907–2912. doi:10.1073/pnas.96.6.2907

Tamura, T., Takemoto, K., & Akutsu, T. (2009). Measuring structural robustness of metabolic networks under a Boolean model using integer programming and feedback vertex sets. *2nd International Workshop on Intelligent Informatics in Biology and Medicine (IIBM2009), A Part of Proceedings of 3rd International Conference on Complex, Intelligent and Software Intensive Systems* (pp. 819–824).

Tanay, A., Sharan, R., Kupiec, M., & Shamir, R. (2004). Revealing modularity and organization in the yeast molecular network by integrated analysis of highly heterogeneous genomewide data. *Proceedings of the National Academy of Sciences of the United States of America*, *101*(9), 2981–2986. doi:10.1073/pnas.0308661100

Tang, A., Jackson, D., Hobbs, J., Chen, W., Smith, J. L., & Patel, H. (2008). A maximum entropy model applied to spatial and temporal correlations from cortical networks in vitro. [PubMed]. *The Journal of Neuroscience*, *28*, 505–518. doi:10.1523/JNEUROSCI.3359-07.2008

Tangelder, J. H., & Veltkamp, R. C. (2008). A survey of content based 3D shape retrieval methods. *Multimedia Tools and Applications*, *39*, 441–471. doi:10.1007/s11042-007-0181-0

Tang, M., Pires, Y., Schultz, M., Duarte, I., Gallegos, M., & Wistuba, I. I. (2003). Microsatellite analysis of synchronous and metachronous tumors: a tool for double primary tumor and metastasis assessment. *Diagnostic Molecular Pathology*, *12*, 151–159.

Tapinos, N., Ohnishi, M., & Rambukkana, A. (2006). ErbB2 receptor tyrosine kinase signaling mediates early demyelination induced by leprosy bacilli. *Nature Medicine*, *12*, 961–966. doi:10.1038/nm1433

Tata, S., Hankins, R. A., & Patel, J. M. (2004). Practical suffix tree construction. In *VLDB'04* (pp. 36–47).

Teng, L., & Chan, L. (2008). Discovering biclusters by iteratively sorting with weighted correlation coefficient in gene expression data. *J Signal Process Syst*, *50*(3), 267–280. doi:10.1007/s11265-007-0121-2

Thadani, R., & Tammi, M. T. (2006). MicroTar: predicting microRNA targets from RNA duplexes. *BMC Bioinformatics*, *7*(5), S20. doi:10.1186/1471-2105-7-S5-S20

The Gene Ontology Consortium. (2008). The Gene Ontology project in 2008. *Nucleic Acids Research*, *36*, D440–D444. doi:10.1093/nar/gkm883

Thieffry, D., Huerta, A. M., Prez-Rueda, E., & Collado-Vides, J. (1998). From specific gene regulation to genomic networks: a global analysis of transcriptional regulation in escherichia coli. *BioEssays*, *20*(5), 433–440. doi:10.1002/(SICI)1521-1878(199805)20:5<433::AID-BIES10>3.0.CO;2-2

Thoma, M., Cheng, H., Gretton, A., Han, J., Kriegel, H., Smola, A., et al. (2009). Near-optimal supervised feature selection among frequent subgraphs. In *Proceedings of the SDM*.

Thomson, R. (2008). Developing markers for multilocus phylogenetics in non-model organisms: A test case with turtles. *Molecular Phylogenetics and Evolution*, *49*, 514–525. doi:10.1016/j.ympev.2008.08.006

Thornton, J. M., Todd, A. E., Milburn, D., Borkakoti, N., & Orengo, C. A. (2000). From structure to function: approaches and limitations. *Nature Structural Biology*, *7*(Suppl), 991–994. doi:10.1038/80784

Trabucchi, M., Briata, P., Garcia-Mayoral, M., Haase, A. D., Filipowicz, W., & Ramos, A. (2009). The RNA-binding protein KSRP promotes the biogenesis of a subset of microRNAs. *Nature*, *459*, 1010–1014. doi:10.1038/nature08025

Tseng, Y. Y., Dundas, J., & Liang, J. (2009). Predicting Protein Function and Binding Profile via Matching of Local Evolutionary and Geometric Surface Patterns. *J. Mol. Biol.*

Uetz, P. (2000). A comprehensive analysis of protein-protein interactions in Saccharomyces cerevisiae. *Nature*, *403*, 623–627. doi:10.1038/35001009

Ullmann, J. R. (1976). An algorithm for sub-graph isomorphism. *Journal of the Association for Computing Machinery*, *23*(1), 31–42.

Ulrich, S. (2009). *SPCC(TDM): Catalogue for the content analysis of summary of product characteristics (SPCs) with reference to therapeutic drug monitoring (TDM)*. Retrieved May 2009, from http://www.agnp.de

Ulrich, S., Hiemke, C., Laux, G., Mueller-Oerlinghausen, B., Havemann-Reinecke, U., & Riederer, P. (2007). Value and actuality of the prescription information for therapeutic drug monitoring of psychopharmaceuticals: A comparison with the medico-scientific evidence. *Pharmacopsychiatry*, *40*(3), 121–127. doi:10.1055/s-2007-977712

Ulrich, S., Laux, G., Mueller-Oerlinghausen, B., Havemann-Reinecke, U., Riederer, P., & Zernig, G. (2008). Nutzen und Aktualität der Fachinformationen für das therapeutische Drug Monitoring von Psychopharmaka. *Psychopharmakotherapie*, *15*(2), 67–74.

Ultsch, A. (2003). Maps for the visualization of high-dimensional data spaces. *Proc.WSOM'03*, 225-230.

Ultsch, A. (2005). ESOM-Maps: tools for clustering, visualization, and classification with Emergent SOM. *Technical report.Dept.of Mathematics and Computer Sciece, Univeristy of Marburg.*

Vaisman, I. I., Tropsha, A., & Zheng, W. (1998, May). *Compositional preferences in quadruplets of nearest neighbor residues in protein structures: statistical geometry analysis.* Paper presented at the IEEE Symposium on Intelligence and Systems, Washington, DC.

van Vliet, M., Klijn, C., Wessels, L., & Reinders, M. (2007). Module-based outcome prediction using breast cancer compendia. *PLoS ONE*, *2*(10), 1047. doi:10.1371/journal.pone.0001047

Vapnik, V. (1998). *Statistical learning theory*. New York: Wiley.

Vert, J.-P., & Yamanishi, Y. (2005). Supervised graph inference. In L. K. Saul, Y. Weiss, & L. Bottou (Eds.), *Advances in neural information processing systems 17* (pp. 1433-1440). Cambridge, MA: MIT Press.

Vert, J. P., Qiu, J., & Noble, W. S. (2007). A new pairwise kernel for biological network inference with support vector machines. *BMC Bioinformatics*, *10*(Suppl 10), S8. doi:10.1186/1471-2105-8-S10-S8

Verzilli, C. J., Whittaker, J. C., Stallard, N., & Chasman, D. (2005). A hierarchical Bayesian model for predicting the functional consequences of amino acid polymorphisms. *Journal of the Royal Statistical Society. Series C, Applied Statistics*, *54*(1), 191–206. doi:10.1111/j.1467-9876.2005.00478.x

Vesanto, J., & Alhoniemi, E. (2000). Clustering of the Self-Organizing Map. *IEEE Transactions on Neural Networks*, *11*, 586–600. doi:10.1109/72.846731

Vetterli, M., & Herley, C. (1992). Wavelets and filter banks: theory and design. *IEEE Transactions on Signal Processing*, *40*, 2207–2232. doi:10.1109/78.157221

Vieille, C., & Zeikus, G. J. (2001). Hyperthermophilic enzymes: sources, uses, and molecular mechanisms for thermostability. *Microbiology and Molecular Biology Reviews*, *65*(1), 1–43. doi:10.1128/MMBR.65.1.1-43.2001

Vinga, S., & Almeida, J. (2004). Renyi continuous entropy of DNA sequences. *Journal of Theoretical Biology*, *231*, 377–388. doi:10.1016/j.jtbi.2004.06.030

Vogel, K., & Horz, W. (1989). The two positively acting regulatory proteins PHO2 and PHO4 physically interact with PHO5 upstream activation regions. *Molecular and Cellular Biology*, *9*(5), 2050–2057.

von Kamp, A., & Schuster, S. (2006). Metatool 5.0: Fast and flexible elementary modes analysis. *Bioinformatics (Oxford, England)*, *22*(15), 1930–1931. doi:10.1093/bioinformatics/btl267

von Mering, C., Krause, R., Snel, B., Cornell, M., Oliver, S. G., Fields, S., & Bork, P. (2002). Comparative assessment of large-scale data sets of protein-protein interactions. *Nature*, *417*(6887), 399–403. doi:10.1038/nature750

von Mering, D., & Krause, C. (2002). Comparative assessment of a large-scale data sets of protein-protein interactions. *Nature*, *417*(6887), 399–403. doi:10.1038/nature750

von Roretz, C., & Gallouzi, I. E. (2008). Decoding ARE-mediated decay: is microRNA part of the equation? *The Journal of Cell Biology*, *181*, 189–194. doi:10.1083/jcb.200712054

Wagner, A., & Fell, D. (2001). The small world inside large metabolic networks. *Proceedings. Biological Sciences*, *268*, 1803–1810. doi:10.1098/rspb.2001.1711

Wagsta, K., Basu, S., et al. (2006). When is constrained clustering beneficial, and why? In *Proceedings of the AAAI.*

Wagsta, K., Cardie, C., et al. (2001). *Constrained k-means clustering with background knowledge*. Paper presented at the 18th International Conference on Machine Learning (ICML-01).

Wang, L., Dong, L., & Fan, H. (2004). Randomized Algorithms for Motif Detection. In *Proceedings of the 15th Annual International Symposium on Algorithms and Computation (ISAAC'04)* (pp. 884-895).

Wang, M., Shang, X. Q., Diao, J. N., & Li, Z. H. (2010). WIBE: Mining frequent closed patterns without candidate maintenance in microarray dataset. In *Proceedings of the 6th International Conference on Data Mining.*

Wang, X., Wirth, A., & Wang, L. (2007). Structure-based statistical features and multivariate time series clustering. In *Proceedings of the 7th IEEE ICDM*, Omaha, NE (pp. 351-360).

Wang, G., & Dunbrack, R. L. Jr. (2003). PISCES: a protein sequence culling server. *Bioinformatics (Oxford, England)*, *19*(12), 1589–1591. doi:10.1093/bioinformatics/btg224

Wang, J. C. (1996). DNA topoisomerases. *Annual Review of Biochemistry*, *65*, 635–692..doi:10.1146/annurev.bi.65.070196.003223

Wang, J. Z., Du, Z., Payattakool, R., Yu, P. S., & Chen, C.-F. (2007). A new method to measure the semantic similarity of GO terms. *Bioinformatics (Oxford, England)*, *23*(10). doi:10.1093/bioinformatics/btm087

Waterman, M., Arratia, R., & Galas, E. (1984). Pattern recognition in several sequences: consenus and alignment. *Bulletin of Mathematical Biology*, *46*, 515–527.

Watts, D. J., & Strogatz, S. H. (1998). Collective dynamics of 'small-world' networks. *Nature*, *393*, 440–442. doi:10.1038/30918

Webb, E. C. (1992). *Recommendations of the Nomenclature Committee of the International Union of Biochemistry and Molecular Biology on the Nomenclature and Classification of Enzymes*. Oxford University Press.

Weigelt, B., Glas, A. M., Wessels, L. F., Witteveen, A. T., Peterse, J. L., & Vant Veer, L. J. (2003). Gene expression profiles of primary breast tumors maintained in distant metastases. In *Proceedings of the National Academy of Sciences of the United States of America* (Vol. 100, pp. 15901-15905). doi:10.1073/pnas.2634067100

Wernicke, S. (2006). Efficient detection of network motifs. *IEEE. ACM Transactions On Computational Biology And Bioinformatics*, *3*(4), 347–359. doi:10.1109/TCBB.2006.51

Westbrook, J., Feng, Z., Chen, L., Yang, H., & Berman, H. M. (2003). The Protein Data Bank and structural genomics. *Nucleic Acids Research*, *31*, 489–491. doi:10.1093/nar/gkg068

Whitaker, J., McConkey, G., & Westhead, D. (2009). The transferome of metabolic genes explored: analysis of the horizontal transfer of enzyme encoding genes in unicellular eukaryotes. *Genome Biology*, *10*, R36. doi:10.1186/gb-2009-10-4-r36

Willy, H. (2009). *Discovering interaction motifs from protein interaction networks*. (pp. 99–116). Hershey, PA, USA: IGI Global.

Wilusz, C. J., & Wilusz, J. (2004). Bringing the role of mRNA decay in the control of gene expression into focus. *Trends in Genetics*, *20*, 491–497. doi:10.1016/j.tig.2004.07.011

Wilusz, C. J., Wormington, M., & Peltz, S. W. (2001). The cap-to-tail guide to mRNA turnover. *Nature Reviews. Molecular Cell Biology*, *2*, 237–246. doi:10.1038/35067025

Wojcik, J., Boneca, I. G., & Legrain, P. (2002). Prediction, assessment and validation of protein interaction maps in bacteria. *Journal of Molecular Biology*, *323*, 763–770. doi:10.1016/S0022-2836(02)01009-4

Wu, L., & Oviatt, S. L. (1999). Multimodal integration - a statistical view. *IEEE Transactions on Multimedia*, *1*(4), 334–341. doi:10.1109/6046.807953

Wulczyn, F. G., Smirnova, L., Rybak, A., Brandt, C., Kwidzinski, E., & Ninnemann, O. (2007). Post-transcriptional regulation of the let-7 microRNA during neural cell specification. *The FASEB Journal*, *21*, 415–426. doi:10.1096/fj.06-6130com

Wu, X., Zhang, C., & Zhang, S. (2004). Efficient mining of both positive and negative association rules. *ACM Transactions on Information Systems*, *22*, 381–405. doi:10.1145/1010614.1010616

Xenarios, I., Rice, D. W., Salwinski, L., Baron, M. K., Marcotte, E. M., & Eisenberg, D. (2000). DIP: The database of interacting proteins. *Nucleic Acids Research*, *28*, 289–291. doi:10.1093/nar/28.1.289

Xia, Y., Turken, U., Whitfield-Gabrieli, S. L., & Gabrieli, J. D. (2005). Knowledge-based classification of neuronal fibers in entire brain. In *Proceedings of the Int. Conf. on Medical Image Computing and Computer Assisted Intervention* (pp. 205-212). New York: Springer.

Xing, E. P., & Ng, A. Y. (2002). Distance metric learning, with application to clustering with side-information. In [Cambridge, MA: MIT.]. *Proceedings of Advances in Neural Information Processing Systems*, *15*, 505–512.

Xu, X., Ester, M., Kriegel, H. P., & Sander, J. (1998). A Distribution-Based Clustering Algorithm for Mining in Large Spatial Databases. In *Proceedings of the 14th Int. Conf. on Data Engineering*, Los Alamitos, CA (pp. 324-331). Washington, DC: IEEE Computer Society Press.

Xulvi-Brunet, R., & Li, H. (2009). Co-expression networks: graph properties and topological comparisons. *Bioinformatics (Oxford, England)*, *26*(2), 205–214. doi:10.1093/bioinformatics/btp632

Xu, N., Chen, C. Y., & Shyu, A. B. (2001). Versatile role for hnRNP D isoforms in the differential regulation of cytoplasmic mRNA turnover. *Molecular and Cellular Biology*, *21*, 6960–6971. doi:10.1128/MCB.21.20.6960-6971.2001

Yamamoto, N., Yang, M., Jiang, P., Xu, M., Tsuchiya, H., & Tomita, K. (2003). Determination of clonality of metastasis by cell-specific color-coded fluorescent-protein imaging. *Cancer Research*, *63*, 7785.

Yamanishi, Y., Vert, J.-P., & Kanehisa, M. (2005). Supervised enzyme network inference from the integration of genomic data and chemical information. *Bioinformatics (Oxford, England)*, *21*(Suppl 1), i468–i477. doi:10.1093/bioinformatics/bti1012

Yan, J., Liu, N., Zhang, B., Yang, Q., Yan, S., & Chen, Z. (2006, December). *A novel scalable algorithm for supervised subspace learning.* Paper presented at the IEEE International Conference on Data Mining, Hong Kong, China.

Yan, X., & Han, J. (2002). gSpan: graph-based substructure pattern mining. In *Proceedings of the ICDM* (pp. 721-724).

Yan, X., Cheng, H., Han, J., & Yu, P. S. (2008). Mining significant graph patterns by leap search. In *Proceedings of the SIGMOD* (pp. 433-444).

Yandell, M. D., & Majoros, W. H. (2002). Genomics and natural language processing. *Nature Reviews. Genetics*, *3*(8), 601–610.

Yang, J., Wang, H., Wang, W., & Yu, P. (2003). Enhanced biclustering on expression data. In *Proceedings of the Third IEEE Symposium on Bioinformatics and Bioengineering (BIBE'03)* (pp. 1-7).

Yang, X., Wang, M., & Fitzgerald, M. C. (2004). Analysis of protein folding and function using backbone modified proteins. *Bioorganic Chemistry*, *32*(5), 438–449. doi:10.1016/j.bioorg.2004.06.011

Yanikoglu, B., & Erman, B. (2002). Minimum Energy Configurations of the 2-Dimensional HP-Model of Proteins by Self-Organizing Networks. *Journal of Computational Biology*, *9*(4), 613–620. doi:10.1089/10665270276027 7345

Yeger-Lotem, E., Sattath, S., Kashtan, N., Itzkovitz, S., Milo, R., & Pinter, R. Y. (2004). Network motifs in integrated cellular networks of transcriptionregulation and proteinprotein interaction. *Proceedings of the National Academy of Sciences of the United States of America*, *101*(16), 5934–5939. doi:10.1073/pnas.0306752101

Ye, J., McGinnis, S., & Madden, T. L. (2008). BLAST: improvements for better sequence analysis. *Nucleic Acids Research*, *34*, W6–W9. doi:10.1093/nar/gkl164

Yeturu, K., & Chandra, N. (2008). PocketMatch: a new algorithm to compare binding sites in protein structures. *BMC Bioinformatics*, *9*, 543. doi:10.1186/1471-2105-9-543

Yeung, K. Y., & Medvedovic, M. (2003). Clustering gene-expression data with repeated measurements. *Genome Biology*, *4*(5), R34. doi:10.1186/gb-2003-4-5-r34

Yeung, K., Medvedovic, M., & Bumgarner, R. (2004). From co-expression to co-regulation: how many microarray experiments do we need? *Genome Biology*, *5*(7). doi:10.1186/gb-2004-5-7-r48

Young, D., Stark, J., & Kirschner, D. (2008). Systems biology of persistent infection: tuberculosis as a case study. *Nature Reviews Microbiology*, *6*, 520–528. doi:10.1038/nrmicro1919

Yuan, X., Buckles, B. P., Yuan, Z., & Zhang, J. (2002). Mining negative association rules. In *Proceedings of the 7th International Symposium on Computers and Communications*.

Zahn, J. M., & Poosala, S. (2007). AGEMAP: A gene expression database for aging in mice. *PLOS Genetics*, *3*(11), 2326–2337. doi:10.1371/journal.pgen.0030201

Zahrt, T. C. (2003). Molecular mechanisms regulating persistent Mycobacterium tuberculosis infection. *Microbes and Infection*, *5*, 159–167. doi:10.1016/S1286-4579(02)00083-7

Zaki, M., & Ramakrishnan, N. (2005). Reasoning about sets using redescription mining. In *Proceedings of the ACM SIGKDD Intl. Conf. on Knowledge Discovery and Data Mining.*

Zaki, M. J., & Hsiao, C.-J. (2005). Efficient algorithms for mining closed itemsets and their lattice structure. *IEEE Transactions on Knowledge and Data Engineering*, *17*, 462–478. doi:10.1109/TKDE.2005.60

Zare, H., & Khodursky, A. B. (2006). Gene clustering and gene function prediction using multiple sources of data. In. *Proceedings of the IEEE Genomic Signal Processing and Statistics*, *2006*, 113–114.

Zhang, S., & Laidlaw, D. H. (2005). DTI fiber clustering and cross-subject cluster analysis. In: International Society for Magnetic Resonance in Medicine (Ed.), *Proceedings of the 13th Scientific Meeting ISMRM*, Miami, FL (pp. 2727).

Zhang, C., & Kim, S. H. (2003). Overview of structural genomics: from structure to function. *Current Opinion in Chemical Biology*, *7*, 28–32. doi:10.1016/S1367-5931(02)00015-7

Zhang, R., & Su, B. (2009). Small but influential: the role of microRNAs on gene regulatory network and 3'UTR evolution. *Journal of Genetics and Genomics = Yi Chuan Xue Bao*, *36*, 1–6. doi:10.1016/S1673-8527(09)60001-1

Zhang, S., Demiralp, C., & Laidlaw, D. H. (2003). Visualizing diffusion tensor MR images using streamtubes and streamsurfaces. *IEEE Transactions on Visualization and Computer Graphics*, *9*(4), 454–462. doi:10.1109/TVCG.2003.1260740

Zhao, L., Zaki, M., & Ramakrishnan, N. (2006). Blosom: A framework for mining arbitrary boolean expressions. In *Proceedings of the ACM SIGKDD Intl. Conf. on Knowledge Discovery and Data Mining.*

Zhao, L., & Zaki, M. (2005). MicroCluster: Efficient deterministic biclustering of Microarray data. *IEEE Intelligent Systems*, *20*(6), 40–49. doi:10.1109/MIS.2005.112

Zheng, J., Close, T. J., Jiang, T., & Lonardi, S. (2003). Efficient selection of unique and popular oligos for large EST databases. *Lecture Notes in Computer Science*, *2676*, 384–401. doi:10.1007/3-540-44888-8_28

Zhong, S., & Ghosh, J. (2003). A comparative study of generative models for document clustering. In *Proceedings of the workshop on Clustering High Dimensional Data and Its Applications in SIAM Data Mining Conference.*

Zhong, W., & Sternberg, P. W. (2006). Genome-wide prediction of C. elegans genetic interactions. *Science*, *311*, 1481–1484. doi:10.1126/science.1123287

Zhou, H., & Zhou, Y. (2002). Distance-scaled, finite ideal-gas reference state improves structure-derived potentials of mean force for structure selection and stability prediction. *Protein Science*, *11*(11), 2714–2726. doi:10.1110/ps.0217002

Zineh, I., Pebanco, G. D., Aquilante, C. L., Gerhard, T., Beitelshees, A. L., Beasley, B. N., & Hartzema, A. G. (2006). Discordance between availability of pharmacogenetic studies and pharmacogenetics-based prescribing information for the top 200 drugs. *The Annals of Pharmacotherapy*, *40*(4), 639–644. doi:10.1345/aph.1G464

Zubiaga, A. M., Belasco, J. G., & Greenberg, M. E. (1995). The nonamer UUAUUUAUU is the key AU-rich sequence motif that mediates mRNA degradation. *Molecular and Cellular Biology*, *15*, 2219–2230.

Zuker, M. (2003). Mfold web server for nucleic acid folding and hybridization prediction. *Nucleic Acids Research*, *31*, 3406–3415. doi:10.1093/nar/gkg595

About the Contributors

Xiao-Li Li is currently a principal investigator in the Data Mining Department at the Institute for Infocomm Research, A*STAR, Singapore. He also holds an appointment of adjunct assistant professor in Nanyang Technological University (NTU). His research interests include data mining, machine learning, and bioinformatics. He has been serving as the members of technical program committees in numerous data mining related conferences (including KDD, ICDM, SDM, PKDD/ECML, PAKDD, AAAI, ACML and CIKM), and bioinformatics conferences (a book editor for Biological Data Mining in Protein Interaction Networks, PC members for ACM BCB, IEEE BIBE, IEEE BIBM, InCob, BIOCOMP etc.). In 2005, he received Best Paper Award in the 16th International Conference on genome informatics (GIW 2005). In 2008, he received the Best Poster Award in the 12th Annual International Conference Research in computational molecular biology (RECOMB 2008). In 2011, he received Best Paper Runner-Up Award of the 16th International Conference on Database Systems for Advanced Applications (DASFAA 2011).

See-Kiong Ng is currently Principal Scientist and the head of the Data Mining Department at the Institute for Infocomm Research, A*STAR. The department focuses on applied research and development in machine learning, text mining, semantic technologies, privacy-preserving data mining, and bioinformatics. See-Kiong obtained both his BS and PhD in computer science from Carnegie Mellon University (CMU), with an MS from University of Pennsylvania. His primary research is in data mining and machine learning, with applications in text mining, bioinformatics, privacy-preserving data mining, and social network mining. See-Kiong has a long-standing interest in cross-disciplinary applied computer science research. He wrote the TrueAllele software when he was a graduate student at CMU. The program was subsequently used by a biotech company in Iceland to genotype the Icelandic population -- thus began his cross-disciplinary journey into the exciting field of genomics as a computer scientist. See-Kiong has since been able to apply what he had learned from bioinformatics to a wide array of application domains from social network analysis to sensor stream data mining. He has published close to 100 papers in leading peer-reviewed journals and conferences to-date. See-Kiong also holds a joint appointment as an Associate Professor of Information Systems Technology and Design at the Singapore University of Technology and Design, the latest university in Singapore.

* * *

Tatsuya Akutsu received B.Eng. and M.Eng. in Aeronautics and D.Eng. in Information Engineering from University of Tokyo, 1984, 1986 and 1989, respectively. From 1989 to 1994, he was with Mechanical

Engineering Laboratory, Japan. From 1994 to 1996, he was an Associate Professor in the Department of Computer Science at Gunma University. From 1996 to 2001, he was an Associate Professor in Human Genome Center, Institute of Medical Science, University of Tokyo. Since 2001, he has been a Professor in Bioinformatics Center, Institute for Chemical Research, Kyoto University. His research interests include bioinformatics and the design and analysis of algorithms.

Paramjeet Singh Bagga received his Bachelors, Masters, and Ph.D. in Biology from Guru Nanak Dev University, Amritsar, India, in 1982, 1983 and 1988 respectively. His postdoctoral research at University of Medicine and Dentistry of New Jersey (UMDNJ), USA, focused on DNA repair in human Xeroderma Pigmentosum cells. As a research associate he studied mechanisms of RNA processing at UMDNJ. He joined as Assistant Professor at Montclair State University, New Jersey, USA, in 1996 and then moved to Ramapo College of New Jersey (RCNJ) in 1999. At present he is working as Professor of Biology and Convener (Head) of Bioinformatics at RCNJ. His current research interests include regulation of post-transcriptional eukaryotic gene expression from a genomics perspective. His research lab develops and uses computational tools for studying mammalian genomes for conserved sequence motifs which are involved in regulatory processes through interactions with cellular proteins.

Pierre Baumann, Dipl. Chem., Ph.D. has been Professor emeritus, since 2007, in psychopharmacology, Dept of Psychiatry (CHUV), University of Lausanne. His research interests are metabolism, pharmacokinetics and pharmacogenetics of psychotropic drugs, phase II – IV studies, analytical methods, therapeutic drug monitoring, pharmacovigilance in psychiatry, neuropsychopharmacology, and clinical psychopharmacology. He has been published in more than 300 books and international journals.

Richipal Singh Bindra received his BE degree from Dr. BAMU University, India. He is currently a graduate student in the Bioinformatics Program and Department of Computer Science at the New Jersey Institute of Technology, Newark, New Jersey, USA. His research interests include computational biology, databases, and data mining.

Shveta Bisht obtained her Bachelor of Science degree from Panjab University in the subject biotechnology in the year 2005. She then joined Indian Institute of Science under the integrated PhD program and obtained her Master in Science degree in 2008. Currently, she is pursuing her PhD under the supervision of Prof M.R.N. Murthy and her current research interests include crystallographic and computational analyses of pathogen proteins.

Christian Boehm is currently professor of informatics at Ludwig-Maximilians-Universit Universität München, Germany (LMU) and visiting professor at Florida State University, Tallahassee (FSU). His previous affiliations include Technische Universität München (TUM) and the University for Medical Informatics and Technology in Innsbruck, Austria (UMIT). He has currently more than 90 reviewed publications in database research (with focus on indexing) and data mining (with focus on clustering). He holds one patent and has received the SIGMOD best paper award 1997 and the best paper honorable mention award at the SIAM Int. Conf. on Data Mining (SDM) 2008.

Francesco Bruno received his master degree in computer engineering at University of Calabria on July, 2008. His master thesis concerned motif extraction from biological networks. Currently, he is working at Value Team as a software consultant.

Young-Rae Cho received his Ph.D. degree in computer science and engineering from State University of New York at Buffalo in 2009, and the Master's degree in computer science from University of Illinois at Urbana-Champaign in 2003. He is currently an assistant professor in the Department of Computer Science at Baylor University. His research interests include bioinformatics, computational biology, systems biology, and data mining. He has authored or coauthored more than 20 publications in journals, conferences, and book chapters. He served as the program committee in several international conferences such as BIBM 2009 and ACM-BCB 2010. He received the Young Investigator Development Program Award from the Vice Provost for Research at Baylor University in 2010.

Frank Collart received his M.S, in Chemistry from Bowling Green State University in 1982 and his Ph.D in Medical Sciences from the Medical College of Ohio in 1984. . He is a Molecular Biologist and Principle Investigator in the Biosciences Division, Argonne National Laboratory. He is a Principal Investigator and Research Coordinator for the Experimental Annotation of Function. Research interests include informatics and molecular approaches for elucidation of sensory and response pathways in bacteria and fungi.

Andreas Conca M.D. is Head of the Department of Psychiatry, Regional Hospital Bolzano (I), Associate Professor, Department of Psychiatry, Medical University Innsbruck (A), and Chairman of the Task force for Special Treatment Modalities in Psychiatry of the Austrian Society of Psychiatry and Psychotherapy (A). He has been published in more than 100 papers, participated in over 250 national and international talks, and received multiple awards due to his expertise.

Yang Dai received Ph.D degree in Management Science and Engineering from University of Tsukuba, Japan, in 1991. From 1991 to 1997 she was a research associate, an assistant professor at the Department of Management Science of Kobe University of Commerce. From 1997 -2001 she was an assistant professor of Department of Mathematical and Computing Sciences, Tokyo Institute of Technology, Japan. Since 2001 she is on faculty of Department of Bioengineering, the University of Illinois at Chicago. Her recent research interests include the analysis of gene expression and pathways, inference of biological networks, and data analysis on genome-wide association study of complex diseases. Professor Dai has published more than 70 peer reviewed papers in various journals and conferences proceedings.

Anna Dominiczak is a British Heart Foundation Chair of Cardiovascular Medicine and heads the BHF Glasgow Cardiovascular Research Centre at the University of Glasgow. She is also a consultant physician and endocrinologist at the Western Infirmary and associate dean for Research at the Faculty of Medicine. She has a major research interest in cardiovascular genomics and systems medicine and holds a BHF Programme Grant. "Genomics and proteomics of hypertension and its vascular complications: the pathwayomic strategies." Her total research income in the last 3 years is in excess of £35 million. She is a Fellow of the Royal College of Physicians, the Academy of Medical Sciences, the Royal Society of Edinburgh and the American Heart Association. In addition to membership of several editorial boards, Professor Dominiczak was editor-in-chief of Clinical Science between 2004 and 2008. Her scientific and clinical contributions have been recognised by invitations to serve on the Medical Research Council Physiological Medicine Board, British Heart Foundation Project Grant Committee, the Wellcome Trust Physiological Science Committee and the Scientific Advisory Board of the Leducq Foundation. Professor Dominiczak is an author of 230 publications in peer-reviewed journals. In 2005 she was awarded OBE for services to medicine.

Susan Fairley completed her PhD in computing science at the University of Glasgow in the area of bioinformatics. She is currently employed at the Wellcome Trust Sanger Institute.

Wangsen Feng got a Ph.D degree in 2008 from Peking University. His research focus is bioinformatics and algorithms.

Klaus R. Hahn has a permanent scientist-position at the institute for Biomathematics and Biometrics of the Helmholtz-Zentrum Munich, Germany. His main scientific contributions are in the fields: Nuclear Physics, Radiation Biology and Neuro Imaging. After studying theoretical physics at the LMU University Munich he achieved his Doctor degree at University Tübingen. His contributions cover theoretical, mathematical, numerical and statistical investigations and led to about 70 refereed papers. He is also engaged in the education of young scientists and CoEditor of the International Journal of Biomathematics and Biostatistics.

Xiaoxu Han received his Ph.D. degree in applied mathematical and computational sciences (AMCS) from the University of Iowa in 2004. He did his postdoctoral research in computational biology in Roy J. Carver Center for Genomics at the University of Iowa. He is currently an associate professor in the Department of Mathematics and Bioinformatics Program at Eastern Michigan University, Ypsilanti MI 47197 USA. His current research interests include bioinformatics, systems biology, and machine learning. His recent research projects include investigating clinical-level cutting-edge pattern recognition methodologies for serum proteomic profiles, high-performance data reduction techniques for petascale data, and privacy data mining in bioinformatics.

Neil Hanlon received his PhD from the Department of Biochemistry at Imperial College. Since then he has completed post-doctoral work at the University of Oxford, the University of Glasgow and the University of Edinburgh. He is currently employed by the Centre for Systems Biology at Edinburgh University.

Christoph Hiemke Is head of the therapeutic drug monitoring group of the AGNP http://www.agnp.de/, managing director of the psiac GmbH, an internet based program on drug interactions in psychiatry http://www.psiac.de/, and member of the scientific boards of pharmacopsychiatry and psychopharmakotherapie. His research interests are clinical and experimental psychopharmacology, therapeutic monitoring of psychoactive drugs, metabolism of psychoactive drugs, metabolism of psychoactive drugs, drug-drug interactions, and drug transport.

Ela Hunt is a lecturer in computer and information sciences at the University of Strathclyde in Glasgow, UK. She graduated with a PhD in computing science from the University of Glasgow, Glasgow, UK in 2002. She has been a Medical Research Council Fellow in bioinformatics 2001-2005, at the University of Glasgow, and an EU-funded Marie Curie fellow at the ETH Zurich 2006-2008. Her research interests include approximate pattern matching, persistence, indexing, suffix trees, XML, bioinformatics, data visualisation and usability.

Robert W. Irving is a senior lecturer in the Department of Computing Science at the University of Glasgow. He has BSc and PhD degrees in mathematics from Glasgow University. He was previously employed as a faculty member in the Department of Mathematics and Computer Science at the University of Salford, and as a communications specialist at the UK Government Communications Headquarters in Cheltenham. His research interests are in the design and analysis of algorithms, data structures, computational complexity, combinatorial optimization, and discrete mathematics, with particular expertise in matching algorithms and string algorithms, and he has published over 70 research papers in these fields.

Ning Jin is a graduate student in the Department of Computer Science at the University of North Carolina at Chapel Hill. He received his bachelor's degree in computer science at the Peking University in 2007. His research interests include bioinformatics and data mining.

Hisashi Kashima received the ME and PhD degrees in informatics from Kyoto University, Kyoto, Japan, in 1999 and 2007, respectively. He has been a researcher at the Tokyo Research Laboratory, IBM Research, Tokyo, Japan, since 1999. His research interests include machine learning and its applications to bioinformatics, autonomic computing, and business intelligence.

Tsuyoshi Kato received the BE, ME, and PhD degree from Tohoku University, Sendai, Japan, in 1998, 2000, and 2003. From 2003 to 2005, he was with the National Institute of Advanced Industrial Science Technology (AIST) as a postdoctoral fellow in the Computational Biology Research Center (CBRC) at Tokyo. From 2005 to 2008, he was an assistant professor at the Graduate School of Frontier Sciences, University of Tokyo. Since 2008, he has been an associate professor at the Center for Informational Biology, Ochanomizu University. Currently, he is also a guest researcher at CBRC. His current scientific interests include bioinformatics and statistical pattern recognition.

Imtiaz A. Khan (IAK) has been active in the fields of bioinformatics and computational biology and has focused lately (over 5 yrs) on image based oncology research. IAK started his professional career in pharmaceutical sciences; and then actively converted to gaining training in bioinformatics inspired by the challenges facing biology and medicine in embracing bioinformatics science and technology. IAK gained post-graduate degrees in Bioinformatics from UK and US institutions along with associated training elements at both NCBI and EBI. IAK has introduced novel bioinformatics infrastructure (PMID: 17387278) that aims to transform images to knowledge and thus augmenting our understanding about the dynamics of cellular behaviour. He is a senior researcher within the Cardiff-Swansea Systems Cytometry Group, an interdisciplinary group bringing together biologists, engineers and mathematicians alike. Currently, as a Marie Curie Fellow he is working at the Broad Institute of Harvard and MIT,

Cambridge, USA. Primary objective of his current research is to develop infrastructure enabling the interoperability of data generated from different cytometric platforms.

Oruganty Krishnadev obtained his Bachelor of Science degree from Delhi University in the subject biochemistry in 2002. He then joined Indian Institute of Science under the integrated PhD program and obtained the Master in Science degree in 2005. Currently, he is pursuing his PhD under the supervision of Prof Srinivasan and his current research interests include development of algorithms for protein-protein interactions and structure prediction using remote homology detection methods.

Hans-Joachim Kuss is dealing with analytical chemistry and drug analysis for about 30 years in Psychiatric Clinic, University Hospital Munich (Germany). He is very experienced in Gas Chromatography and Mass Spectrometry. He was investigating the pharmacokinetics of catechol amines and metabolites. He was analysing psychopharmaceuticals by HPLC in body fluids. He provides university seminars in Drug Interactions of Psychopharmaceuticals, Statistical Analysis, and Chromatography in Biomedical Analysis. He is author of about 30 publications in the area of drug analysis, chromatography and psychopharmacology.

Peter E. Larsen received his MS in Bioengineering from the University of Illinois at Chicago in 2006. He is currently at the Molecular and Systems Biology group at Argonne National Laboratory, Biosciences Division. His research interests include analysis of next generation transcriptomic sequencing data, generating novel analysis methods for structural annotation of gene models, and generating regulatory and metabolic biological interaction networks.

Tao Li is currently an Associate Professor in the School of Computing and Information Sciences at Florida International University. He received his PhD in Computer Science in 2004 from the University of Rochester. His research interests are in data mining, machine learning, bioinformatics, and information retrieval. He is a recipient of NSF CAREER Award and IBM Faculty Research Awards.

Zhanhuai Li is a professor at The School of Computer Science and Engineering at the Northwestern Polytechnical University, Xi'An, China. He is a senior member of China Computer Federation. He is vice dean of the school and vice chairman of Database Technical Committee, China Computer Federation. He received his Master's and Ph.D. degrees from Northwestern Polytechnical University in 1987 and 1996 respectively. His research interest is database technology, software engineering and data mining.

Michael G. Madden is a Senior Lecturer in the College of Engineering and Informatics in the National University of Ireland Galway, which he joined in 2000. In 2006/07, he spent a year as a Visiting Research Scientist in the University of Helsinki, the University of California Irvine, and the University of California Berkeley. He previously worked in professional software R&D. He established and leads the Machine Learning & Data Mining Group in NUI Galway (http://datamining.it.nuigalway.ie). His interdisciplinary research focuses on the development of new machine learning methods (Bayesian, SVMs, and others) and their application to scientific, medical and other domains, and includes contributions in theoretical algorithms and practical applications. He has published over 60 papers and has received seven publication awards. He has three patent filings, and he co-founded a spin-out company, Analyze IQ Limited, based on research that he led.

Afra Maria Wohlschläger currently is the head of the neuroimaging group in the department of Neuroradiology, Klinikum rechts der Isar of the Technical University in Munich, Germany. She received her master and PhD degree in Physics at University of Cologne, Germany. Her research fields include resting state activity of the brain, functional connectivity analysis, multimodal integration: functional magnetic resonance imaging (fMRI), Diffusion Tensor Imaging (DTI), Voxel Based Morphometry (VBM), etc.

Majid Masso received a B.S and an M.S. in Mathematics from the University of Delaware and a Ph.D. in Bioinformatics from George Mason University (GMU) under the direction of Iosif I. Vaisman, where in 2005 he was awarded the inaugural Navachat Tongvichit Memorial Fellowship for Excellence in Bioinformatics. He currently conducts research and mentors students at GMU as a postdoctoral fellow. His main research interest is in modeling and prediction of the functional consequences to proteins upon amino acid replacements, which merges diverse disciplines including structural biology, biochemistry, biophysics, machine learning, probability theory, hypothesis testing, and computer programming.

Martin McBride currently holds a lectureship in systems biology in the BHF Glasgow Cardiovascular Research Centre at the University of Glasgow. He completed his undergraduate and postgraduate training in Genetics at the University of Glasgow and received post-doctoral research training in various laboratories before joining Professor Dominiczak's group. His principle research interests include the identification of genetic determinants of complex disease, including hypertension and stroke, in animal models of disease and subsequent translation to human.

John McClure currently holds a lectureship in genetic statistics within the BHF Glasgow Cardiovascular Research Centre with a major interest in complex quantitative traits such as blood pressure, insulin resistance and obesity. He is working with a large group of molecular biologists with large volumes of data on human and experimental high through put genetic analysis. He facilitates statistical analyses within several large collaborative programmes funded by the Wellcome Trust and European Union producing large volume of data ready for statistical analyses.

Nicholas Myers studied Neuroscience at Columbia University, where he assisted in using brain imaging techniques to study human language comprehension. While working at the Technical University in Munich he has collaborated on using brain imaging to identify the relationship between structural and functional alterations and attentional impairments in early Alzheimer's disease using diffusion tensor imaging, resting state fMRI, and PET. He is currently a student of Neuroscience at the University of Oxford, working on brain imaging studies of attention and memory.

Giri Narasimhan received his B.Tech degree from the Indian Institute of Technology, Mumbai, India, and his PhD from the University of Wisconsin - Madison. He was a faculty member in the Mathematical Sciences Department at the University of Memphis until 2001. Since then he has been a faculty member in the School of Computing and Information Sciences at Florida International University. His research interests include the design and analysis of algorithms and bioinformatics.

Kinya Okada received the BS degree in Agriculture from the University of Tohoku, Sendai, Japan in 2003, and the MS degree in information science from Nara Institute of Science and Technology, Nara, Japan in 2004, and the PhD degree in science from the University of Tokyo, Kashiwa, Japan in 2008. Currently, he is in KO Institute for Medical Bioinformatics, Kanagawa, Japan. His research interest is bioinformatics.

Luigi Palopoli is a Professor of Computer Science at DEIS of University of Calabria and research associate with ICAR-CNR. He is chairing the School of Computer Engineering at University of Calabria. He was a visiting scholar at UCLA, AT&T Laboratories, Oxford University and Technical University of Wien. His main research interests are in the areas of information systems and databases, artificial intelligence, game theory and bioinformatics. He authored about 150 papers appearing in refereed journals and conference proceedings. Palopoli serves in the editorial board of AI Communications – The European J. on AI, and in several conference program committees.

Claudia Plant currently is visiting professor at the department of Scientific Computing at the Florida State University sponsored by a Feodor Lynen fellowship of the Alexander von Humboldt Foundation. She received her PhD in biomedical informatics in 2004 from the University of Health Sciences Medical Informatics and Technology (UMIT), Austria. She has been working on several biomedical data mining projects at UMIT, University of Munich and Technische Universität München, Germany with applications in metabolomics, proteomics and neuroscience. Besides mining biomedical data she has been contributing to basic research especially in clustering, high-performance data mining and information-theoretic data mining.

Naren Ramakrishnan is a professor and the associate head for graduate studies in the Department of Computer Science at Virginia Tech. His research interests include computational science (esp. computational biology), mining scientific data, problem solving environments, and information personalization. He received his PhD in computer sciences from Purdue University in 1997. Ramakrishnan currently serves on the editorial boards of IEEE Computer, Knowledge and Information Systems, and the Journal of Intelligent Information Systems. He is an ACM Distinguished Scientist.

Simona E. Rombo is a post-doc fellow at DEIS, University of Calabria. She obtained her PhD in "computer science, biomedicine and telecommunications" at University "Mediterranea" of Reggio Calabria. She was a visiting scholar at the Department of Computer Science, Purdue University. Her research interests include bioinformatics, data mining, combinatorial algorithms.

Ricardo Santiago-Mozos is a Spanish Foundation for Science and Technology post-doctoral fellow (since September 2010) at the Machine Learning & Data Mining Group in National University of Ireland, Galway where he has just completed a Marie Curie Fellowship that commenced in October 2009. In September 2009 he obtained his PhD in Signal Processing and Communications at the Carlos III University of Madrid, where he developed a screening system for tuberculosis diagnosis. His research interests include machine learning, medical applications, genetic algorithms and communications, and he has co-authored more than 20 papers appearing in refereed journals and conference proceedings.

Xuequn Shang is an association professor at The School of Computer Science and Engineering at the Northwestern Polytechnical University, Xi'an, China. She completed her PhD degree from University of Magdeburg, Germany in 2005. She is head of computer software and engineering department. She is a member of China Computer Federation. Since 2001 she has been researching in data mining, database technology and bioinformatics.

Junming Shao is currently a PhD student of Insititute of informatics at University of Munich, Germany (LMU). He received his bachelor and master degree in Computer Science from Northwest A&F University, China, in 2005 and 2008 respectively. Currently, he is also working with neuroimaging group in the department of Neuroradiology, Klinikum rechts der Isar of the Technical University to investigate the relationship between structural and functional connectivity in Alzheimer's disease using diffusion tensor imaging and fMRI. Besides his basic research of data mining, he has been applying data mining techniques in diverse fields, such as biology, neuroscience and environment.

Narayanaswamy Srinivasan obtained his Bachelor's and Master's degrees in physics and biophysics respectively from University of Madras. He received his PhD from Indian Institute of Science for his thesis work on computational analysis of protein conformation. He had three postdoctoral stints in England; in Birkbeck College, London, Ludwig Institute for Cancer Research, London and Department of Biochemistry, University of Cambridge. He is currently leading a research team in the Molecular Biophysics Unit, Indian Institute of Science and his team focuses on computational biology with problems centered around genome-wide analysis, prediction of protein structure, function and interactions and biological pathways. He is the Fellow of two of the National Academies of India and received Shanti Swroop Bhatnagar Prize and National Bioscience Award. He is a Senior Fellow of the Manchester University, UK and a visiting Professor at Reunion University, France. He serves in the editorial boards of several journals including IJKDB.

Viktoria Stieffenhofer née Tatusch studied Pharmacy at the University of Mainz from 2001-2006, and wrote her dissertation at the Neurochemical Laboratory, Department of Psychiatry and Psychotherapy, University Medical Centre Mainz.

Masashi Sugiyama received the BE, ME, and PhD degrees in computer science from Tokyo Institute of Technology, Tokyo, Japan, in 1997, 1999, and 2001, respectively. In 2001, he was appointed as a research associate at the Tokyo Institute of Technology, where since 2003, he has been an associate professor. His research interests include machine learning and signal/image processing.

Kazuhiro Takemoto received his Doctoral degree in Informatics from Kyoto University, Japan, in 2008 after earned his Bachelor's and Master's degrees in Computer Science from Kyushu Institute of Technology, Japan, in 2004 and 2006, respectively. He is currently belonging to the Department of Computational Biology, University of Tokyo, Japan, as a Postdoctoral Fellow after served as a JSPS Research Fellow for two years from 2007. His research interests are network science including theory, bioinformatics, and mathematical biology. In particular, he recently focuses on metabolism, and attempts to elucidate the adaptive mechanisms of metabolic networks in extremophiles and the diversity of secondary metabolites in plants by a combination of their scientific fields.

Takeyuki Tamura received B.E., M.E. and Ph.D. degrees in informatics from Kyoto University, Japan, in 2001, 2003, and 2006, respectively. He joined Bioinformatics Center, Institute for Chemical Research, Kyoto University as a postdoctoral fellow in April, 2006. He works as an assistant professor from December, 2007. His research interests are bioinformatics and the theory of combinatorial optimization for graphs and networks. The current research topics include algorithms related to fixed points of Boolean networks, finding reaction cuts of metabolic networks, completing networks using observed data and calculating impact degree of deleting reactions of metabolic networks. His approach for these topics includes mathematically proving computational complexity, developing efficient algorithms, applying integer programming.

Sven Ulrich studied at the Organic Chemistry Technical University Merseburg, Germany and Moscow State University, Russia. He wrote his Ph.D. Theses in organic and analytical chemistry in 1992. From 1993-2002 Sven Ulrich studied therapeutic drug monitoring, clinical psychopharmacology, and drug analysis in body fluids at the Institute of Clinical Pharmacology, University Hospital, Magdeburg, Germany. From 2003 to the present date, Sven Ulrich is active in the pharmaceutical industry studying regulatory, pharmacovigilance, medical-scientific, and clinical psychopharmacology. He has been published in about 30 clinical psychopharmacology, therapeutic drug monitoring and drug analysis articles.

Jason T.L. Wang received a B.S. in Mathematics from National Taiwan University, Taipei, Taiwan, and a Ph.D. in Computer Science from the Courant Institute of Mathematical Sciences at New York University, New York, USA. He is a Professor of computer science and bioinformatics at the New Jersey Institute of Technology, Newark, New Jersey, USA. He has published over 120 peer-reviewed papers, 5 books, has been a program committee member of over 100 conferences, and is on the editorial board of 15 journals as well as the Executive Editor of the World Scientific book series on science, engineering and biology informatics. His research interests include data mining, knowledge discovery, computational biology and bioinformatics.

Lusheng Wang received the Ph.D. degree from McMaster University, Hamilton, ON, Canada, in 1995. Currently, he is a Professor in the Department of Computer Science, City University of Hong Kong. His research interests include algorithms, bioinformatics, and computational biology.

Miao Wang is a doctoral student at The School of Computer Science and Engineering at the Northwestern Polytechnical University, Xi'an, China. He completed his Master Degree from University of Northwestern Polytechnique in 2008. Since 2006 he has been researching in data mining and bioinformatics.

Wei Wang is a professor in the Department of Computer Science and a member of the Carolina Center for Genomic Sciences at the University of North Carolina at Chapel Hill. She received a MS degree from the State University of New York at Binghamton in 1995 and a PhD degree in Computer Science from the University of California at Los Angeles in 1999. She was a research staff member at the IBM T. J. Watson Research Center between 1999 and 2002. Dr. Wang's research interests include data mining, bioinformatics, and databases. She has filed seven patents, and has published one monograph and more than one hundred research papers in international journals and major peer-reviewed conference proceedings. She serves on the program committees of prestigious international conferences such as ACM SIGMOD, ACM SIGKDD, VLDB, ICDE, EDBT, ACM CIKM, IEEE ICDM, and SSDBM.

Chengyong Yang is currently leading the algorithm team to develop image analysis and base-calling algorithms for next-gen sequencing platforms at the Life Technologies Inc. Chengyong received his PhD degree in Computer Science from Florida International University (2006) and his dissertation focused on knowledge discovery from homogenous and heterogeneous sources of biological data. His research interests include image processing, high performance computing, data mining, and statistical modelling.

Qinli Yang obtained her bachelor degree from Northwest A&F University in the subject Environmental Science in 2007. Currently, she is pursuing her PhD in the field of Environment Engineering in the University of Edinburgh. Her research interests include water engineering, flood risk control and data mining.

Calvin Young completed his bachelor's degree in computer science at the University of North Carolina at Chapel Hill in 2010. He is currently working as a software engineer at Google. His research interests include bioinformatics, data mining, and graph mining.

Mohammed J. Zaki is a Professor of Computer Science at RPI. He received his Ph.D. degree in computer science from the University of Rochester in 1998. His research interests focus on developing novel data mining techniques, especially in bioinformatics. He has published over 175 papers and book-chapters on data mining and bioinformatics. He was the founding co-chair for the BIOKDD series of workshops. He is currently an Executive Editor for Statistical Analysis and Data Mining, and an Associate Editor for Data Mining and Knowledge Discovery, ACM Transactions on Knowledge Discovery from Data, Knowledge and Information Systems, ACM Transactions on Intelligent Systems and Technology, Social Networks and Mining, and International Journal of Knowledge Discovery in Bioinformatics. He was the program co-chair for SDM'08, SIGKDD'09 and PAKDD'10. He received the National Science Foundation CAREER Award in 2001 and the Department of Energy Early Career Principal Investigator Award in 2002. He also received the ACM Recognition of Service Award in 2003 & 2009, and an IEEE Certificate of Appreciation in 2005.

Erliang Zeng is currently a research faculty member in the Department of Computer Science and Engineering, and the Managing Director of Bioinformatics Core Facility at the University of Notre Dame. Prior to joining Notre Dame, he worked as a post-doctoral researcher with professor Mitsunori Ogihara at the University of Miami. He received his PhD degree in Computer Science from Florida International University in 2008 under the supervision of professor Giri Narasimhan, and a MS degree in Biochemistry and Molecular Biology from Shanghai Jiao Tong University in 2001. His research interests are in the general areas of bioinformatics, computational biology, and data mining.

Aidong Zhang received her Ph.D. degree in computer science from Purdue University in 1994. She is currently a Professor and the Chair in the Department of Computer Science and Engineering, State University of New York at Buffalo. Her research interests include bioinformatics, data mining, multimedia and database systems, and content-based image retrieval. She has authored or coauthored more than

200 publications in these areas. She is currently a member of several journal editorial boards such as the International Journal of Bioinformatics Research and Applications and the ACM Multimedia Systems. She is the author of two books: Protein Interaction Networks: Computational Analysis (Cambridge University Press, 2009) and Advanced Analysis of Gene Expression Microarray Data (World Scientific Publishing, 2006). She was the Chair or a member of more than 100 program committees of international conferences and workshops. She is a recipient of the National Science Foundation CAREER Award and the SUNY Chancellor's Research Recognition Award. She is an IEEE fellow.

Shaohua Zhang is a graduated student at The School of Computer Science and Engineering at the Northwestern Polytechnical University, Xi'an, China. He completed his Bachelor's Degree from Xin-JiangUniversity in 2008. Since 2008 he has been researching in data mining and bioinformatics.

Lizhuang Zhao is a software design engineer at Microsoft Inc. His research interests involve developing novel large-scale software using data mining techniques, with applications in clustering, and data logical-relationship analysis, and so on. He received his PhD in computer science from Rensselaer Polytechnic Institute.

Index

N

O

P

Q

R

S

T

U

V

W

Y

www.ingramcontent.com/pod-product-compliance
Lightning Source LLC
LaVergne TN
LVHW060357300926
849162LV00006B/1287
* 9 7 8 1 4 6 6 6 1 7 8 5 8 *